A Heat Transfer Textbook

JOHN H. LIENHARD

Professor of Mechanical Engineering
University of Houston
Houston, Texas

(with Chapter 6 by **Roger Eichhorn**,
Dean, College of Engineering,
University of Kentucky,
Lexington, Kentucky)

PRENTICE-HALL, INC., *Englewood Cliffs, New Jersey 07632*

Library of Congress Cataloging in Publication Data

LIENHARD, JOHN H   1930–
    A heat transfer textbook.

    Includes bibliographical references and index.
    1. Heat—Transmission. I. Title.
TJ260.L445   621.402′2   80-17380
ISBN 0-13-385112-5

Editorial/production supervision by Ros Herion and Gary Samartino
Interior design by Gary Samartino
Cover design by Edsal Enterprises
Manufacturing buyer: Anthony Caruso

Printed in the United States of America

10 9 8 7 6 5 4 3 2 1

PRENTICE-HALL INTERNATIONAL, INC., *London*
PRENTICE-HALL OF AUSTRALIA PTY. LIMITED, *Sydney*
PRENTICE-HALL OF CANADA, LTD., *Toronto*
PRENTICE-HALL OF INDIA PRIVATE LIMITED, *New Delhi*
PRENTICE-HALL OF JAPAN, INC., *Tokyo*
PRENTICE-HALL OF SOUTHEAST ASIA PTE. LTD., *Singapore*
WHITEHALL BOOKS LIMITED, WELLINGTON, *New Zealand*

For my father, the third **Johann Heinrich.**

The first two emigrated from a house on a hill in Bilten
where, by chance, the opening pages of this book
were finished. The fifth one wrote its closing pages.

But my father, born before the airplane and dying as
closeup photos of Saturn reached the earth, told me
on some authority what language and technology were
all about.

Contents

Preface

This book is intended to serve undergraduate engineering students in their introductory heat transfer course. It has been adjusted specifically to match the needs of chemical and mechanical engineering students at the University of Kentucky. The anticipated prerequisites for the course are engineering thermodynamics, two years of calculus, and an introduction to fluid mechanics. The material can be taught to students who lack a fluid mechanics background, but the pace should be considerably slower.

The organization of material has a few unconventional aspects, and the intent of these features should be explained:

- Heat exchangers are discussed in the beginning of the book rather than toward the end. This is easy to do, once the students have learned what a convective heat transfer coefficient is, even though they do not yet know how to predict its value. We have found that students have much more confidence in the worth of subsequent analytical work once they have encountered a major area of application of the subject. Furthermore, heat exchanger design provides a natural link between heat transfer and the previous subject of thermodynamics.

- The first three chapters make up a mini-course in heat transfer—an attempt to acquaint the students rather broadly with the whole subject. In the subsequent development, we try to keep the students aware of the interactions among different modes of heat transfer.

- Dimensional analysis is introduced, without reference to the cumbersome method of indices, in connection with heat conduction. It is then used throughout the book to simplify the mathematics and to explain data correlations.

- Radiation, though it appears as the final chapter, can be dealt with at any point following Chapter 3.

- The role of the Biot number as a diagnostic device is introduced in the first chapter along with the related notion of a lumped-capacity solution. We view these simple notions as being quite important, and we return to them often.

- Natural convection and film condensation appear in the same chapter. They can be dealt with as separate issues, but we suggest that they be treated together. There is a strong kinship between the subjects, and it is useful to capitalize on this kinship.

- It has been conventional to give short shrift to boiling heat transfer. But modern processes more and more often employ boiling to remove heat. We therefore provide a great deal of material for the instructor who has time to present it.

- R. Eichhorn's chapter on the numerical analysis of heat transfer is rather detailed and self-explanatory. It can be dealt with at any point after Chapter 3—either all at once, or as the material is needed.

It should be easy to teach the entire book in two, 3-credit "quarters." About 80% of the material is normally treated in a conventional 3-credit, one-semester course at Kentucky by passing lightly over selected portions of Chapters 5 through 11.

I owe an enormous debt of gratitude to many colleagues and students for their generous contributions to this effort. At the University of Kentucky these include: Roger Eichhorn, who in addition to writing Chapter 6 provided extremely useful and detailed criticism of all the other chapters; James Elliott, Md. Alamgir, Robert Altenkirch, and Shiva Singh, who taught from earlier versions of the text and helped greatly in correcting it; Graeme Fairweather, who provided very helpful criticism of Chapter 6; Clifford Cremers, Richard Birkebak, Eralp Özil, Robert Peck, Margaret Somers, and Hans Züst, who provided critical readings and commentary to other portions of the text; Annaliese Griffin, who provided programming assistance for Chapter 6; and the many students who have responded constructively to the effort (three—Frank Loxley, Kenneth Allen, and Ronald Foster—were so energetic in tracking errors that I must mention them by name).

Among colleagues at other institutions who have been of great help, the first has been Ephriam M. Sparrow who very graciously provided a line-by-line reading of the entire first draft. With inexorable thoroughness, he identified pedagogical lapses, errors, and better sources of information. My debt to him is very great. I also wish to thank Jerry Taborek for his important advice on heat exchanger problems; Stewart Churchill for his help in providing the best information on convective heat transfer data correlations; Vijay K. Dhir for his critical reading of Chapters 9 and 10; and Prof. J. R. Mahan, whose suggestions improved Chapter 11.

Beth Altenkirch—one of the best technical typists in the business—prepared the manuscript. She not only ate through three drafts of the 875-page manuscript, but she also contributed materially to its consistency and correctness and made the job look easy. The graphical work was done at the University of Kentucky by Pati Marks, Mack Mosely, Jr., and Cindy Taulbee, under the direction of R. William DeVore.

Special thanks are also due to the Prentice-Hall editors and staff. The support and positive attitude of Mike Melody, John Cross, Hank Kennedy, and Riley Humler, who initiated this project, and of Chuck Iossi, Gary Samartino, and Ros Herion, who saw it through, has been unrelenting.

Finally, I am really grateful to my nice family: Carol, Andrew, and John, for the cheerful way in which they suffered their absentee *paterfamilias* during the writing of the book. John also indexed the book and did some of the computer programming for it.

University of Kentucky   JOHN LIENHARD
March 1980

Part I

The General Problem of Heat Exchange

The General Problem
of Heat Exchange

Chapter 1

Introduction

The radiation of the sun in which the planet is incessantly plunged, penetrates the air, the earth, and the waters; its elements are divided, change direction in every way, and, penetrating the mass of the globe, would raise its temperature more and more, if the heat acquired were not exactly balanced by that which escapes in rays from all points of the surface and expands through the sky.

The Analytical Theory of Heat, J. Fourier

1.1 Heat transfer

People have always understood that something flows from hot objects to cold ones. We call that flow "heat." In the eighteenth and early nineteenth centuries, scientists imagined that all bodies contained an invisible fluid which they called "caloric." Caloric was assigned a variety of properties, some of which proved to be inconsistent with nature (e.g., it had weight and it could not be created nor destroyed). But its most important feature was that it flowed from hot bodies into cold ones. It was a very useful way to think about heat. Later we shall explain the flow of heat in terms more satisfactory to the modern ear; however, it will seldom be wrong to imagine caloric flowing from a hot body to a cold one.

The flow of heat is all-pervasive. It is active to some degree or another in everything. Heat flows constantly from your bloodstream to the air around you. The warmed air buoys off your body to warm the room you are in. If you leave the room, some small buoyancy-driven (or "convective") motion of the air will continue because the walls can never be perfectly isothermal. Such processes go on in all plant and animal life and in the air around us. They occur throughout the earth, which is hot at its core and cooled around its surface. The only conceivable domain free from heat flow would have to be isothermal and totally isolated from any other region. It would be "dead" in the fullest sense of the word—devoid of any process of any kind.

The overall driving force for these heat flow processes is the cooling—or leveling of thermal gradients—within our universe. The heat flows that result from the cooling of the sun are the primary processes that we experience naturally. The conductive cooling of the earth's center and the radiative cooling of the other stars are processes of secondary importance in our lives.

The life forms on our planet have necessarily evolved to match the magnitude of these natural energy flows. But while "natural" man is in balance with these heat flows, technological man[1] has used his mind, his back, and his will to harness and control energy flows that are far more intense than those he experiences naturally. To emphasize this point we suggest that the reader make an experiment.

Experiment 1.1

Generate as much power as you can, in some way that permits you to measure your own work output. You might lift a weight, or run your own weight up a stairwell, against a stopwatch. Express the result in watts (W). Perhaps you might collect the results in your class. They should generally be less than 1 kW or even 1 horsepower (746 W). How much less might be surprising.

Thus, when we do so small a thing as turning on a 150-W light bulb, we are manipulating a quantity of energy substantially greater than a human being could produce in sustained effort. The energy consumed by an oven, toaster, or hot water heater is an order of magnitude beyond human capacity. The energy consumed by an automobile can easily be three orders of magnitude greater. If all the people in the United States worked continuously like galley slaves, they could barely equal the power output of even a single city power plant.

Our voracious appetite for energy has steadily driven the intensity of actual heat transfer processes upward until they are far greater than those normally involved with life forms on the earth. Until the middle of the sixteenth century, man's energy was drawn indirectly from the sun using comparatively gentle processes—animal power, wind and water power, and the combustion of wood. Then population growth and deforestation drove the English to using coal. By the end of the seventeenth century, England had almost completely converted to coal in place of wood. At the turn of the eighteenth century, the first commercial steam engines were developed, and that set the stage for enormously increased

---

[1]Some anthropologists think that the term *homo technologicus* (technological man) serves to define human beings, as apart from animals, better than the older term *homo sapiens* (man, the wise). We may not be as much wiser than the animals as we think we are, but only we are serious toolmakers.

consumption of coal. Europe and America followed England in these developments.

The development of fossil energy sources has been a bit like Jules Verne's description in *Around the World in Eighty Days* in which, to win a race, a crew burns the inside of a ship to power the steam engine. The combustion of nonrenewable fossil energy sources (and, more recently, the fission of uranium) has led to remarkably intense energy releases in power-generating equipment. The energy transferred as heat in a nuclear reactor is on the order of *one million watts per square meter*.

A complex system of heat and work transfer processes is invariably needed to bring these concentrations of energy back down to human proportions. We must understand and control the processes that divide and diffuse intense heat flows down to the level on which we can interact with them. To see how this works, consider a specific situation. Suppose that you live in a town where coal is processed into fuel-gas and coke. Such power supply systems used to be common and they will probably return as natural gas supplies dwindle. Let us list a few of the process heat transfer problems that must be solved before you can enjoy a glass of iced tea.

- A variety of high-intensity heat transfer processes are involved with combustion and chemical reaction in the gasifier unit itself.
- The gas goes through various cleanup and pipe-delivery processes to get to your stove. The heat transfer processes involved in these stages are generally less intense.
- The gas is burned in the stove. Heat is transferred from the flame to the bottom of your teakettle. While this process is small, it is intense because boiling is a very efficient way to remove heat.
- The coke is burned in a steam power plant. The heat transfer rates from the combustion chamber to the boiler, and from the wall of the boiler to the water inside, are very intense.
- The steam passes through a turbine, where it is involved with many heat transfer processes, including some condensation in the last stages. The spent steam is then condensed in any of a variety of heat transfer devices.
- Cooling must be provided in each stage of the electrical supply system: the windings and bearings of the generator, the transformers, the switches, the power lines, and the wiring in your house.
- The ice cubes for your tea are made in an electrical refrigerator. It involves three major heat exchange processes and several lesser ones. The major ones are the condensation of refrigerant at room temperature to reject heat, the absorption of heat from within the refrigerator by evaporating the refrigerant, and the balancing heat leakage from the room to the inside.

- Drink your iced tea quickly because heat transfer from the room to the water and from the water to the ice will first dilute, and then warm, your drink if you linger.

A society based on power technology teems with heat transfer problems. Our aim is to learn the principles of heat transfer so that we can solve these problems and design the equipment that is needed to transfer thermal energy from one substance to another. In a broad sense, all of these problems resolve themselves into collecting and focusing large quantities of energy for the use of people; and then distributing and interfacing this energy with people in such a way that they can use it on their own puny level.

We begin our study by recollecting how heat transfer was treated in the study of thermodynamics, and by seeing why thermodynamics is not adequate to the task of solving heat transfer problems.

1.2 Relation of heat transfer to thermodynamics

The First Law with work equal to zero

The subject of thermodynamics, as taught in engineering programs, makes constant reference to the heat transfer between systems. The First Law of Thermodynamics for a closed system takes the following form on a rate basis:

$$Q = Wk + \frac{dU}{dt} \tag{1.1}$$

$$\underbrace{}_{\substack{\text{positive toward} \\ \text{the system}}} \quad \underbrace{}_{\substack{\text{positive away} \\ \text{from the system}}} \quad \underbrace{\phantom{\frac{dU}{dt}}}_{\substack{\text{positive when} \\ \text{the system's} \\ \text{energy increases}}}$$

where Q is the heat transfer rate and Wk is the work transfer rate. They may be expressed in joules per second (J/s) or watts (W). The derivative dU/dt is the rate of change of internal thermal energy, U, with time, t. This statement is sketched schematically in Fig. 1.1a.

The analysis of heat transfer processes can generally be done without reference to any work processes, although heat transfer might subsequently be combined with work in the analysis of real systems. Therefore, our starting point for almost all subsequent analysis will be the truncated First Law statement (see Fig. 1.1b):

$$\boxed{Q = \frac{dU}{dt}} \tag{1.2}$$

We shall often refer back to this equation.

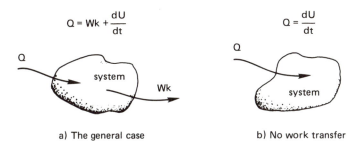

a) The general case   b) No work transfer

FIGURE 1.1   The First Law of Thermodynamics for a closed system.

The reader who has studied engineering thermodynamics will remember that eqn. (1.1) takes the following form[2] for reversible processes:

$$\underbrace{\frac{d}{dt}\int T\,dS}_{Q_{\text{rev}}} = \underbrace{\frac{d}{dt}\int p\,dV}_{Wk_{\text{rev}}} + \frac{dU}{dt} \tag{1.3}$$

This seems to suggest that Q can be evaluated independently for inclusion in either eqn. (1.1) or (1.2). Actually, it cannot be evaluated using $\int T\,dS$, because all heat transfer processes are irreversible and S is not defined as a function of T in an irreversible process. The reader will recall that engineering thermodynamics might better be named thermo*statics*, because it only describes the equilibrium states on either end of irreversible processes.

If the rate of heat transfer cannot be predicted using $\int T\,dS$, then how can it be determined? If $U(t)$ were known, then (when $Wk=0$) eqn. (1.2) would give Q, but $U(t)$ is seldom known a priori.

The answer is that a new set of physical principles must be introduced to predict Q. These principles are *transport laws*, which are not a part of the subject of thermodynamics. They include Fourier's law, Newton's law of cooling, and the Stefan–Boltzmann law. We introduce these laws later in the chapter. The important thing to remember is that a description of heat transfer requires that additional principles be combined with the First Law of Thermodynamics.

Reversible heat transfer
as the temperature gradient vanishes

Consider a wall connecting two thermal reservoirs as shown in Fig. 1.2. As long as $T_1 > T_2$, heat will flow *spontaneously* and *irreversibly* from 1 to 2. In accordance with our understanding of the Second Law of Thermodynamics, we expect the entropy of the universe to increase as a consequence of this process. If $T_2 \rightarrow T_1$, the process will approach being quasistatic and reversible. But the rate

[2]T = absolute temperature, S = entropy, V = volume, p = pressure, and "rev" denotes a reversible process.

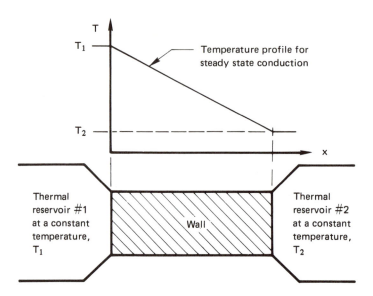

FIGURE 1.2 Irreversible heat flow between two thermal reservoirs through an intervening wall.

of heat transfer will also approach zero if there is no temperature difference to drive it. Thus all real heat transfer processes generate entropy.

Now we come to a dilemma: If the irreversible process occurs at steady state, the properties of the wall do not vary with time. We know that the entropy of the wall depends on its state and must therefore be constant. How, then, does the entropy of the universe increase? We turn to this question next.

Entropy production

The entropy increase of the universe as the result of a process is the sum of the entropy changes of *all* elements that are involved in that process. The *rate of entropy production* of the universe, $\dot{S}_{Un}$, resulting from the preceding heat transfer process through a wall is

$$\dot{S}_{Un} = \dot{S}_{res\,1} + \underbrace{\dot{S}_{wall}}_{\substack{=\,0,\text{ since } S_{wall} \\ \text{must be constant}}} + \dot{S}_{res\,2} \qquad (1.4)$$

where the dots denote time derivatives (i.e., $\dot{x} \equiv dx/dt$). Since the reservoir temperatures are constant,

$$\dot{S}_{res} = \frac{\dot{Q}}{T_{res}} \qquad (1.5)$$

Now $Q_{\text{res }1}$ is negative and equal in magnitude to $Q_{\text{res }2}$, so eqn. (1.4) becomes

$$\dot{S}_{\text{Un}} = |Q_{\text{res }1}|\left(\frac{1}{T_2} - \frac{1}{T_1}\right) \qquad (1.6)$$

The term in parentheses *must* be positive, so $\dot{S}_{\text{Un}} > 0$. This agrees with Clausius's statement of the Second Law of Thermodynamics.

Notice an odd fact here: The rate of heat transfer Q, and hence $\dot{S}_{\text{Un}}$, is determined by the wall's resistance to heat flow. Although the wall is the agent that causes the entropy of the universe to increase, its own entropy does not change. Only the entropies of the reservoirs change.

1.3 Modes of heat transfer

Figure 1.3 shows an analogy that might be useful in fixing the concepts of heat conduction, convection, and radiation as we proceed to look at each in some detail.

Heat conduction

Fourier's law. Joseph Fourier[3] (see Fig. 1.4) published his remarkable book *Théorie Analytique de la Chaleur* in 1822. In it he formulated a very complete exposition of the theory of heat conduction.

He began his treatise by stating the empirical law that bears his name: *The heat flux,[4] $q W/m^2$, resulting from thermal conduction is proportional to the magnitude of the temperature gradient and opposite to it in sign*. If we call the constant of proportionality, k, then

$$q = -k\frac{dT}{dx} \qquad (1.7)$$

The constant, k, is called the *thermal conductivity*. It obviously must have the

---

[3]Joseph Fourier lived a remarkable double life. He served as a high government official in Napoleonic France and he was also an applied mathematician of great importance. He was with Napoleon in Egypt between 1798 and 1801, and he was subsequently Prefect of the administrative area (or "Department") of Isére in France until Napoleon's first fall in 1814. During the latter period he worked on the theory of heat flow and in 1807 submitted a 234 page monograph on the subject. It was given to such luminaries as Lagrange and Laplace for review. They found faults with his adaptation of a series expansion suggested by Daniel Bernoulli in the eighteenth century. Fourier's theory of heat flow, his governing differential equation, and the now-famous "Fourier series" solution of that equation did not emerge in print from the ensuing controversy until 1822.

[4]The heat flux, q, is a heat rate per unit area and can be expressed as Q/A, where A is an appropriate area.

Help! The barn is on fire.

Let the *water* be analogous to *heat,* and let the *people* be analogous to the *heat transfer medium.* Then:

Case 1 The hose directs water from (W) to (B) independently of the medium. This is analogous to *thermal radiation* in a vacuum or in most gases.

Case 2 In the bucket brigade, water goes from (W) to (B) through the medium. This is analogous to *conduction.*

Case 3 A single runner, representing the medium, carries water from (W) to (B) . This is analogous to *convection.*

FIGURE 1.3   An analogy for the three modes of heat transfer.

FIGURE 1.4   Baron Jean Baptiste Joseph Fourier (1768–1830). (Photo courtesy *Appl. Mech. Revs.*, vol. 26, Feb 1973.)

dimensions J/m-s-°C, or W/m-°C, or Btu/ft-hr-°F if eqn. (1.7) is to be dimensionally correct.

The heat flux is a vector quantity. Equation (1.7) tells us that if temperature decreases with x, then q will be positive—it will flow in the x-direction. If T increases with x, q will be negative—it will flow opposite to the x-direction. In either case, q will flow from higher temperatures to lower temperatures. Equation (1.7) is the one-dimensional form of Fourier's law. We develop its three-dimensional form in Chapter 2, namely:

$$\vec{q} = -k\vec{\nabla}T$$

Example 1.1

The front of a slab of lead ($k = 35$ W/m-°C) is kept at 110°C and the back is kept at 50°C. If the area of the slab is 0.4 m^2 and it is 0.03 m thick, compute the heat flux, q, and the heat transfer, Q.

SOLUTION.   For the moment we presume that dT/dx is a constant equal to $(T_{\text{back}} - T_{\text{front}})/(x_{\text{back}} - x_{\text{front}})$. (This will be verified in Chapter 2.) Thus eqn. (1.7) becomes

$$q = -35\left(\frac{50-110}{0.03}\right) = +70{,}000 \text{ W/m}^2 = \underline{70 \text{ kW/m}^2}$$

and

$$Q = qA = 70(0.4) = \underline{28 \text{ kW}}$$

In one-dimensional heat conduction problems, there is never any real problem in deciding which way the heat should flow. It is therefore sometimes convenient to write Fourier's law in simple scalar form:

$$q = k \frac{\Delta T}{L} \tag{1.8}$$

where L is the thickness in the direction of heat flow and q and ΔT are both written as positive quantities. When we use eqn. (1.8), we must remember that q always flows from high to low temperatures.

Thermal conductivity values.   It will help if we first consider how conduction occurs in, for example, a gas. We know that the molecular velocity depends on temperature. Thus during conduction from a hot wall to a cold one, as shown in Fig. 1.5, the molecules near the hot wall collide with it and are agitated by the molecules of wall. They leave with generally higher speed and collide with their neighbors to the right, increasing the speed of those neighbors. This process continues until the molecules on the right pass their kinetic energy to those in the cool wall. Within solids, comparable processes occur as the molecules vibrate within their lattice structure and as the lattice vibrates as a whole. This sort of process also occurs, to some extent, in the electron "gas" that moves through the solid. The processes are more efficient in solids than they are in

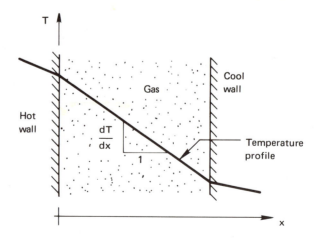

FIGURE 1.5   Heat conduction through the gas separating two solid walls.

gases. Notice that

$$-\frac{dT}{dx} = \underbrace{\frac{q}{k}}_{\substack{\text{since, in steady} \\ \text{conduction, } q \text{ is} \\ \text{constant}}} \propto \frac{1}{k} \tag{1.9}$$

Thus solids, with generally higher thermal conductivities than gases, yield smaller temperature gradients for a given heat flux. In a gas, by the way, it can be shown that k is proportional to molecular speed and molar specific heat, and inversely proportional to the cross-sectional area of the molecules.

This book deals almost exclusively in S.I. units, or *Système International d'Unités*. Since much reference material will continue to be available in English units, we should have at hand a conversion factor for thermal conductivity:

$$1 = \frac{J}{0.0009478 \text{ Btu}} \cdot \frac{hr}{3600 \text{ s}} \cdot \frac{ft}{0.3048 \text{ m}} \cdot \frac{1.8°F}{°C}$$

Thus the conversion factor from W/m-°C to its English equivalent, Btu/ft-hr-°F, is

$$\boxed{1 = 1.731 \frac{\text{W/m-°C}}{\text{Btu/ft-hr-°F}}} \tag{1.10}$$

Consider, for example, copper—the common substance with the highest conductivity at ordinary temperatures:

$$k_{\text{Cu at room temp.}} = (383 \text{ W/m-°C}) \Big/ 1.731 \frac{\text{W/m-°C}}{\text{Btu/ft-hr-°F}} = 221 \text{ Btu/ft-hr-°F}$$

The range of thermal conductivities is enormous. As we can see from Fig. 1.6, k varies by a factor of about 10^5 between gases and diamond at room temperature. This variation can be increased to about 10^7 if we include the effective conductivity of various cryogenic "superinsulations." (These involve powders, fibers, or multilayered materials that have been evacuated of all air.) The reader should study and remember the order of magnitude of the thermal conductivities of different types of materials. This will be a help in avoiding mistakes in future computations, and it will be a help in making assumptions during problem solving. Actual numerical values of the thermal conductivity are given in Appendix A (which is a broad listing of many of the physical properties you might need in this course) and in Figs. 2.2 and 2.3.

Example 1.2

A copper slab ($k = 372$ W/m-°C) is 3 mm thick. It is protected from corrosion by 2-mm-thick layers of stainless steel ($k = 17$ W/m-°C) on both sides. The temperature is

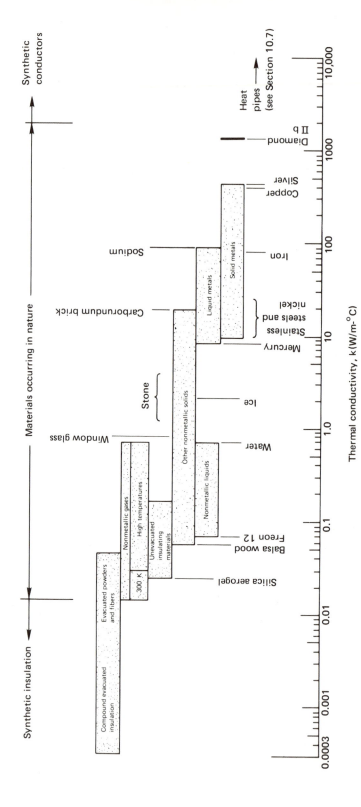

FIGURE 1.6 The approximate ranges of thermal conductivity of various substances. (All values are for the neighborhood of room temperature unless noted.)

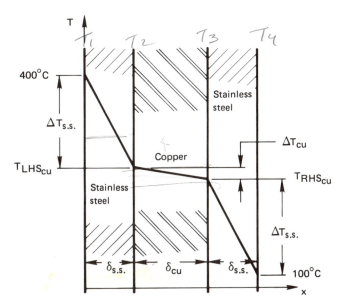

FIGURE 1.7 Temperature drop through a copper wall protected by stainless steel (Example 1.2).

400°C on one side and 100°C on the other. Find the temperature distribution in the copper slab and the heat conduction through the wall (see Fig. 1.7).

If we recall Fig. 1.5 and eqn. (1.9), it should be clear that the temperature drop will take place almost entirely in the stainless steel, where k is less than $\frac{1}{20}$ of k in the copper. Thus the copper will be virtually isothermal at the average temperature of $(400 + 100)/2 = 250°C$. Furthermore, the heat conduction can be calculated in a 4-mm slab of stainless steel as though the copper were not even there. With the help of Fourier's law in the form of eqn. (1.7), we get

$$q = -k\frac{dT}{dx} \simeq 17 \text{ W/m-°C}\left(\frac{400 - 100}{0.004}\right) °C/m = \underline{1275 \text{ kW/m}^2}$$

The accuracy of this rough calculation can be improved by considering the copper. To do this we first solve for $\Delta T_{s.s.}$ and ΔT_{Cu} (see Fig. 1.7). Conservation of energy requires that the steady heat flux through all three slabs must be the same. Therefore,

$$q \simeq \left[k\frac{\Delta T}{L}\right]_{s.s.} = \left[k\frac{\Delta T}{L}\right]_{Cu}$$

but

$$(400 - 100)°C \equiv \Delta T_{Cu} + 2\Delta T_{s.s.} = \Delta T_{Cu}\left[1 + 2\frac{(k/L)_{Cu}}{(k/L)_{s.s.}}\right] = 30.18\Delta T_{Cu}$$

Solving this, we obtain $\Delta T_{Cu} = 9.94°C$. So $\Delta T_{s.s.} = (300 - 9.94)/2 = 145.03°C$. It follows that $\underline{T_{left_{Cu}} = 254.97°C}$ and $\underline{T_{right_{Cu}} = 245.03°C}$.

The heat flux can be obtained by applying Fourier's law to any of the three layers. We consider either stainless steel layer and get

$$q = 17 \frac{W}{m\text{-}°C} \frac{145.03°C}{0.002 \text{ m}} = \underline{1233 \text{ kW/m}^2}$$

Thus our initial approximation was accurate within a few percent.

One-dimensional heat diffusion equation.   In Example 1.2 we had to deal with a major problem that arises in heat conduction problems. The problem is that Fourier's law involves two dependent variables, T and q. To eliminate q, and first solve for T, we introduced the First Law of Thermodynamics implicitly: Conservation of energy required that q was the same in each metallic slab.

The elimination of q from Fourier's law must now be done in a more general way. Consider a one-dimensional element as shown in Fig. 1.8. From this we see that during general unsteady heat flow, Fourier's law gives

$$q_{net} A = Q_{net} = kA \frac{\partial^2 T}{\partial x^2} \delta x \tag{1.11}$$

To eliminate Q_{net} we use the general First Law statement for closed, nonworking systems, eqn. (1.2):

$$Q_{net} = \frac{dU}{dt} = \rho c A \frac{d(T - T_{ref})}{dt} \delta x = \rho c A \frac{dT}{dt} \delta x \tag{1.12}$$

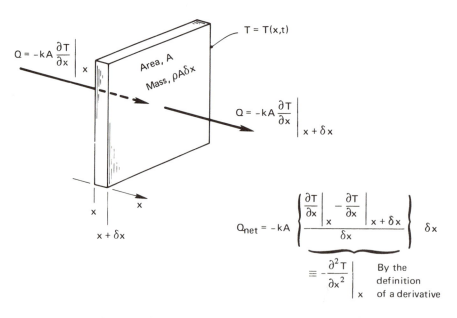

FIGURE 1.8   One-dimensional heat conduction through a differential element.

where ρ is the density of the slab and c is its specific heat.[5] Equations (1.11) and (1.12) can be combined to give

$$\frac{\partial^2 T}{\partial x^2} = \frac{\rho c}{k} \frac{\partial T}{\partial t} \equiv \frac{1}{\alpha} \frac{\partial T}{\partial t} \qquad (1.13)$$

This is the *one-dimensional heat diffusion equation*. Its importance is this: By combining the First Law with Fourier's law, we have eliminated the unknown Q and obtained a differential equation that can be solved for the temperature distribution, $T(x)$. It is the primary equation upon which all of heat conduction theory is based.

The heat diffusion equation includes a new property which is as important to transient heat conduction as k is to steady-state conduction. This is the thermal diffusivity, α:

$$\alpha = \frac{k}{\rho c} \frac{\dfrac{J}{m\text{-}s\text{-}°C}}{\dfrac{kg}{m^3} \dfrac{J}{kg\text{-}°C}} = \alpha \; m^2/s \qquad (\text{or } \alpha \, ft^2/hr)$$

The thermal diffusivity is a measure of how quickly a material can carry heat away from a hot source. Since a material does not just transmit heat but must be warmed by it as well, α involves both the conductivity, k, and the volumetric heat capacity, ρc.

Heat convection

The physical process. Consider a typical convective cooling situation. Cool gas flows past a warm body as shown in Fig. 1.9. The fluid immediately adjacent to the body forms a thin slowed-down region called a boundary layer. Heat is conducted into this layer, which sweeps it away and farther downstream mixes it into the stream. We call such processes of carrying heat away by a moving fluid, convection.

In 1701, Isaac Newton considered the convective process and suggested that the cooling would be such that

$$\frac{dT_{body}}{dt} \propto T_{body} - T_\infty \qquad (1.14)$$

where T_∞ is the temperature of the oncoming fluid. This implies that energy is flowing from the body. If the energy of the body is constantly replenished, the

---

[5]The reader might wonder if c should be c_p or c_v. This is a strictly incompressible equation and $c_p = c_v = c$ for an incompressible material. The compressible equation involves additional terms, and this particular term emerges with c_p in it, in the conventional rearrangements of terms.

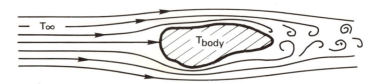

FIGURE 1.9   The convective cooling of a heated body.

body temperature need not change. Then with the help of eqn. (1.2) we get, from eqn. (1.14) (Problem 1.2),

$$Q \propto T_{\text{body}} - T_\infty \tag{1.15}$$

This can be rephrased in terms of $q = Q/A$ as

$$\boxed{q = \bar{h}(T_{\text{body}} - T_\infty)} \tag{1.16}$$

This is the steady-state form of Newton's law of cooling, as it is usually quoted, although Newton never wrote such an expression.

The constant $\bar{h}$ is the "unit surface conductance" or "heat transfer coefficient." The bar over h indicates that it is an average over the surface of the body. Without the bar, h denotes the "local" value of the heat transfer coefficient at a point on the surface. The units of h and $\bar{h}$ are W/m²-°C or J/s-m²-°C. The conversion factor for English units is:

$$1 = \frac{0.0009478 \text{ Btu}}{\text{J}} \quad \frac{°C}{1.8°F} \quad \frac{3600 \text{ s}}{\text{hr}} \quad \frac{(0.3048 \text{ m})^2}{\text{ft}^2}$$

or

$$1 = 0.1761 \; \frac{\text{Btu/ft}^2\text{-hr-}°F}{\text{W/m}^2\text{-}°C} \tag{1.17}$$

It turns out that Newton oversimplified the process in making his conjecture. Heat convection is very complicated and $\bar{h}$ can depend on the temperature difference $T_{\text{body}} - T_\infty \equiv \Delta T$. In Chapter 7 we find that $\bar{h}$ really *is* independent of ΔT in situations in which fluid is forced past a body and ΔT is not too large. This is called *forced convection.*

When fluid buoys up from a hot body or down from a cold one, h varies as some weak power of ΔT—typically as $\Delta T^{1/4}$ or $\Delta T^{1/3}$. This is called *free* or *natural convection.* If the body is hot enough to boil a liquid surrounding it, h will typically vary as ΔT^2.

For the moment we restrict consideration to situations in which Newton's law is either true or at least a reasonable approximation to real behavior.

We should have some idea of how big $\bar{h}$ might be in a given situation. Table 1.1 provides some illustrative values of $\bar{h}$ that have been observed or calculated

Table 1.1 Some illustrative values of convective heat transfer coefficients

Situation	$\bar{h}$ (W/m²-°C)
Natural convection, single phase	
• 0.3-m vertical wall in air, $\Delta T = 30°C$	4.33
• 4-cm-O.D. horizontal pipe in H_2O, $\Delta T = 30°C$	570
• 0.025-cm-diameter wire in methanol, $\Delta T = 50°C$	4,000
Forced convection	
• Air at 30 m/s over a 1-m plate, $\Delta T = 70°C$	80
• Water at 2 m/s over a 0.06-m plate, $\Delta T = 15°C$	590
• Aniline–alcohol mixture at 3 m/s in a 2.5-cm-I.D. tube, $\Delta T = 80°C$	2,600
• Liquid sodium at 5 m/s in a 1.3-cm-I.D. tube, at 370°C	75,000
Boiling water	
• During film boiling at 1 atm	300
• In a teakettle	4,000
• A peak-pool boiling heat flux at 1 atm	40,000
• A peak-flow boiling heat flux at 1 atm	100,000
• Approximate maximum convective heat flux under optimum conditions	900,000
Condensing	
• Typical value in a horizontal cold-water-tube steam condenser	15,000
• Same but condensing benzene	1,700
• Dropwise condensation of water	100,000

for different situations. They are only illustrative and should not be used in calculations because the situations for which they apply have not been fully described. Most of the values in the table could be changed a great deal by varying quantities (such as surface roughness or geometry) that have not been specified. The determination of h or $\bar{h}$ is a fairly complicated task and one that will receive a great deal of our attention. Notice too that $\bar{h}$ can change dramatically from one situation to the next. Reasonable values of h range over about six orders of magnitude.

Example 1.3

The heat flux, q, is 6000 W/m² at the surface of an electrical resistance heater. The heater temperature is 120°C when it is cooled by air at 70°C. What is the average convective heat transfer coefficient, $\bar{h}$? What will the heater temperature be if q is reduced to 2000 W/m²?

SOLUTION.

$$\bar{h} = \frac{q}{\Delta T} = \frac{6000}{120 - 70} = \underline{120 \text{ W/m}^2\text{-°C}}$$

If the heat flux is reduced, $\bar{h}$ should remain unchanged during forced convection. Thus

$$\Delta T = T_{\text{heater}} - 70°C = \frac{q}{\bar{h}} = \frac{2000 \text{ W/m}^2}{120 \text{ W/m}^2\text{-°C}} = 16.67°C$$

so

$$T_{\text{heater}} = 70 + 16.67 = 86.67°C$$

Lumped-capacity solution.   We now wish to deal with a very simple, but extremely important, kind of convective heat transfer problem. The problem is that of predicting the transient cooling of a convectively cooled object such as is shown in Fig. 1.9. We begin with our now-familiar First Law statement, eqn. (1.2):

$$\underbrace{Q}_{-\bar{h}A(T-T_\infty)} = \underbrace{\frac{dU}{dt}}_{\frac{d}{dt}[\rho c V(T-T_{\text{ref}})]} \tag{1.18}$$

where A and V are the surface area and volume of the body, T is the temperature of the body, $T = T(t)$, and T_{ref} is the arbitrary temperature at which U is defined equal to zero. Thus[6]

$$\frac{d(T-T_\infty)}{dt} = -\frac{\bar{h}A}{\rho c V}(T-T_\infty) \tag{1.19}$$

The general solution of this equation is

$$\ell n(T-T_\infty) = -t \bigg/ \frac{\rho c V}{\bar{h}A} + C \tag{1.20}$$

The group $\rho c V / \bar{h}A$ is the *time constant*, **T**. If the initial temperature is $T(t=0) \equiv T_i$, then $C = \ell n(T_i - T_\infty)$, and the cooling of the body is given by

$$\boxed{\frac{T-T_\infty}{T_i-T_\infty} = e^{-t/T}} \tag{1.21}$$

All of the physical parameters in the problem have now been "lumped" into the time constant. It represents the time required for a body to cool to $1/e$, or 37% of its initial temperature difference above (or below) T_∞. The ratio t/T can also be interpreted as

$$\frac{t}{T} = \frac{\bar{h}At\ (\text{J}/°\text{C})}{\rho c V\ (\text{J}/°\text{C})} = \frac{\text{capacity for convection from surface}}{\text{heat capacity of the body}} \tag{1.22}$$

Notice that the thermal conductivity is missing from eqns. (1.21) and (1.22). The reason is that we have assumed that the temperature of the body is nearly

---

[6]Is it clear why $(T - T_{\text{ref}})$ has been changed to $(T - T_\infty)$ under the derivative? Remember that the derivative of a constant (like T_{ref} or T_∞) is zero. We can therefore introduce $(T - T_\infty)$ to get the same dependent variable on either side of the equation.

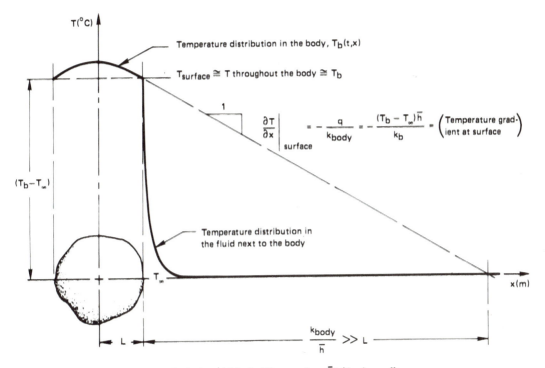

FIGURE 1.10   The cooling of a body for which the Biot number, $\bar{h}L/k_b$, is small.

uniform, and this means that internal conduction is not important. We see in Fig. 1.10 that, if $L/(k_b/\bar{h}) \ll 1$, the temperature of the body, T_b, is almost constant within the body at any time. Thus

$$\frac{\bar{h}L}{k_b} \ll 1 \quad \text{implies that} \quad T_b(x,t) \simeq T(t) \simeq T_{\text{surface}}$$

and the thermal conductivity, k_b, becomes irrelevant to the cooling process. This condition must be satisfied or the lumped-capacity solution will not be accurate.

We call the group $\bar{h}L/k_b$ the *Biot*[7] *number,* Bi. If Bi were large, of course, the situation would be reversed, as shown in Fig. 1.11. In this case Bi$= hL/k_b \gg 1$ and the convection process offers little resistance to heat transfer. We

---

[7]Pronounced Bée-oh. (Later we encounter a similar dimensionless group called the Nusselt number, Nu$\equiv hL/k_{\text{fluid}}$. The latter relates only to the boundary layer and not to the body being cooled. We deal with it extensively in the study of convection.) J. B. Biot, although younger than Fourier, worked on the analysis of heat conduction even earlier—in 1802 or 1803. He came to grips with the problem of including external convection in heat conduction analyses in 1804 but could not see how to do it. Fourier read Biot's work and by 1807 had determined how to analyze the problem.

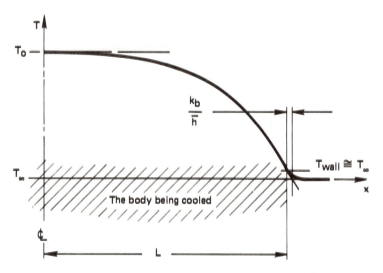

FIGURE 1.11   The cooling of a body for which the Biot number, $\bar{h}L/k_b$, is large.

could solve the heat diffusion equation

$$\frac{\partial^2 T}{\partial x^2} = \frac{1}{\alpha}\frac{\partial T}{\partial t}$$

subject to the simple boundary condition $T(x,t) = T_\infty$ when $x = 0$, to determine the temperature in the body and its rate of cooling, in this case. The Biot number will therefore be the basis for determining what sort of problem we have to solve.

To calculate the rate of entropy production in a lumped-capacity system, we note that the entropy change of the universe is the sum of the entropy decrease of the body and the more rapid entropy increase of the surroundings. The source of irreversibility is heat flow through the boundary layer. Accordingly, we write the time rate of change of entropy of the universe, $dS_{Un}/dt \equiv \dot{S}_{Un}$, as

$$\dot{S}_{Un} = \dot{S}_b + \dot{S}_{surroundings} = \frac{-Q_{rev}}{T_b} + \frac{Q_{rev}}{T_\infty}$$

or

$$\dot{S}_{Un} = -\rho c V \frac{dT_b}{dt}\left(\frac{1}{T_\infty} - \frac{1}{T_b}\right)$$

We can multiply both sides of this equation by dt and integrate the right-hand side from $T_b(t=0) \equiv T_{b_0}$ to T_b at the time of interest:

$$\Delta S = -\rho c V \int_{T_{b_0}}^{T_b}\left(\frac{1}{T_\infty} - \frac{1}{T_b}\right)dT_b \tag{1.23}$$

Example 1.4

A thermocouple bead is largely solder, 1 mm in diameter. It is initially at room temperature and is suddenly placed in a 200°C gas flow. $\bar{h}$ is 250 W/m²-°C, and the effective values of k, ρ, and c are 45 W/m-°C, 9300 kg/m³, and $c = 0.18$ kJ/kg-°C, respectively. Evaluate the response of the thermocouple.

SOLUTION. The time constant, T, is

$$T = \frac{\rho c V}{\bar{h} A} = \rho c \frac{\pi}{6} \frac{D^3}{\bar{h}\pi D^2} = \frac{\rho c D}{6\bar{h}}$$

$$= \frac{(9300)(0.18)(0.001)}{6(250)} \frac{\text{kg}}{\text{m}^2} \frac{\text{kJ}}{\text{kg-°C}} \text{m} \frac{\text{m}^2\text{-°C}}{\text{W}} \frac{1000 \text{ W}}{\text{kJ/s}}$$

$$= \underline{1.116 \text{ s}}$$

Therefore, eqn. (1.21) becomes

$$\frac{T - 200°C}{(20 - 200)°C} = e^{-t/1.116} \qquad \text{or} \qquad \underline{T = 200 - 180e^{-t/1.116}}$$

This result is plotted in Fig. 1.12, where we see that this thermocouple catches up with the gas stream in less than 5 s. Indeed, it should be apparent that any such system will come within 95% of the signal in three time constants. Notice too that if the response could continue at its initial rate, the thermocouple would reach the signal temperature in one time constant.

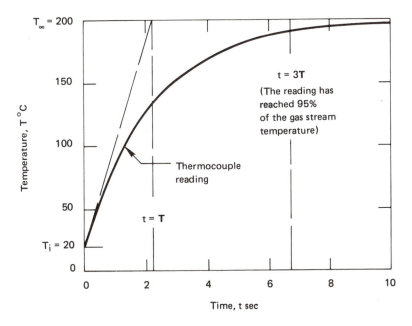

FIGURE 1.12 Thermocouple response to hot gas flow.

This calculation is based entirely on the assumption that Bi≪1 for the thermocouple. We must check that assumption:

$$\text{Bi} \equiv \frac{\bar{h}L}{k} = \frac{(250 \text{ W/m}^2\text{-}°\text{C})(0.001 \text{ m}/2)}{45 \text{ W/m-}°\text{C}} = 0.00278$$

This is very small indeed, so the assumption is valid.

Experiment 1.2

Invent, and carry out, a simple procedure for evaluating the time constant of a fever thermometer in your mouth.

Radiation

Heat transfer by thermal radiation. All bodies constantly emit energy by a process of electromagnetic radiation. The intensity of such energy flux depends upon the temperature of the body and the nature of its surface. Most of the heat that reaches you when you sit in front of a fire is radiant energy. Radiant energy browns your toast in an electric toaster and it warms you when you walk in the sun. Objects that are cooler than the fire, the toaster, or the sun emit much less energy because the energy emission varies as the fourth power of absolute temperature. Very often, the emission of energy, or *radiant heat transfer*, from cooler bodies can be neglected in comparison with convection and conduction. But heat transfer processes that occur at high temperature, or with conduction and convection suppressed by evacuated insulations, usually involve a significant fraction of radiation.

Experiment 1.3

Open the freezer door of your refrigerator. Put your face near it but stay far enough away to avoid the downwash of cooled air. This way you cannot be cooled by convection and, because the air between you and the freezer is a fine insulator, you cannot be cooled by conduction. Still, your face will feel cooler. The reason is that you radiate heat directly into the cold region and it radiates very little heat to you. Consequently, your face cools perceptibly.

The electromagnetic spectrum. Thermal radiation occurs in a range of the electromagnetic spectrum of energy emission. Accordingly, it exhibits the same wavelike properties as light or radio waves. Each quantum of radiant energy has a wavelength, λ, and a frequency, ν, associated with it.

The full electromagnetic spectrum includes an enormous range of energy-bearing waves, of which heat is only a small part. Table 1.2 lists the various forms over a range of wavelengths that spans 24 orders of magnitude. Only the tiniest "window" exists in this spectrum through which we can *see* the world around us. Heat radiation, whose main component is usually the spectrum of infrared radiation, passes through a much larger window—about three orders of magnitude in λ or ν.

Table 1.2   Forms of the electromagnetic wave spectrum

Character of Radiation	Wavelength,[a] $\lambda = c_\ell / \nu$	Frequency,[a] $\nu = (c_\ell / \lambda)$ Hz
Cosmic rays	Up to 4×10^{-3} Å	Above 0.75×10^{21}
γ-radiation	4×10^{-3}–1.4 Å	0.75×10^{21}–2×10^{18}
X-rays	0.1–200 Å	3×10^{19}–1.5×10^{16}
Ultraviolet light	100 Å–0.0004 mm	3×10^{16}–0.75×10^{15}
Visible light	0.0004–0.0008 mm	0.75×10^{15}–0.4×10^{15}
{ Infrared rays	0.0008–1 mm	0.4×10^{15}–3×10^{11}
{ *Heat*	0.0001–0.1 mm	3×10^{15}–3×10^{12}
Microwave (including radar)	1 mm–10 m	3×10^{11}–30,000
Radio waves	10 m–30 km	3×10^{7}–10,000
Electric transmission in the United States	5000 km	60

[a]The velocity of light, c_ℓ, is 3×10^5 km/s, and 1 Å $= 10^{-10}$ m $= 10^{-7}$ mm.

The model for the perfect thermal radiator is a so-called *black body*. This is a body which absorbs all energy that reaches it, and reflects nothing. The term can be a little confusing, since such bodies *emit* energy. Thus, if we possessed infrared vision, a black body would glow with the "color" appropriate to its temperature. Of course, perfect radiators *are* "black" in the sense that they absorb all visible light (and all other radiation) that reaches them.

It is necessary to have an experimental method for making a perfectly black body. The conventional device for approaching this ideal is called by the German term *hohlraum*, which literally means "hollow space." Figure 1.13 shows

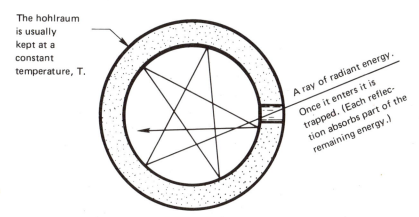

The hohlraum is usually kept at a constant temperature, T.

A ray of radiant energy. Once it enters it is trapped. (Each reflection absorbs part of the remaining energy.)

FIGURE 1.13   Cross-section of a spherical hohlraum. The hole has the attributes of a nearly perfect thermal black body.

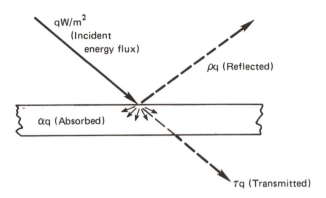

FIGURE 1.14   The distribution of energy incident on a translucent slab.

how a hohlraum is arranged. It is simply a device that traps all the energy that reaches the aperture.

What are the important features of a thermally black body? First consider a distinction between *heat* and infrared radiation. *Infrared radiation* refers to a particular range of wavelengths, while heat involves a distribution of radiant energy in wavelength peculiar to a given surface at a given temperature. Suppose that a radiant heat flux, q, falls upon a translucent plate that is not black, as shown in Fig. 1.14. A fraction, α, of the total incident energy, called the *absorptance*, is absorbed in the body; a fraction, ρ, called the *reflectance*, is reflected from it; and a fraction, τ, called the *transmittance*, passes through. Thus

$$1 = \alpha + \rho + \tau \tag{1.24}$$

This relation can also be written for the energy carried by each wavelength in the distribution of wavelengths that makes up *heat* from a source at any temperature:

$$1 = \alpha_\lambda + \rho_\lambda + \tau_\lambda \tag{1.25}$$

All radiant energy incident on a black body is absorbed (so that α or $\alpha_\lambda = 1$ and $\rho = \tau = 0$). Furthermore, the energy emitted from such a body reaches a theoretical maximum, which is given by the Stefan–Boltzmann law. We look at this next.

The Stefan–Boltzmann law.   The flux of energy radiating from a black body is commonly designated $e(T) W/m^2$. The symbol $e_\lambda(\lambda, T)$ designates the distribution function of radiative flux in λ, or the *monochromatic emissive power*:

$$e_\lambda(\lambda, T) = \frac{de(\lambda, T)}{d\lambda} \qquad \text{or} \qquad e(\lambda, T) = \int_0^\lambda e_\lambda(\lambda, T)\, d\lambda \tag{1.26}$$

Thus

$$e(T) \equiv e(\infty, T) = \int_0^\infty e_\lambda(\lambda, T)\, d\lambda$$

The dependence of $e(T)$ on T was established experimentally by Stefan in 1879 and explained by Boltzmann on the basis of thermodynamic arguments in 1884. The Stefan–Boltzmann law is

$$e(T) = \sigma T^4 \tag{1.27}$$

where the Stefan–Boltzmann constant, σ, is 5.6697×10^{-8} W/m²-°K⁴ or 1.714×10^{-9} Btu/ft²-hr-°R⁴, and T is the absolute temperature.

e_λ *vs.* λ.   Nature requires that, at a given temperature, a black body will emit a unique distribution of energy in wavelength. Thus when you heat a poker in the fire, it first glows a dull red—emitting most of its energy at long wavelengths and just a little bit in the visible regime. When it is white-hot the energy distribution has been both greatly increased and shifted toward the shorter-wavelength visible range. A black body yields the highest attainable value of e_λ at any temperature.

The very accurate measurements of the energy spectrum by Lummer and Pringsheim (1899) are shown in Fig. 1.15. The locus of maxima of the curves is also plotted. It obeys the relation

$$(\lambda T)_{e_\lambda = \text{max}} = 0.2898 \text{ cm-°K} \tag{1.28}$$

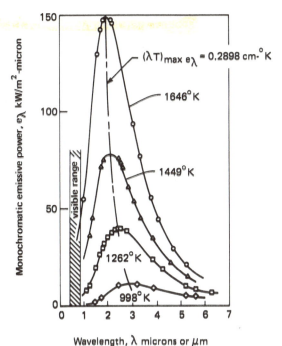

FIGURE 1.15   Emissive power of a black body at several temperatures—predicted and observed.

About three-fourths of the radiant energy of a black body lies to the right of this line in Fig. 1.15. Notice that, while the locus of maxima leans toward the visible range at higher temperatures, only a small fraction of the radiation is visible even at the highest temperature.

Predicting how the monochromatic emissive power of a black body depends on λ was an increasingly serious problem at the close of the nineteenth century. The prediction was a keystone of the most profound scientific revolution the world has seen. In 1901, Max Planck made the prediction, and his work included the initial formulation of quantum mechanics. He found that

$$e_\lambda = \frac{2\pi h c_\ell^2}{\lambda^5 \left[\exp(hc_\ell / kT\lambda) - 1 \right]} \tag{1.29}$$

where c_ℓ is the speed of light; h is Planck's constant, 6.6256×10^{-34} J-s; and k is Boltzmann's constant, 1.3805×10^{-23} J/°K.

Radiant heat exchange. Suppose that a heated object (1 in Fig. 1.16a) radiates to some other object (2). Then if both objects are thermally black, the net heat transferred from object 1 to object 2, Q_{net}, is the difference between Q_{1-2} and Q_{2-1}, or

$$\boxed{Q_{net} = A_1 \left[e_1(T) - e_2(T) \right] = \sigma A_1 \left(T_1^4 - T_2^4 \right)} \tag{1.30}$$

If the first object "sees" other objects in addition to object 2, as indicated in Fig. 1.16b, then a *view factor* (sometimes called an "angle factor" or a "shape

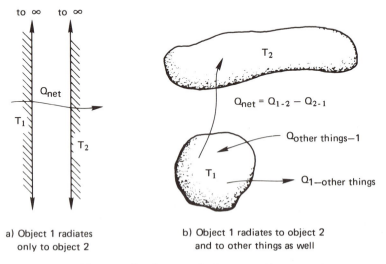

a) Object 1 radiates
   only to object 2

b) Object 1 radiates to object 2
   and to other things as well

FIGURE 1.16   The net radiant heat transfer from one object to another.

factor"), F_{1-2}, must be included in eqn. (1.30):

$$Q_{\text{net}} = F_{1-2}\sigma A_1 \left(T_1^4 - T_2^4 \right) \tag{1.31}$$

where F_{1-2} is the fraction of energy leaving object 1 that is intercepted by object 2. Finally, if the bodies are not black, then the view factor, F_{1-2}, must be replaced by a new factor, $\mathscr{F}_{1-2}$, which depends on surface properties of the various objects as well as the geometrical "view."

Example 1.5

A black thermocouple measures the temperature in a chamber with black walls. If the air around the thermocouple is at 20°C, the walls are at 100°C, and the heat transfer coefficient between the thermocouple and the air is 15 W/m²-°C, what temperature will the thermocouple read?

SOLUTION.   The heat convected away from the thermocouple by the air must exactly balance that radiated to it by the hot walls if the system is steady. Furthermore, F_{1-2} is unity since the thermocouple is enclosed:

$$\bar{h} A (T_{\text{tc}} - T_{\text{air}}) = \sigma A \left(T_{\text{wall}}^4 - T_{\text{tc}}^4 \right)$$

or

$$15(T_{\text{tc}} - 20) \text{ W/m}^2 = 5.6697 \times 10^{-8} \left[373^4 - (T_{\text{tc}} + 273)^4 \right] \text{ W/m}^2$$

Trial-and-error solution of this equation yields $T_{\text{tc}} = 51°C$.

Radiation shielding.   The preceding example points out an important practical problem that can be solved with radiation shielding. The idea is as follows: If we want to measure the true air temperature, we can place a thin foil casing, or shield, around the thermocouple. The casing is shaped to obstruct the thermocouple's "view" of the room but to permit the free flow of air around the thermocouple. Then the casing (or shield) will be closer to 50°C than to 100°C, and the thermocouple will be influenced by this much cooler radiator. If the shield is highly reflecting on the outside, it will assume a temperature still closer to that of the air and the error will be still less. Multiple layers of shielding can further reduce the error.

Radiation shielding can take many forms and serve many purposes. It is an important element in superinsulations. A glass firescreen in a fireplace serves as a radiation shield because it is largely opaque to radiation. It absorbs energy and reradiates (ineffectively) at a temperature much lower than that of the fire.

Example 1.6

A crucible of molten metal at 1800°C is placed on the foundry floor. The foundryman covers it with a metal sheet to reduce heat loss to the room. If $\mathscr{F}$ is 0.4 between the melt and the plate and 0.8 between either the melt or the top of the plate and the room, how much will the heat loss to the room be reduced by the sheet?

SOLUTION.   First find the sheet temperature:

$$q = (0.8)\sigma\left[T_s^4 - (20+273)^4 \right] = (0.4)\sigma\left[(1800+273)^4 - T_s^4 \right]$$

This gives $T_{\text{sheet}} = 1575^\circ\text{K}$, so

$$\frac{q_{\text{with sheet}}}{q_{\text{without sheet}}} = \frac{0.8\sigma(1575^4 - 293^4)}{0.8\sigma(2073^4 - 293^4)} = 0.333$$

The heat loss is therefore reduced by 66.7% by the shield.

Experiment 1.4

Find a small open flame that produces a fair amount of soot. A candle, kerosene lamp, or a cutting torch with a rich mixture should work well. A clean blue flame will not work well because such gases do not radiate much heat. First place your finger in a position about 1 to 2 cm to one side of the flame, where it becomes uncomfortably hot. Now take a piece of fine mesh screen and dip it in soapy water, which will fill up the holes. Put it between your finger and the flame. You will see that your finger is protected from heating until the water evaporates.

Water is relatively transparent to light. What does this experiment show you about the transmittance of water to infrared wavelengths?

1.4 A look ahead

What we have done up to this point has been no more than to reveal the tip of the iceberg. The basic mechanisms of heat transfer have been explained and some quantitative relations have been presented. However, this information will barely get you started when you are faced with a real heat transfer problem. Three tasks, in particular, must be completed to solve actual problems:

- The heat diffusion equation must be solved subject to appropriate boundary conditions if the problem involves heat conduction of any complexity.
- The convective heat transfer coefficient, h, must be determined if convection is important in a problem.
- The factor F_{1-2} or $\mathscr{F}_{1-2}$ must be determined to calculate radiative heat transfer.

Any of these determinations can involve a great deal of complication, and most of the chapters that lie ahead are devoted to these three basic problems.

Before becoming engrossed in these three questions, we shall first look at the archetypical applied problem of heat transfer—namely the design of a heat exchanger. Chapter 2 sets up the elementary analytical apparatus that is needed for this, and Chapter 3 shows how to do such design if $\bar{h}$ is already known. This will make it easier to see the importance of undertaking the three basic problems in subsequent parts of the book.

1.5 Problems

We have noted that this book is set down in S.I. units almost exclusively. The student who has problems with dimensional conversion will find Appendix B to be helpful. The only use of English units appears in certain of the problems at the end of each chapter. A few such problems are included to provide experience in converting back into English units, since such units will undoubtedly persist in this country for many more years. You will almost certainly find such problems to be annoying and not as simple as those in S.I. units.

There is another matter that often lends to some discussion between students and teachers in heat transfer courses. That is the question as to whether a problem is "theoretical" or "practical." Quite often the student is inclined to view as "theoretical" a problem that does not involve numbers, or that requires him to develop algebraic results.

The problems assigned in this book are all intended to be useful in that they do one or more of five things:

1. They involve a calculation of a type that actually arises in practice (e.g., Problems 1.1, 1.3, 1.8 to 1.18, and 1.21 through 1.25).

2. They illustrate a physical principle (e.g., Problems 1.2, 1.4 to 1.7, 1.9, and 1.20). These are probably closest to having a "theoretical" objective.

3. They ask you to use methods developed in the text, to develop other results which would be needed in certain applied problems (e.g., Problems 1.10, 1.16, 1.17, and 1.21). Such problems are usually the most difficult and the most valuable to you.

4. They anticipate developments that will appear in subsequent chapters (e.g., Problems 1.16 and 1.20).

5. They require that you develop your ability to handle numerical and algebraic computation effectively. (This is the case with most of the problems in Chapter 1, but it is especially true of Problems 1.6 to 1.9, 1.15, and 1.17).

Partial numerical answers to some of the problems follow them in brackets.

Actually, we wish to look at the *theory*, *analysis*, and *practice* of heat transfer —all three—where we accept Webster's definitions of the terms:

Theory: "a systematic statement of principles; a formulation of apparent relationships or underlying principles of certain observed phenomena."

Analysis: "the solving of problems by the means of equations; the breaking up of any whole into its parts so as to find out their nature, function, relationship, etc."

Practice: "the *doing* of something as an application of knowledge."

PROBLEMS

1.1. A composite wall consists of alternate layers of fir (5 cm thick), aluminum (1 cm thick), lead (1 cm thick), and corkboard (6 cm thick). The temperature is 60°C on the outside of the fir and 10°C on the outside of the corkboard. Plot the temperature gradient through the wall. Does the temperature profile suggest any simplifying assumptions that might be made in subsequent analysis of the wall?

1.2. Verify eqn. (1.15).

1.3. $q = 5000$ W/m^2 in a 1-cm slab and $T = -40$°C on the cold side. Tabulate the temperature drop through the slab if it is made of

- Silver
- Aluminum
- Mild steel (0.5% C)
- Ice
- Spruce
- Insulation (85% magnesia)
- Silica aerogel

Indicate which situations would be unreasonable, and why.

1.4. Explain in words why the heat diffusion equation, eqn. (1.13), shows that in transient conduction the temperature depends on the thermal diffusivity, α, but we can solve steady conduction problems using just k (as in Example 1.1).

1.5. A 1-m rod of pure copper 1 cm^2 in cross section connects a 200°C thermal reservoir, with a 0°C thermal reservoir. The system has already reached steady state. What are the rates of change of entropy of (a) the first reservoir, (b) the second reservoir, (c) the rod, and (d) the whole universe, as a result of the process? Explain whether or not your answer satisfies the Second Law of Thermodynamics. [(d): +0.0120 W/°K.]

1.6. Two thermal energy reservoirs at temperatures of 27°C and −43°C, respectively, are separated by a slab of material 10 cm thick and 930 cm² in cross-sectional area. The slab has a thermal conductivity of 0.14 W/m-°C. The system is operating at steady-state conditions. What are the rates of change of entropy of (a) the higher temperature reservoir, (b) the lower temperature reservoir, (c) the slab, and (d) the whole universe as a result of this process? (e) Does your answer satisfy the Second Law of Thermodynamics?

1.7. (a) If the thermal energy reservoirs in Problem 1.6 are suddenly replaced with adiabatic walls, determine the final equilibrium temperature of the slab. (b) What is the entropy change for the slab for this process? (c) Does your answer satisfy the Second Law of Thermodynamics in this instance? Explain. The density of the slab is 26 lb/ft³ and the specific heat is 0.65 Btu/lb-°F. [(b): 30.81 J/°K.]

1.8. A copper sphere 2.5 cm in diameter has a uniform temperature of 40°C. The sphere is suspended in a slow-moving air stream at 0°C. The air stream produces a convection heat transfer coefficient of 15 W/m²-°C. Radiation can be neglected. Since copper is highly conductive, temperature gradients in the sphere will smooth out rapidly, and its temperature at any instant during the cooling process can be taken as uniform (i.e., Bi≪1). Write the instantaneous energy balance between the sphere and the surrounding air. Solve this equation and plot the resulting temperature as a function of time between 40 and 0°C.

1.9. Determine the total heat transfer in Problem 1.8 as the sphere cools from 40 to 0°C. Plot the net entropy increase resulting from the cooling process above, ΔS, vs. T°K. [Total heat transfer = 1123 J.]

1.10. A truncated cone 30 cm high is constructed of portland cement. The diameter at the top is 15 cm and at the bottom is 7.5 cm. The lower surface is maintained at 6°C and the top at 40°C. The other surface is insulated. Assume one-dimensional heat transfer and calculate the rate of heat transfer in watts from top to bottom. To do this, note that the heat transfer, Q, must be the same at every cross section. Write Fourier's law locally, and integrate it from top to bottom to get a relation between this unknown Q and the known end temperatures. [$Q = -1.70$ W.]

1.11. A hot water heater contains 100 kg of water at 75°C in a 20°C room. Its surface area is 1.3 m². Select an insulating material, and specify its thickness, to keep the water from cooling more than 3°C/h. (Notice that this problem will be greatly simplified if the temperature drop in the steel casing and the temperature drop in the convective boundary layers are negligible. Can you make such assumptions? Explain.)

1.12. What is the temperature at the left-hand wall shown in Fig. P1.12? Both walls are thin, very large in extent, highly conducting, and thermally black. [$T_{right} = 42.5$°C.]

1.13. Develop S.I. to English conversion factors for:

- The thermal diffusivity, α
- The heat flux, q

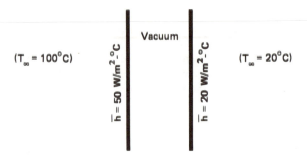

FIGURE P1.12

- The density, ρ
- The Stefan–Boltzmann constant, σ
- The view factor, F_{1-2}
- The molar entropy
- The specific heat per unit mass, c

In each case begin with basic dimensions J, m, kg, s, and °C, and check your answers against Appendix B if possible.

1.14. Three infinite, parallel, black, opaque plates transfer heat by radiation, as shown in Fig. P1.14. Find T_2.

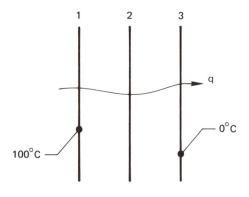

FIGURE P1.14

1.15. Four infinite, parallel, black, opaque plates transfer heat by radiation, as shown in Fig. P1.15. Find T_2 and T_3. [$T_2 = 75.53$°C.]

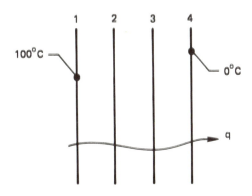

FIGURE P1.15

1.16. Two large black horizontal plates are spaced a distance L from one another. The top one is warm at a controllable temperature, T_h, and the bottom one is cool at a specified temperature, T_c. A gas separates them. The gas is stationary because it is warm on top and cold on the bottom. Write the equation $q_{rad}/q_{cond} =$ fn$(N, \Theta \equiv T_h/T_c)$, where N is a dimensionless group containing σ, k, L, and T_c. Plot N as a function of Θ for $q_{rad}/q_{cond} = 1$, 0.8, and 1.2 (and for other values if you wish).

Now suppose that you have a system in which $L = 10$ cm, $T_c = 100°$K, and the gas is hydrogen with an average k of 0.1 W/m-°C. Further suppose you wish to operate in such a way that the conduction and radiation heat fluxes are identical. Identify the operating point on your curve and report the value of T_h that you must maintain.

1.17. A blackened copper sphere 2 cm in diameter and uniformly at 200°C is introduced into an evacuated black chamber which is maintained at 20°C.

- Write the differential equation that expresses $T(t)$ for the sphere, assuming lumped thermal capacity.

- Identify a dimensionless group, analogous to the Biot number, which can be used to tell whether or not the lumped-capacity solution is valid.

- Show that the lumped-capacity solution is valid.

- Integrate your differential equation and plot the temperature response for the sphere.

1.18. As part of a space experiment, a small instrumentation package is released from a space vehicle. It can be approximated as a solid aluminum sphere, 4 cm in diameter. The sphere is initially at 30°C and it contains a pressurized hydrogen component that will condense and misfunction at 30°K. If we take the surrounding space to be at 0°K, how long may we expect the instrumentation package to function properly? Is it legitimate to use the lumped-capacity method in solving the problem? (*Hint:* See the directions for Problem 1.17.) [Time = 5.8 weeks.]

1.19. Consider heat conduction through the wall as shown in Fig. P1.19. Calculate q and the temperature of the right-hand side of the wall.

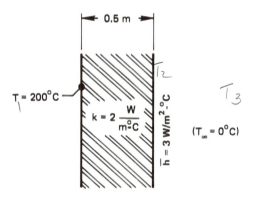

FIGURE P1.19

1.20. Throughout Chapter 1 we have assumed that the steady temperature distribution in a plane uniform wall is linear. To prove this, simplify the heat diffusion equation to the form appropriate for steady flow. Then integrate it twice and eliminate the two constants using the known outside temperatures T_{left} and T_{right} at $x=0$ and $x=$ wall thickness, L.

1.21. The thermal conductivity in a particular plane wall depends as follows on the wall temperature: $k=A+BT$, where A and B are constants. The temperatures are T_1 and T_2 on either side of the wall and its thickness is L. Develop an expression for q.

1.22. Find k for the wall shown in Fig. P1.22. What might it be made of?

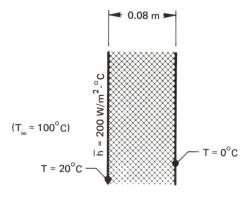

FIGURE P1.22

1.23. What are T_i, T_j, and T_r in the wall shown in Fig. P1.23? [$T_j=16.44°C.$]

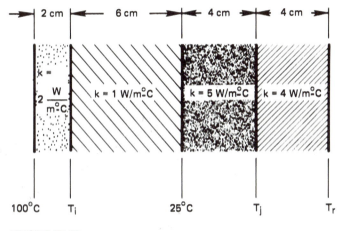

FIGURE P1.23

1.24.   An aluminum can of beer or soda pop is removed from the refrigerator and set on the table. If $\bar{h}$ is 13.5 W/m²-°C, estimate when the beverage will be at 15°C. State all of your assumptions.

1.25.   One large black wall at 27°C faces another whose surface is at 127°C. The gap between the two walls is evacuated. If the second wall is 0.1 m thick and has a thermal conductivity of 17.5 W/m-°C, what is its temperature on the back side? (Assume steady state.)

A SHORT BIBLIOGRAPHY

Subsequent chapters have lists of specific references appended to them. This, however, is a bibliography of general readings in heat transfer for the readers who wish to supplement their study of any broad area of the subject. This list is only a sampling of much good material that is available.

Reissued Books of Historical Importance

[1.1]   J. FOURIER, *The Analytical Theory of Heat*, Dover Publications, Inc., New York, 1955.

[1.2]   L. M. K. BOELTER, V. H. CHERRY, H. A. JOHNSON, and R. C. MARTINELLI, *Heat Transfer Notes*, McGraw-Hill Book Company, New York, 1965. (Originally issued as class notes at the University of California at Berkeley between 1932 and 1941.)

General Introductory Textbooks Reflecting the Influence of [*1.2*]

[1.3]   J. P. HOLMAN, *Heat Transfer*, 4th ed., McGraw-Hill Book Company, New York, 1976.

[1.4]   F. KREITH, *Principles of Heat Transfer*, 3rd ed., Intext Press, Inc., New York, 1973.

[1.5]   W. H. GIEDT, *Engineering Heat Transfer*, D. Van Nostrand Co., Inc., Princeton, N.J., 1957.

Some Good Intermediate-Level Books

[1.6]   W. M. ROHSENOW and H. Y. CHOI, *Heat, Mass and Momentum Transfer*, Prentice-Hall, Inc., Englewood Cliffs, N.J., 1961.

[1.7]   E. R. G. ECKERT and R. M. DRAKE, JR., *Analysis of Heat and Mass Transfer*, McGraw-Hill Book Company, New York, 1972.

[1.8]   S. T. I. HSU, *Engineering Heat Transfer*, D. Van Nostrand Co. Inc., Princeton, N.J., 1963.

[1.9]   B. GEBHART, *Heat Transfer*, 2nd ed., McGraw-Hill Book Company, New York, 1971.

[1.10]   M. N. ÖZIŞIK, *Basic Heat Transfer*, McGraw-Hill Book Company, New York, 1977.

Some Books of Solved Problems

[1.11]   J. SUCEC, *Heat Transfer*, Simon and Shuster Technical Outlines, New York, 1975.

[1.12]   D. R. PITTS and L. E. SISSOM, *Heat Transfer*, Schaum's Outline Series, McGraw-Hill Book Company, New York, 1977.

Some Treatises and Advanced Texts in Certain Areas of Heat Transfer

[1.13]   W. M. ROHSENOW and J. P. HARTNETT, *Handbook of Heat Transfer*, McGraw-Hill Book Company, New York, 1973.

[1.14]   H. S. CARSLAW and J. C. JAEGER, *Conduction of Heat in Solids*, 2nd ed., Oxford University Press, New York, 1959.

[1.15]   P. J. SCHNEIDER, *Conduction Heat Transfer*, Addison-Wesley Publishing Co., Inc., Reading, Mass., 1955.

[1.16]   V. S. ARPACI, *Conduction Heat Transfer*, Addison-Wesley Publishing Co., Reading, Mass., 1966.

[1.17]   G. E. MYERS, *Analytical Methods in Conduction Heat Transfer*, McGraw-Hill Book Company, New York, 1971.

[1.18]   W. M. KAYS, *Convective Heat and Mass Transfer*, McGraw-Hill Book Company, New York, 1966.

[1.19]   H. C. HOTTEL and A. F. SAROFIM, *Radiative Heat Transfer*, McGraw-Hill Book Company, New York, 1967.

[1.20]   E. M. SPARROW and R. D. CESS, *Radiation Heat Transfer*, Hemisphere Publishing Corp./McGraw-Hill Book Company, Washington, D.C., 1978.

[1.21]   R. SIEGEL and J. R. HOWELL, *Thermal Radiative Heat Transfer*, McGraw-Hill Book Company, New York, 1972.

[1.22]   J. G. COLLIER, *Convective Boiling and Condensation*, McGraw-Hill Book Company, New York, 1972.

[1.23]   Y. Y. HSU and R. W. GRAHAM, *Transport Processes in Boiling and Two-Phase Systems*, Hemisphere Publishing Corp.: McGraw-Hill Book Company, Washington, D.C., 1976.

[1.24]   W. M. KAYS and A. L. LONDON, *Compact Heat Exchangers*, McGraw-Hill Book Company, New York, 1964.

[1.25]   A. H. P. SKELLAND, *Diffusional Mass Transfer*, Wiley–Interscience, New York, 1974.

Chapter 2

Heat conduction concepts, thermal resistance, and the overall heat transfer coefficient

---

2.1 The heat diffusion equation

Objective

We must now develop some ideas that will be needed for the design of heat exchangers. The most important of these is the notion of an overall heat transfer coefficient. This is a measure of the general resistance of a heat exchanger to the flow of heat, and usually it must be built up from analyses of component resistances. In particular, we must know how to predict $\bar{h}$ and how to evaluate the conductive resistance of bodies more complicated than plane passive walls. The evaluation of $\bar{h}$ is a matter that must be deferred to Chapters 7 and 8. For the present, $\bar{h}$ values must be considered given information in any problem.

The heat conduction component of most heat exchanger problems is more complex than the simple planar analysis discussed in Chapter 1. To do such analyses we must next derive the heat conduction equation and learn to solve it.

Consider the general temperature distribution in a three-dimensional body as depicted in Fig. 2.1: For some reason (heating from one side, in this case), there is a space- and time-dependent temperature field in the body. This field,

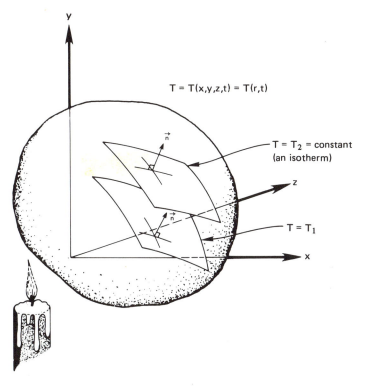

T = T(x,y,z,t) = T(r,t)

T = T_2 = constant
(an isotherm)

T = T_1

FIGURE 2.1 A three-dimensional, transient, temperature field.

$T = T(x,y,z,t)$ or $T(\vec{r},t)$, defines instantaneous isothermal surfaces, T_1, T_2, and so on.

We next consider a very important vector associated with the scalar, T. The vector that has both the magnitude and direction of maximum increase of temperature at each point is called the *temperature gradient*, $\vec{\nabla} T$:

$$\vec{\nabla} T \equiv \vec{i}\, \frac{\partial T}{\partial x} + \vec{j}\, \frac{\partial T}{\partial y} + \vec{k}\, \frac{\partial T}{\partial z} \tag{2.1}$$

Fourier's law

"Experience"—that is, physical observation—suggests two things about the heat that results from temperature nonuniformities in a body. These are:

$$\frac{\vec{q}}{|\vec{q}|} = - \frac{\vec{\nabla} T}{|\vec{\nabla} T|} \qquad \left\{ \begin{array}{l} \text{This says that } \vec{q} \text{ and } \vec{\nabla} T \text{ are exactly} \\ \text{opposite one another in direction} \end{array} \right.$$

and

$$|\vec{q}| \propto |\vec{\nabla} T|$$ $\left\{\begin{array}{l}\text{This says that the magnitude of the heat}\\\text{flux is directly proportional to the tem-}\\\text{perature gradient}\end{array}\right.$

Notice that the heat flux is now written as a quantity that has a specified direction as well as a specified magnitude. Fourier's law summarizes this physical experience succinctly as

$$\boxed{\vec{q} = -k\vec{\nabla} T} \tag{2.2}$$

which resolves itself into three components:

$$q_x = -k\frac{\partial T}{\partial x} \qquad q_y = -k\frac{\partial T}{\partial y} \qquad q_z = -k\frac{\partial T}{\partial z}$$

The "constant" k—the thermal conductivity—also depends on position and temperature in the most general case:

$$k = k[\vec{r}, T(\vec{r}, t)] \tag{2.3}$$

Fortunately, most materials (although not all of them) are homogeneous. Thus we can usually write $k = k(T)$. The assumption that we really want to make is the assumption that k is constant. Whether or not that is legitimate must be determined in each case. As is apparent from Figs. 2.2 and 2.3, k almost always varies with temperature. It always rises with T in gases at low pressures, but it may rise or fall in metals and liquids. The problem is that of assessing whether or not k is approximately constant in the range of interest. We could safely take k to be a constant for iron between 0 and 40°C (see Fig. 2.2), but we would incur error between -100 and 800°C.

It is easy to prove (Problem 2.1) that if k varies linearly with T, and if heat transfer is plane and steady, then $q = k\Delta T/L$, with k evaluated at the average temperature in the plane. If heat transfer is not planar or if $k \neq A + BT$, it can be much more difficult to specify a single accurate effective value of k. If ΔT is not large, one can still make a reasonably accurate approximation using a constant average value of k.

Fourier's law has some extremely important analogies in other kinds of physical behavior. These analogous processes provide us with some guidance in the solution of heat transfer problems, and heat conduction analyses can often be adapted to describe such processes as well. First consider *Ohm's law*:

$$\text{flux of electrical charge} = \frac{\vec{I}}{A} \equiv \vec{i} = -\gamma\vec{\nabla} V \tag{2.4}$$

where $\vec{I}$ amperes is the vectorial electrical current, A is an area normal to $\vec{I}$, $\vec{i}$ is the current flux, γ cm/ohm-cm^2 is the electrical conductivity, and V is the voltage.

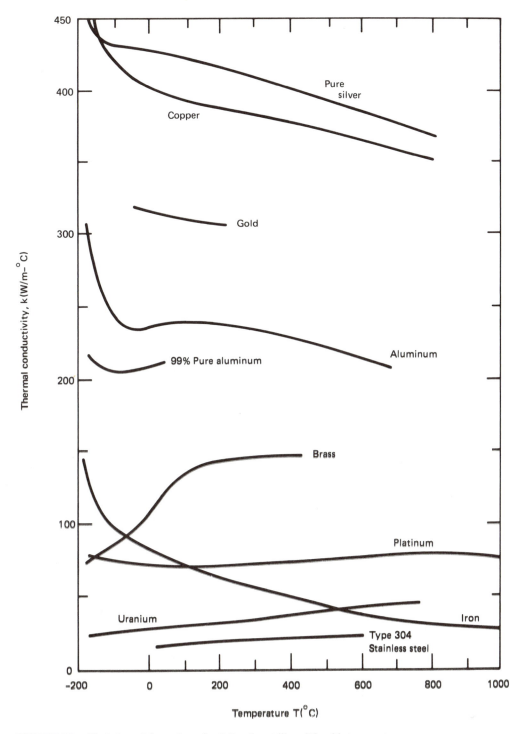

FIGURE 2.2 Variation of thermal conductivity of metallic solids with temperature.

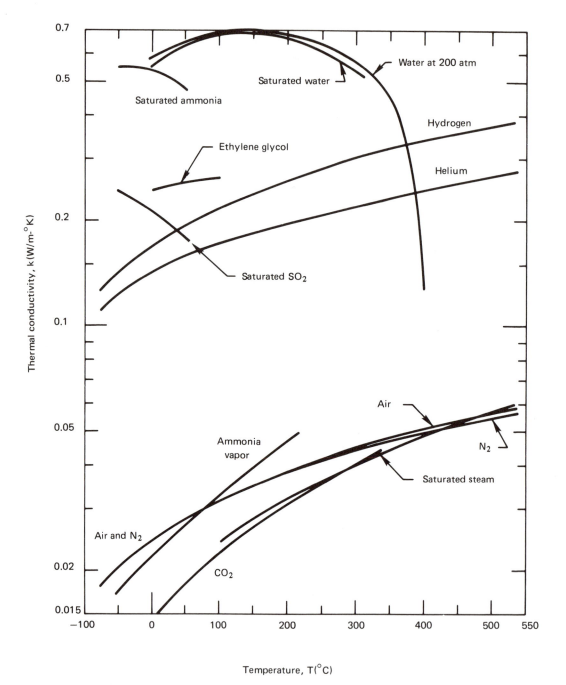

FIGURE 2.3 The temperature dependence of the thermal conductivity of liquids, and of gases which are either saturated or at 1 atm. pressure.

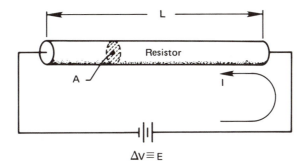

FIGURE 2.4   The one-dimensional flow of current.

Consider the application of eqn. (2.4) to a one-dimensional current flow situation pictured in Fig. 2.4. In this case eqn. (2.4) becomes

$$i = -\gamma \frac{dV}{dx} = \gamma \frac{\Delta V}{L} \tag{2.5}$$

but ΔV is the applied voltage, E, and $\gamma \equiv L/RA$, where R is the resistance of the wire. Then, since $I \equiv iA$, eqn. (2.5) becomes

$$I = \frac{E}{R} \tag{2.6}$$

which is the restrictive one-dimensional statement of Ohm's law that is familiar.

Fick's law is another analogous relation. It states that during mass diffusion, the flux, $\vec{J}_1$, of a dilute component, 1, is proportional to the gradient of its mass concentration, c_1. Thus

$$\vec{J}_1 = -\mathscr{D}_{12} \vec{\nabla} c_1 \tag{2.7}$$

where the constant $\mathscr{D}_{12}$ is the coefficient of mutual diffusion.

Example 2.1

Air fills a tube 1 m in length. There is a small water leak at one end where the water vapor concentration builds to a mole fraction of 0.01. A desiccator maintains the concentration at zero on the other side. What is the steady flux of water from one side to the other if $\mathscr{D}_{12}$ is 0.000284 m^2/s?

SOLUTION.

$$\vec{J}_{\text{water vapor}} = 0.000284 \, \frac{\text{m}^2}{\text{s}} \, \frac{0.01 \text{ mole H}_2\text{O/mole mixture}}{1 \text{ m}}$$

$$= 0.00000284 \text{ m}^3/\text{m}^2\text{-s} = \underline{10.2 \text{ liters/m}^2\text{-min}}$$

We have now revisited Fourier's law in three dimensions and seen that there is a little more to it than we saw in Chapter 1. Next we write the heat conduction equation in three dimensions.

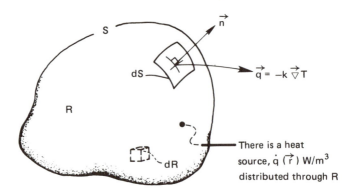

FIGURE 2.5   Control volume in a heat flow field.

Three-dimensional heat diffusion equation
with heat generation

We begin as we did in Chapter 1 with the First Law statement, eqn. (1.2):

$$Q = \frac{\partial U}{\partial t} \qquad (1.2)$$

where we write the derivative in partial form to distinguish between changes of U with respect to $\vec{r}$ and t. This time we apply eqn. (1.2) to a three-dimensional control volume[1] as shown in Fig. 2.5. The control volume is a finite region of a conducting body which we set aside for analysis. The surface is denoted as S and the volume of the region as R. An element of the surface, dS, is identified and two vectors are shown on dS: one is the unit normal vector, $\vec{n}$ (with $|\vec{n}| = 1$), and the other is the heat flux vector, $\vec{q} = - k\vec{\nabla} T$, at that point on the surface.

We also allow the possibility that there is a volumetric heat release of $\dot{q}(\vec{r})$ W/m^3 distributed through the region. This might be the result of chemical or nuclear reaction, of electrical resistance heating, of external radiation into the region, or of still other causes. With reference to Fig. 2.5, we can write the heat flux, dQ, out of dS, as

$$dQ = (- k\vec{\nabla} T) \cdot (\vec{n}\, dS) \qquad (2.8)$$

If heat is also being generated (or consumed) within the region, R, it must be added to eqn. (2.8) to get the *net* heat rate to R:

$$Q = - \int_S (- k\vec{\nabla} T) \cdot (\vec{n}\, dS) + \int_R \dot{q}\, dR \qquad (2.9)$$

---

[1] Figure 2.5 is the three-dimensional version of the control volume shown in Fig. 1.8.

The rate of energy increase of the region, R, is

$$\frac{\partial U}{\partial t} = \int_R \left(\rho c \frac{\partial T}{\partial t} \right) dR \tag{2.10}$$

Finally, we combine Q as given by eqn. (2.9), and $\partial U/\partial t$ as given by eqn. (2.10), into eqn. (1.2). After rearranging the terms, we obtain

$$\int_S k \vec{\nabla} T \cdot \vec{n} \, dS = \int_R \left[\rho c \frac{\partial T}{\partial t} - \dot{q} \right] dR \tag{2.11}$$

To get the left-hand side into convenient form we introduce Gauss's theorem, which converts a surface integral into a volume integral. Gauss's theorem says that if $\vec{A}$ is any continuous function of position, then

$$\int_S \vec{A} \cdot \vec{n} \, dS = \int_R \vec{\nabla} \cdot \vec{A} \, dR \tag{2.12}$$

Therefore, if we identify $\vec{A}$ with $(k \vec{\nabla} T)$, eqn. (2.11) reduces to

$$\int_R \left(\vec{\nabla} \cdot k \vec{\nabla} T - \rho c \frac{\partial T}{\partial t} + \dot{q} \right) dR = 0 \tag{2.13}$$

Finally, since the region, R, is arbitrary, the integrand must vanish identically.[2] We therefore get the *heat diffusion equation* in three dimensions:

$$\boxed{\vec{\nabla} \cdot k \vec{\nabla} T + \dot{q} = \rho c \frac{\partial T}{\partial t}} \tag{2.14}$$

The limitations on this equation are:

*   Incompressible medium. (This was implied when no expansion work term was included.)
*   No convection. (The medium cannot undergo any relative motion. However, it *can* be a liquid or gas as long as it sits still.)

If the variation of k with T is small, k can be factored out in eqn. (2.14) to get

$$\boxed{\nabla^2 T + \frac{\dot{q}}{k} = \frac{1}{\alpha} \frac{\partial T}{\partial t}} \tag{2.15}$$

This is a more complete version of the heat conduction equation [recall eqn. (1.13)] and α is the thermal diffusivity which was discussed after eqn. (1.13). The term $\nabla^2 T \equiv \vec{\nabla} \cdot \vec{\nabla} T$ is called the "Laplacian." It arises thus in a Cartesian

---

[2]Consider $\int f(x)\,dx = 0$. If $f(x)$ were, say, $\sin x$, then this could only be true over intervals of $x = 2\pi$ or multiples of it. For eqn. (2.13) to be true for *any* range of integration one might choose, the terms in parentheses must be zero everywhere.

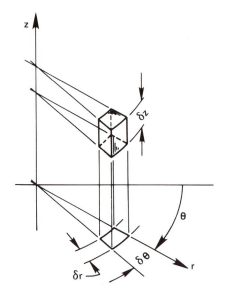

Polar coordinates

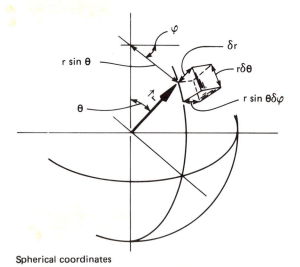

Spherical coordinates

FIGURE 2.6 Cylindrical and spherical coordinate schemes.

coordinate system:

$$\vec{\nabla} \cdot k\vec{\nabla} T \simeq k\vec{\nabla} \cdot \vec{\nabla} T = k\left(\vec{i}\,\frac{\partial}{\partial x} + \vec{j}\,\frac{\partial}{\partial y} + \vec{k}\,\frac{\partial}{\partial z}\right) \cdot \left(\vec{i}\,\frac{\partial T}{\partial x} + \vec{j}\,\frac{\partial T}{\partial y} + \vec{k}\,\frac{\partial T}{\partial z}\right)$$

or

$$\boxed{\nabla^2 T = \frac{\partial^2 T}{\partial x^2} + \frac{\partial^2 T}{\partial y^2} + \frac{\partial^2 T}{\partial z^2}}$$ (2.16)

The Laplacian can also be expressed in cylindrical or spherical coordinates. The results are:

- Cylindrical:

$$\nabla^2 T \equiv \frac{1}{r}\frac{\partial}{\partial r}\left(r\frac{\partial T}{\partial r}\right) + \frac{1}{r^2}\frac{\partial^2 T}{\partial \theta^2} + \frac{\partial^2 T}{\partial z^2}$$ (2.17)

- Spherical:

$$\nabla^2 T \equiv \frac{1}{r}\frac{\partial^2 (rT)}{\partial r^2} + \frac{1}{r^2 \sin\theta}\frac{\partial}{\partial \theta}\left(\sin\theta\frac{\partial T}{\partial \theta}\right) + \frac{1}{r^2 \sin^2\theta}\frac{\partial^2 T}{\partial \phi^2}$$ (2.18)

where the coordinates are described in Fig. 2.6.

2.2 Solutions of the heat diffusion equation

We are now in a position to calculate the temperature distribution and/or heat flux in bodies with the help of the heat diffusion equation. In every case, we first calculate $T(\vec{r}, t)$. Then, if we want the heat flux as well, we differentiate T to get q from Fourier's law.

The heat diffusion equation is a partial differential equation (p.d.e.) and the task of solving it may seem difficult, but we can actually do a lot with fairly elementary mathematical tools. For one thing, in one-dimensional steady-state situations the heat diffusion equation becomes an ordinary differential equation (o.d.e.); for another, the equation is linear and therefore not too formidable, in any case. Our procedure can be laid out, step by step, with the help of the following example.

Example 2.2 Basic Methodology

A large, thin concrete slab of thickness L is "setting." Setting is an exothermic process that releases $\dot{q}$ W/m^3. The outside surfaces are kept at the ambient temperature, so $T_w = T_\infty$. What is the maximum internal temperature?

Step 1. Pick the coordinate scheme that best fits the problem and identify the independent variables that determine T.

In the example, T will probably vary only along the thin dimension, which we will call the x-direction. (We would want to know that the edges were insulated and that L was much smaller than the width or height. If they are, this assumption should be quite good.) Since the interior temperature will reach its maximum value when the process becomes steady, we write $T = T(x, \text{only})$.

Step 2. Write the appropriate d.e., starting with one of the forms of eqn. (2.15).

$$\frac{\partial^2 T}{\partial x^2} + \underbrace{\frac{\partial^2 t}{\partial y^2}}_{\substack{=0, \text{ since} \\ T \neq T(y \text{ or } z)}} + \frac{\partial^2 T}{\partial z^2} + \frac{\dot{q}}{k} = \underbrace{\frac{1}{\alpha}\frac{\partial T}{\partial t}}_{\substack{=0, \text{ since} \\ \text{steady}}}$$

Therefore, since $T = T(x \text{ only})$, the equation reduces to the ordinary d.e.

$$\frac{d^2 T}{dx^2} = -\frac{\dot{q}}{k}$$

Step 3. Obtain the general solution of the d.e. (This is usually the easiest step.) We simply integrate the d.e. twice and get

$$T = -\frac{\dot{q}}{2k}x^2 + C_1 x + C_2$$

Step 4. Write the "side conditions" on the d.e.—the initial and boundary conditions. This is always the hardest part for beginning students; it is the part that most seriously tests their physical or "practical" understanding of problems.

Normally, we have to make two specifications of temperature on each position coordinate, and one on the time coordinate, to get rid of the constants of integration in the general solution. (These matters are discussed at greater length in Chapter 4.)

In this case there are two boundary conditions:

$$T(x=0) = T_w \qquad T(x=L) = T_w$$

Very Important Warning: Never, never introduce inaccessible information in a boundary or initial condition. Always stop and ask yourself, "Would I have access to a numerical value of the temperature (or other data) that I specify at a given position or time?" If the answer is "no," then your result will be useless.

Step 5. Substitute the general solution in the boundary and initial conditions and solve for the constants. This process gets very complicated in the transient and multidimensional cases. Fourier and Bessel-function series methods are typically needed to solve the problem. However, the steady one-dimensional problems are usually easy.

In the example we get:

$$T_w = -0 + 0 + C_2 \qquad C_2 = T_w$$

$$T_w = -\frac{\dot{q}L^2}{2k} + C_1 L + \underbrace{C_2}_{=T_w} \qquad C_1 = \frac{\dot{q}L}{2k}$$

Step 6. Put the calculated constants back in the general solution to get the particular solution to the problem.

In the example problem we obtain

$$T = -\frac{\dot{q}}{2k}x^2 + \frac{\dot{q}}{2k}Lx + T_w$$

This should be put in neat dimensionless form:

$$\frac{T - T_w}{\dot{q}L^2/k} = \frac{1}{2}\left[\frac{x}{L} - \left(\frac{x}{L}\right)^2\right] \tag{2.19}$$

Step 7. Play with the solution—look it over—see what it has to tell you. Make any checks you can think of to be sure it is correct.

In this case we plot eqn. (2.19) in Fig. 2.7. The resulting temperature distribution is parabolic and, as we would expect, symmetrical. It satisfies the boundary conditions at the wall and maximizes in the center. By nondimensionalizing the result we have succeeded in representing all situations with a simple curve. That is most desirable when the calculations are not simple, as they are here. (Notice that T actually depends on *five* different things, yet the solution is a single curve on a two-coordinate graph.)

Finally, we check to see if the heat flux at the wall is correct:

$$q_{\text{wall}} = -k\frac{\partial T}{\partial x}\bigg|_{x=0} = k\left[\frac{\dot{q}}{k}x - \frac{\dot{q}L}{2k}\right]_{x=0} = -\frac{\dot{q}L}{2}$$

Thus half of the total energy generated in the slab comes out of the front side, as we would expect. The solution appears to be correct.

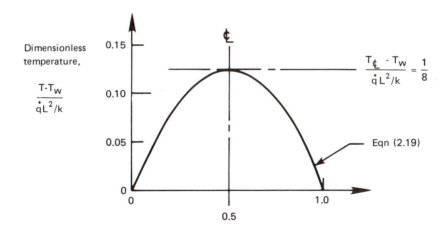

FIGURE 2.7   Temperature distribution in the setting concrete slab of Example 1.2.

Step 8.   If the temperature field is now correctly established, you can, if you wish, calculate the heat flux at any point in the body by substituting $T(\vec{r},t)$ back into Fourier's law. We did this already to check our solution in the example, in step 7.

Now we shall run through additional examples and, in the process, develop some important results for future use.

Example 2.3   The Simple Slab

A slab shown in Fig. 2.8 is at steady state with dissimilar temperatures on either side and no internal heat generation. We want the temperature distribution and the heat flux through it. These can be found quickly by following the steps set down in Example 2.2

Step 1. $T = T(x)$

Step 2. $\dfrac{d^2T}{dx^2} = 0$

Step 3. $T = C_1 x + C_2$

Step 4. $T(x=0) = T_1$; $T(x=L) = T_2$

Step 5. $T_1 = 0 + C_2$, so $\underline{C_2 = T_1}$; $T_2 = C_1 x + C_2$, so $\underline{C_1 = \dfrac{T_2 - T_1}{L}}$

Step 6. $\underline{T = T_1 + \dfrac{T_2 - T_1}{L} x}$ or $\underline{\dfrac{T - T_1}{T_2 - T_1} = \dfrac{x}{L}}$

Step 7.   We note that the solution satisfies the b.c.'s and that the temperature profile is linear.

Step 8. $q = -k\dfrac{dT}{dx} = -k\dfrac{d}{dx}\left(T_1 - \dfrac{T_1 - T_2}{L} x\right)_{x \text{ of interest}}$

$\qquad \underline{= k\dfrac{\Delta T}{L}}$

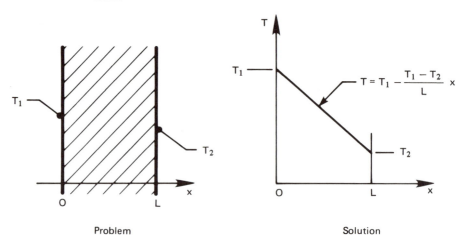

Problem	Solution

FIGURE 2.8   Example 2.3, heat conduction in a slab.

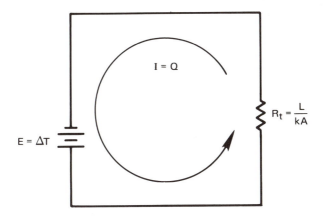

FIGURE 2.9 Ohm's-Law analogy to plane conduction.

This result, which is the simplest heat conduction solution, calls to mind Ohm's law. Thus, if we rearrange it:

$$Q = \frac{\Delta T}{L/kA} \quad \text{is like} \quad I = \frac{E}{R}$$

and L/kA assumes the role of a "thermal resistance," R_t (see Fig. 2.9). R_t has the dimensions of $(W/°C)^{-1}$.

Example 2.4 Radial Heat Conduction in a Tube

Find the temperature distribution and the heat flux for the long hollow cylinder shown in Fig. 2.10.

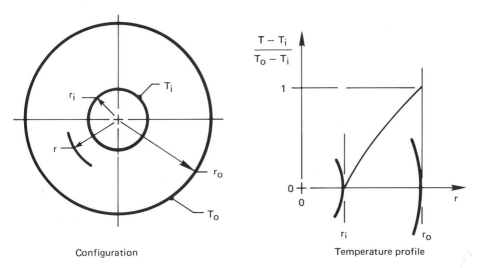

Configuration

Temperature profile

FIGURE 2.10 Example 2.4, heat transfer through a cylinder with fixed wall temperature.

Step 1. $T = T(r)$

Step 2. $\dfrac{1}{r}\dfrac{\partial}{\partial r}\left(r\dfrac{\partial T}{\partial r}\right) + \underbrace{\dfrac{1}{r^2}\dfrac{\partial^2 T}{\partial \phi^2} + \dfrac{\partial^2 T}{\partial z^2}}_{\substack{=0,\text{ since}\\ T\neq T(\phi,z)}} + \underset{=0}{\dfrac{\dot{q}}{k}} = \underset{\substack{=0,\text{ since}\\ \text{steady}}}{\dfrac{1}{\alpha}\dfrac{\partial T}{\partial t}}$

Step 3.   Integrate once: $r\dfrac{\partial T}{\partial r} = C_1$;

integrate again: $\underline{T = C_1\ln r + C_2}$

Step 4. $T(r = r_1) = T_i$ and $T(r = r_o) = T_o$

Step 5. $\left.\begin{array}{l} T_i = C_1\ln r_i + C_2 \\ T_o = C_1\ln r_o + C_2 \end{array}\right\}$ $\underline{C_1 = \dfrac{T_i - T_o}{\ln(r_i/r_o)} = -\dfrac{\Delta T}{\ln(r_o/r_i)}}$

$\underline{C_2 = T_i + \dfrac{\Delta T}{\ln(r_o/r_i)}\ln r_i}$

Step 6. $T = T_i - \dfrac{\Delta T}{\ln r_o/r_i}(\ln r - \ln r_i)$

or $\boxed{\dfrac{T - T_i}{T_o - T_i} = \dfrac{\ln(r/r_i)}{\ln(r_o/r_i)}}$ (2.20)

Step 7.   The solution is plotted in Fig. 2.10. We see that the temperature profile is logarithmic and that it satisfies both boundary conditions. Furthermore, it is instructive to see what happens when the wall of the cylinder is very thin, or when $r_i/r_o \to 1$. In this case:

$$\ln(r/r_i) \to \frac{r}{r_i} - 1 = \frac{r - r_i}{r_i}$$

and

$$\ln(r_o/r_i) \to \frac{r_o - r_i}{r_i}$$

Thus eqn. (2.20) becomes

$$\frac{T - T_i}{T_o - T_i} = \frac{r - r_i}{r_o - r_i} \qquad \text{a simple linear profile}$$

This is the same solution that we would get in a plane wall.

Step 8.   At any station, r:

$$q_{\text{radial}} = -k\frac{\partial T}{\partial r} = +\frac{k\Delta T}{\ln(r_o/r_i)}\frac{1}{r}$$

So the heat *flux* falls off inversely with radius. That is reasonable, since the same heat flow must pass each radial surface. Let us see if this is the case for a cylinder of length l:

$$Q(W) = q[2\pi rl] = \frac{2\pi kl\,\Delta T}{\ln(r_o/r_i)} \neq f(r) \qquad (2.21)$$

Finally, we again recognize Ohm's law in this result and write the thermal resistance for a cylinder:

$$R_t = \frac{\ell n(r_o/r_i)}{2\pi lk} \qquad (2.22)$$

This can be compared with the resistance of a plane wall:

$$R_{t_{wall}} = \frac{L}{kA} \left(\frac{°C}{W}\right) \qquad R_{t_{cyl}} = \frac{\ell n(r_o/r_i)}{2\pi kl} \left(\frac{°C}{W}\right)$$

Both resistances are inversely proportional to k, but each reflects a different geometry.

In the preceding examples, the boundary conditions were all the same—a temperature specified at an outer edge. Next let us suppose that the temperature is specified in the environment away from a body, with a heat transfer coefficient between the environment and the body:

Example 2.5 Heat Conduction with a Convective Boundary Condition

A convective heat transfer coefficient around the outside of the cylinder in Example 2.4 provides thermal resistance between the cylinder and an environment at $T = T_\infty$ as shown in Fig. 2.11. Find the temperature distribution and heat flux in this case.

Steps 1 through 3.   These are the same as in Example 2.4.

Step 4.   The first boundary condition is $T(r = r_i) = T_i$. The second boundary condition must be expressed as an energy balance at the outer wall (recall Section 1.3).

$$q_{\text{convection}} = q_{\text{conduction at the wall}}$$

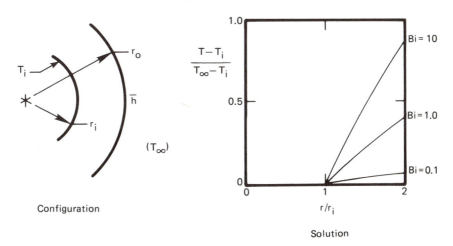

Configuration

Solution

FIGURE 2.11   Example 2.5, heat transfer through a cylinder with a convective boundary condition.

or

$$\bar{h}(T - T_\infty)_{r=r_o} = -k\frac{\partial T}{\partial r}\bigg|_{r=r_o}$$

Step 5.   From the first boundary condition we obtain $T_i = C_1 \ln r_i + C_2$. It is easy to make mistakes when we substitute the general solution into the second boundary condition, so we will do it in detail:

$$\bar{h}[(C_1 \ln r + C_2) - T_\infty]_{r=r_o} = -k\left[\frac{\partial}{\partial r}(C_1 \ln r + C_2)\right]_{r=r_o} \qquad (2.23)$$

A common error is to substitute $T = T_o$ on the left-hand side instead of substituting the entire general solution. That will do no good because T_o is not an accessible piece of information. Equation (2.23) reduces to

$$\bar{h}(T_\infty - C_1 \ln r_o - C_2) = \frac{kC_1}{r_o}$$

We combine this with the result of the first boundary condition to eliminate C_2:

$$C_1 = -\frac{T_i - T_\infty}{k/\bar{h}r_o + \ln(r_o/r_i)} = \frac{T_\infty - T_i}{1/\text{Bi} + \ln(r_o/r_i)}$$

Then

$$C_2 = T_i - \frac{T_\infty - T_i}{1/\text{Bi} + \ln(r_o/r_i)}\ln r_i$$

Step 6. $T = \dfrac{T_\infty - T_i}{1/\text{Bi} + \ln(r_o/r_i)}\ln(r/r_i) + T_i$

This can be rearranged in fully dimensionless form:

$$\frac{T - T_i}{T_\infty - T_i} = \frac{\ln(r/r_i)}{1/\text{Bi} + \ln(r_o/r_i)} \qquad (2.24)$$

Step 7.   Let us fix a value of r_o/r_i—say, 2—and plot eqn. (2.24) for several values of the Biot number. The results are included in Fig. 2.11. Some very important things show up in this plot. When $\text{Bi} \gg 1$ the solution reduces to the solution given in Example 2.4. It is as though the convective resistance to heat flow were not there. That is exactly what we anticipated in Section 1.3 for large Bi. When $\text{Bi} \ll 1$, the opposite is true: $(T - T_i)/(T_\infty - T_i)$ remains on the order of Bi and internal conduction can be neglected.

How big is big and how small is small? We do not really have to specify exactly. But in this case $\text{Bi} < 0.1$ signals constancy of temperature inside the cylinder within about $\pm 3\%$. $\text{Bi} > 20$ means that we can neglect convection within about 5% error.

Step 8. $q_{\text{radial}} = -k\dfrac{\partial T}{\partial r} = k\dfrac{T_i - T_\infty}{1/\text{Bi} + \ln(r_o/r_i)}\dfrac{1}{r}$

This can be written in terms of $Q(W) = q_{\text{radial}}(2\pi r l)$ for a cylinder of length l:

$$Q = \frac{T_i - T_\infty}{\dfrac{1}{2\pi\bar{h}r_o l} + \dfrac{\ln(r_o/r_i)}{2\pi k l}} = \frac{T_i - T_\infty}{R_{t_{\text{conv}}} + R_{t_{\text{cond}}}} \qquad (2.25)$$

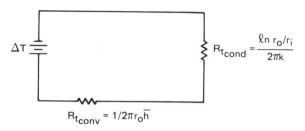

FIGURE 2.12   Thermal circuit with two resistances.

Equation (2.25) is once again analogous to Ohm's law. But this time the denominator is the sum of two thermal resistances, as would be the case in a series circuit. We accordingly present the analogous electrical circuit in Fig. 2.12.

Example 2.6   Critical Radius of Insulation

An interesting consequence of the preceding result can be brought out with a specific example. Suppose that we insulate a 0.5-cm-O.D. copper steam line with 85% magnesium to prevent the steam from condensing too rapidly. The steam is under pressure and stays at 150°C. The copper is thin and highly conductive—obviously a tiny resistance in series with the convective and insulation resistances, as we see in Fig. 2.13. The condensation of steam in the tube also offers very little resistance.[3] But a heat transfer coefficient of $\bar{h} = 20$ W/m²-°C offers fairly high resistance on the outside.

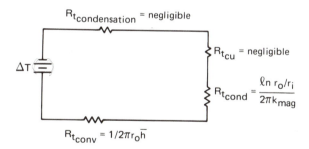

FIGURE 2.13   Thermal circuit for an insulated tube.

It turns out that insulation can actually *improve* heat transfer in this case. Figure 2.14 is a plot of the two significant resistances and their sum. A very interesting thing occurs here. $R_{t_{conv}}$ falls off rapidly when r_o is increased, because the outside area is increasing. Accordingly, the total resistance passes through a minimum in this case. Will it always do

---

[3]The question of how much resistance to heat transfer is offered by condensation inside the tube is the subject of Chapter 9. It turns out that $\bar{h}$ is generally enormous during condensation and that $R_{t_{condensation}}$ is tiny.

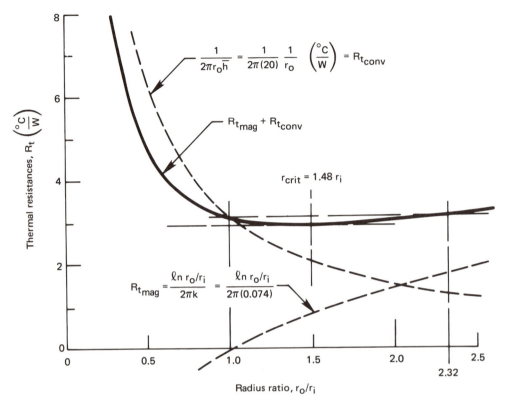

FIGURE 2.14   Example 2.6, the critical radius of insulation. Written for a cylinder of unit length ($l=1$ m).

so? To find out, we differentiate eqn. (2.25), setting l equal to a unit length of 1 m:

$$\frac{dQ}{dr_o} = \frac{1(T_i - T_\infty)}{\left(\dfrac{1}{2\pi r_o \bar{h}} + \dfrac{\ln(r_o/r_i)}{2\pi k}\right)^2}\left(-\frac{1}{2\pi r_o^2 \bar{h}} + \frac{1}{2\pi k r_o}\right) = 0$$

We solve this for the value of $r_o = r_{crit}$ at which R_t is minimum. Thus we obtain

$$\text{Bi} = 1 = \frac{\bar{h} r_{crit}}{k} \tag{2.26}$$

at the maximum heat flux. In the present example, added insulation will increase heat loss instead of reducing it, until $r_{crit} = k/\bar{h} = 0.0037$ m or $r_{crit}/r_i = 1.48$. Indeed, insulation will not even start to do any good until $r_o/r_i = 2.32$ or $r_o = 0.0058$ m. We call r_{crit} the "critical radius" of insulation.

There is an interesting catch here. For most cylinders $r_{crit} < r_i$ and the critical radius idiosyncrasy is of no concern. If our steam line had had a 1-cm O.D., the critical radius difficulty would not have arisen. The problem of cooling electrical wiring must be undertaken with this problem in mind, but one need not worry about the critical radius in the design of most large process equipment.

2.3 Overall heat transfer coefficient, U^4

Definition

We often want to transfer heat through composite resistances, as shown in Fig. 2.15. It is very convenient to have a number, U, which works like this:

$$\boxed{Q = UA\,\Delta T} \tag{2.27}$$

This number, called the *overall heat transfer coefficient,* is defined largely by the system, and in many cases it proves to be insensitive to the operating conditions of the system.

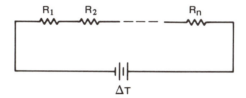

FIGURE 2.15   A thermal circuit with many resistances.

In Example 2.5, for example, we can use the value of Q given by eqn. (2.25) to get

$$U = \frac{Q(W)}{\left[2\pi r_o l(m^2)\right]\Delta T(^\circ C)} = \frac{1}{\dfrac{1}{\bar{h}} + \dfrac{r_o \ell n(r_o/r_i)}{k}}\left(\frac{W}{m^2\text{-}^\circ C}\right) \tag{2.28}$$

We have based U on the outside area—on r_o—in this case. We might also have based it on inside area and obtained

$$U = \frac{1}{\dfrac{r_i}{\bar{h}r_o} + \dfrac{r_i \ell n(r_o/r_i)}{k}} \tag{2.29}$$

It is therefore important to remember which area an overall heat transfer

---

[4]This U must not be confused with internal energy. The two terms should always be distinct in context.

coefficient is based on. It is particularly important that A and U are consistent when we write $Q = UA\,\Delta T$.

Example 2.7

Estimate the overall heat transfer coefficient for the teakettle shown in Fig. 2.16. Note that the flame convects heat to the thin aluminum. The heat is then conducted through the aluminum and finally convected by boiling into the water.

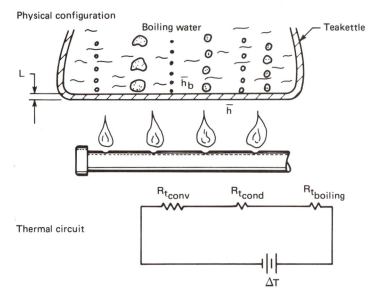

FIGURE 2.16   Heat transfer through the bottom of a teakettle.

SOLUTION.   We need not worry about deciding which area to base A on because the area normal to the heat flux vector does not change. We simply write the heat flux

$$\frac{\Delta T}{\sum R_{\mathrm{t}}} = \frac{T_{\mathrm{flame}} - T_{\mathrm{boiling\ water}}}{\dfrac{1}{\bar{h}A} + \dfrac{L}{k_{\mathrm{Al}}A} + \dfrac{1}{\bar{h}_{\mathrm{b}}A}}$$

Then

$$U = \frac{Q}{A\,\Delta T} = \frac{1}{\dfrac{1}{\bar{h}} + \dfrac{L}{k_{\mathrm{Al}}} + \dfrac{1}{\bar{h}_{\mathrm{b}}}}$$

Let us see what typical numbers would look like in this example. $\bar{h}$ might be around 200 W/m²-°C, L/k_{Al} might be 0.001 m/(160 W/m-°C) or 1/160,000 W/m²-°C, and $\bar{h}_{\mathrm{b}}$ is quite large—perhaps about 5,000 W/m²-°C. Thus

$$U \simeq \frac{1}{\dfrac{1}{200} + \dfrac{1}{160,000} + \dfrac{1}{5000}} = \underline{192.1\ \mathrm{W/m^2\text{-}°C}}$$

It is clear that the first resistance is dominant, as we have shown it in Fig. 1.5. Notice that in such cases

$$U \Rightarrow 1/R_{t_{\text{dominant}}} \tag{2.30}$$

where we now express R_t on a unit area basis, $(°C/W)/(m^2$ of heat exchanger area).

Experiment 2.1

Boil water in a paper cup over an open flame and explain why you can do so. [Recall eqn. (2.30) and see Problem 2.12.]

Example 2.8

A wall consists of alternating layers of pine and sawdust, as shown in Fig. 2.17. The sheathes on the outside have negligible resistance and $\bar{h}$ is known on the sides. Compute Q and U for the wall.

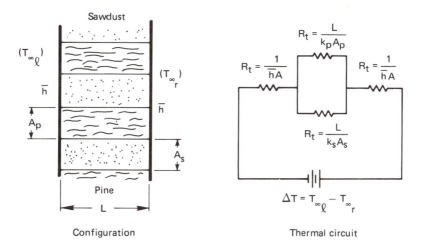

Configuration	Thermal circuit

FIGURE 2.17   Heat transfer through a composite wall.

SOLUTION. As long as the wood and sawdust do not differ dramatically from one another in thermal conductivity, we can approximate the wall as a parallel resistance circuit, as shown in the figure.[5] The total thermal resistance of such a circuit is

$$R_{t_{\text{total}}} = R_{t_{\text{conv}}} + \cfrac{1}{\cfrac{1}{R_{t_{\text{pine}}}} + \cfrac{1}{R_{t_{\text{sawdust}}}}} + R_{t_{\text{conv}}}$$

---

[5]For this approximation to be exact, the resistances must be equal. If they differ radically, the problem must be treated as two-dimensional.

Thus

$$Q = \frac{\Delta T}{R_{t_{total}}} = \frac{T_{\infty_l} - T_{\infty_r}}{\dfrac{1}{\bar{h}A} + \dfrac{1}{\dfrac{1}{L/k_p A_p} + \dfrac{1}{L/k_s A_s}} + \dfrac{1}{\bar{h}A}}$$

and

$$U = \frac{Q}{A\,\Delta T} = \frac{1}{\dfrac{2}{\bar{h}} + \dfrac{1}{\dfrac{k_p}{L}\dfrac{A_p}{A} + \dfrac{k_s}{L}\dfrac{A_s}{A}}}$$

Typical values of U

In a fairly general use of the word, a heat exchanger is anything that lies between two fluid masses at different temperatures. In this sense a heat exchanger might be designed either to impede or to enhance heat exchange. Consider some typical values of U in Table 2.1. These data were assembled from

Table 2.1   Typical values or ranges of U

Heat Exchange Configuration	$U(\text{W/m}^2\text{-}°\text{C})$
Walls and roofs of dwellings with a 24-km/hr exterior wind velocity:	
Frame walls	0.8–5
Finished masonry walls	0.5–6
Uninsulated roofs	1.2–4
Insulated roofs	0.3–2
Single-pane windows	~6, but main heat loss is by infiltration
Air to heavy tars and oils	{ As low as 45
Air to low-viscosity liquids	{ As high as 600
Air to various gases	60–550
Air condensers	350–780
Steam or water to oil	60–340
Liquids in coils immersed in liquids	110–2000
Steam-jacketed, agitated vessels	500–1900
Shell-and-tube, ammonia condensers	800–1400
Steam condensers with 25°C water	1500–5000
Feedwater heaters	110–8500
Condensing steam to high-pressure boiling water	θ (7000)
Heat pipes:	
cryogenic	<1000
water	3000
liquid metal	50,000

[2.1], [2.2], various manufacturers' literature, [1.13, Chap. 18], and other general sources listed at the end of Chapter 1. If the exchanger is intended to improve heat exchange, U will generally be much greater than 40 W/m²-°C. If it is intended to impede heat flow, it will be less than 10 W/m²-°C—anywhere down to almost perfect insulation. One should have some numerical concept of the relative values of U, so we recommend that the reader scrutinize the numbers in Table 2.1. Some things worth bearing in mind are:

- The fluids with low thermal conductivities, such as tars, oils, or any of the gases, usually yield low values of $\bar{h}$. When such a fluid flows on one side of an exchanger, U will generally be pulled down.

- Condensing and boiling are very effective heat transfer processes. They greatly improve U but they cannot override one very small value of $\bar{h}$ on the other side of the exchanger. (Recall Example 2.7.) In fact:

- For a high U, *all* resistances in the exchanger must be low.

- The highly conducting liquids, such as water and liquid metals, give high values of $\bar{h}$ and U.

Fouling resistance

Figure 2.18 shows one of the simplest forms of a heat exchanger—a pipe. The inside is new and clean on the left, but on the right it has built up a layer of scale. In conventional freshwater preheaters, for example, this scale is typically $MgSO_4$ (magnesium sulfate) or $CaSO_4$ (calcium sulfate), which precipitates onto the pipe wall after a time. To account for the resistance offered by these buildups, we must include an additional, highly empirical resistance when we

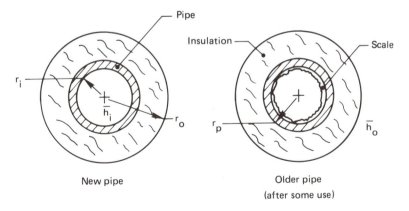

FIGURE 2.18   The fouling of a pipe.

calculate U. Thus for the pipe shown in Fig. 2.18,

$$U\big|_{\substack{\text{older pipe} \\ \text{based on } r_i}} = \frac{1}{\dfrac{1}{h_i} + \dfrac{r_i \ln(r_o/r_p)}{k_{\text{insul}}} + \dfrac{r_i \ln(r_p/r_i)}{k_{\text{pipe}}} + \dfrac{r_i}{r_o h_o} + R_f}$$

And clearly

$$R_f \equiv \frac{1}{U_{\text{old}}} - \frac{1}{U_{\text{new}}} \tag{2.31}$$

Some typical values of R_f are given in Table 2.2. These values have been adapted from [1.13, Chap. 18] and [2.3]. Notice that fouling has the effect of adding a resistance on the order of 10^{-4} (m²-°C/W) in series. It is rather like another heat transfer coefficient, $\bar{h}_f$, on the order of 10,000, in series with the other resistances in the exchanger.

The tabulated values of R_f are given to only one significant figure because they are very approximate. Clearly, exact values would have to be referred to specific heat exchanger materials, to fluid velocities, to operating temperatures, and to age. The resistance generally drops with increased velocity and increases with temperature and age. The values given in the table are based on reasonable maintenance and the use of conventional heat exchangers. With misuse, a given heat exchanger can yield much higher values of R_f.

Table 2.2   Some typical fouling resistances

Fluid and Situation	Fouling Resistance R_f (m²-°C/W)
Distilled water	0.0001
Seawater	0.0001–0.0002
Clean river and lake water	0.0002–0.0006
About the worst waters used in heat exchangers	<0.0020
Treated boiler feedwater	0.0001–0.0002
Transformer or lubricating oil	0.0002
Fuel oil	0.0010
Most industrial liquids	0.0002
Most refinery liquids	0.0002–0.0008
Non-oil-bearing steam	0.0001
Oil-bearing steam (e.g., turbine exhaust)	0.0002
Most stable gases	0.0005
Engine exhaust gases	0.0020
Fuel gases	0.0020

Notice, too, that if $U \ll 10,000$ W/m²-°C, fouling will be unimportant, because it will introduce small resistances in series. Thus in a water-to-water heat exchanger, in which U is on the order of 2000, fouling might be important; but in a finned-tube heat exchanger with hot gas in the tubes and cold gas passing across them, U might be around 200 and fouling should be insignificant.

Example 2.9

You have unpainted aluminum siding on your house and the engineer has based a heat loss calculation on $U = 5$ W/m²-°C. You discover that air pollution levels are such that R_f is 0.0005 m²-°C/W on the siding. Should the engineer redesign the siding?

SOLUTION.   From eqn. (2.31) we get

$$\frac{1}{U_{corrected}} = \frac{1}{U_{uncorrected}} + R_f = 0.2000 + 0.0005$$

Therefore, fouling is irrelevant to the calculation of domestic heat loads.

Example 2.10

Since the engineer did not fail you on this calculation, you entrust him with the installation of a heat exchanger at your plant. He installs a water-cooled steam condenser with $U = 4000$ W/m²-°C. You discover that he used a water-side fouling resistance for distilled water, but that the water flowing in the tubes is not clean at all. Should you go back to him this time?

SOLUTION.   Equation (2.31) and Table 2.2 give

$$\frac{1}{U_{corrected}} = \frac{1}{4000} + (0.0006 \text{ to } 0.0020)$$

$$= 0.00085 \text{ to } 0.00225 \text{ m}^2\text{-°C/W}$$

Thus U is reduced from 4000 to between 444 and 1176 W/m²-°C. Fouling is crucial in this case, and the engineer was in serious error.

Contact resistance

When two solid surfaces are pressed together, they will not form perfect *thermal* contact, owing to the air gaps that result from unavoidable roughnesses in the interface. A typical plane of contact between two surfaces is shown in Fig. 2.19 with an enormously exaggerated vertical scale. Heat transfer follows two paths in such an interface. The heat conduction path through points of solid-to-solid contact is very effective, but the path through the interstices which contain air (or some other low-conductivity gas) is ineffective.

The contact resistance correction, although it can be rationalized somewhat further than the fouling factor, is still quite similar to it. It is usually expressed as an interfacial conductance, h_c W/m²-°C—like an inverse of R_f. Thus for a

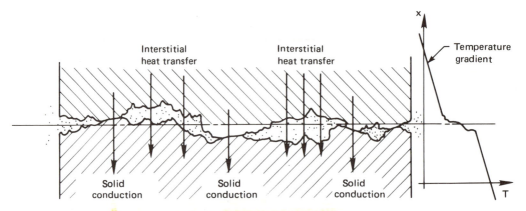

FIGURE 2.19   Heat transfer through the contact plane between two solid surfaces.

compound wall with one plane of thermal contact in it, we would have

$$U = \frac{1}{\dfrac{1}{\bar{h}_i} + \dfrac{L_1}{k_1} + \dfrac{1}{h_c} + \dfrac{L_2}{k_2} + \dfrac{1}{\bar{h}_o}}$$

It follows that

$$\frac{1}{h_c} = \frac{1}{U_{\text{w/o contact res}}} - \frac{1}{U_{\text{w/contact res}}} \qquad (2.32)$$

The interfacial conductance, h_c, depends on the following factors:

- The surface finish and cleanliness of the contacting solids.
- The materials that are in contact.
- The pressure with which the surfaces are forced together.
- The substance (or lack of it) in the interstitial spaces.
- The temperature at the contact plane.

The influence of pressure is usually a modest one up to around 10 atm in most metals. Then there is increasing plastic deformation of the local contact points and h_c increases more dramatically at high pressure. Table 2.3 gives typical values of contact resistances which bear out most of the preceding points. These values have been adapted from [1.13, Chap. 3] and [2.4].

Example 2.11

Heat flows through two stainless steel slabs ($k = 18$ W/m-°C) pressed together. How thin must the slabs be before contact resistance is important? With reference to Fig. 2.20 we

Table 2.3   Some typical interfacial conductances

Situation	h_c (W/m²-°C)
Ceramic/ceramic (moderate pressure and normal finishes)	500–3000
Ceramic/metals (moderate pressure and normal finishes)	1500–8500
Graphite/metals (moderate pressure and normal finishes)	3000–6000
Stainless steel/stainless steel (moderate pressure and normal finishes)	1700–3700
Aluminum/aluminum (moderate pressure and normal finishes)	2200–12,000
Copper/copper (moderate pressure and normal finishes)	10,000–25,000
Rough aluminum/aluminum (low pressure and evacuated interstices)	~150
Iron/aluminum (70 atm pressure)	45,000

can write

$$R_{total} = \frac{L}{18} + \frac{1}{h_c} + \frac{L}{18}$$

but h_c is about 2500. Therefore,

$$\frac{2L}{18} \quad \text{must be} \quad \gg \frac{1}{2500} = 0.0004$$

or L must be much greater than 0.0036 m if contact resistance is to be ignored. A thickness of 4 cm would reduce the error to about 10%.

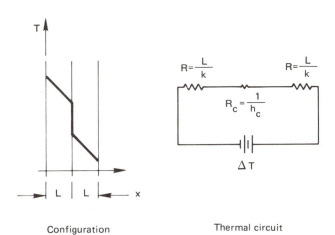

Configuration   Thermal circuit

FIGURE 2.20   Conduction through two stainless steel slabs with a contact resistance.

Four things have been done in this chapter:

- The heat diffusion equation has been established. A methodology has been erected for solving it in simple problems, and some important results have been presented. (There is much more to the matter of solving the heat diffusion equation. A more detailed treatment is the subject of Part II of this book.)

- The overall heat transfer coefficient has been defined and we have seen how to build it up out of component resistances.

- The electric analogy to steady heat flow has been exploited to solve heat transfer problems in the same way as electrical circuit problems.

- Some practical problems encountered in the evaluation of overall heat transfer coefficients have been discussed.

There are three very important things that we have not considered in Chapter 2. They are:

- In all evaluations of U that involve values of $\bar{h}$, we have taken these values as given information. In any real situation we must determine correct values of $\bar{h}$ for the specific situation. Part III deals with such determinations.

- When fluids flow through heat exchangers, they give up, or gain, heat. Thus the driving temperature difference varies through the exchanger. (Problem 2.14 asks you to consider this difficulty in its simplest form.) Accordingly, the design of an exchanger is complicated. We deal with this problem in Chapter 3.

- The heat transfer coefficients themselves vary with position inside many types of heat exchangers, causing U to be position-dependent.

PROBLEMS

2.1. Prove that if k varies linearly with T in a slab, and if heat transfer is one-dimensional and steady, then q may be evaluated precisely using k evaluated at the mean temperature in the slab.

2.2. Invent a numerical method for calculating the steady heat flux through a plane wall when $k(T)$ is an arbitrary function. Use the method to predict q in an iron slab 1 cm thick if the temperature varies from $-100°C$ on the left to $400°C$ on the right. How far would you have erred if you had taken $k_{\text{average}} = (k_{\text{left}} + k_{\text{right}})/2$?

2.3. The steady heat flux at one side of a slab is a known value, q_o. The thermal conductivity varies with temperature in the slab, and the variation can be expressed with a power series as

$$k = \sum_{i=0}^{i=n} A_i T^i$$

(a) Start with eqn. (2.14) and derive an equation that relates T to position in the slab, x. (b) Calculate the heat flux at any position in the wall from this expression using Fourier's law. Is the resulting q a function of x?

2.4. Combine Fick's law with the principle of conservation of mass (of the dilute species) in such a way as to eliminate J_1, and obtain a second-order differential equation in c_1. Discuss the importance and the use of the result.

2.5. Solve for the temperature distribution in a thick-walled pipe if the bulk interior temperature and the exterior air temperature, T_{∞_i} and T_{∞_o}, are known. The interior and exterior heat transfer coefficients are $\bar{h}_i$ and $\bar{h}_o$, respectively. Follow the methodology in Example 2.2 and put your result in dimensionless form so that

$$\frac{T - T_{\infty_i}}{T_{\infty_i} - T_{\infty_o}} = \text{fn}(\text{Bi}_i, \text{Bi}_o, r/r_i, r_o/r_i)$$

2.6. Put the boundary conditions from Problem 2.5 into dimensionless form so that the Biot numbers appear in them. Let the Biot numbers approach infinity. This should get you back to the boundary conditions for Example 2.4. Therefore, the solution that you obtain in Problem 2.5 should reduce to the solution of Example 2.4 when the Biot numbers approach infinity. Show that this is the case.

2.7. Write an accurate explanation of the idea of the "critical radius of insulation" that your kid brother, who is still in grade school, could understand. (If you do not have a kid brother, borrow one, to see if your explanation really works.)

2.8. The slab shown in Fig. P2.8 is embedded on five sides in insulating material. The sixth side is exposed to an ambient temperature through a heat transfer coefficient. Heat is generated in the slab at the rate of 1.0 kW/m³. The thermal conductivity of the slab is 0.2 W/m-°C. (a) Solve for the temperature distribution in the slab, noting any assumptions you must make. Be careful to clearly identify the boundary conditions. (b) Evaluate T at the front and back faces of the slab. (c) Show that your solution gives the expected heat fluxes at the back and front faces.

2.9. Consider the composite wall shown in Fig. P2.9. The concrete and brick sections are of equal thickness. Determine T_1, T_2, q, and the percentage of q that flows through the brick. To do this, approximate the heat flow as one-dimensional. Draw the thermal circuit for the wall and identify all four resistances before you begin.

2.10. Compute Q and U for Example 2.8 if the wall is 0.3 m thick. Five (each) pine and sawdust layers are 5 and 8 cm thick, respectively; and the heat transfer coefficients are 10 on the left and 18 on the right. $T_{\infty_l} = 30°C$ and $T_{\infty_r} = 10°C$.

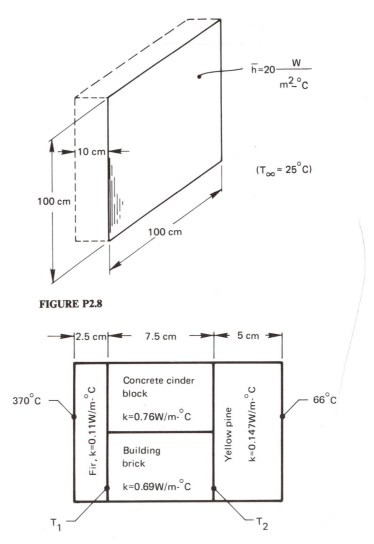

FIGURE P2.8

FIGURE P2.9

2.11. Compute U for the wall in Example 1.2.

2.12. Consider the teakettle in Example 2.7. Suppose that the kettle holds 1 kg of water (about 1 liter) and that the flame impinges on 0.02 m² of the bottom. (a) Find out how fast the water temperature is increasing when it reaches its boiling point and calculate the temperature of the bottom of the kettle immediately below the water if the gases from the flame are at 500°C when they touch the bottom of the kettle. (b) There is an old parlor trick in which one puts a *paper* cup of water over an open flame and boils the water without burning the paper (see Experiment 2.1). Explain this *using an electrical analogy*. [(a): $dT/dt = 0.37$ °C/s.]

2.13. Copper plates 2 mm and 3 mm in thickness are pressed rather lightly together. Non-oil-bearing steam condenses under pressure at $T_{sat}=200°C$ on one side ($\bar{h}=12{,}000$ W/m²-°C) and methanol boils under pressure at 130°C on the other ($\bar{h}=9000$ W/m²-°C). Estimate U and q initially and after extended service. List the relevant thermal resistances in order of decreasing importance and suggest whether or not any of them can be ignored.

2.14. 0.5 kg/s of air at 20°C moves along a 1-m-wide channel. One side of the channel is a heat exchange surface ($U=300$ W/m²-°C) with steam condensing at 120°C on the other side. Determine:

- q at the entrance.

- The rate of increase of temperature of the fluid with x at the entrance.

- The temperature and heat flux 2 m downstream. [$T_{2m}=89.7°C$.]

2.15. An isothermal sphere 3 cm in diameter is kept at 80°C in a large clay region. The temperature of the clay far from the sphere is kept at 10°C. How much heat must be supplied to the sphere to maintain its temperature if $k_{clay}=1.28$ W/m-°C? (*Hint:* You must solve the boundary value problem not in the sphere but in the clay surrounding it.) [$Q=16.9$ W.]

2.16. Is it possible to increase the heat transfer from a convectively cooled isothermal sphere by adding insulation? Explain fully.

2.17. A wall consists of layers of metals and plastic with heat transfer coefficients on either side. U is 225 W/m²-°C and the overall temperature difference is 200°C. One layer in the wall is stainless steel ($k=18$ W/m-°C) 3 mm thick. What is ΔT across the stainless steel?

2.18. A 1% carbon-steel sphere 20 cm in diameter is kept at 250°C on the outside. It has an 8-cm-diameter cavity containing boiling water ($\bar{h}_{inside}$ is very high) which is vented to the atmosphere. What is Q through the shell?

2.19. A slab is insulated on one side and exposed to a surrounding temperature, T_∞, through a heat transfer coefficient on the other. There is nonuniform heat generation in the slab such that $\dot{q}=[A\text{ (W/m}^4)][x\text{ (m)}]$, where $x=0$ at the insulated wall and $x=L$ at the cooled wall. Derive the temperature distribution in the slab.

2.20. 800 W/m³ of heat is generated within a 10-cm-diameter nickel-steel sphere for which $k=10$ W/m-°C. The environment is at 20°C and there is a natural convection heat transfer coefficient of 10 W/m²-°C around the outside of the sphere. What is its center temperature at steady state? [21.37°C.]

2.21. The outside of a pipe is insulated and we measure its temperature with a thermocouple. The pipe serves as an electrical resistance heater and $\dot{q}$ is known from resistance and current measurements. The inside of the pipe is cooled by the flow of liquid with a known bulk temperature. Evaluate the heat transfer coefficient, $\bar{h}$, in terms of known information. The pipe dimensions and properties are known. [*Hint:* Remember that $\bar{h}$ is not known and we cannot use a b.c. of the third kind at the inner wall to get $T(r)$.]

2.22. Consider the hot water heater in Problem 1.11. Suppose that it is insulated with 2 cm of a material for which $k = 0.12$ W/m-°C, and suppose that $\bar{h} = 16$ W/m²-°C. Find: (a) **T** for the tank, neglecting the steel casing; (b) the initial rate of cooling in °C/hr; (c) the time required for the water to cool from its initial temperature of 75°C to 40°C; (d) the percentage of additional heat loss that would result if an outer casing for the insulation were held on by eight steel rods, 1 cm in diameter, between the inner and outer casings.

2.23. A slab of thickness L is subjected to a constant heat flux, q_l, on the left side. The right-hand side is cooled convectively by an environment at T_∞. (a) Develop a dimensionless equation for the temperature of the slab. (b) Present dimensionless equations for the left- and right-hand wall temperatures, as well. (c) If the wall is firebrick, 10 cm thick, q_l is 400 W/m², $\bar{h}$ is 20 W/m²-°C, and $T_\infty = 20$°C, compute the left- and right-hand temperatures.

2.24. Heat flows steadily through a stainless steel wall of thickness $L_{ss} = 0.06$ m, with a variable thermal conductivity of $k_{ss} = 16.7 + 0.0143(T°C)$. It is partially insulated on the right side with glass wool of thickness $L_{gw} = 0.1$ m, with a thermal conductivity of $k_{gw} = 0.04$. The temperature on the left-hand side of the stainless steel is 400°C and on the right-hand side of the glass wool is 100°C. Evaluate q and T_i.

REFERENCES

Most of the ideas in Chapter 2 are dealt with at various levels in the general references following Chapter 1. The specific references introduced in this chapter follow.

[2.1]   *Handbook of Air Conditioning, Heating and Ventilating*, C. Strock, ed., The Industrial Press, New York, 1959.

[2.2]   *Chemical Engineer's Handbook*, 2nd ed., J. H. Perry, ed., McGraw-Hill Book Company, New York, 1941.

[2.3]   *Standards of Tubular Exchanger Manufacturer's Association*, 4th and 6th eds., Tubular Exchanger Manufacturer's Association, Inc., New York, 1959 and 1978.

[2.4]   R. F. WHEELER, "Thermal Conductance of Fuel Element Materials," USAEC Rep. HW-60343, April 1959.

Chapter 3

Heat exchanger
design

3.1 Function and configuration of heat exchangers

The archetypical problem that any heat exchanger solves is that of getting energy from one fluid mass to another, as we see in Fig. 3.1. A simple or composite wall of some kind divides the two flows and provides an element of thermal resistance between them. There is an exception to this in the direct-contact form of heat exchanger. Figure 3.2 shows one such arrangement in which steam is bubbled into water which condenses it and heats up. In other arrangements, immiscible fluids might contact each other, or noncondensible gases might be bubbled through liquids.

This discussion will be restricted to heat exchangers with a dividing wall between the two fluids. There is an enormous variety of such configurations, but there are three basic types to which most commercial exchangers can be reduced. Figure 3.3 shows these types in schematic form. They are:

- The simple parallel or counterflow configuration. These arrangements are basic and they are rather versatile. Figure 3.4 shows how the

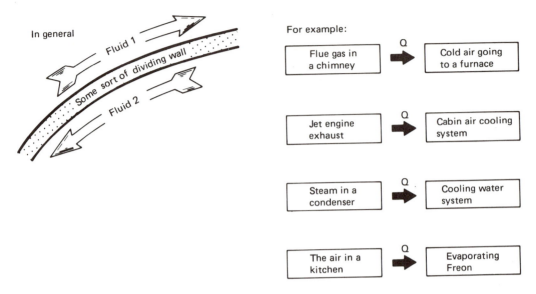

FIGURE 3.1   Heat exchange.

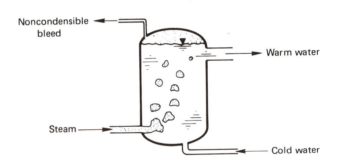

FIGURE 3.2   A direct contact heat exchanger.

counterflow arrangement is bent around in a so-called Heliflow compact heat exchanger configuration.

- The shell-and-tube configuration. Figure 3.5 shows the U-tubes of a
  two-tube-pass, one-shell-pass exchanger being installed in the support-
  ing baffles. The shell is yet to be added. Most of the really large heat
  exchangers are of the shell-and-tube form.

- The cross-flow configuration. Figure 3.6 shows typical cross-flow units.
  In Fig. 3.6a and c both flows are *unmixed*. Each flow must stay in a
  prescribed path through the exchanger and it is not allowed to "mix" to
  the right or left. Figure 3.6b shows a typical plate-fin cross-flow element. Here the flows are also unmixed.

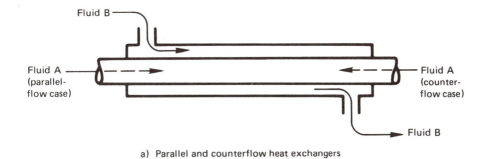

a) Parallel and counterflow heat exchangers

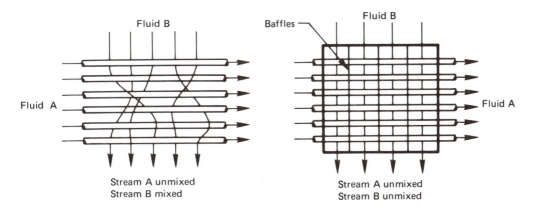

Baffles

Fluid A

Fluid B

Fluid B

Fluid A →

Simplest form:
2 tube-passes
1 shell-pass

4 tube-passes
2 shell-passes

b) Two kinds of shell-and-tube heat exchangers

Fluid B

Baffles

Fluid B

Fluid A

Fluid A

Stream A unmixed
Stream B mixed

Stream A unmixed
Stream B unmixed

c) Two kinds of cross-flow exchangers

FIGURE 3.3   The three basic types of heat exchanger.

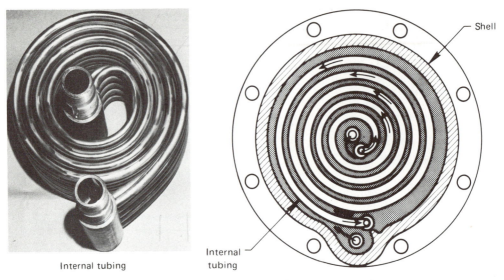

Internal tubing

Internal tubing

Shell

Plan view showing flow patterns

FIGURE 3.4   Heliflow compact counterflow heat exchanger. (Photograph courtesy of Graham Manufacturing Co., Inc., Batavia, New York.)

Figure 3.7, taken from the standards of the Tubular Exchanger Manufacturer's Association, TEMA [2.3], shows four typical single-shell-pass heat exchangers and it establishes nomenclature for such units. These pictures also show some of the complications that arise in translating simple concepts into hardware. Figure 3.7a shows an exchanger with a single tube pass. Although the shell flow is baffled so that it criss-crosses the tubes, it still proceeds from the hot to cold (or cold to hot) end of the shell. Therefore, it is like a simple parallel (or counterflow) unit. The kettle reboiler in Fig. 3.7d involves a divided shell-pass flow configuration over two tube passes (from right to left and back to the "channel header"). In this case the isothermal shell flow could be flowing in any direction—it makes no difference to the tube flow. Therefore, it is also equivalent to either the simple parallel or counterflow configuration.

Notice that a salient feature of the shell-and-tube exchangers is the presence of baffles. Baffles serve to direct the flow normal to the tubes. We find in Part III that heat transfer from a tube to a flowing fluid is usually better when the flow moves across the tube than it is when the flow moves along the tube. This augmentation of heat transfer gives the complicated shell-and-tube exchanger an advantage over the simpler single-pass parallel and counterflow exchangers.

However, baffles bring with them a variety of problems. The flow patterns are very complicated and almost defy analysis. A good deal of the shell-side fluid might unpredictably leak through the baffle holes in the axial direction, or it might bypass the baffles near the wall. In certain shell-flow configurations,

Above and left: A very large feed-water preheater. Tubes are shown withdrawn from the shell on the left. Inset above shows baffles before tubes are inserted. (Photos courtesy of Southwest Engineering Co., Subsidiary of Cronus Industries, Inc., Los Angeles, Calif.)

Below: Small "Swinglok" exchanger with tube-bundle removed from shell. (Photo courtesy of Graham Manufacturing Co. Inc., Batavia, New York.)

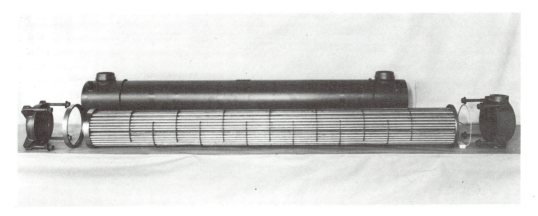

FIGURE 3.5   Typical commercial one-shell-pass, two-tube-pass heat exchangers.

a) A 1980 Chevette radiator. Cross-flow exchanger with neither flow mixed. Vertical tubes cannot be seen.

b) A section of an automotive air conditioning condenser. The flow through the horizontal wavy fins is allowed to mix with itself while the two-pass flow through the U-tubes remains unmixed

c) The basic 1 ft. × 1 ft. × 2 ft. module for a waste heat recuperator. It is a plate-fin, gas-to-air cross-flow heat exchanger with neither flow mixed.

FIGURE 3.6   Several commercial cross-flow compact heat exchangers. (Photographs courtesy of Harrison Radiator Division, G.M. Corp.)

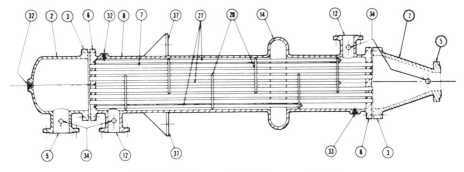

a) Single shell-pass, single tube-pass exchanger

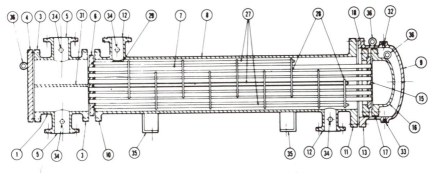

b) One shell-pass, two tube-pass exchanger

1. Stationary head-channel
2. Stationary head-bonnet
3. Stationary head-flange-channel or bonnet
4. Channel cover
5. Stationary head nozzle
6. Stationary tube sheet
7. Tubes
8. Shell
9. Shell cover
10. Shell flange-stationary head end
11. Shell flange-rear head end
12. Shell nozzle
13. Shell cover flange

14. Expansion joint
15. Floating tube sheet
16. Floating head cover
17. Floating head flange
18. Floating head backing device
19. Split shear ring
20. Slip-on backing flange
21. Floating head cover-external
22. Floating tube sheet skirt
23. Packing box
24. Packing
25. Packing gland

26. Lantern ring
27. Tie rods and spacers
28. Transverse baffles or support plates
29. Impingement plate
30. Longitudinal baffle
31. Pass partition
32. Vent connection
33. Drain connection
34. Instrument connection
35. Support saddle
36. Lifting lug
37. Support bracket
38. Weir
39. Liquid level connection

FIGURE 3.7   Four typical heat exchanger configurations (continued on next page). (Photos courtesy of the Tubular Exchanger Manufacturer's Association.)

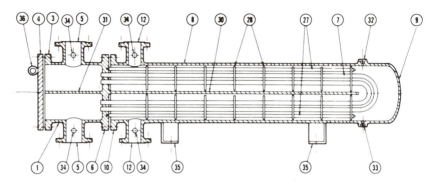

c) Two tube-pass, two shell-pass exchanger

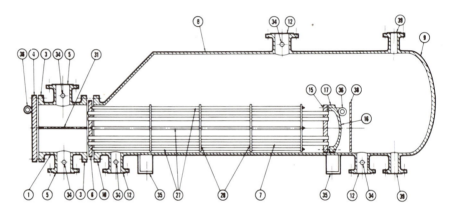

d) One split shell-pass, two tube-pass, kettle type of exchanger

FIGURE 3.7 Continued

unanticipated vibrational modes of the tubes might be excited. Many of the cross-flow configurations also baffle the fluid so as to move it across a tube bundle. The plate-and-fin configuration (Fig. 3.6b) is such a cross-flow heat exchanger.

In all of these heat exchanger arrangements it becomes clear that a dramatic investment of human ingenuity is directed toward the task of augmenting the amount of heat that can be transferred from one flow to another. The variations are endless, as you will quickly see if you try Experiment 3.1.

Experiment 3.1

Carry a notebook with you for a day and mark down every heat exchanger you encounter in home, university, or automobile. Classify each according to type and note any special augmentation features.

The analysis of heat exchangers first becomes complicated when we account for the fact that the two flow streams change one another's temperature. It is to the problem of predicting an appropriate mean temperature difference that we address ourselves in Section 3.2. Section 3.3 then presents a strategy to use when this mean cannot be determined initially.

3.2 Evaluation of the mean temperature difference in a heat exchanger

Logarithmic mean temperature difference, LMTD

To begin with, we take U to be a constant value. This is fairly reasonable in compact single-phase heat exchangers. In larger exchangers, particularly in shell-and-tube configurations and large condensers, U is apt to vary with position in the exchanger and/or with local temperature. But in situations in which U is fairly constant, the varying temperatures of the fluid streams can be dealt with by writing the overall heat transfer in terms of the mean temperature difference between the two fluid streams:

$$Q = UA\,\Delta T_{\mathrm{mean}} \tag{3.1}$$

Our problem is then reduced to that of finding the appropriate mean temperature difference that will make this equation true. This can be done by considering the simple parallel and counterflow configurations as sketched in Fig. 3.8.

The temperature of both streams is plotted in Fig. 3.8 for both single-pass arrangements—the parallel and counterflow configurations—as a function of the length of travel (or area passed over). Notice that in the parallel-flow configuration, temperatures tend to change more rapidly with position, and less length is required. But the counterflow arrangement achieves generally more complete heat exchange from one flow to the other.

Figure 3.9 shows another variation on the single-pass configuration. This is a condenser in which one stream flows through with its temperature changing, but the other simply condenses at constant temperature. This arrangement has some special characteristics which we shall point to shortly.

The determination of ΔT_{mean} for such arrangements proceeds as follows. The differential heat transfer within either arrangement is (see Fig. 3.8)

$$dQ = U(dA)\,\Delta T = -\left(\dot{m}c_{\mathrm{p}}\right)_{\mathrm{h}} dT_{\mathrm{h}} = \left(\dot{m}c_{\mathrm{p}}\right)_{\mathrm{c}} dT_{\mathrm{c}} \tag{3.2}$$

where the subscripts h and c denote the hot and cold streams, respectively. We give symbols to the total heat capacities of the hot and cold streams:

$$C_{\mathrm{h}} \equiv \left(\dot{m}c_{\mathrm{p}}\right)_{\mathrm{h}} \ \mathrm{W}/^{\circ}\mathrm{C} \qquad \text{and} \qquad C_{\mathrm{c}} \equiv \left(\dot{m}c_{\mathrm{p}}\right)_{\mathrm{c}} \ \mathrm{W}/^{\circ}\mathrm{C} \tag{3.3}$$

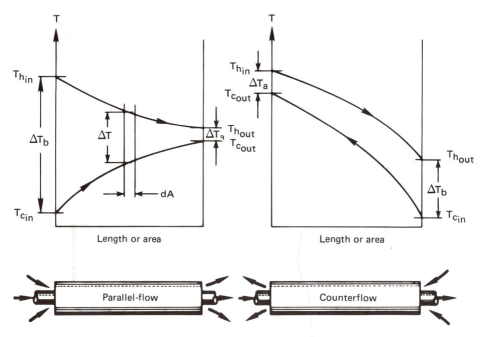

FIGURE 3.8   The temperature variation through single-pass heat exchangers.

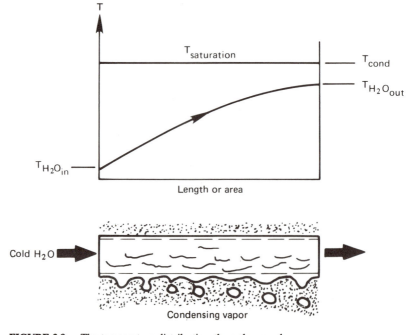

FIGURE 3.9   The temperature distribution through a condenser.

Thus, for either heat exchanger, $-C_h dT_h = C_c dT_c$. This equation can be integrated from $T_h(T_c = T_{c_{in}}) = T_{h_{in}}$ for parallel flow [or from $T_h(T_c = T_{c_{out}}) = T_{h_{in}}$ for counterflow] to $T_h(T_c)$. The results are:

$$
\text{parallel flow:} \quad T_h = T_{h_{in}} - \frac{C_c}{C_h}(T_c - T_{c_{in}}) \doteq T_{h_{in}} - \frac{Q}{C_h}
$$

$$
\text{counterflow:} \quad T_h = T_{h_{in}} - \frac{C_c}{C_h}(T_{c_{out}} - T_c) = T_{h_{in}} - \frac{Q}{C_h}
$$

(3.4)

where Q is the total heat transfer from the entrance to the point of interest. Equations (3.4) can be solved for the local temperature differences:

$$
\Delta T_{parallel} = T_h - T_c = T_{h_{in}} - \left(1 + \frac{C_c}{C_h}\right)T_c + \frac{C_c}{C_h}T_{c_{in}}
$$

$$
\Delta T_{counter} = T_h - T_c = T_{h_{in}} - \left(1 - \frac{C_c}{C_h}\right)T_c - \frac{C_c}{C_h}T_{c_{out}}
$$

(3.5)

Substitution of these in $dQ = C_c dT_c = UdA\,\Delta T$ yields

$$
\left.\frac{UdA}{C_c}\right|_{parallel} = \frac{dT_c}{\left[-\left(1 + \dfrac{C_c}{C_h}\right)T_c + \dfrac{C_c}{C_h}T_{c_{in}} + T_{h_{in}}\right]}
$$

$$
\left.\frac{UdA}{C_c}\right|_{counter} = \frac{dT_c}{\left[-\left(1 - \dfrac{C_c}{C_h}\right)T_c - \dfrac{C_c}{C_h}T_{c_{out}} + T_{h_{in}}\right]}
$$

(3.6)

Equation (3.6) can be integrated from one side of the exchanger to an arbitrary point within it:

$$
\int_0^A \frac{U}{C_c}\,dA = \int_{T_{c_{in}}}^{T_{c_{out}}} \frac{dT_c}{[---]}
$$

(3.7)

If U and C_c can be treated as constant, this integration gives

$$
\text{parallel:} \quad \ln\left[\frac{-\left(1 + \dfrac{C_c}{C_h}\right)T_{c_{out}} + \dfrac{C_c}{C_h}T_{c_{in}} + T_{h_{in}}}{-\left(1 + \dfrac{C_c}{C_h}\right)T_{c_{in}} + \dfrac{C_c}{C_h}T_{c_{in}} + T_{h_{in}}}\right] = -\frac{UA}{C_c}\left(1 + \frac{C_c}{C_h}\right)
$$

$$
\text{counter:} \quad \ln\left[\frac{-\left(1 - \dfrac{C_c}{C_h}\right)T_{c_{out}} - \dfrac{C_c}{C_h}T_{c_{out}} + T_{h_{in}}}{-\left(1 - \dfrac{C_c}{C_h}\right)T_{c_{in}} - \dfrac{C_c}{C_h}T_{c_{out}} + T_{h_{in}}}\right] = -\frac{UA}{C_c}\left(1 - \frac{C_c}{C_h}\right)
$$

(3.8)

If U were variable, the integration leading from eqn. (3.7) to eqn. (3.8) is where its variability would have to be considered. Any such variability of U can complicate eqns. (3.8) terribly. Presuming that eqns. (3.8) are valid, we can simplify them with the help of the definitions of ΔT_a and ΔT_b, given in Fig. 3.8:

$$\text{parallel:} \quad \ell n \left[\frac{(1 + C_c/C_h)(T_{c_{in}} - T_{c_{out}}) + \Delta T_b}{\Delta T_b} \right] = -UA \left(\frac{1}{C_c} + \frac{1}{C_h} \right)$$

$$\text{counter:} \quad \ell n \frac{\Delta T_a}{(-1 + C_c/C_h)(T_{c_{in}} - T_{c_{out}}) + \Delta T_a} = -UA \left(\frac{1}{C_c} - \frac{1}{C_h} \right) \tag{3.9}$$

Conservation of energy $(Q_c = Q_h)$ requires that

$$\boxed{\frac{C_c}{C_h} = -\frac{T_{h_{out}} - T_{h_{in}}}{T_{c_{out}} - T_{c_{in}}}} \tag{3.10}$$

Then eqns. (3.9) and (3.10) give

$$\text{parallel:} \quad \ell n \left[\frac{\overbrace{\left[(T_{c_{in}} - T_{c_{out}}) + (T_{h_{out}} - T_{h_{in}}) \right]}^{\Delta T_a - \Delta T_b} + \Delta T_b}{\Delta T_b} \right]$$

$$= \ell n \frac{\Delta T_a}{\Delta T_b} = -UA \left(\frac{1}{C_c} + \frac{1}{C_h} \right)$$

$$\text{counter:} \quad \ell n \left(\frac{\Delta T_a}{\Delta T_b - \Delta T_a + \Delta T_a} \right) = \ell n \frac{\Delta T_a}{\Delta T_b} = -UA \left(\frac{1}{C_c} - \frac{1}{C_h} \right) \tag{3.11}$$

Finally, we write $1/C_c = (T_{c_{out}} - T_{c_{in}})/Q$ and $1/C_h = (T_{h_{in}} - T_{h_{out}})/Q$ on the right-hand side of either of eqns. (3.11), and get for *either* parallel or counter-flow,

$$\boxed{Q = UA \left(\frac{\Delta T_a - \Delta T_b}{\ell n (\Delta T_a/\Delta T_b)} \right)} \tag{3.12}$$

The appropriate ΔT_{mean} for use in eqn. (3.1) is thus the *logarithmic mean temperature difference*, or LMTD:

$$\boxed{\Delta T_{mean} = \text{LMTD} \equiv \left(\frac{\Delta T_a - \Delta T_b}{\ell n (\Delta T_a/\Delta T_b)} \right)} \tag{3.13}$$

Example 3.1

The idea of a logarithmic mean difference is not new to us. We have already encountered it in Chapter 2. Suppose that we had asked, "What mean radius of pipe would have allowed us to compute the conduction through the wall of a pipe as though it were a slab of thickness $L = r_o - r_i$ (see Fig. 3.10)?" To answer this we compare

$$Q = kA\frac{\Delta T}{L} = 2\pi k l \Delta T \frac{r_{\text{mean}}}{r_o - r_i}$$

with eqn. (2.21):

$$Q = 2\pi k l \Delta T \frac{1}{\ell n(r_o/r_i)}$$

It follows that

$$\underline{r_{\text{mean}} = \frac{r_o - r_i}{\ell n(r_o/r_i)} = \text{logarithmic mean radius}}$$

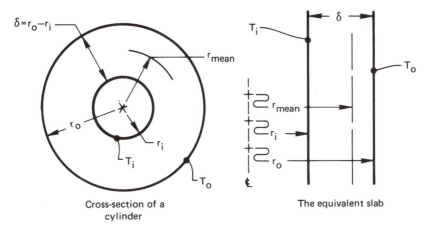

Cross-section of a   The equivalent slab
cylinder

FIGURE 3.10   Calculation of the mean radius for heat conduction through a slab.

Example 3.2

Suppose that the temperature differences on either end of a heat exchanger, ΔT_a and ΔT_b, are equal. Clearly, the effective ΔT must equal ΔT_a and ΔT_b, in this case. Does the LMTD reduce to this value?

SOLUTION.   If we substitute $\Delta T_a = \Delta T_b$ in eqn. (3.13), we get

$$\text{LMTD} = \frac{\Delta T_b - \Delta T_b}{\ell n(\Delta T_b/\Delta T_b)} = \frac{0}{0} = \text{indeterminate}$$

Therefore, it is necessary to use L'Hospital's rule:

$$\lim_{\Delta T_a \to \Delta T_b} \frac{\Delta T_a - \Delta T_b}{\ell n(\Delta T_a/\Delta T_b)} = \frac{\dfrac{\partial}{\partial \Delta T_a}(\Delta T_a - \Delta T_b)_{\Delta T_a = \Delta T_b}}{\dfrac{\partial}{\partial \Delta T_a}\ell n\left(\dfrac{\Delta T_a}{\Delta T_b}\right)_{\Delta T_a = \Delta T_b}}$$

$$= \left(\frac{1}{1/\Delta T_a}\right)_{\Delta T_a = \Delta T_b} = \Delta T_a = \Delta T_b$$

It follows that the LMTD reduces to the intuitively obvious result in the limit.

Example 3.3

Water enters the tubes of a small single-pass heat exchanger at 20°C and leaves at 40°C. On the shell side, 25 kg/min of steam condenses at 60°C. Calculate the overall heat transfer coefficient and the required flow rate of water if the area of the exchanger is 12 m². (The latent heat, h_{fg}, is 2358.7 kJ/kg at 60°C.)

SOLUTION.

$$Q = \dot{m}_{condensate}h_{fg_{60°C}} = \frac{25(2358.7)}{60}$$

$$= 983 \text{ kJ/s}$$

and, with reference to Fig. 3.9, we can calculate the LMTD without naming the exchanger "parallel" or "counterflow," since the condensate temperature is constant.

$$\text{LMTD} = \frac{(60-20)-(60-40)}{\ell n \dfrac{60-20}{60-40}} = 28.85°C$$

Then

$$U = \frac{Q}{A(\text{LMTD})}$$

$$= \frac{983(1000)}{12(28.85)} = 2839 \text{ W/m}^2\text{-°C}$$

and

$$\dot{m}_{H_2O} = \frac{Q}{c_p \Delta T} = \frac{983,000}{4174(20)} = 11.78 \text{ kg/s}$$

Extended use of the LMTD

Limitations. There are two basic limitations on characterizing heat exchangers with a LMTD. The first is that it is restricted to the single-pass parallel and counterflow configurations. This restriction can be overcome by adjusting the LMTD for other configurations—a matter that we take up in the following subsection.

The second limitation is more serious—that is our use up to this point of a constant value of U. U must be negligibly dependent on T to complete the integration of eqn. (3.7). If $U \neq f(T)$, the changing flow configuration and the

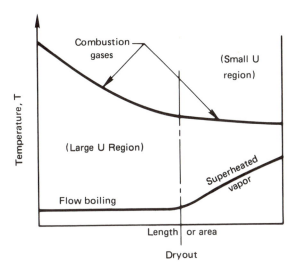

FIGURE 3.11   A typical case of a heat exchanger in which U varies dramatically.

variation of temperature can still give rise to serious variations of U within a given heat exchanger. Figure 3.11 shows a typical situation in which the variation of U within a heat exchanger might be very great. In this case the mechanism of heat exchange on the water side is completely altered when the liquid is finally boiled away. If we had constant values of U for the two portions of the heat exchanger, then we could treat it as two different exchangers in series.

However, the more common difficulty that we face is that of designing heat exchangers in which U varies continuously with position within it. This problem is most severe in large[1] industrial shell-and-tube configurations (see, e.g., Fig. 3.5 or 3.12) and less serious in compact heat exchangers with less surface area. If U depends on the location, analyses such as we have just completed [eqns. (3.1) to (3.13)] must be done using an average U defined as $(\int_0^A U\,dA)/A$.

LMTD correction factor, **F.**   Suppose that we have a heat exchanger in which U can reasonably be taken constant, but one that involves such configurational complications as multiple passes and/or cross-flow. In such cases it is necessary to rederive the appropriate mean temperature difference in the same way as we derived the LMTD. Each configuration must be analyzed separately and the results are generally more complicated than eqn. (3.13).

This task had been undertaken on an ad hoc basis during the early twentieth century. In 1940, Bowman, Mueller, and Nagle [3.1] organized such calculations

---

[1]Actual heat exchangers can have areas well in excess of 10,000 m^2. Large power plant condensers and other large heat exchangers are often remarkably big pieces of equipment.

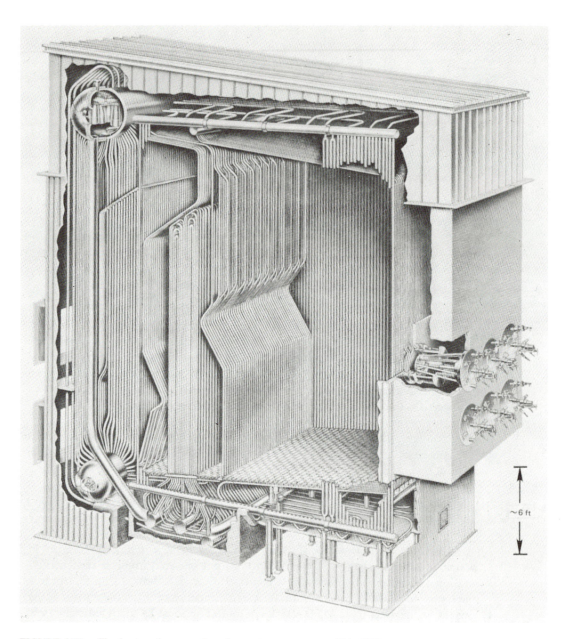

FIGURE 3.12 The heat exchange surface for a steam generator. This **PFT** type Integral-Furnace boiler, with a surface area of 4560 m², is not particularly large. About 88% of the area is in the furnace tubing and 12% is in the boiler. (Photograph courtesy of Babcock and Wilcox Co.)

for the common range of heat exchanger configurations. In each case they wrote

$$Q = UA(\text{LMTD}) \cdot F\left[\underbrace{\frac{T_{t_{out}} - T_{t_{in}}}{T_{s_{in}} - T_{t_{in}}}}_{P} , \quad \underbrace{\frac{T_{s_{in}} - T_{s_{out}}}{T_{t_{out}} - T_{t_{in}}}}_{R} \right] \tag{3.14}$$

where T_t and T_s are temperatures of tube and shell flows, respectively. The factor F is an LMTD correction which varies from unity to zero, depending on conditions. The dimensionless groups P and R have the following physical significance:

- P is the relative influence of the overall temperature difference ($T_{s_{in}} - T_{t_{in}}$) on the tube flow temperature. It must obviously be less than unity.
- R, according to eqn. (3.10), equals the heat capacity ratio C_t / C_s.
- If one flow remains at constant temperature (as, for example, in Fig. 3.9), then either P or R will equal zero. In this case the simple LMTD will be the correct ΔT_{mean} and F must go to unity.

The factor F is defined in such a way that *the LMTD should always be calculated for the equivalent counterflow single-pass exchanger with the same hot and cold temperatures*. This is explained in Fig. 3.13.

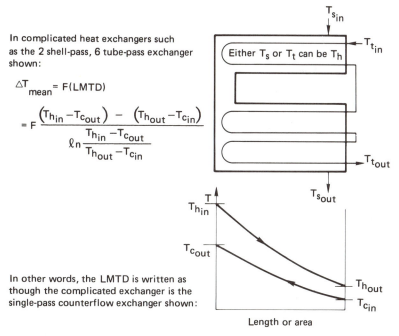

FIGURE 3.13   The basis of the LMTD in a multi-pass exchanger, prior to correction.

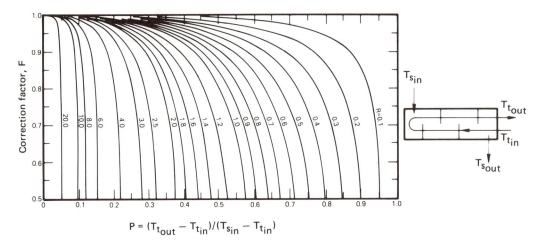

FIGURE 3.14a   LMTD correction factor, F, for a 1 shell-pass, 2, 4, 6, ... tube-passes, exchanger.

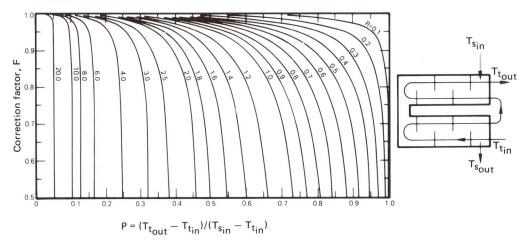

FIGURE 3.14b   LMTD correction factor, F, for a 2 shell-pass, 4 or more tube-passes, exchanger.

Bowman et al. [3.1] summarized the previously derived equations for F, in various configurations, and presented them graphically in not-very-accurate figures that have been widely copied. The TEMA [2.3] version of these curves has been recalculated for shell-and-tube heat exchangers and it is more accurate. We include two of these curves in Fig. 3.14a and b. TEMA presents many additional curves for more complex shell-and-tube configurations. Figure 3.14c and d are the Bowman et al. curves for the simplest cross-flow configurations.

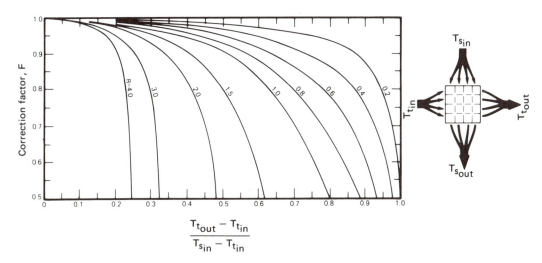

FIGURE 3.14c LMTD correction factor, F, for a one-pass cross-flow exchanger with both passes unmixed.

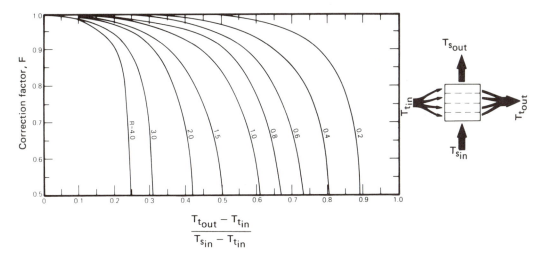

FIGURE 3.14d LMTD correction factor, F, for a one-pass cross-flow exchanger with one pass mixed.

Gardner and Taborek [3.2] redeveloped Fig. 3.14c over a different range of parameters. They also showed how Fig. 3.14a and b must be modified if the number of baffles in a tube-in-shell heat exchanger is small enough to make it behave like a series of cross-flow exchangers.

Example 3.4

5.795 kg/s of oil flows through the shell side of a two-shell-pass, four-tube-pass oil cooler. It enters at 181°C and leaves at 38°C. Water flows in the tubes, entering at 32°C and leaving at 49°C. $c_{p_{oil}} = 2,282$ J/kg-°C. $U = 416$ W/m²-°C. Find how much area the heat exchanger must have.

SOLUTION.

$$\text{LMTD} = \frac{\left(T_{h_{in}} - T_{c_{out}}\right) - \left(T_{h_{out}} - T_{c_{in}}\right)}{\ell n \dfrac{T_{h_{in}} - T_{c_{out}}}{T_{h_{out}} - T_{c_{in}}}} = \frac{(181 - 49) - (38 - 32)}{\ell n \dfrac{181 - 49}{38 - 32}} = 40.76°C$$

$$R = \frac{181 - 38}{49 - 32} = 8.412 \quad P = \frac{49 - 32}{181 - 32} = 0.114$$

so Fig. 3.14b gives[2] $F = 0.92$. It follows that

$$Q = UAF(\text{LMTD})$$

$$5.795(2282)(181 - 38) = 416(A)(0.92)(40.76)$$

$$\underline{A = 121.2 \text{ m}^2}$$

3.3 Heat exchanger effectiveness

We are now in position to predict the performance of an exchanger once we know its configuration *and* the imposed temperature differences. Unfortunately, we do not often know that much about a system before the design is complete.

Often we begin with information such as is shown in Fig. 3.15. If we sought to calculate Q in such a case, we would have to do so by guessing an exit temperature such as to make $Q_h = Q_c = C_h \Delta T_h = C_c \Delta T_c$. Then we could calculate Q from $UA(\text{LMTD})$ or $UAF(\text{LMTD})$ and check it against Q_h. The answers would differ, so we would have to guess new exit temperatures and try again.

Such problems can be greatly simplified with the help of the so-called *effectiveness method*. This method was first developed in full detail by Kays and London [1.24] in 1955, in a book titled *Compact Heat Exchangers*. We should take particular note of the title. It is with compact heat exchangers that the

---

[2]Notice that, for a 1 shell-pass exchanger, these R and P lines do not quite intersect (see Fig. 3.14a). Therefore, one could not obtain these temperatures with any single shell exchanger.

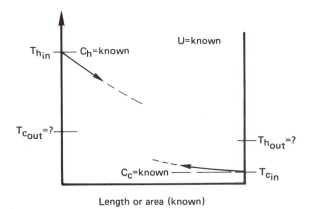

FIGURE 3.15   A design problem in which the LMTD cannot be calculated a priori.

present methods can reasonably be used since the overall heat transfer coefficient is far more likely to remain fairly constant.

The heat exchanger effectiveness is defined as

$$\varepsilon \equiv \frac{C_h(T_{h_{in}} - T_{h_{out}})}{C_{min}(T_{h_{in}} - T_{c_{in}})} = \frac{C_c(T_{c_{out}} - T_{c_{in}})}{C_{min}(T_{h_{in}} - T_{c_{in}})} \tag{3.15}$$

where C_{min} is the smaller of C_c and C_h. The effectiveness can be interpreted as

$$\varepsilon = \frac{\text{actual heat transferred}}{\substack{\text{maximum heat that could possibly be} \\ \text{transferred from one stream to the other}}} \tag{3.16}$$

It follows that

$$Q = \varepsilon C_{min}(T_{h_{in}} - T_{c_{in}}) \tag{3.17}$$

A second definition that we will need was originally made by W. Nusselt, whom we shall meet again in Part III. This is the *number of transfer units*, or NTU:

$$NTU \equiv \frac{UA}{C_{min}} \tag{3.18}$$

This dimensionless group can be viewed as a comparison of the heat capacity of the heat exchanger, expressed in W/°C, with the heat capacity of the flow.

We can immediately reduce the parallel-flow result from eqns. (3.9) to the following equation, based on these definitions:

$$-\left(\frac{C_{min}}{C_c} + \frac{C_{min}}{C_h}\right)NTU = \ell n\left[-\left(1 + \frac{C_c}{C_h}\right)\varepsilon\frac{C_{min}}{C_c} + 1\right] \tag{3.19}$$

We solve this for ε and, regardless of whether C_{min} is associated with the hot or cold flow, obtain for the parallel single-pass heat exchanger:

$$\varepsilon = \frac{1 - \exp[-(1 + C_{min}/C_{max})\text{NTU}]}{1 + C_{min}/C_{max}} = \text{fn}\left(\frac{C_{min}}{C_{max}}, \text{NTU only}\right) \quad (3.20)$$

The corresponding expression for the counterflow case is

$$\varepsilon = \frac{1 - \exp[-(1 - C_{min}/C_{max})\text{NTU}]}{1 - (C_{min}/C_{max})\exp[-(1 - C_{min}/C_{max})\text{NTU}]} \quad (3.21)$$

Equations (3.20) and (3.21) are given in graphical form in Fig. 3.16. Similar calculations give the effectiveness for the other heat exchanger configurations (see [1.24]) and we include some of the resulting effectiveness plots in Fig. 3.17. To see how the effectiveness can conveniently be used to complete a design, consider the following two examples.

Example 3.5

Consider the following parallel-flow heat exchanger specification:

cold flow enters at 40°C: $C_c = 20{,}000$ W/°C

hot flow enters at 150°C: $C_h = 10{,}000$ W/°C

$$A = 30 \text{ m}^2 \qquad U = 500 \text{ W/m}^2\text{-°C}$$

Determine the heat transfer and the exit temperatures.

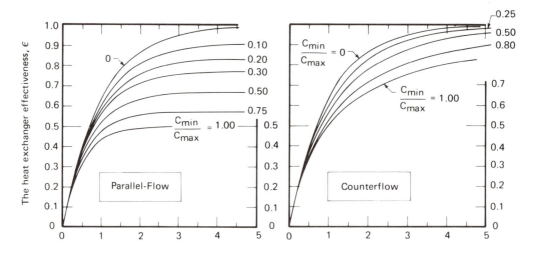

FIGURE 3.16   The effectiveness of parallel and counterflow heat exchangers. (Data provided by A. D. Kraus.)

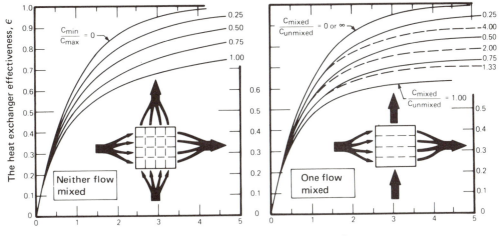

a.) Cross-flow exchanger, neither fluid mixed

b.) Cross-flow exchanger, one fluid mixed

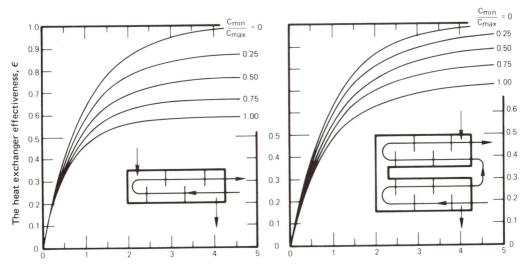

The number of transfer units, NTU = UA/C_{min}

c.) One shell pass, two tube pass exchanger. (Can also be used for 4, 6, 8, 10, 12 tube passes with a maximum error in ϵ, of 0.040 at C_{min}/C_{max} = 1 and large NTU.)

d.) Two shell pass 4 tube pass exchanger. (Can also be used for 4, 6, 8, ... tube passes with reasonable accuracy if there are equal numbers of tube passes in each shell pass.)

FIGURE 3.17 The effectiveness of some other heat exchanger configurations. (Data provided by A. D. Kraus.)

SOLUTION.   In this case we do not know the exit temperatures, so it is not possible to calculate the LMTD. Instead, we can either go to the parallel-flow effectiveness chart in Fig. 3.16, or to eqn. (3.20), using

$$\text{NTU} = \frac{UA}{C_{min}} = \frac{500(30)}{10,000} = 1.5$$

$$\frac{C_{min}}{C_{max}} = 0.5$$

and we obtain $\varepsilon = 0.596$. Now from eqn. (3.17), we find that

$$Q = \varepsilon C_{min}(T_{h_{in}} - T_{c_{in}}) = 0.596(10,000)(110)$$

$$= 655,600 \text{ W} = \underline{655.6 \text{ kW}}$$

Finally, from energy balances such as are expressed in eqn. (3.4), we get

$$T_{h_{out}} = T_{h_{in}} - \frac{Q}{C_h} = 150 - \frac{655,600}{10,000} = \underline{84.44°C}$$

$$T_{c_{out}} = T_{c_{in}} + \frac{Q}{C_c} = 40 + \frac{655,600}{20,000} = \underline{72.78°C}$$

Example 3.6

Suppose that we had the same kind of exchanger as we considered in Example 3.5, but that the area remained unspecified as a design variable. Then calculate the area that would bring the hot flow out at 90°C.

SOLUTION.   Once the exit cold fluid temperature is known, the problem can be solved with equal ease by either the LMTD or the effectiveness approach.

$$T_{c_{out}} = T_{c_{in}} + \frac{C_h}{C_c}(T_{h_{in}} - T_{h_{out}}) = 40 + \tfrac{1}{2}(150 - 90) = 70°C$$

Then, using the effectiveness method,

$$\varepsilon = \frac{C_h(T_{h_{in}} - T_{h_{out}})}{C_{min}(T_{h_{in}} - T_{c_{in}})} = \frac{10,000(150 - 90)}{10,000(150 - 40)} = 0.5455$$

so from Fig. 3.16 we read $\text{NTU} \simeq 1.15 = UA/C_{min}$. Thus

$$A = \frac{10,000(1.15)}{500} = \underline{23.00 \text{ m}^2}$$

We could also have calculated the LMTD:

$$\text{LMTD} = \frac{(150 - 40) - (90 - 70)}{\ell n(110/20)} = 52.79°C$$

so from $Q = UA(\text{LMTD})$, we obtain

$$A = \frac{10,000(150 - 90)}{500(52.79)} = \underline{22.73 \text{ m}^2}$$

The answers differ by 1%, which reflects graph-reading inaccuracy.

When the temperature of either fluid in a heat exchanger stays constant, the problem of analyzing heat transfer is greatly simplified. We have already noted that no *F*-correction is needed to adjust the LMTD in this case. The reason is that, when only one fluid changes in temperature, the configuration of the exchanger becomes irrelevant. Any exchanger is equivalent to a single fluid flowing in an isothermal pipe.[3]

Since all heat exchangers are equivalent in this case, it follows that the equation for the effectiveness in any configuration must reduce to the same common expression as C_{max} approaches infinity. The volumetric heat capacity rate might *approach* infinity because the flow rate or specific heat is very large, or it might *be* infinite because the flow is absorbing or giving up latent heat (as in Fig. 3.9). The limiting effectiveness expression can also be derived directly from energy-balance considerations (see Problem 3.11), but we obtain it here by letting $C_{max} \to \infty$ in either eqn. (3.20) or (3.21). The result is

$$\text{Limit}_{C_{max} \to \infty} \; \varepsilon = 1 - e^{-\text{NTU}} \tag{3.22}$$

Equation (3.22) defines the curve for $C_{min}/C_{max} = 0$ in all six of the effectiveness graphs in Figs. 3.16 and 3.17.

3.4 Heat exchanger design

The preceding sections have provided means for designing heat exchangers that generally are valuable in the design of smaller exchangers—typically, the kind of compact cross-flow exchanger used in transportation equipment. Larger shell-and-tube exchangers pose two kinds of difficulty in relation to *U*. The first is the variation of *U* through the exchanger, which we have already discussed. The second difficulty is that convective heat transfer coefficients are very hard to predict for the complicated flows that move through a baffled shell.

We shall achieve considerable success in using analysis to predict $\bar{h}$'s for various convective flows in Part III. The determination of $\bar{h}$ in a baffled shell remains a problem that cannot be solved analytically. Instead, it is normally computed with the help of empirical correlations or with the aid of large commercial computer programs that include relevant experimental correlations. The problem of predicting $\bar{h}$ when the flow is boiling or condensing is even more complicated. There is presently a great deal of research under way aimed at perfecting such empirical predictions.

There is also a host of considerations that must be made in designing heat exchangers in addition to the prediction of heat transfer. The primary ones are the minimization of pumping power and the minimization of fixed costs.

---

[3]We make use of this notion in Section 8.6 when we analyze heat convection in pipes and tubes.

The pumping power calculation, which we do not treat here in any detail, is based on the principles discussed in a first course on fluid mechanics. It generally takes the following form for each stream of fluid through the heat exchanger:

$$\text{pumping power} = \left(\dot{m}\ \frac{\text{kg}}{\text{s}}\right)\left(\frac{\Delta p}{\rho}\ \frac{\text{N/m}^2}{\text{kg/m}^3}\right) = \frac{\dot{m}\Delta p}{\rho}\left(\frac{\text{N-m}}{\text{s}}\right) = \frac{\dot{m}\Delta p}{\rho}(\text{W}) \quad (3.23)$$

where $\dot{m}$ is the mass flow rate of the stream, Δp the pressure drop of the stream as it passes through the exchanger, and ρ the fluid density.

Determining the pressure drop is a matter that can be relatively simple in a single-pass pipe-in-tube heat exchanger; or extremely difficult in, say, a shell-and-tube exchanger. The pressure drop in a straight run of pipe, for example, is given by

$$\Delta p = \frac{L}{D}\ \frac{\rho u_{av}^2}{2} f \quad (3.24)$$

where L is the length of the pipe and D is its diameter. u_{av} is the mean velocity of the flow in the pipe and f is the Darcy–Weisbach friction factor (see Fig. 8.6).

Optimizing the design of an exchanger is not just a matter of making Δp small as possible. Often heat exchange can be augmented by employing fins or roughening elements in an exchanger. (We discuss such elements in Chapter 4; see, e.g., Fig. 4.6.) Such augmentation will invariably increase the pressure drop, but it can also reduce the fixed cost of an exchanger by increasing U and reducing the required area. Furthermore, it can reduce the required flow rate of, say, a coolant, by increasing the effectiveness and thus balance the increase of Δp in eqn. (3.23).

To better understand the course of the design process, faced with such an array of trade-offs of advantages and penalties, we follow Taborek's [3.3] list of design considerations for a large shell-and-tube exchanger:

- Decide which fluid should flow on the shell side and which should flow in the tubes. Normally, this decision will be made to minimize the pumping cost. If, for example, water is being used to cool oil, the more viscous oil would flow in the shell. Corrosion behavior, fouling, and the problems of cleaning fouled tubes also weigh heavily in this decision.

- Early in the process, the designer should assess the cost of the calculation in comparison with:
  (a) The converging accuracy of the computation.
  (b) The investment in the exchanger.
  (c) The cost of a miscalculation.

- Make a rough estimate of the size of the heat exchanger using, for example, U values from Table 2.1 and/or anything else that might be known from experience. This serves to circumscribe the subsequent

trial-and-error calculations; it will help to size flow rates and to anticipate temperature variations; and it will help to avoid subsequent errors.

- Evaluate the heat transfer, pressure drop, and cost of various exchanger
  configurations which appear reasonable for the application. This is
  usually done with large-scale computer programs that have been developed, and are constantly being improved, as new research is included in
  them.

The computer runs suggested by this procedure are normally very complicated
and might typically involve 200 successive redesigns, even when relatively
efficient optimization procedures are used.

However, most students of heat transfer will not have to deal with such
designs. Many, if not most, *will* be called upon at one time or another to design
smaller exchangers in the range 0.1 to 10 m^2. The heat transfer calculation can
usually be done effectively with the methods described in this chapter. Some
useful sources of guidance in the pressure drop calculation are Kern's classical
treatment of *Process Heat Transfer* [3.4], the TEMA design book [2.3], and
Perry's Chemical Engineer's Handbook [3.5].

In such a calculation we start off with one fluid to heat and one to cool.
Perhaps we know the flow heat capacities (the C's), certain temperatures,
and/or the amount of heat that is to be transferred. The problem can be
annoyingly wide open and nothing can be done until it is somehow delimited.
The normal starting point is the specification of an exchanger configuration, and
to make this choice one needs experience. The descriptions in this chapter
provide a kind of first level of experience. References [3.4] to [3.7] provide a
second level. Manufacturer's catalogs are an excellent source of more advanced
information.

Once the exchanger configuration is set, U will be approximately set and the
area becomes the basic design variable. The design can then proceed along the
lines of Sections 3.2 or 3.3. If it is possible to begin with a complete specification
of inlet and outlet temperatures,

$$\underbrace{Q}_{C\Delta T} = \underbrace{U}_{\text{known}} \underbrace{A\,F(\text{LMTD})}_{\text{calculable}}$$

Then A can be calculated and the design completed. Usually, a reevaluation of
U and some iteration of the calculation is needed.

More often, we begin without full knowledge of the outlet temperatures. In
such cases we normally have to invent an appropriate trial-and-error method to
get the area, and a more complicated sequence of trials if we seek to optimize
pressure drop and cost by varying the configuration as well. If the C's are design
variables, then U will change significantly, because the $\bar{h}$'s are generally
velocity-dependent and more iteration will be needed.

We conclude Part I facing a variety of incomplete issues. Most notably, we face a serious need to be able to determine convective heat transfer coefficients. The prediction of $\bar{h}$ depends upon a knowledge of heat conduction. We therefore turn attention in Part II to a much more thorough study of heat conduction analysis than was undertaken in Section 1.3. In addition to setting up the methodology ultimately needed to predict $\bar{h}$'s, Part II will also deal with a variety of issues that have great practical importance in their own right.

PROBLEMS

3.1. Can you have a cross-flow exchanger in which both flows are mixed? Discuss.

3.2. Find the appropriate mean radius, $\bar{r}$, which will make $Q = kA(\bar{r})\Delta T/(r_o - r_i)$ valid for the one-dimensional heat conduction through a thick spherical shell (cf. Example 3.1).

3.3 Rework Problem 2.14, using the methods of Chapter 3.

3.4. 2.4 kg/s of a fluid having a specific heat of 0.8 kJ/kg-°K enters a counterflow heat exchanger at 0°C and is heated to 400°C by 2 kg/s of a fluid having a specific heat of 0.96 kJ/kg-°K entering the unit at 700°C. Show that to heat the cooler fluid to 500°C, all other conditions remaining unchanged, would require the surface area for heat transfer to be increased by 87.5%.

3.5. A cross-flow heat exchanger with both fluids unmixed is used to heat water ($c_p = 4.18$ kJ/kg-°K) from 40°C to 80°C, flowing at the rate of 1.0 kg/s. What is the overall heat transfer coefficient if hot engine oil ($c_p = 1.9$ kJ/kg-°K), flowing at the rate of 2.6 kg/s, enters at 100°C? The heat transfer area is 20 m². (Notice that you can use either an effectiveness or an LMTD method. It would be wise to use both as a check.)

3.6. Saturated non-oil-bearing steam at 1 atm enters the shell pass of a two-tube-pass shell condenser. There are thirty 20-ft tubes in each tube pass. They are made of schedule 160, $\frac{3}{4}$-in. steel pipe (nominal diameter). A volume flow rate of 0.01 ft³/s of water entering at 60°F enters each tube. The condensing heat transfer coefficient is 2000 Btu/ft²-hr-°F, and we calculate $\bar{h} = 1380$ Btu/ft²-hr-°F for the water in the tubes. Estimate the exit temperature of the water and mass rate of condensate. [$\dot{m}_c \simeq 8393$ lb$_m$/hr.]

3.7. Consider a counterflow heat exchanger which must cool 3000 kg/hr of mercury from 150°F to 128°F. The coolant is 100 kg/hr of water, supplied at 70°F. If U is 300 W/m²-°C, complete the design by determining reasonable values for the area and the exit-water temperature. [$A = 0.147$ m².]

3.8. An automobile air-conditioner condenser gives up 18 kW at 65 km/hr if the outside temperature is 35°C. The refrigerant temperature is constant at 65°C under these conditions, and the air rises 6°C in temperature as it flows across the heat exchanger tubes. The heat exchanger is of the finned-tube type shown in Fig. 3.6b with $U \simeq 200$ W/m²-°C. If $U \sim$(air velocity)$^{0.7}$ and the mass flow rate increases directly with the velocity, plot the percentage reduction of heat transfer in the condenser as a function of air velocity between 15 and 65 km/hr.

3.9. Derive eqn. (3.21).

3.10. Derive the infinite NTU limit of the effectiveness of parallel and counterflow heat exchangers at several values of C_{min}/C_{max}. Use common sense and the First Law of Thermodynamics and make reference to eqns. (3.20) and (3.21) only to check your results.

3.11. Derive the equation $\varepsilon = (NTU, C_{min}/C_{max})$ for the heat exchanger depicted in Fig. 3.9.

3.12. A single-pass heat exchanger condenses steam at 1 atm on the shell side and heats water from 10 to 30°C on the tube side. $^*U = 2500$ W/m²-°C. The tubing is thin-walled, 5 cm in diameter and 2 m in length. (a) Your boss asks whether the exchanger should be counterflow or parallel-flow. How do you advise him? Evaluate: (b) the LMTD; (c) $\dot{m}_{H_2O}$; (d) ε [$\varepsilon \approx 0.222$].

3.13. 2 kg/s of air at 27°C and 1.5 kg/s of water at 60°C each enter a heat exchanger. Evaluate the exit temperatures if $A = 12$ m², $U = 185$ W/m²-°C, and
(a) The exchanger is parallel flow.
(b) The exchanger is counterflow [$T_{h_{out}} \approx 54.0°C$].
(c) The exchanger is cross-flow, one stream mixed.
(d) The exchanger is cross-flow, neither stream mixed. [$T_{h_{out}} = 53.62°C$.]

3.14. 0.25 kg/s of air at 0°C enters a cross-flow heat exchanger. It is to be warmed to 20°C by 0.14 kg/s of air at 50°C. The streams are unmixed. As a first step in the design process, plot U against A and identify the approximate range of area for the exchanger.

3.15. A particular two-shell-pass, four-tube-pass heat exchanger uses 20 kg/s of river water at 10°C on the shell side to cool 8 kg/s of processed water from 80°C to 25°C on the tube side. At what temperature will the coolant be returned to the river? If U is 800 W/m²-°C, how large must the exchanger be?

3.16. A particular cross-flow process heat exchanger operates with the fluid mixed on one side only. When it is new, $U = 2000$ W/m²-°C, $T_{c_{in}} = 25°C$, $T_{c_{out}} = 80°C$, $T_{h_{in}} = 160°C$, and $T_{h_{out}} = 70°C$. After 6 months of operation the plant manager reports that the hot fluid is only being cooled to 90°C and that he is suffering a 30% reduction in total heat transfer. What is the fouling resistance after 6 months of use? (Assume no reduction of cold-side flow rate by fouling.)

3.17. Water at 15°C is supplied to a one-shell-pass, two-tube-pass heat exchanger, to cool 10 kg/s of liquid ammonia from 120°C to 40°C. You anticipate a U on the order of 1500 W/m²-°C when the water flows in the tubes. If A is to be 90 m², choose the correct flow rate of water.

3.18. Suppose that the heat exchanger in Example 3.5 had been a two-shell-pass, four-tube-pass exchanger with the hot fluid moving in the tubes. (a) What would be the exit temperatures in this case? [$T_{c_{out}} = 75.09°C$.] (b) What would the area be if we wanted the hot fluid to leave at the same temperature that it does in the example?

3.19. Plot the maximum tolerable fouling resistance as a function of U_{new} for a counterflow exchanger, with given inlet temperatures, if a 30% reduction in U is the maximum that can be tolerated.

REFERENCES

[3.1] R. A. BOWMAN, A. C. MUELLER, and W. M. NAGLE, "Mean Temperature Difference in Design," *Trans. ASME*, vol. 62, 1940, pp. 283–294.

[3.2] K. GARDNER and J. TABOREK, "Mean Temperature Difference: A Reappraisal," *AIChE J.*, vol. 23, no. 6, 1977, pp. 770–786.

[3.3] J. TABOREK, "Evolution of Heat Exchanger Design Techniques," *Heat Transfer Engineering*, vol. 1, no. 1, 1979, pp. 15–29.

[3.4] D. Q. KERN, *Process Heat Transfer*, McGraw-Hill Book Company, New York, 1950.

[3.5] *Perry's Chemical Engineering Handbook*, 5th ed. (H. H. Perry and C. H. Chilton, eds.), McGraw-Hill Book Company, New York, 1973.

[3.6] D. M. CONSIDINE, *Energy Technology Handbook*, McGraw-Hill Book Company, New York, 1975.

[3.7] A. P. FRAAS and M. N. ÖZIŞIK, *Heat Exchanger Design*, John Wiley & Sons, Inc., New York, 1965.

Part II

Analysis
of Heat Conduction

Chapter 4

Analysis of heat conduction and some steady one-dimensional problems

The effects of heat are subject to constant laws which cannot be discovered without the aid of mathematical analysis. The object of the theory which we are about to explain is to demonstrate these laws; it reduces all physical researches on the propagation of heat to problems of the calculus whose elements are given by experiment.

***The Analytical Theory of Heat**,* **J. Fourier**

4.1 The well-posed problem

The heat diffusion equation was derived in Section 2.1 and some attention was given to its solution. Before we go further with heat conduction problems we must indicate how to state such problems so that they can really be solved. This is particularly important in approaching the more complicated problems of transient and multidimensional heat conduction that we have avoided up to now.

A well-posed heat conduction problem is one in which all of the relevant information needed to obtain a unique solution is stated. A well-posed and hence solvable heat conduction problem will always read as follows:

Find $T(x,y,z,t)$ such that:[1]

1.

$$\vec{\nabla} \cdot (k\vec{\nabla} T) + \dot{q} = \rho c \frac{\partial T}{\partial t}$$

---

[1] x,y,z might be any three coordinates. $T(x,y,z,t) = T(\vec{r},t)$.

for $0 < t < \mathfrak{T}$ (where $\mathfrak{T}$ can $\rightarrow \infty$), and for x, y, z belonging to some region, R, which might extend to infinity.

2.
$$T = T_i(x, y, z) \text{ at } t = 0$$

(this is called an *initial condition* or *i.c.*)

(a) Condition 1 above is not imposed at $t = 0$.

(b) Only one i.c. is required.

(c) The i.c. is not needed:

   (1) In the steady-state case: $\vec{\nabla} \cdot (k \vec{\nabla} T) + \dot{q} = 0$.

   (2) For "periodic" heat transfer, where $\dot{q}$ or the boundary conditions vary periodically with time, and where we ignore the starting transient behavior.

3. T must also satisfy two *boundary conditions*, or *b.c.'s*, for each coordinate. The b.c.'s are very often of three common types.

(a) "Dirichlet conditions," or b.c.'s of the *first kind*:

   T is specified on the boundary of R for $t > 0$.

   We saw such b.c.'s in Examples 2.2, 2.3, and 2.4.

(b) "Neumann conditions" or b.c.'s of the *second kind*:

   The derivative of T normal to the boundary is specified on the boundary of R for $t > 0$.

   Such a condition arises when the heat flux,[2] $k(\partial T / \partial x)$, is specified on a boundary or when, with the help of insulation, we set $\partial T / \partial x$ equal to zero.

(c) b.c.'s of the *third kind*:

   A derivative of T in a direction normal to a boundary is proportional to the temperature on that boundary.

   Such a condition most commonly arises when there is convection at a boundary and is typically expressed[2] as

$$-k \frac{\partial T}{\partial x}\bigg|_{\text{bdy}} = \bar{h}(T - T_\infty)_{\text{bdy}}$$

   when the body lies to the left of the boundary on the x-coordinate. We have already used such a b.c. in step 4 of Example 2.5, and we have discussed it in Section 1.3 as well.

This list of b.c.'s is not complete, by any means, but it includes a great number of important cases. Missing from this list are time-dependent, or other more complicated, b.c.'s.

---

[2]Although we write $\partial T / \partial x$ here, we understand that this might be $\partial T / \partial z$, or $\partial T / \partial r$, or any other derivative in a direction locally normal to the surface on which the b.c. is specified.

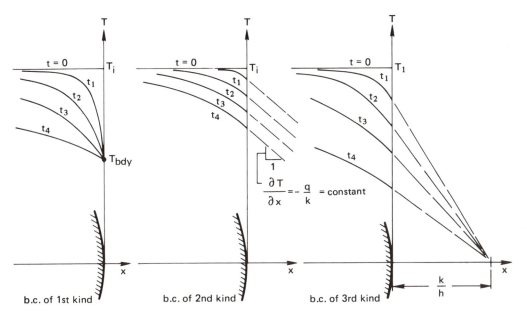

FIGURE 4.1   The transient cooling of a body as it might occur, subject to boundary
conditions of the first, second, and third kinds.

Next let us see what the three b.c.'s presented above might look like
graphically. Figure 4.1 shows the transient cooling of body from a constant
initial temperature, subject to each of the three b.c.'s we have described. Notice
that the initial temperature is not subject to the boundary condition, as was
pointed out previously under 2(a).

The eight-point procedure that was outlined in Section 2.2 for solving the
heat diffusion equation was contrived in part to assure that a problem will meet
the preceding requirements and be well posed.

4.2 The general solution

Once the heat conduction problem has been "posed" properly, the first step in
solving it is to find the general solution of the heat diffusion equation. We have
remarked that this is usually the easiest part of the problem. We consider next
some examples of general solutions.

One-dimensional steady heat conduction

Problem 4.1 emphasizes the simplicity of finding the general solutions of linear
ordinary differential equations, by asking for a table of all general solutions of

one-dimensional heat conduction problems. We shall work out some of those results to show what is involved. We begin with the heat diffusion equation with constant k and $\dot{q}$:

$$\nabla^2 T + \frac{\dot{q}}{k} = \frac{1}{\alpha}\frac{\partial T}{\partial t} \tag{2.15}$$

Cartesian coordinates: Steady conduction in the y-*direction.*   Equation (2.15) reduces as follows:

$$\underbrace{\frac{\partial^2 T}{\partial x^2}}_{=0} + \frac{\partial^2 T}{\partial y^2} + \underbrace{\frac{\partial^2 T}{\partial z^2}}_{=0} + \frac{\dot{q}}{k} = \underbrace{\frac{1}{\alpha}\frac{\partial T}{\partial t}}_{=0,\ \text{since steady}}$$

Therefore,

$$\frac{d^2 T}{dy^2} = -\frac{\dot{q}}{k}$$

which we integrate twice to get

$$T = -\frac{\dot{q}}{2k}y^2 + C_1 y + C_2$$

or, if $\dot{q} = 0$,

$$T = C_1 y + C_2$$

Cylindrical coordinates with a heat source: Tangential conduction.   This time we express eqn. (2.15) in cylindrical coordinates with the help of eqn. (2.17):

$$\underbrace{\frac{1}{r}\frac{\partial}{\partial r}\left(r\frac{\partial T}{\partial r}\right)}_{=0} + \underbrace{\frac{1}{r^2}\frac{\partial^2 T}{\partial \phi^2}}_{r\ \text{constant}} + \underbrace{\frac{\partial^2 T}{\partial z^2}}_{=0} + \frac{\dot{q}}{k} = \underbrace{\frac{1}{\alpha}\frac{\partial T}{\partial t}}_{=0,\ \text{since steady}}$$

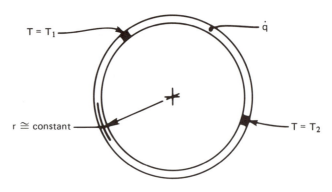

FIGURE 4.2   One-dimensional heat conduction in a ring.

Two integrations give

$$T = -\frac{r^2 \dot{q}}{2k}\phi^2 + C_1\phi + C_2 \tag{4.1}$$

This would describe, for example, the temperature distribution in the thin ring shown in Fig. 4.2. Here the b.c.'s might consist of temperatures specified at two angular locations as shown.

T = T(t *only*)

If T is spatially uniform, it can still vary with time. In such cases

$$\underbrace{\nabla^2 T}_{=0} + \frac{\dot{q}}{k} = \frac{1}{\alpha}\frac{\partial T}{\partial t}$$

and $\partial T/\partial t$ becomes an ordinary derivative. Then, since $\alpha = k/\rho c$,

$$\frac{dT}{dt} = \frac{\dot{q}}{\rho c} \tag{4.2}$$

This result is actually consistent with the lumped capacity solution described in Section 1.3. If the Biot number is low and internal resistance is unimportant, the convective removal of heat from the boundary of a body can be prorated over the volume of the body and interpreted as

$$\dot{q}_{\text{effective}} = -\frac{\bar{h}(T_{\text{body}} - T_\infty)A}{\text{volume}} \ \ \text{W/m}^3 \tag{4.3}$$

and the heat diffusion equation for this case, eqn. (4.2), becomes

$$\frac{dT}{dt} = -\frac{\bar{h}A}{\rho c V}(T - T_\infty) \tag{4.4}$$

The general solution in this situation was given in eqn. (1.20). [A particular solution was also written in eqn. (1.21).]

Separation of variables:
The general solution of multidimensional problems

Suppose that the physical situation does not permit us to throw away all but one of the derivatives in the heat diffusion equation. Suppose, for example, that we wish to predict the transient cooling in a slab as a function of the location within it. If there is no heat generation, the heat diffusion equation is

$$\frac{\partial^2 T}{\partial x^2} = \frac{1}{\alpha}\frac{\partial T}{\partial t} \tag{4.5}$$

A common trick is to ask: "Can we find a solution in the form of a product of functions of t and x: $T = \mathcal{T}(t) \cdot \mathcal{X}(x)$?" To find the answer we substitute this in eqn. (4.5) and get

$$\mathcal{X}''\mathcal{T} = \frac{1}{\alpha}\mathcal{T}'\mathcal{X} \tag{4.6}$$

where each prime denotes one differentiation of a function with respect to its argument. Thus $\mathcal{T}' = d\mathcal{T}/dt$ and $\mathcal{X}'' = d^2\mathcal{X}/dx^2$. Rearranging eqn. (4.6), we get

$$\frac{\mathcal{X}''}{\mathcal{X}} = \frac{1}{\alpha}\frac{\mathcal{T}'}{\mathcal{T}} \tag{4.7}$$

This is an interesting result in that the left-hand side depends only upon x and the right-hand side depends only upon t. Thus we set *both* sides equal to the same constant, which for convenience we call $-\lambda^2$:

$$\frac{\mathcal{X}''}{\mathcal{X}} = \frac{1}{\alpha}\frac{\mathcal{T}'}{\mathcal{T}} = -\lambda^2 \qquad \text{a constant} \tag{4.7a}$$

It follows that the differential equation (4.7) can be resolved into two ordinary differential equations:

$$\mathcal{X}'' = -\lambda^2\mathcal{X} \qquad \text{and} \qquad \mathcal{T}' = -\alpha\lambda^2\mathcal{T} \tag{4.8}$$

The general solutions of both of these equations are well known and among the first ones dealt with in any study of differential equations. They are:

$$\left.\begin{array}{ll} \mathcal{X}(x) = A\sin\lambda x + B\cos\lambda x & \text{for } \lambda \neq 0 \\ \mathcal{X}(x) = Ax + B & \text{for } \lambda = 0 \end{array}\right\} \tag{4.9}$$

and

$$\left.\begin{array}{ll} \mathcal{T}(t) = C\exp(-\alpha\lambda^2 t) & \text{for } \lambda \neq 0 \\ \mathcal{T}(t) = C & \text{for } \lambda = 0 \end{array}\right\} \tag{4.10}$$

where we use capital letters to denote constants of integration. [In either case, these solutions can be verified by substituting them back into eqns. (4.8).] Thus the general solution of eqn. (4.5) can indeed be written in the form of a product, and that product is

$$\left.\begin{array}{ll} T = \mathcal{X}\mathcal{T} = e^{-\alpha\lambda^2 t}(D\sin\lambda x + E\cos\lambda x) & \text{for } \lambda \neq 0 \\ T = \mathcal{X}\mathcal{T} = Dx + E & \text{for } \lambda = 0 \end{array}\right\} \tag{4.11}$$

Whether or not this result will prove useful depends on whether or not we can fit it to the b.c.'s and the i.c. It turns out that it is often possible to do so using the methods of Fourier series. But that is a somewhat difficult business, and a matter that we shall put aside for the present.

This very simple method for obtaining general solutions of linear partial d.e.'s is called the method of *separation of variables*. It can be applied to all kinds of linear d.e.'s. Consider, for example, two-dimensional steady heat conduction without heat sources:

$$\frac{\partial^2 T}{\partial x^2} + \frac{\partial^2 T}{\partial y^2} = 0 \tag{4.12}$$

Set $T = \mathcal{X}\mathcal{Y}$ and get

$$\frac{\mathcal{X}''}{\mathcal{X}} = -\frac{\mathcal{Y}''}{\mathcal{Y}} = -\lambda^2$$

where λ can be an imaginary number. Then

$$\left. \begin{array}{l} \mathcal{X} = A\sin\lambda x + B\cos\lambda x \\ \mathcal{Y} = Ce^{\lambda y} + De^{-\lambda y} \end{array} \right\} \quad \text{for } \lambda \neq 0$$

or

$$\left. \begin{array}{l} \mathcal{X} = Ax + B \\ \mathcal{Y} = Cy + D \end{array} \right\} \quad \text{for } \lambda = 0$$

and the general solution is

$$\left. \begin{array}{ll} T = (E\sin\lambda x + F\cos\lambda x)(e^{-\lambda y} + Ge^{\lambda y}) & \text{for } \lambda \neq 0 \\ T = (Ex + F)(y + G) & \text{for } \lambda = 0 \end{array} \right\} \quad (4.13)$$

Example 4.1

A long slab is cooled to 0°C on the sides and a blowtorch is turned on the top edge, giving an approximately sinusoidal temperature distribution along the top as shown in Fig. 4.3. Find the temperature distribution within the slab.

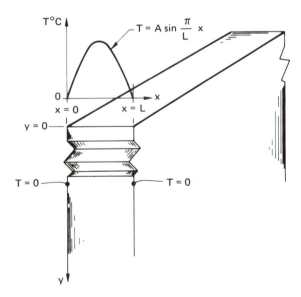

FIGURE 4.3   A two-dimensional slab maintained at a constant temperature on the sides and subjected to a sinusoidal variation of temperature on one face.

SOLUTION. The general solution is given by eqn. (4.13). We must therefore identify the appropriate b.c.'s and then fit the general solution to it. Those b.c.'s are:

on the top surface: $T(x,0) = A \sin \pi \dfrac{x}{L}$

on the sides: $T(0 \text{ or } L, y) = 0$

as $y \to \infty$: $T(x, y \to \infty) = 0$

Substitute eqn. (4.13) in the third b.c.:

$$(E \sin \lambda x + F \cos \lambda x)(0 + G[\infty]) = 0$$

The only way that this can be true is if $G = 0$. Substitute eqn. (4.13), with $G = 0$, into the second b.c.:

$$(0 + F)e^{-\lambda y} = 0$$

so F also $= 0$. Substitute eqn. (4.13), with $G = F = 0$, into the first b.c.:

$$E(\sin \lambda x) = A \sin \pi \frac{x}{L}$$

It follows that $A = E$ and $\lambda = \pi/L$. Then eqn. (4.13) becomes the particular solution that satisfies the b.c.'s:

$$T = A\left(\sin \pi \frac{x}{L}\right)e^{-\pi y/L}$$

Thus the sinusoidal variation of temperature at the top of the slab is attenuated exponentially at lower positions in the slab. At a position $y = 2L$ below the top, T will be $0.0019A \sin \pi x/L$. The temperature distribution in the x-direction will still be sinusoidal, but it will have less than $\frac{1}{500}$ of the amplitude at $y = 0$.

Consider some important features of this and other solutions:

- The b.c. at $y = 0$ is a special one which works very well with this particular general solution. If we had tried to fit the equation to a general temperature distribution, $T(x, y = 0) = \text{fn}(x)$, it would not have been obvious how to proceed. Actually, this is the kind of problem that Fourier solved with the help of his "Fourier series" method. We discuss this matter in more detail in Chapter 5.

- Not all forms of general solutions lend themselves to a particular set of boundary and/or initial conditions. In this example, we made the process look simple, but more often than not, *it is in fitting a general solution to a set of boundary conditions that we get stuck.*

- Normally, in formulating a problem, we must approximate real behavior in stating the b.c.'s. It is advisable to consider what kind of assumption will put the b.c.'s in a form compatible with the general solution. The temperature distribution imposed on the slab by the blowtorch in Example 4.1 might just as well have been approximated as a parabola. But as small as the difference between a parabola and a sine function might be, the latter b.c. was far easier to accommodate.

- The twin issues of existence and uniqueness of solutions require a comment here: It has been established that solutions to all well-posed heat diffusion problems are unique. Furthermore, we know from our experience that if we describe a physical process correctly, a unique outcome exists. Therefore, we are normally safe to leave these issues to the mathematician—at least in the sort of problems we discuss here.

- Given that a unique solution exists, we accept any solution as correct once we have carved it to fit the boundary conditions. In this sense the solution of differential equations is often more of an inventive than a formal operation. The person who does it best is often the person who has done it before and so has a large assortment of tricks up his or her sleeve.

4.3 Dimensional analysis

Introduction

Most college curricula place the first course in heat transfer after an introduction to fluid mechanics, and most fluid mechanics courses involve a unit on dimensional analysis. The reader with such a background can probably scan lightly over this section, although we shall deal with illustrations drawn largely from heat transfer instead of from fluid mechanics. The reader who has not been through such a course will find a rather nice introduction in Streeter's *Fluid Mechanics* text [4.1].

Part I included many illustrations in which a problem involving many variables was reduced into a much simpler form by combining variables into dimensionless groups. In Example 2.5, for example, we had a problem involving seven variables: They included a dependent variable which was the temperature,[3] $(T - T_i)$; an independent variable which was the radius, r; and five system parameters, r_i, r_o, $\bar{h}$, k, and $(T_\infty - T_i)$. By reorganizing the solution into dimensionless groups [eqn. (2.24)] we reduced the total number of variables to only four:

$$\underbrace{\frac{T - T_i}{T_\infty - T_i}}_{\substack{\text{dependent} \\ \text{variable}}} = \text{fn}\left(\underbrace{\frac{r}{r_i}}_{\substack{\text{indep.} \\ \text{var.}}}, \; \underbrace{\frac{r_o}{r_i}, \text{Bi}}_{\substack{\text{two system} \\ \text{parameters}}} \right) \qquad (2.24a)$$

This solution offered a number of advantages over the dimensional solution. For one thing, it permitted us to plot *all* conceivable solutions for a particular

---

[3]Notice that we do not call T_i a variable. It is simply the reference temperature against which the problem is worked. If it happened to be 0°C, we would not notice its subtraction from the other temperatures.

shape of cylinder, (r_o/r_i), in a single figure, Fig. 2.11. For another, it allowed us to study the simultaneous roles of $\bar{h}$, k, and r_o in defining the character of the solution. By combining them as a Biot number, we were able to say—even before we had solved the problem—whether or not external convection really had to be considered.

Nondimensionalization made it possible for us to consider, simultaneously, the behavior of all *similar* systems of heat conduction through cylinders. Thus a large, highly conducting cylinder might be *similar* in its behavior to a small cylinder with a low thermal conductivity.

Finally, we shall discover that, by nondimensionalizing a problem *before* we solve it, we can often greatly simplify the process of solving it.

Our next aim is to map out a method for nondimensionalizing problems before we have solved them or, indeed, before we have even written the equations that must be solved. The key to the method is a result called the Buckingham pi-theorem.

The Buckingham pi-theorem

The attention of scientific workers was apparently drawn very strongly toward the question of similarity at about the beginning of World War I. Buckingham first organized previous thinking and developed his famous theorem in 1914 in the *Physical Review* [4.2] and he expanded upon the idea in the *Transactions of the ASME* one year later [4.3]. Lord Rayleigh almost simultaneously discussed the problem with great clarity in 1915 [4.4]. To understand Buckingham's theorem, we begin with an example.

Example 4.2

Consider the heat exchanger problem described in Fig. 3.15. The "unknown," or dependent variable, in the problem is either of the exit temperatures. Without any knowledge of heat exchanger analysis, we can write the functional equation on the basis of our physical understanding of the problem:

$$\underbrace{T_{c_{out}} - T_{c_{in}}}_{°C} = \text{fn}\left[\underbrace{C_{max}}_{\frac{W}{°C}}, \underbrace{C_{min}}_{\frac{W}{°C}}, \underbrace{\left(T_{h_{in}} - T_{c_{in}}\right)}_{°C}, \underbrace{U,}_{\frac{W}{m^2 \cdot °C}} \underbrace{A}_{m^2} \right] \qquad (4.14)$$

where the dimensions of each term are noted under the equation.

We want to know how many dimensionless groups the variables in eqn. (4.14) should reduce to. To do this we pick one of the variables and arbitrarily divide or multiply it into each of the variables that has its units, in such a way as to eliminate those units from all but one term. Doing this will in no way change the validity of the dimensional functional equation. First let us do this with the term $(T_{h_{in}} - T_{c_{in}})$.

$$\underbrace{\frac{T_{c_{out}} - T_{c_{in}}}{T_{h_{in}} - T_{c_{in}}}}_{\text{dimensionless}} = \text{fn}\left[\underbrace{C_{max}(T_{h_{in}} - T_{c_{in}})}_{W}, \underbrace{C_{min}(T_{h_{in}} - T_{c_{in}})}_{W}, \underbrace{\left(T_{h_{in}} - T_{c_{in}}\right)}_{°C}, \underbrace{U(T_{h_{in}} - T_{c_{in}})}_{W/m^2}, \underbrace{A}_{m^2} \right]$$

The interesting thing about the equation in this form is that the only term in it with the units of °C is $(T_{h_{in}} - T_{c_{in}})$. No such term *can* exist in the equation because it is impossible to achieve dimensional homogeneity without another term in °C to balance it. Therefore, we must remove it.

$$\underbrace{\frac{T_{c_{out}} - T_{c_{in}}}{T_{h_{in}} - T_{c_{in}}}}_{\text{dimensionless}} = \text{fn}\left[\underbrace{C_{max}(T_{h_{in}} - T_{c_{in}})}_{\text{W}}, \underbrace{C_{min}(T_{h_{in}} - T_{c_{in}})}_{\text{W}}, \underbrace{U(T_{h_{in}} - T_{c_{in}})}_{\text{W/m}^2}, \underbrace{A}_{\text{m}^2} \right]$$

Now the equation has only two dimensions in it—W and m². Next we multiply $U(T_{h_{in}} - T_{c_{in}})$ by A to get rid of m² in the second-to-last term. Accordingly, the term A m² can no longer stay in the equation, and we have

$$\underbrace{\frac{T_{c_{out}} - T_{c_{in}}}{T_{h_{in}} - T_{c_{in}}}}_{\text{dimensionless}} = \text{fn}\left[\underbrace{C_{max}(T_{h_{in}} - T_{c_{in}})}_{\text{W}}, \underbrace{C_{min}(T_{h_{in}} - T_{c_{in}})}_{\text{W}}, \underbrace{UA(T_{h_{in}} - T_{c_{in}})}_{\text{W}} \right]$$

Next we divide the first and third terms on the right by the second. This leaves only $C_{min}(T_{h_{in}} - T_{c_{in}})$, with the dimensions of W. That term must then be removed and we are left with the completely dimensionless result:

$$\frac{T_{c_{out}} - T_{c_{in}}}{T_{h_{in}} - T_{c_{in}}} = \text{fn}\left(\frac{C_{max}}{C_{min}}, \frac{UA}{C_{min}} \right) \tag{4.15}$$

Equation (4.15) has exactly the same functional form as eqn. (3.21), which we obtained by direct analysis.

Notice that we removed one variable from eqn. (4.14) for each dimension in which the variables are expressed. If there are n variables—including the dependent variable—expressed in m dimensions, we then expect to be able to express the equation in $(n-m)$ dimensionless groups, or pi-groups as Buckingham called them.

This fact is expressed by the *Buckingham pi-theorem*, which we state formally in the following way:

> A physical relationship among n variables, which can be expressed in *a minimum* of m dimensions, can be rearranged into a relationship among $(n-m)$ *independent* dimensionless groups of the original variables.

Two important qualifications have been italicized. They will be explained in detail in subsequent examples.

Buckingham called the dimensionless groups pi-groups and identified them as $\Pi_1, \Pi_2, \ldots, \Pi_{n-m}$. Normally, we call Π_1 the dependent variable and retain $\Pi_{2\rightarrow(n-m)}$ as independent variables. Thus the dimensional functional equation reduces to a dimensionless functional equation of the form

$$\Pi_1 = \text{fn}(\Pi_2, \Pi_3, \ldots, \Pi_{n-m}) \tag{4.16}$$

Example 4.3

Is eqn. (2.24) consistent with the pi-theorem? To find out, we first write the dimensional functional equation for Example 2.5:

$$T - T_i = \text{fn}[\underbrace{r}_{\text{m}}, \underbrace{r_i}_{\text{m}}, \underbrace{r_o}_{\text{m}}, \underbrace{\bar{h}}_{\text{W/m}^2\text{-}°\text{C}}, \underbrace{k}_{\text{W/m-}°\text{C}}, \underbrace{(T_\infty - T_i)}_{°\text{C}}]$$

There are seven variables ($n = 7$) in three dimensions, °C, m, and W ($m = 3$). Therefore, we look for $7 - 3$, or 4 pi-groups. There *are* four pi-groups in eqn. (2.24):

$$\Pi_1 = \frac{T - T_i}{T_\infty - T_i} \qquad \Pi_2 = \frac{r}{r_i} \qquad \Pi_3 = \frac{r_o}{r_i} \qquad \Pi_4 = \frac{\bar{h} r_o}{k} \equiv \text{Bi}$$

Consider two features of this result. First, the minimum number of dimensions was three. If we had written watts as J/s, we would have had four dimensions, instead. But joules never appear in that particular problem independently of seconds. They always appear as a ratio and should not be separated. (If we had worked in English units, this would have seemed more confusing, since there is no name for a Btu/sec unless we first convert it to horsepower.) The failure to combine dimensions that are grouped together is one of the major errors that the beginner makes in using the pi-theorem.

The second feature is the *independence* of the groups. This means that we may pick any four dimensionless arrangements of variables as long as no group or groups can be made into any other group by mathematical manipulation. For example, suppose that someone suggested that there was a fifth pi-group in Example 4.3:

$$\Pi_5 = \sqrt{\frac{\bar{h} r}{k}}$$

It is easy to see that Π_5 can be written as

$$\Pi_5 = \sqrt{\frac{\bar{h} r_o}{k}} \sqrt{\frac{r}{r_i}} \sqrt{\frac{r_i}{r_o}} = \sqrt{\text{Bi} \, \Pi_2 / \Pi_3}$$

Therefore Π_5 is not independent of the existing groups, nor will we ever find a fifth grouping that is.

Another matter, which is frequently made much of, is the problem of finding sets of independent pi-groups once the variables of a problem are known. There exists an arithmetic strategy for doing this called the method of indices (see, for example, [4.1]). This method is a little cumbersome and it requires that a number of judgments must be made. Nevertheless, it is perfectly correct to use it.

We shall suggest, in its place, either of two alternative methods: (1) repeat the elimination-of-dimensions strategy that was used to derive the pi-theorem in

Example 4.2; or (2) simply arrange the variables into the required number of independent dimensionless groups by inspection. In any method, one must make judgments in the process of combining variables, and these decisions can lead to different arrangements of the pi-groups. Therefore, if the problem can be solved by inspection, there is no advantage to be gained by the use of a more formal method.

The methods of dimensional analysis can be used to help find the solution of many physical problems. We offer the following example, not entirely with tongue in cheek; and we work it formally, although it could be done just as easily by inspection.

Example 4.4

Suppose that Einstein had been clever enough to recognize that the energy equivalent, e, of a rest mass, m_o, depended on the velocity of light, c_ℓ; and that he did so before he knew any mathematics beyond the pi-theorem. Then he would have had the three variables in the following dimensional functional equation:

$$e \text{ N-m or } e \frac{\text{kg-m}^2}{\text{s}^2} = \text{fn}(c_\ell \text{ m/s}, m_o \text{ kg})$$

The minimum number of dimensions is only two: kg and m/s, so we look for $3-2$, or 1 pi-group. To find it formally, we eliminate the dimension of mass from e by dividing it by m_o kg. Thus

$$\frac{e}{m_o} \frac{\text{m}^2}{\text{s}^2} = \text{fn}(c_\ell \text{ m/s}, \underbrace{m_o \text{ kg}}_{})$$

this must be removed because it
is the only term with mass in it

Then we eliminate the dimension of velocity (m/s) by dividing e/m_o by c_ℓ^2:

$$\frac{e}{m_o c_\ell^2} = \text{fn}(c_\ell \text{ m/s})$$

This time c_ℓ must be removed from the function on the right since it is the only term with the dimensions m/s. This gives the result (which could have been written by inspection once it was known that there could only be one pi-group):

$$\Pi_1 = \frac{e}{m_o c_\ell^2} = \text{fn(no other groups)} = \text{constant}$$

or

$$e = \text{constant}(m_o c_\ell^2)$$

Of course, it required Einstein's relativity theory to tell us that the constant is unity.

Example 4.5

What is the velocity of efflux of liquid from the tank shown in Fig. 4.4?

SOLUTION. In this case we can guess that the velocity, V, might depend on gravity, g, and the head, H. We might be tempted to include the density as well until we realize that g is already a *force per unit mass*. To understand this, we can use English units and divide

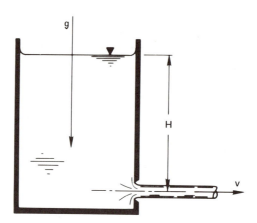

FIGURE 4.4   Efflux of liquid from a tank.

g by the conversion factor,[4] g_c. Thus $(g \text{ ft/s}^2)/(g_c \text{ lb}_m\text{-ft/lb}_f\text{-s}^2) = g \text{ lb}_f/\text{lb}_m$. Then

$$\underbrace{V}_{m/s} = \text{fn}(\underbrace{H}_{m}, \underbrace{g}_{m/s^2})$$

so there are three variables in two dimensions, and we look for $3-2$, or 1 pi-group. It would have to be

$$\Pi_1 = \frac{V}{\sqrt{gH}} = \text{fn(no other pi-groups)} = \text{constant}$$

or

$$V = \text{constant}\sqrt{gH}$$

The analytical study of fluid mechanics tells us that this form is correct and that the constant is $\sqrt{2}$. The group, V^2/gh, by the way, is called a Froude (pronounced "Frood") number, Fr. It compares inertial forces to gravitational forces. Fr is about 1000 for a pitched baseball and it is between 1 and 10 for the water flowing over the spillway of a dam.

Example 4.6

Obtain the dimensionless functional equation for the temperature distribution during steady conduction in a slab with a heat source, $\dot{q}$.

SOLUTION.   In such a case there might be one or two specified temperatures in the problem: T_1 or T_2. Thus the dimensional functional equation is

$$\underbrace{T - T_1}_{°C} = \text{fn}\,[\underbrace{(T_2 - T_1)}_{°C}, \underbrace{x, L}_{m}, \underbrace{\dot{q}}_{W/m^3}, \underbrace{k}_{W/m\text{-}°C}, \underbrace{\bar{h}}_{W/m^2\text{-}°C}]$$

[4]One can always divide any variable by a conversion factor without changing it.

where we presume that there is a convective b.c. involved and we identify a characteristic length, L, in the x-direction. There are seven variables in three dimensions, or $7-3=4$ pi-groups. Three of these groups are ones that we have dealt with in the past, in one form or another:

$$\Pi_1 = \frac{T - T_1}{T_2 - T_1}$$ dimensionless temperature which we will give the name Θ

$$\Pi_2 = \frac{x}{L}$$ dimensionless length which we call ξ

$$\Pi_3 = \frac{\bar{h}L}{k}$$ which we recognize as the Biot number, Bi

The fourth group is new to us:

$$\Pi_4 = \frac{\dot{q}L^2}{k(T_2 - T_1)}$$ which compares the rate of generation of heat to the rate at which it is carried away; we call it Γ

Thus the solution is

$$\Theta = \text{fn}(\xi, \text{Bi}, \Gamma) \tag{4.17}$$

In Example 2.2 we undertook such a problem, but it differed in two respects. There was no convective boundary condition and hence no $\bar{h}$, and only one temperature was specified in the problem. In this case, the dimensional functional equation was

$$(T - T_1) = \text{fn}(x, L, \dot{q}, k)$$

so there were only five variables in the same three dimensions. The resulting dimensionless functional equation therefore involved only two pi-groups. One was $\xi = x/L$ and the other is a new one equal to Θ/Γ. We call it Φ:

$$\Phi \equiv \frac{T - T_1}{\dot{q}L^2/k} = \text{fn}\left(\frac{x}{L}\right) \tag{4.18}$$

And this is exactly the form of the analytical result, eqn. (2.19).

A final issue that we must consider is that of dealing with dimensions that convert into one another. For example, kg and N are defined in terms of one another through Newton's Second Law of Motion. Therefore, they cannot be identified as separate dimensions. The same would appear to be true of J and N-m, since both are the dimensions of energy. However, we must discern whether or not a mechanism exists for interchanging them in a given problem. If mechanical energy remains distinct from thermal energy in a given problem, then J should not be interpreted as N-m.

Additional examples of the use of dimensional analysis appear through this book. Dimensional analysis is, indeed, our court of first resort in solving most of the new problems that we undertake.

4.4 Use of dimensional analysis in the solution of steady heat conduction problems

Heat conduction problems with convective boundary conditions can rapidly become very complicated, even if they start out simple. We therefore look for tricks to avoid making mistakes. One such trick is to take great care that the dimensions are consistent at each stage of the solution. The best way to do this, and to eliminate a great deal of algebra at the same time, is to nondimensionalize the equation before we apply the b.c.'s. This nondimensionalization should be consistent with the pi-theorem. We illustrate this idea with a fairly complex example.

Example 4.7

A slab shown in the center of Fig. 4.5 has different temperatures and different heat transfer coefficients on either side and heat is generated within it. Calculate the temperature distribution in the slab.

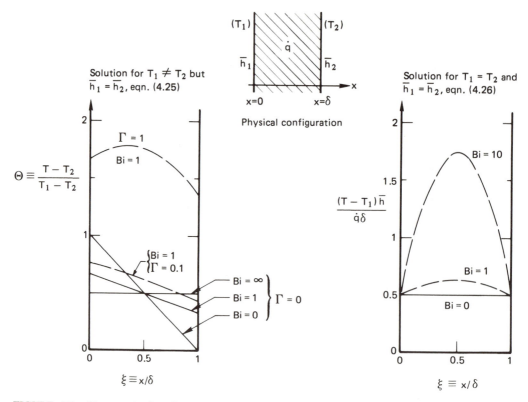

FIGURE 4.5   Heat conduction through a heat-generating slab with asymmetric boundary conditions.

SOLUTION. The differential equation is

$$\frac{d^2 T}{dx^2} = -\frac{\dot{q}}{k}$$

and its general solution is

$$T = -\frac{\dot{q}x^2}{2k} + C_1 x + C_2 \tag{4.19}$$

with b.c.'s

$$\bar{h}_1(T_1 - T)_{x=0} = -k\frac{dT}{dx}\Big|_{x=0} \qquad \bar{h}_2(T - T_2)_{x=L} = -k\frac{dT}{dx}\Big|_{x=L} \tag{4.20}$$

There are eight variables involved in the problem: $(T - T_2)$, $(T_1 - T_2)$, x, L, k, $\bar{h}_1$, $\bar{h}_2$, and $\dot{q}$; and there are three dimensions: °C, W, and m. This results in $8 - 3$, or 5 pi-groups. For these we choose

$$\Pi_1 \equiv \Theta = \frac{T - T_2}{T_1 - T_2} \qquad \Pi_2 \equiv \xi = \frac{x}{L} \qquad \Pi_3 \equiv Bi_1 = \frac{\bar{h}_1 L}{k}$$

$$\Pi_4 \equiv Bi_2 = \frac{\bar{h}_2 L}{k} \qquad \Pi_5 \equiv \Gamma = \frac{\dot{q}L^2}{2k(T_2 - T_1)}$$

where Γ can be interpreted as a comparison of the heat generated in the slab to that which could flow through it.

Under this nondimensionalization, eqn. (4.19) becomes[5]

$$\Theta = -\Gamma\xi^2 + C_3\xi + C_4 \tag{4.21}$$

and the b.c.'s become

$$Bi_1(1 - \Theta_{\xi=0}) = -\Theta'_{\xi=0} \qquad Bi_2\Theta_{\xi=1} = -\Theta'_{\xi=1} \tag{4.22}$$

where the primes denote differentiation with respect to ξ. Substituting eqn. (4.21) in eqns. (4.22), we obtain

$$Bi_1(1 - C_4) = -C_3 \qquad Bi_2(-\Gamma + C_3 + C_4) = 2\Gamma - C_3 \tag{4.23}$$

Substituting the first of eqns. (4.23) in the second, we get

$$C_4 = 1 + \frac{-Bi_1 + 2(Bi_1/Bi_2)\Gamma + Bi_1\Gamma}{Bi_1 + Bi_1^2/Bi_2 + Bi_1^2}$$

$$C_3 = Bi_1(C_4 - 1)$$

Thus eqn. (4.21) becomes

$$\Theta = 1 + \Gamma\left[\frac{2(Bi_1/Bi_2) + Bi_1}{1 + Bi_1/Bi_2 + Bi_1}\xi - \xi^2 + \frac{2(Bi_1/Bi_2) + Bi_1}{Bi_1 + Bi_1^2/Bi_2 + Bi_1^2}\right]$$

$$-\frac{Bi_1}{1 + Bi_1/Bi_2 + Bi_1}\xi - \frac{Bi_1}{Bi_1 + Bi_1^2/Bi_2 + Bi_1^2} \tag{4.24}$$

[5]The rearrangement of the dimensional equations into dimensionless form is straightforward algebra. If the results shown here are not immediately obvious to you, sketch the calculation on another piece of paper.

This is a complicated result and one that would have required enormous patience and accuracy to obtain without first simplifying the problem statement as we did. If the heat transfer coefficients were the same on either side of the wall, then $Bi_1 = Bi_2 \equiv Bi$ and eqn. (4.24) would reduce to

$$\Theta = 1 + \Gamma(\xi - \xi^2 + 1/Bi) - \frac{\xi + 1/Bi}{1 + 2/Bi} \qquad (4.25)$$

which is a very great simplification.

Equation (4.25) is plotted on the left-hand side of Fig. 4.5 for Bi equal to 0, 1, and ∞ and for Γ equal to 0, 0.1, and 1. The following features should be noted:

- When $\Gamma \ll 0.1$, the heat generation can be ignored.
- When $\Gamma \gg 1$, $\Theta \rightarrow \dfrac{Bi + \Gamma}{Bi} + \Gamma(\xi - \xi^2)$. This is a simple parabolic temperature distribution displaced upward an amount that depends on the relative external resistance as reflected in the Biot number.
- If both Γ and $1/Bi$ become large, $\Theta \rightarrow \Gamma/Bi$. This means that, when internal resistance is low and the heat generation is great, the slab temperature is constant and quite high.

If T_2 were equal to T_1 in this problem, Γ would go to infinity. In such a situation we should redo the dimensional analysis of the problem. The dimensional functional equation now shows $(T - T_1)$ to be a function of x, L, k, $\bar{h}$, and $\dot{q}$. There are six variables in three dimensions, so there are three pi-groups:

$$\frac{T - T_1}{\dot{q}L/\bar{h}} = \text{fn}(\xi, Bi)$$

where the dependent variable is like Φ [recall eqn. (4.18)] multiplied by Bi. We can put eqn. (4.25) in this form by multiplying both sides of it by $\bar{h}(T_1 - T_2)/\dot{q}\delta$. The result is

$$\frac{(T - T_1)\bar{h}}{\dot{q}L} = \tfrac{1}{2}Bi(\xi - \xi^2) + \tfrac{1}{2} \qquad (4.26)$$

This result is plotted on the right-hand side of Fig. 4.5. The following features of that graph are of interest:

- Heat generation is the only "force" giving rise to temperature nonuniformity. Since it is symmetrical, the graph is also symmetrical.
- When $Bi \ll 1$, the slab temperature approaches a uniform value equal to $T_1 + \dot{q}L/2\bar{h}$. (In this case we could have solved the problem with far greater ease by using a simple heat balance, since it is no longer a heat conduction problem.)

- When $\mathrm{Bi} > 100$, the temperature distribution is a very large parabola with $\frac{1}{2}$ added to it. In this case the problem could have been solved using boundary conditions of the first kind because the surface temperature stays very close to T_∞ (recall Fig. 1.11).

4.5 Fin design

The purpose of fins

The convective removal of heat from a surface can be substantially improved if we put extensions on that surface to increase its area. These extensions can take a variety of forms. Figure 4.6, for example, shows many different ways in which the surface of commercial heat exchanger tubing can be "extended" with protrusions of a kind that we call "fins."

Figure 4.7 shows another very interesting application of fins in a heat exchanger design. This picture is taken from an issue of *Science* magazine [4.5], which presents an intriguing argument by Farlow, Thompson, and Rosner. They showed that the strange rows of fins on the back of the *Stegosaurus* were probably used to shed excess body heat after strenuous activity. This supports a recent suspicion that *Stegosaurus* was actually warm-blooded.

These examples involve some rather complicated fins. But the analysis of a straight fin protruding from a wall will show us the essential features of all fin behavior. This analysis will have considerable direct application to a host of problems.

Analysis of a one-dimensional fin

The equations. Figure 4.8 shows a one-dimensional fin protruding from a wall. The wall—and the root of the fin—are at a temperature, T_o, which is either greater or less than the ambient temperature, T_∞. The length of the fin is cooled or heated through a heat transfer coefficient, $\bar{h}$, by the ambient fluid. The heat transfer coefficient will be assumed constant, although (as we shall see in Part III) that can introduce serious error in boiling, condensing, or other natural convection situations.

The tip may or may not exchange heat with the surroundings through a heat transfer coefficient, $\bar{h}_L$, which would generally differ from $\bar{h}$. The length of the fin is L, its constant cross-sectional area is A, and its circumferential perimeter is P.

The characteristic dimension of the fin in the transverse direction (normal to the x-axis) is taken to be A/P. Thus for a circular cylindrical fin, $A/P = \pi(\text{radius})^2/(2\pi \text{ radius}) = (\text{radius}/2)$. We define a Biot number for conduction in

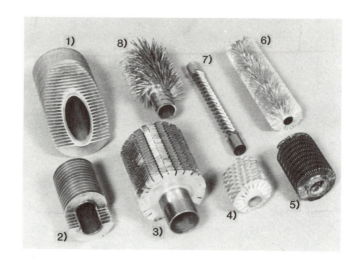

a) Eight examples of externally finned tubing.
   1) and 2) Typical commercial circular fins of constant thickness;
   3) and 4) Serrated circular fins and dimpled spirally-wound circular fins, both intended to improve convection.
   5) Spirally-wound copper coils outside and inside.
   6) and 8) Bristle fins, spirally wound and machined from base metal.
   7) A spirally indented tube to improve convection as well as to increase surface area.

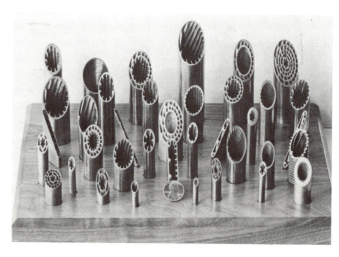

b) An array of commercial internally finned tubing (photo courtesy of Noranda Metal Industries, Inc.)

FIGURE 4.6   Some of the many varieties of finned tubes.

124

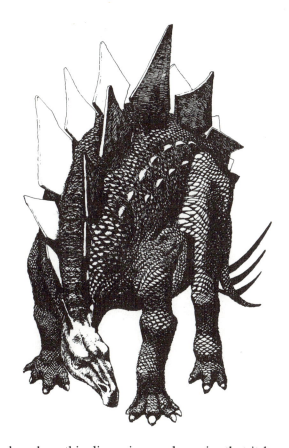

Figure 4.7 The *stegosaurus* with what might have been cooling fins. (Etching by Daniel Rosner.)

the transverse direction, based on this dimension, and require that it be small:

$$\mathrm{Bi}_{\mathrm{fin}} = \frac{\bar{h}(A/P)}{k} \ll 1 \qquad (4.27)$$

This condition means that the transverse variation of T at any axial position, x, is much less than $(T_{\mathrm{surface}} - T_\infty)$. Thus $T \simeq T(x$ only) and the heat flow can be treated as one-dimensional.

An energy balance on the thin slice of the fin shown in Fig. 4.8 gives

$$-kA\left.\frac{dT}{dx}\right|_{x+\delta x} + kA\left.\frac{dT}{dx}\right|_{x} + \bar{h}(P\delta x)(T - T_\infty)_x = 0 \qquad (4.28)$$

but

$$\frac{dT/dx|_{x+\delta x} - dT/dx|_x}{\delta x} \rightarrow \frac{d^2 T}{dx^2} = \frac{d^2(T - T_\infty)}{dx^2} \qquad (4.29)$$

so

$$\frac{d^2(T - T_\infty)}{dx^2} = \frac{\bar{h}P}{kA}(T - T_\infty) \qquad (4.30)$$

125

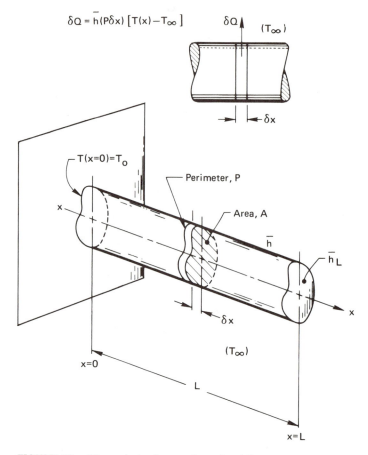

$$\delta Q = \bar{h}(P\delta x)\left[T(x) - T_\infty\right]$$

FIGURE 4.8   The analysis of a one-dimensional fin.

The b.c.'s for this equation are

$$\left.\begin{array}{l}(T - T_\infty)_{x=0} = T_o - T_\infty \\[2mm] -kA\dfrac{d(T - T_\infty)}{dx}\bigg|_{x=L} = \bar{h}_L A(T - T_\infty)_{x=L}\end{array}\right\} \qquad (4.31\text{a})$$

Alternatively, if the tip is insulated, or if we can guess that $\bar{h}_L$ is small enough to be unimportant, the b.c.'s are

$$(T - T_\infty)_{x=0} = T_o - T_\infty \qquad \text{and} \qquad \frac{d(T - T_\infty)}{dx}\bigg|_{x=L} = 0 \qquad (4.31\text{b})$$

Before we solve this problem it will pay to do a dimensional analysis of it. The dimensional functional equation is

$$T - T_\infty = \text{fn}\left[(T_o - T_\infty),\, x,\, L,\, kA,\, \bar{h}P,\, \bar{h}_L A\right] \qquad (4.32)$$

Notice that we have written kA, $\bar{h}P$, and $\bar{h}_L A$ as single variables. The reason for doing this is subtle, but important. Setting $\bar{h}(A/P)/k \ll 1$, erases any geometrical detail of the cross section from the problem. The *only* place where P and A enter the problem is as products of k, $\bar{h}$, or $\bar{h}_L$. If they showed up elsewhere, they would have to do so in a physically incorrect way. Thus, we have just seven variables in W, °C, and m. This gives four pi-groups if the tip is uninsulated:

$$\frac{T-T_\infty}{T_o-T_\infty} = \text{fn}\left[\frac{x}{L}, \sqrt{\frac{\bar{h}P}{kA}L^2}, \underbrace{\frac{\bar{h}_L AL}{kA}}_{=\frac{\bar{h}_L L}{k}} \right]$$

or if we rename the groups,

$$\Theta = \text{fn}(\xi, mL, \text{Bi}_{\text{axial}}) \tag{4.33a}$$

where $\sqrt{\bar{h}PL^2/kA} \equiv mL$ because that terminology is common in the literature on fins.

If the tip of the fin is insulated, $\bar{h}_L$ will not appear in eqn. (4.32). There is one less variable but the same number of dimensions; hence there will be only three pi-groups. The one that is removed is Bi_{axial}, which involves $\bar{h}_L$. Thus, for the insulated fin,

$$\Theta = \text{fn}(\xi, mL) \tag{4.33b}$$

We put eqn. (4.30) in these terms by multiplying it by $L^2/(T_o - T_\infty)$. The result is

$$\boxed{\frac{d^2\Theta}{d\xi^2} = (mL)^2\Theta} \tag{4.34}$$

This equation is satisfied by $\Theta = Ce^{\pm(mL)\xi}$. The sum of two such particular solutions forms the general solution of eqn. (4.34):

$$\boxed{\Theta = C_1 e^{mL\xi} + C_2 e^{-mL\xi}} \tag{4.35}$$

Temperature distribution in a one-dimensional fin with the tip insulated. The b.c.'s (4.31b) can be written as

$$\Theta_{\xi=0} = 1 \quad \text{and} \quad \left.\frac{d\Theta}{d\xi}\right|_{\xi=1} = 0 \tag{4.36}$$

Substituting eqn. (4.35) into both of eqns. (4.36), we get

$$C_1 + C_2 = 1 \quad \text{and} \quad C_1 e^{mL} - C_2 e^{-mL} = 0 \tag{4.37}$$

Mathematical Digression 4.1

To put the solution of eqns. (4.37) for C_1 and C_2 in the simplest form, we need to recall a few properties of hyperbolic functions. The four basic functions that we need are defined as

$$
\left.
\begin{aligned}
\sinh x &\equiv \frac{e^x - \bar{e}^x}{2} \\[4pt]
\cosh x &\equiv \frac{e^x + e^{-x}}{2} \\[4pt]
\tanh x &\equiv \frac{\sinh x}{\cosh x} = \frac{e^x - e^{-x}}{e^x + e^{-x}} \\[4pt]
\coth x &\equiv \frac{e^x + e^{-x}}{e^x - e^{-x}}
\end{aligned}
\right\}
\tag{4.38}
$$

where x is the argument of the function. Additional functions are defined by analogy to the trigonometric counterparts. The differential relations can be written out formally, and they also resemble their trigonometric counterparts.

$$
\left.
\begin{aligned}
\frac{d}{dx} \sinh x &= \tfrac{1}{2}[e^x - (-e^{-x})] = \cosh x \\[4pt]
\frac{d}{dx} \cosh x &= \tfrac{1}{2}[e^x + (-e^{-x})] = \sinh x
\end{aligned}
\right\}
\tag{4.39}
$$

Notice that the latter result is positive, while $d\cos x / dx = -\sin x$.

The solution of eqns. (4.37) is then

$$
C_1 = \frac{e^{-mL}}{2\cosh mL} \quad \text{and} \quad C_2 = 1 - \frac{e^{-mL}}{2\cosh mL}
\tag{4.40}
$$

Therefore, eqn. (4.35) becomes

$$
\Theta = \frac{e^{-mL(1-\xi)} + (2\cosh mL)e^{-mL\xi} - e^{-mL(1+\xi)}}{2\cosh mL}
$$

which simplifies to

$$
\boxed{\Theta = \frac{\cosh mL(1-\xi)}{\cosh mL}}
\tag{4.41}
$$

for the one-dimensional fin with the tip insulated.

The most important design variable for a fin is the amount of heat that it removes from (or delivers to) the wall. To calculate this, we write Fourier's law for the heat flow into the base of the fin:[6]

$$
Q = -kA \left. \frac{d(T - T_\infty)}{dx} \right|_{x=0}
\tag{4.42}
$$

---

[6]We could also integrate $\bar{h}(T - T_\infty)$ over the outside area of the fin to get Q. The answer would be the same, but the calculation would be a little more complicated.

We multiply eqn. (4.42) by $L/kA(T_o - T_\infty)$ and obtain, after substituting eqn. (4.41) on the right-hand side,

$$\frac{QL}{kA(T_o - T_\infty)} = mL\frac{\sinh mL}{\cosh mL} = mL\tanh mL \tag{4.43}$$

which can be rewritten

$$\frac{Q}{\sqrt{(kA)(\bar{h}P)}\,(T_o - T_\infty)} = \tanh mL \tag{4.44}$$

Figure 4.9 includes two graphs showing the behavior of a one-dimensional fin with an insulated tip. The top graph shows how the heat removal increases

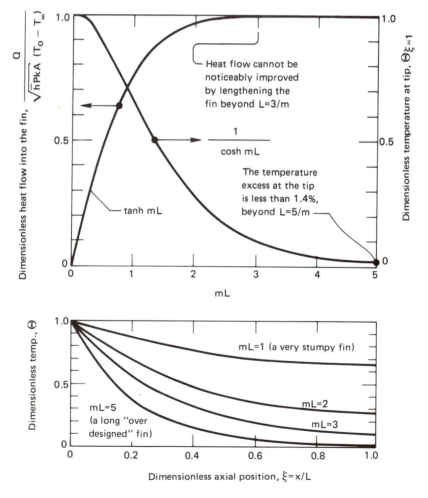

FIGURE 4.9 The temperature distribution, tip temperature, and heat flux in a straight one-dimensional fin with the tip insulated.

with mL to a virtual maximum at $mL \simeq 3$. This means that no such fin should have a length in excess of 2 or $3/m$ if it is being used to cool (or heat) a wall. Additional length would simply increase the cost without doing any good.

Also shown in the top graph is the temperature of the tip of such a fin. Setting $\xi = 1$ in eqn. (4.41), we discover that

$$\Theta_{\text{tip}} = \frac{1}{\cosh mL} \qquad (4.45)$$

This dimensionless temperature drops to about 0.014 at the tip when mL reaches 5. This means that the end is $0.014(T_o - T_\infty)°\text{C}$ above T_∞ at the end. Thus if the "fin" is actually functioning as a holder for a thermometer or a thermocouple which is intended to read T_∞, the reading will be in error if mL is not significantly greater than five.

The lower graph in Fig. 4.9 shows how the temperature is distributed in insulated-tip fins of increasing dimensionless length mL.

Experiment 4.1

Clamp a 20-cm-or-so length of copper rod by one end in a horizontal position. Put a candle flame very near the other end and let the arrangement come to steady state. Run your finger along the rod. How does what you feel correspond to Fig. 4.9? (The diameter for the rod should not exceed about 3 mm. A larger rod of metal with a lower conductivity will also work.)

Exact temperature distribution in a fin with an uninsulated tip. The b.c.'s given in eqn. (4.31a) take the following dimensionless form:

$$\Theta_{\xi=0} = 1 \qquad \text{and} \qquad -\frac{d\Theta}{d\xi}\bigg|_{\xi=1} = \text{Bi}_{\text{ax}}\Theta_{\xi=1} \qquad (4.46)$$

Substitution of the general solution, eqn. (4.35), in these b.c.'s yields

$$C_1 + C_2 = 1$$

and

$$-mL\left(C_1 e^{mL} + C_2 e^{-mL}\right) = \text{Bi}_{\text{ax}}\left(C_1 e^{mL} + C_2 e^{-mL}\right)$$

$$\left.\right\} \qquad (4.47)$$

It requires a fair amount of manipulation to solve eqns. (4.47) for C_1 and C_2 and to substitute the results in eqn. (4.35). We leave this exercise for the student to complete (Problem 4.11). The result is

$$\Theta = \frac{\cosh mL(1-\xi) + (\text{Bi}_{\text{ax}}/mL)\sinh mL(1-\xi)}{\cosh mL + (\text{Bi}_{\text{ax}}/mL)\sinh mL} \qquad (4.48)$$

which is in the form of eqn. (4.33a), as we anticipated. The corresponding heat flux equation is

$$\frac{Q}{\sqrt{(kA(\bar{h}P))}\,(T_o - T_\infty)} = \frac{(\text{Bi}_{\text{ax}}/mL) + \tanh mL}{1 + (\text{Bi}_{\text{ax}}/mL)\tanh mL} \qquad (4.49)$$

We have seen that mL is not too much greater than unity in a well-designed fin with an insulated tip. Furthermore, when $\bar{h}_L$ is small (as it might be in natural convection) Bi_{ax} is normally much less than unity. Therefore, in such cases, we expect to be justified in neglecting terms multiplied by Bi_{ax}. Then eqn. (4.48) reduces to

$$\Theta = \frac{\cosh mL(1-\xi)}{\cosh mL} \tag{4.41}$$

which we obtained by analyzing an insulated fin.

It is worth pointing out that we are in serious difficulty if $\bar{h}_L$ is very large and we cannot assume the tip to be insulated. The reason is that $\bar{h}_L$ is next to impossible to predict in most practical cases.

Example 4.8

A 2-cm-diameter aluminum rod with $k=205$ W/m-°C, 8 cm in length, protrudes from a 150°C wall. Air at 26°C flows by it, and $\bar{h}=120$ W/m²-°C. Determine whether or not tip conduction is important in this problem. To do this make the very crude assumption that $\bar{h} \simeq \bar{h}_L$. Then compare the tip temperatures as calculated with and without considering heat transfer from the tip.

SOLUTION.

$$mL = \sqrt{\frac{\bar{h}PL^2}{kA}} = \sqrt{\frac{120(0.08)^2}{205(0.010/2)}} = 0.8656$$

$$\text{Bi}_{ax} = \frac{\bar{h}L}{k} = \frac{120(0.08)}{205} = 0.0468$$

Therefore, eqn. (4.48) becomes

$$\Theta(\xi=1) = \Theta_{tip} = \frac{\cosh 0 + (0.0468/0.8656)\sinh 0}{\cosh(0.8656) + (0.0468/0.8656)\sinh(0.8656)}$$

$$= \frac{1}{1.3986 + 0.0529} = 0.6886$$

so the exact tip temperature is

$$T_{tip} = T_\infty + 0.6886(T_o - T_\infty)$$

$$= 26 + 0.6886(150 - 26) = \underline{111.43°C}$$

Equation (4.41) or Fig. 4.9, on the other hand, gives

$$\Theta_{tip} = \frac{1}{1.3986} = 0.7150$$

so the approximate tip temperature is

$$T_{tip} = 26 + 0.715(150 - 26) = \underline{114.66°C}$$

Thus the insulated-tip approximation is entirely adequate for the computation in this case.

Very long fin. If a fin is so long that $mL \gg 1$, then eqn. (4.41) becomes

$$\lim_{mL \to \text{large}} \Theta = \lim_{mL \to \text{large}} \frac{e^{mL(1-\xi)} + e^{-mL(1-\xi)}}{e^{mL} + e^{-mL}} = \frac{e^{mL(1-\xi)}}{e^{mL}}$$

or

$$\lim_{mL \to \text{large}} \Theta = e^{-mL\xi} \tag{4.50}$$

Substituting this result in eqn. (4.42), we obtain [cf. eqn. (4.44)]

$$Q = \sqrt{(kA)(\bar{h}P)}\,(T_o - T_\infty) \tag{4.51}$$

These results are only good for terribly overdesigned fins—that is, fins with mL much larger than it ought to be for heating or cooling fins.

Physical significance of mL. The group mL has thus far proved to be extremely useful in the analysis of fins. We should therefore say a brief word about its physical significance. Notice that

$$(mL)^2 = \frac{L/kA}{1/\bar{h}(PL)} = \frac{\text{internal resistance in } x\text{-direction}}{\text{gross external resistance}}$$

Thus $(mL)^2$ is a hybrid Biot number. When it is big, $\Theta_{\xi=1} \to 0$ and we can neglect tip convection. When it is small, the temperature drop along the axis of the fin becomes small (see the lower graph in Fig. 4.9).

The group $(mL)^2$ also has a peculiar similarity to the NTU (Chapter 3) and the dimensionless time, t/T, that appeared in the lumped-capacitance solution (Chapter 1). Thus

$$\frac{\bar{h}(PL)}{kA/L} \quad \text{is like} \quad \frac{UA}{C_{\min}} \quad \text{is like} \quad \frac{\bar{h}A}{\rho c V/t}$$

In each case a convective heat rate is compared with a heat rate that characterizes the capacity of a system; and in each case the system temperature asymptotically approaches its limit as the numerator becomes large. This was true in eqns. (1.21), (3.21) and (3.22), and (4.50).

The problem of specifying the root temperature

The root temperature of a fin has been taken as known information in all the problems that we have discussed. There really are many circumstances in which it might be known; however, if a fin protrudes from a wall of the same material, as sketched in Fig. 4.10a, it is clear that for heat to flow there must be a temperature gradient in the neighborhood of the root.

Consider the situation in which the surface of a wall is kept at a temperature T_s. Then a fin is placed on the wall as shown in the figure. If $T_\infty < T_s$, the wall temperature will be depressed in the neighborhood of the root as heat flows into the fin. The fin performance should then be predicted using the lowered root temperature, T_{root}.

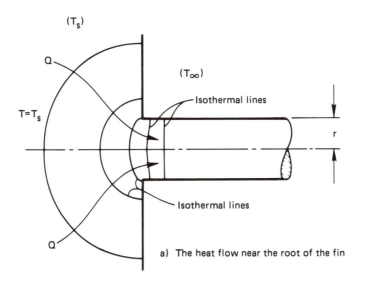

a) The heat flow near the root of the fin

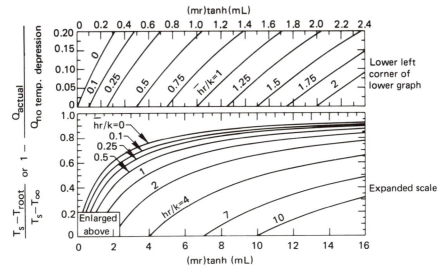

b) Predicted deviations of root temperature and heat flux resulting
from local temperature distortion near the root

FIGURE 4.10 The influence of heat flow into the root of circular cylindrical fins
[4.6].

This heat conduction problem has been analyzed for several fin arrangements by Sparrow and co-workers. Figure 4.10b is the result of Sparrow and Hennecke's [4.6] analysis for a single circular cylinder. They give

$$1 - \frac{Q_{\text{actual}}}{Q_{\text{no temp. depression}}} = \frac{T_s - T_{\text{root}}}{T_s - T_\infty} = \text{fn}\left[\frac{\bar{h}r}{k}, (mr)\tanh(mL) \right] \quad (4.52)$$

where r is the radius of the fin. From the figure we see that the actual heat flow into the fin, Q_{actual}, and the actual root temperature are both reduced when the Biot number, $\bar{h}r/k$, is large and the fin constant, m, is small.

Example 4.9

Neglect the tip convection from the fin in Example 4.8 and suppose that it is embedded in a wall of the same material. Calculate the error in Q and the actual temperature of the root if the wall is kept at 150°C.

SOLUTION. From Example 4.8 we have $mL = 0.8656$ and $\bar{h}r/k = 120(0.010)/205 = 0.00586$. Then $(mr)\tanh(mL) = 0.8656(0.010/0.080)\tanh(0.8656) = 0.0756$. The lower portion of Fig. 4.10b then gives

$$1 - \frac{Q_{\text{actual}}}{Q_{\text{no temp. depression}}} = \frac{T_s - T_{\text{root}}}{T_s - T_\infty} = 0.05$$

so the heat flow is reduced by 5% and the actual root temperature is

$$T_{\text{root}} = 150 - (150 - 26)0.05 = 143.8 \, °C$$

The correction is fairly small in this case.

Fin design

There are two basic measures of fin performance that are particularly useful in fin design. The first is called the "efficiency," η_f.

$$\text{fin efficiency, } \eta_f \equiv \frac{\substack{\text{actual heat} \\ \text{transferred from a wall}}}{\substack{\text{heat that would be} \\ \text{transferred if the en-} \\ \text{tire fin were at } T = T_o}} \quad (4.53)$$

To see how this works, we evaluate it for a one-dimensional fin with an insulated tip:

$$\eta_f = \frac{\sqrt{(\bar{h}P)(kA)} \, (T_o - T_\infty)\tanh mL}{\bar{h}(PL)(T_o - T_\infty)} = \frac{\tanh mL}{mL} \quad (4.54)$$

This says that, under the definition of efficiency, a very long fin will give $(\tanh mL)/mL \to 1/\text{large number}$, so the fin will be inefficient. On the other hand, the efficiency goes to 100% as the length is reduced to zero, because $\tanh(mL)_{\text{small}} \to mL$.

It is therefore clear that, while η_f provides some useful information as to how well a fin is contrived, it is not possible to design toward any particular value of η_f. A second measure of fin performance is called the "effectiveness," ε:

$$\text{effectiveness, } \varepsilon \equiv \frac{\begin{array}{c}\text{heat flux from}\\ \text{the wall with the fin}\end{array}}{\begin{array}{c}\text{heat flux from the}\\ \text{wall without the fin}\end{array}} \tag{4.55}$$

This can easily be computed from the efficiency:

$$\varepsilon = \eta_f \frac{\text{surface area of the fin}}{\text{cross-sectional area of the fin}} \tag{4.56}$$

Normally, we want the effectiveness to be as high as possible. But this can always be done by extending the length of the fin, and that—as we have seen—rapidly becomes a losing proposition.

The measures η_f and ε probably attract the interest of designers not because their absolute values guide the designs, but because they are useful in characterizing fins of more complex shape. In such cases the solutions are often so complex that η_f and ε plots serve as labor-saving graphical solutions. We deal with some of these curves in the following section.

The design of a fin thus becomes an open-ended matter of optimizing, subject to many factors. Some of the factors that have to be considered include:

- The weight of material added by the fin. This might be a cost factor or it might be an important consideration in its own right.
- The possible dependence of $\bar{h}$ on $(T - T_\infty)$, flow velocity past the fin, or other influences.
- The influence of the fin (or fins) on the heat transfer coefficient, $\bar{h}$, as the fluid moves around it (or them).
- The geometrical configuration of the channel that the fin lies in.
- The cost and complexity of manufacturing the fins.
- The pressure drop introduced by the fins.

Fins of variable cross section

Let us consider next what is involved in the design of a fin for which A and P are functions of x. Such a fin is shown in Fig. 4.11. We shall restrict such considerations to fins in which

$$\frac{\bar{h}(A/P)}{k} \ll 1 \qquad \text{and} \qquad \frac{d(A/P)}{d(x/L)} \ll 1$$

so the heat flow will be approximately locally one-dimensional.

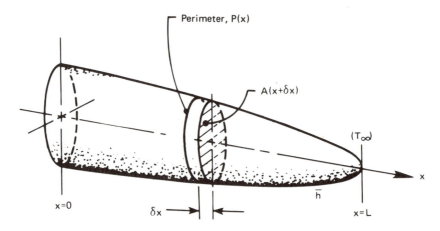

FIGURE 4.11   A general fin of variable cross section.

We begin the analysis, as always, with the First Law statement:

$$Q_{net} = Q_{cond} - Q_{conv} = \frac{dU}{dt}$$

or

$$\underbrace{\left[kA(x+\delta x)\frac{dT}{dx}\Big|_{x+\delta x} - kA(x)\frac{dT}{dx}\Big|_{x} \right]}_{\frac{d}{dx}kA(x)\frac{dT}{dx}\delta x} - \bar{h}P\delta x(T-T_{\infty}) = \underbrace{\rho c A(x)\delta x \frac{dU}{dt}}_{=0,\ \text{since steady}}$$

Therefore,

$$\boxed{\frac{d}{dx}\left[A(x)\frac{d(T-T_{\infty})}{dx} \right] = \frac{\bar{h}P}{k}(T-T_{\infty})} \tag{4.57}$$

If $A(x) = $ constant, this reduces to $\Theta'' = (mL)^2\Theta$, which is the straight fin equation.

To see how eqn. (4.57) works, consider the triangular fin shown in Fig. 4.12. In this case eqn. (4.57) becomes

$$\frac{d}{dx}\left[2\delta\left(\frac{x}{L}\right)b\frac{d(T-T_{\infty})}{dx} \right] = \frac{2\bar{h}b}{k}(T-T_{\infty})$$

or

$$\xi\frac{d^2\Theta}{d\xi^2} + \frac{d\Theta}{d\xi} = \underbrace{\frac{\bar{h}L^2}{k\delta}}_{\substack{\text{a kind} \\ \text{of } (mL)^2}}\Theta \tag{4.58}$$

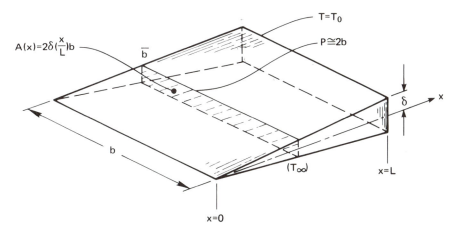

FIGURE 4.12 A two-dimensional wedge-shaped fin.

This second-order ordinary linear differential equation is difficult to solve since it has a variable coefficient. The solution of eqn. (4.58) is expressible in Bessel functions:

$$\Theta = \frac{J_o\left(2\sqrt{-\bar{h}Lx/k\delta}\right)}{J_o\left(2\sqrt{-\bar{h}L^2/k\delta}\right)} \qquad (4.59)$$

where the Bessel function, J_o, with imaginary arguments, can be looked up in appropriate tables.

We shall not deal with the mathematical exercises of solving eqn. (4.57). Instead, we simply present the solution for several geometries in terms of the fin efficiency, η_f, in Fig. 4.13. These curves were given by Schneider [1.15]. Kern and Kraus [4.7] provide the most complete discussion of fins and include a great many additional efficiency curves.

Example 4.10

A thin brass pipe 3 cm in outside diameter carries hot water at 85°C. It is proposed to place 0.8-mm-thick straight circular fins on the pipe to cool it. The fins are 8 cm in diameter and are spaced 2 cm apart. It is determined that $\bar{h}$ will equal 20 W/m²-°C on the pipe and 15 W/m²-°C on the fins, when they have been added. If $T_\infty = 22$°C, compute the heat loss per meter of pipe before and after the fins are added.

SOLUTION. Before the fins are added,

$$Q = \pi(0.03 \text{ m})(20 \text{ W/m}^2\text{-}°\text{C})(85-22)°\text{C} = \underline{119 \text{ W/m}}$$

where we set $T_{wall} = T_{water}$ since the pipe is thin. Notice that, since the wall is constantly heated by the water, we should not have a root-temperature depression problem after the

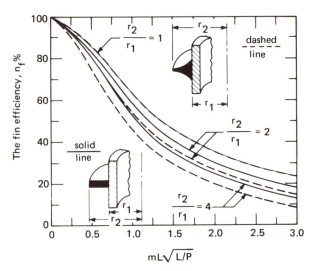

Comparison of a constant thickness circular fin, and a hyperbolic fin with thickness inversely proportional to radius. ($L \equiv r_2 - r_1$ and m is based on the area, A, shown in black.)

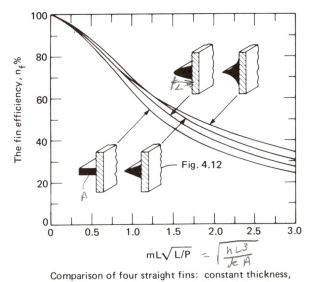

Comparison of four straight fins: constant thickness, triangular, parabolic, and hyperbolic. (m is based on A shown in black.)

FIGURE 4.13   The efficiency of several fins with variable cross section.

$$mL\sqrt{4/P} = \sqrt{\frac{hL^3}{k A}}$$

138

fins are added. Then we can enter Fig. 4.13a with

$$\frac{r_2}{r_1} = 2.67 \quad \text{and} \quad mL\sqrt{\frac{L}{P}} = \sqrt{\frac{\bar{h}L^3}{kA}} = \sqrt{\frac{15(0.04-0.015)^3}{125(0.025)(0.0008)}} = 0.306$$

and we obtain $\eta_f = 89\%$. Thus the actual heat transfer is given by

$$\underbrace{Q_{\text{without fin}}}_{119} \underbrace{\left(\frac{0.02-0.0008}{0.02}\right)}_{\substack{\text{fraction of}\\\text{unfined area}}} + 0.89 \underbrace{[\pi(0.04^2-0.015^2)2]}_{\substack{\text{area per fin}\\\text{(both sides), m}^2}} 50 \,\frac{\text{fins}}{\text{m}}\, 15 \,\frac{\text{W}}{\text{m}^2\text{-}{}^\circ\text{C}}\, (85-22)\,{}^\circ\text{C}$$

so

$$Q_{\text{net}} = 478 \text{ W/m} = 4.02\, Q_{\text{without fins}}$$

PROBLEMS

4.1. Make a table listing the general solutions of all steady, unidimensional, constant-properties heat conduction problems in Cartesian, cylindrical, and spherical coordinates, with and without uniform heat generation. This table should prove to be a very useful tool in future problem solving. It should include a total of 18 solutions. State any restrictions on your solutions. Do not include calculations.

4.2. The left side of a slab of thickness L is kept at 0°C. The right side is cooled by air at T_∞ °C blowing on it. $\bar{h}_{\text{RHS}}$ is known. An exothermic reaction takes place in the slab such that heat is generated at $A(T-T_\infty)$ W/m^3, where A is a constant. Find a fully dimensionless expression for the temperature distribution in the wall.

4.3. A long wide plate of known size, material, and thickness L is connected across the terminals of a power supply and serves as a resistance heater. The voltage, current, and T_∞ are known. The plate is insulated on the bottom and transfers heat out the top by convection. The temperature, T_{tc}, of the bottom is measured with a thermocouple. Obtain expressions for (a) temperature distribution in the plate; (b) $\bar{h}$ at the top; (c) temperature at the top. (Note that your answers must depend on *known information* only.)

4.4. The heat transfer coefficient, $\bar{h}$, resulting from a forced flow over a flat plate, depends on the fluid velocity, viscosity, density, specific heat, and thermal conductivity, as well as on the length of the plate. Develop the dimensionless functional equation for the heat transfer coefficient (cf. Section 7.5).

4.5. Water vapor condenses on a cold pipe and drips off the bottom in regularly spaced nodes as sketched in Fig. 3.9. The wavelength of these nodes, λ, depends on the liquid–vapor density difference, $\rho_f - \rho_g$, the surface tension, σ, and the gravity, g. Find how λ varies with its dependent variables.

4.6. A thin film flows down a vertical wall. The local film velocity at any distance from the wall depends on that distance, gravity, the liquid kinematic viscosity, and the film thickness. Obtain the dimensionless functional equation for the local velocity (cf. Section 9.5).

4.7. A steam preheater consists of a thick electrically conducting cylindrical shell insulated on the outside, with wet steam flowing down the middle. The inside heat transfer coefficient is highly variable, depending on the velocity, quality, and so on, but the flow temperature is constant. Heat is released at $\dot{q}$ J/m^3-s within the cylinder wall. Evaluate the temperature within the cylinder as a function of position. Plot Θ against ρ, where Θ is an appropriate dimensionless temperature and $\rho = r/r_o$. Use $\rho_i = \frac{2}{3}$ and note that Bi will be the parameter of a family of solutions. *On the basis of this plot*, recommend criteria (in terms of Bi) for: (a) replacing the convective boundary condition on the inside with a constant temperature condition; (b) neglecting temperature variations within the cylinder.

4.8. Steam condenses on the inside of a small pipe, keeping it at a specified temperature, T_i. The pipe is heated by electrical resistance at a rate $\dot{q}$ W/m^3. The outside temperature is T_∞ and there is a natural convection heat transfer coefficient, $\bar{h}$, around the outside. (a) Derive an expression for the dimensionless temperature distribution, $\Theta = (T - T_\infty)/(T_i - T_\infty)$, as a function of the radius ratios, $\rho = r/r_o$ and $\rho_i = r_i/r_o$; a heat generation number, $\Gamma = \dot{q}r_o^2/k(T_i - T_\infty)$; and the Biot number. (b) Plot this result for the case $\rho_i = \frac{2}{3}$, Bi = 1, and for several values of Γ. (c) Discuss any interesting aspects of your result.

4.9. Solve Problem 2.5 if you have not already done so, putting it in dimensionless form before you begin. Then let the Biot numbers approach infinity in the solution. You should get the same solution we got in Example 2.4, using b.c.'s of the first kind. Do you?

4.10. Complete the algebra that is missing between eqns. (4.30) and (4.31b), and eqn. (4.41).

4.11. Complete the algebra that is missing between eqns. (4.30) and (4.31a), and eqn. (4.48).

4.12. Obtain eqn. (4.50) from the general solution for a fin [eqn. (4.35)] using the b.c.'s $T(x = 0) = T_o$ and $T(x = L) = T_\infty$. Comment on the significance of the computation.

4.13. What is the minimum length, l, of a thermometer well necessary to ensure an error less than 0.5% of the difference between the pipe wall and the fluid temperature? Assume that the fluid flowing is steam at a temperature of 260°C and that the coefficient between the steam and the tube wall is 300 W/m^2-°C. The well consists of a tube with the end plugged. It has a 2-cm O.D. and a 1.88-cm I.D. The material is type 304 stainless steel. [3.44 cm.]

4.14. Thin fins with a 0.002-m by 0.02-m rectangular cross section and a thermal conductivity of 50 W/m-°C protrude from a wall. $\bar{h} \simeq 600$ W/m^2-°C and $T_o = 170$°C. What is the maximum possible heat flow rate into each fin? $T_\infty = 20$°C.

4.15. A thin rod is anchored at a wall at $T = T_o$ on one end and is insulated at the other end. Plot the dimensionless temperature distribution in the rod as a function of dimensionless length: (a) if the rod is exposed to an environment at T_∞ through a heat transfer coefficient; (b) if the rod is insulated but heat is consumed in it at the uniform rate $-\dot{q} = \bar{h}P(T_o - T_\infty)/A$. Comment on the implications of the comparison.

4.16. A tube of outside diameter d_o and inside diameter d_i carries fluid at $T = T_1$ from a wall at temperature T_1 to another, a distance L away, at T_r. $\bar{h}_o$ is negligible outside the tube and $\bar{h}_i$ is substantial inside the tube. Treat the tube as a fin and plot the dimensionless temperature distribution in it as a function of dimensionless length.

4.17. (If you have had some applied mathematics beyond the usual two years of calculus, this problem will not be difficult.) The shape of the fin in Fig. 4.12 is changed so that $A(x) = 2\delta(x/L)^2 b$ instead of $2\delta(x/L)b$. Calculate the temperature distribution and the heat flux at the base. Plot the temperature distribution and fin thickness against x/L. Derive an expression for η_f.

4.18. Work Problem 2.21, if you have not already done so, nondimensionalizing the problem before you attempt to solve it. It should now be much simpler.

4.19. One end of a copper rod 30 cm long is maintained at 200°C and the other is maintained at 93°C. The heat transfer coefficient in between is 17 W/m²-°C. If $T_\infty = 38$°C and the diameter of the rod is 1.25 cm, what is the net heat removed by the air around the rod? [19.13 W.]

4.20. How much error will the insulated-tip assumption give rise to in the calculation of the heat flow into the fin in Example 4.8?

4.21. A straight cylindrical fin 0.6 cm in diameter and 6 cm long protrudes from a magnesium block at 200°C. Air at 25°C is forced past the fin so that $\bar{h}$ is 130 W/m²-°C. Calculate the heat removed by the fin, considering the temperature depression of the root.

4.22. Work Problem 4.19 considering temperature depression at both roots. To do this, find mL for the two fins with insulated tips, which would give the same temperature gradient at each wall. Base the correction on these values of mL.

4.23. A fin of triangular axial section (cf. Fig. 4.12) 0.1 m in length and 0.02 m wide at its base is used to extend the surface area of a mild steel wall. If the wall is at 40°C and heated gas flows past at 200°C ($\bar{h} = 230$ W/m²-°C), compute the heat removed by the fin, per meter of breadth, b, of the fin. Neglect temperature distortion at the root.

4.24. Consider the concrete slab in Example 2.2. Suppose that the heat generation were to cease abruptly at $t = 0$ and the slab were to start cooling back toward T_w. Predict $T = T_w$ as a function of time, noting that the initial parabolic temperature profile can be nicely approximated as a sine function.

4.25. Steam condenses in a 2-cm-I.D. thin-walled tube of 99% aluminum at 10 atm pressure. There are circular fins of constant thickness, 3.5 cm in diameter every 0.5 cm. The fins are 0.8 mm thick and the heat transfer coefficient, $\bar{h} = 6$ W/m²-°C on the outside. What is the mass rate of condensation if the pipe is 1.5 m in length, the ambient temperature is 18°C, and $\bar{h}$ for condensation is very large? [$\dot{m}_{cond} = 0.802$ kg/hr.]

4.26. How long must a copper fin, 0.4 cm in diameter, be if the temperature of its insulated tip exceeds the surrounding air temperature by 20% of $(T_o - T_\infty)$? $T_{air} = 20$°C, and $\bar{h} = 28$ W/m²-°C.

4.27. A 2 cm ice cube sits on a shelf of aluminum rods, 3 mm in diam., in a refrigerator at 10°C. How rapidly, in mm/min, does the ice cube melt through the wires if $\bar{h}$ between the wires and the air is 20 W/m²-°C. (Be careful that you understand the physical mechanism before you make the calculation.) Check your result experimentally. $h_{fs} = 333,300$ W/kg.

REFERENCES

[4.1] V. L. STREETER, *Fluid Mechanics*, 5th ed., McGraw-Hill Book Company, New York, 1971, Chapter 4.

[4.2] E. BUCKINGHAM, *Phys. Rev.*, vol. 4, 1914, p. 345.

[4.3] E. BUCKINGHAM, "Model Experiments and the Forms of Empirical Equations," *Trans. ASME*, vol. 37, 1915, p. 263–296.

[4.4] LORD RAYLEIGH (JOHN WM. STRUTT), "The Principle of Similitude," *Nature*, vol. 95, 1915, pp. 66–68.

[4.5] J. O. FARLOW, C. V. THOMPSON, and D. E. ROSNER, "Plates of the Dinosaur Stegosaurus: Forced Convection Heat Loss Fins?" *Science*, vol. 192, no. 4244, 1976, pp. 1123–1125 and cover.

[4.6] D. K. HENNECKE and E. M. SPARROW, "Local Heat Sink on a Convectively Cooled Surface—Application to Temperature Measurement Error," *Int. J. Heat Mass Transfer*, vol. 13, 1970, pp. 87–304.

[4.7] D. Q. KERN and A. D. KRAUS, *Extended Surface Heat Transfer*, McGraw Hill Book Company, New York, 1972.

Chapter 5

Transient and multidimensional heat conduction

5.1 Introduction

James Watt, of course, did not invent the steam engine. What he did do was to eliminate a destructive transient heating and cooling process which wasted great amounts of energy. By 1763, the great puffing engines of Savery and Newcomen had been used for over half a century to pump the water out of Cornish mines and to do other tasks. In that year the young instrument maker, Watt, was called upon to renovate the Newcomen engine model at the University of Glasglow. The Glasglow engine was then being used as a demonstration in the course on natural philosophy. Watt did much more than just renovate the machine—he first recognized, and eventually eliminated, its major shortcoming.

The cylinder of Newcomen's engine was cold when steam entered it and nudged the piston outward. A great deal of the steam was wastefully condensed on the cylinder walls until they were warm enough to accommodate it. When the cylinder was filled, the steam valve was closed and jets of water were activated inside the cylinder to cool it again and condense the steam. This created a powerful vacuum which sucked the piston back in on its working stroke. First Watt tried to eliminate the wasteful initial condensation of steam by insulating the cylinder. When that simply reduced the vacuum and cut the power of the working stroke, he thought of leading the steam outside to a *separate condenser*. The cylinder could then stay hot while the vacuum was created.

The separate condenser was the main issue in Watt's first patent (1769) and it immediately doubled the thermal efficiency of steam engines from a maximum of 1.1% to 2.2%. By the time Watt died in 1819, his invention had led to efficiencies of 5.7% and the engine had altered the face of the world by powering the Industrial Revolution. And from 1769 until today the steam power cycles that engineers study in their thermodynamics courses are accurately represented as steady flow processes.

The repeated transient heating and cooling that occurred in Newcomen's engine was the kind of process that today's design engineer might still carelessly ignore, but the lesson that we learn from history is that transient heat transfer can be of overwhelming importance. Today, for example, designers of food storage enclosures know that such systems need relatively little energy to keep food cold at steady conditions. The real cost of operating such systems results from the consumption of energy needed to bring the food down to a low temperature, and the losses resulting from people entering and leaving the system with food. The *transient* heat transfer processes are a dominant concern in the design of food storage units.

We therefore turn our attention first to the analysis of unsteady heat transfer, beginning with a more detailed consideration of the lumped-capacity system that we looked at in Section 1.3.

5.2 Lumped-capacity solutions

We begin by looking briefly at the dimensional analysis of transient conduction in general, and of lumped-capacity systems in particular.

Dimensional analysis of transient heat conduction

We first consider a fairly general problem of one-dimensional transient heat conduction:

$$\frac{\partial^2 T}{\partial x^2} = \frac{1}{\alpha} \frac{\partial T}{\partial t} \qquad \text{i.c.: } T(t=0) = T_i$$

$$\text{b.c.'s: } T(t>0, x=0) = T_1$$

$$-k \frac{\partial T}{\partial x}\bigg|_{x=L} = \bar{h}(T - T_1)_{x=L}$$

where "i.c." and "b.c." denote the initial and boundary conditions. Here we have a situation leading to a dimensional functional equation of the form

$$T - T_1 = \text{fn}\left[(T_i - T_1), x, L, t, \alpha, \bar{h}, k \right]$$

There are eight variables in four dimensions (°C, s, m, W), so we look for $8-4$, or 4 pi-groups. We anticipate, from Section 4.3, that they will include $\Theta \equiv (T - T_1)/(T_i - T_1)$, $\xi \equiv x/L$, and $\text{Bi} \equiv \bar{h}L/k$, and we write

$$\Theta = \text{fn}(\xi, \text{Bi}, \Pi_4) \tag{5.1}$$

One possible candidate for Π_4, which is independent of the other three, is

$$\Pi_4 \equiv \text{Fo, Fourier number, } \alpha t/L^2 \tag{5.2}$$

Other possibilities exist for Π_4. One that we use later is

$$\Pi_4 \equiv \zeta = \frac{x}{\sqrt{\alpha t}} \quad \left(\text{this is exactly } \frac{\xi}{\sqrt{\text{Fo}}} \right) \tag{5.3}$$

If the problem involved only b.c.'s of the first kind, the heat transfer coefficient, $\bar{h}$—and hence the Biot number—would go out of the problem. Then the dimensionless function equation (5.1) is

$$\Theta = \text{fn}(\xi, \text{Fo}) \tag{5.4}$$

By the same token, if the b.c.'s had introduced different values of $\bar{h}$ at $x=0$ and $x=L$, *two* Biot numbers would appear in the solution.

The lumped-capacity problem is particularly interesting from the standpoint of dimensional analysis. In this case neither k nor x enters the problem because we do not retain any features of the internal conduction problem. Therefore, we keep only the denominator of α, namely ρc. Furthermore, we do not have to separate ρ and c because they only appear as a product. Finally, we use the volume-to-external area ratio, V/A, as a characteristic length. Thus for the transient lumped-capacity problem, the dimensional functional equation is

$$T - T_\infty = \text{fn}\left[(T_i - T_\infty), \rho c, V/A, \bar{h}, t \right] \tag{5.5}$$

With six variables in the dimensions J, °C, m, and s, we know that there will only be two pi-groups in the dimensionless function equation.

$$\Theta = \text{fn}\left(\frac{\bar{h}At}{\rho c V} \right) = \text{fn}\left(\frac{t}{\mathcal{T}} \right) \tag{5.6}$$

This is exactly the form of the simple lumped-capacity solution, eqn. (1.21). Notice, too, that the group $t/\mathcal{T}$ becomes

$$\frac{t}{\mathcal{T}} = \frac{hk(V/A)t}{\rho c(V/A)^2 k} = \frac{\bar{h}(V/A)}{k} \cdot \frac{\alpha t}{(V/A)^2} = \text{BiFo} \tag{5.7}$$

The term *capacitance* is adapted from electrical circuit theory to the heat transfer problem. Therefore, we sketch a simple resistance–capacitance circuit in Fig. 5.1. The capacitor is initially charged to voltage, E_0. When the switch is suddenly opened, the capacitor discharges through the resistor and the voltage drops according to the relation

$$\frac{dE}{dt} + \frac{E}{RC} = 0 \qquad (5.8)$$

The solution of eqn. (5.8) with the i.c. $E(t=0) = E_0$ is

$$E = E_0 e^{-t/RC} \qquad (5.9)$$

and the current can be computed from Ohm's law, once $E(t)$ is known.

$$I = \frac{E}{R} \qquad (5.10)$$

Normally, in a heat conduction problem the *thermal* capacitance, $\rho c V$, is *distributed* in space. But when the Biot number is small, $T(t)$ is uniform in the body and we can *lump* the capacitance into a single circuit element. The thermal

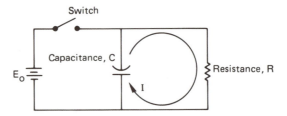

FIGURE 5.1   A simple resistance-capacitance circuit.

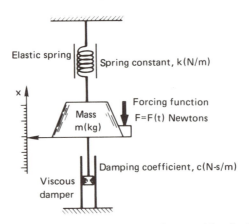

FIGURE 5.2   A spring–mass–damper with a forcing function.

resistance is $1/\bar{h}A$, and the temperature difference $(T-T_\infty)$ is analogous to $E(t)$. Thus the thermal response, analogous to eqn. (5.9), is

$$T - T_\infty = (T_i - T_\infty)\exp\left(-\frac{\bar{h}At}{\rho cV}\right) \qquad (1.21)$$

Notice that the electrical time constant, analogous to $\rho cV/\bar{h}A$, is RC.

Now consider a slightly more complex system. Figure 5.2 shows a spring–mass–damper system. The well-known response equation (actually, a force balance) for this system is

$$m\frac{d^2x}{dt^2} + c\frac{dx}{dt} + kx = F(t) \qquad (5.11)$$

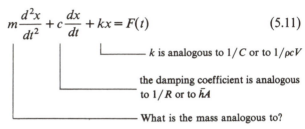

k is analogous to $1/C$ or to $1/\rho cV$

the damping coefficient is analogous to $1/R$ or to $\bar{h}A$

What is the mass analogous to?

A term analogous to mass would arise from electrical inductance, but we did not include it in the electrical circuit. Mass has the effect of carrying the system beyond its final equilibrium point. Thus, in an underdamped mechanical system we might obtain the sort of response shown in Fig. 5.3 if we gave a specified velocity at $x=0$ and provided no forcing function. Electrical inductance provides a similar effect. But the Second Law of Thermodynamics does not permit temperatures to overshoot their equilibrium values spontaneously. *There are no physical elements analogous to mass or inductance in thermal systems.*

Next, consider another mechanical element that does have a thermal analogy, namely the forcing function, F. We consider a (massless) spring–damper system with a forcing function F which probably is time-dependent; and we ask: "What might a thermal forcing function look like?"

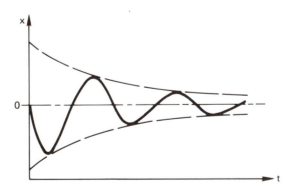

FIGURE 5.3 Response of an unforced spring-mass-damper system with an initial velocity.

To answer the preceding question, let us suddenly immerse an object at a temperature $T = T_i$, with $Bi \ll 1$, into a cool bath whose temperature is rising as $T_\infty(t) = T_i + bt$, where T_i and b are constants. Then eqn. (1.19) becomes

$$\frac{d(T - T_i)}{dt} = -\frac{T - T_\infty}{T} = -\frac{T - T_i - bt}{T}$$

where we have arbitrarily subtracted T_i under the differential. Then

$$\frac{d(T - T_i)}{dt} + \frac{T - T_i}{T} = \frac{bt}{T} \tag{5.12}$$

To solve eqn. (5.12) we must first recall that the general solution to a linear ordinary differential equation with constant coefficients is equal to the sum of any particular integral of the complete equation and the general solution of the homogeneous equation. We know the latter; it is $T - T_i = (\text{constant}) \exp(-t/T)$. A particular integral of the complete equation can often be formed by guessing solutions and trying them in the complete solution. Here we discover that

$$T - T_i = bt - bT$$

satisfies eqn. (5.12). The general solution of the complete eqn. (5.12) is thus

$$T - T_i = C_1 e^{-t/T} + b(t - T) \tag{5.13}$$

Example 5.1

The flow rates of hot and cold water are regulated into a mixing chamber. We measure the temperature of the water as it leaves, using a thermometer with a time constant, T. On a particular day, the system is started with cold water at $T = T_i$ in the mixing chamber. Then hot water is added in such a way that the outflow temperature rises linearly, as shown in Fig. 5.4, with $T_{\text{exit flow}} = T_i + bt$. How will the thermometer report the temperature variation?

SOLUTION. The initial condition in eqn. (5.13), which describes this process, is $T - T_i = 0$ at $t = 0$. Substituting eqn. (5.13) in the i.c., we get

$$0 = C_1 - bT \qquad \text{so} \qquad C_1 = bT$$

and the response equation is

$$T - (T_i + bt) = bT(e^{-t/T} - 1) \tag{5.14}$$

This result is shown graphically in Fig. 5.4. Notice that the thermometer reading reflects a transient portion, $bTe^{-t/T}$, which decays for a few time constants and then can be neglected; and a steady portion, $T_i + b(t - T)$, which persists thereafter. When the steady response is established, the thermometer follows the bath with a temperature lag of bT. This constant error is reduced when either T or the rate of temperature increase, b, is reduced.

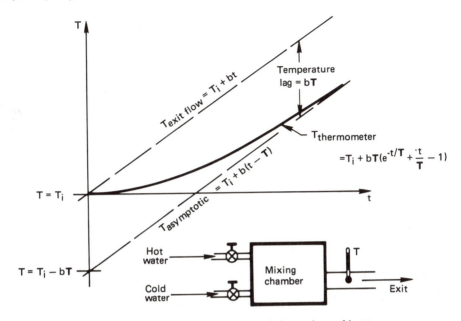

FIGURE 5.4   Response of a thermometer to a linearly increasing ambient temperature.

Second-order lumped-capacity systems

Now we look at situations in which two lumped-thermal capacity systems are connected in series. Such an arrangement is shown in Fig. 5.5. Heat is transferred through two slabs with an interfacial resistance, h_c^{-1} between them. We shall require that $h_c L_1 / k_1$, $h_c L_2 / k_2$, and $\bar{h} L_2 / k_2$ are all much less than unity so

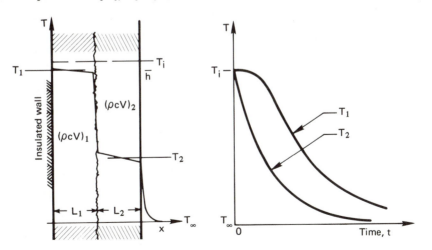

FIGURE 5.5   Two slabs conducting in series through an interfacial resistance.

that it will be legitimate to lump the thermal capacitance of each slab. The differential equations dictating the temperature response of each slab are then

$$\text{slab 1:} \quad h_c A(T_1 - T_2) = -(\rho c V)_1 \frac{dT_1}{dt} \tag{5.15}$$

$$\text{slab 2:} \quad \bar{h} A(T_2 - T_\infty) - h_c A(T_1 - T_2) = -(\rho c V)_2 \frac{dT_2}{dt} \tag{5.16}$$

and the initial conditions on the temperatures T_1 and T_2 are

$$T_1(t=0) = T_2(t=0) = T_i \tag{5.17}$$

We next identify two time constants for this problem:[1]

$$\mathcal{T}_1 \equiv (\rho c V)_1 / h_c A \qquad \text{and} \qquad \mathcal{T}_2 \equiv (\rho c V)_2 / \bar{h} A$$

Then eqn. (5.15) becomes

$$T_2 = \mathcal{T}_1 \frac{dT_1}{dt} + T_1 \tag{5.18}$$

which we substitute in eqn. (5.16) to get

$$\left(\mathcal{T}_1 \frac{dT_1}{dt} + T_1 - T_\infty \right) + \frac{h_c}{\bar{h}} \, \mathcal{T}_1 \frac{dT_1}{dt} = -\mathcal{T}_1 \mathcal{T}_2 \frac{d^2 T_1}{dt^2} - \mathcal{T}_2 \frac{dT_1}{dt}$$

or

$$\frac{d^2 T_1}{dt^2} + \underbrace{\left[\frac{1}{\mathcal{T}_1} + \frac{1}{\mathcal{T}_2} + \frac{h_c}{\bar{h} \mathcal{T}_2} \right]}_{\equiv \, b} \frac{dT_1}{dt} + \underbrace{\frac{T_1 - T_\infty}{\mathcal{T}_1 \mathcal{T}_2}}_{c(T_1 - T_\infty)} = 0 \tag{5.19}$$

if we call $T_1 - T_\infty \equiv \theta$, then eqn. (5.19) can be written as

$$\frac{d^2\theta}{dt^2} + b \frac{d\theta}{dt} + c\theta = 0 \tag{5.19a}$$

Thus we have reduced the pair of first-order equations, (5.15) and (5.16), to a single second-order equation, (5.19a).

The general solution of eqn. (5.19a) is obtained by guessing a solution of the form $\theta = C_1 e^{Dt}$. Substitution of this guess into eqn. (5.19a) gives

$$D^2 + bD + c = 0 \tag{5.20}$$

from which we find that $D = -(b/2) \pm \sqrt{(b/2)^2 - c}$. This gives us two values of D, from which we can get two exponential solutions. By adding them together

---

[1] Notice that we could also have used $(\rho c V)_2 / h_c A$ for $\mathcal{T}_2$ since both h_c and $\bar{h}$ act on slab 2. The choice is arbitrary.

we form a general solution:

$$\theta = C_1 \exp\left[-\frac{b}{2} + \sqrt{\left(\frac{b}{2}\right)^2 - c}\,\right]t + C_2 \exp\left[-\frac{b}{2} - \sqrt{\left(\frac{b}{2}\right)^2 - c}\,\right]t \quad (5.21)$$

To solve for the two constants we first substitute eqn. (5.21) in the first of i.c.'s (5.17) and get

$$T_i - T_\infty = \theta_i = C_1 + C_2 \quad (5.22)$$

The second i.c. can be put into terms of T_1 with the help of eqn. (5.15):

$$-\frac{dT_1}{dt}\bigg|_{t=0} = \frac{h_c A}{(\rho c V)_1}(T_1 - T_2)_{t=0} = 0$$

We substitute eqn. (5.21) in this and obtain

$$0 = \left[-\frac{b}{2} + \sqrt{\left(\frac{b}{2}\right)^2 - c}\,\right]C_1 + \left[-\frac{b}{2} - \sqrt{\left(\frac{b}{2}\right)^2 - c}\,\right]\underbrace{C_2}_{=\theta_i - C_1}$$

so

$$C_1 = -\theta_i\left[\frac{-b/2 - \sqrt{(b/2)^2 - c}}{2\sqrt{(b/2)^2 - c}}\right]$$

and

$$C_2 = \theta_i\left[\frac{-b/2 + \sqrt{(b/2)^2 - c}}{2\sqrt{(b/2)^2 - c}}\right]$$

So we obtain at last:

$$\frac{T_1 - T_\infty}{T_i - T_\infty} \equiv \frac{\theta}{\theta_i} = \frac{b/2 + \sqrt{(b/2)^2 - c}}{2\sqrt{(b/2)^2 - c}}\exp\left[-\frac{b}{2} + \sqrt{\left(\frac{b}{2}\right)^2 - c}\,\right]t$$

$$+ \frac{-b/2 + \sqrt{(b/2)^2 - c}}{2\sqrt{(b/2)^2 - c}}\exp\left[-\frac{b}{2} - \sqrt{\left(\frac{b}{2}\right)^2 - c}\,\right]t \quad (5.23)$$

This is a pretty complicated result—all the more complicated when we remember that b involves three algebraic terms [recall eqn. (5.19)]. Yet there is nothing very sophisticated about it; it is easy to understand. A system involving three capacitances in series would similarly yield a third-order equation of correspondingly higher complexity, and so forth.

5.3 Transient conduction in a one-dimensional slab

Now let us relax the requirement that all thermal resistance is lodged in a convective region outside the body of interest. When the temperature within a body—say, a one-dimensional one—varies with position as well as time, we must solve the heat diffusion equation for $T(x,t)$. We shall do this somewhat complicated task for the simplest case and then look at the results of such calculations in other situations.

A simple slab, shown in Fig. 5.6, is initially at a temperature T_i. The temperature of the surface of the slab is suddenly changed to T_1, and we wish to calculate the interior temperature profile as a function of time. The differential equation is

$$\frac{\partial^2 T}{\partial x^2} = \frac{1}{\alpha} \frac{\partial T}{\partial t} \tag{5.24}$$

with the following b.c.'s and i.c.:

$$T(-L, t>0) = T(L, t>0) = T_1 \qquad \text{and} \qquad T(\text{all } x, 0) = T_i \tag{5.25}$$

In fully dimensionless form eqns. (5.24) and (5.25) are

$$\frac{\partial^2 \Theta}{\partial \xi^2} = \frac{\partial \Theta}{\partial \text{Fo}} \tag{5.26}$$

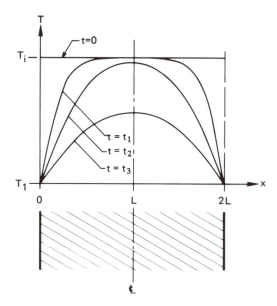

FIGURE 5.6   The transient cooling of a slab. $\xi = (X/L) + 1$.

and

$$\Theta(0, \text{Fo}) = \Theta(2, \text{Fo}) = 0 \quad \text{and} \quad \Theta(\xi, 0) = 1 \qquad (5.27)$$

where we have nondimensionalized the problem in accordance with eqn. (5.4) using $\Theta \equiv (T - T_1)/(T_i - T_1)$ and $\text{Fo} \equiv \alpha t/L^2$; but for convenience in solving the equation we have set ξ equal to $(x/L) + 1$ instead of x/L.

The general solution of eqn. (5.26) is given by eqn. (4.11), which we recast in dimensionless form as

$$\Theta = e^{-(\lambda L)^2 \text{Fo}} \left[G \sin(\lambda L)\xi + E \cos(\lambda L)\xi \right] \qquad (5.28)$$

This solution appears to have introduced a fourth dimensionless group, λL. This needs explanation. The number, λ, which was introduced in the separation-of-variables process, is called an *eigenvalue*.[2] It has the dimensions of $(\text{length})^{-1}$. In the present problem, λL will turn out to be a number—or rather a sequence of numbers—which is independent of system parameters.

Substituting the general solution (5.28) in the first b.c. gives

$$0 = e^{-(\lambda L)^2 \text{Fo}}(0 + E) \quad \text{so} \quad E = 0$$

and substituting it in the second yields

$$0 = e^{-(\lambda L)^2 \text{Fo}} \left[G \sin 2(\lambda L) \right] \quad \text{so} \quad G = 0$$

$$\text{or} \quad 2\lambda L = n\pi, \, n = 0, 1, 2, \ldots$$

In the second case we are presented with two choices. The first, $G = 0$, would give $\Theta \equiv 0$ in all situations, so that the initial condition could never be accommodated. (This is what mathematicians call a "trivial" solution.) The second choice, $\lambda L = n\pi/2$, actually yields a string of solutions, each of the form

$$\Theta = G_n e^{-n^2\pi^2\text{Fo}/4} \sin \frac{n\pi}{2}\xi \qquad (5.29)$$

where G_n is the constant appropriate to the nth one of these solutions.

We still face the problem that no one of eqns. (5.29) will fit the initial condition, $\Theta(\xi, 0) = 1$. To get around this, we remember that the sum of any number of solutions of a linear differential equation is also a solution. Then we write

$$\Theta = \sum_{n=1}^{\infty} G_n e^{-n^2\pi^2\text{Fo}/4} \sin n\frac{\pi}{2}\xi \qquad (5.30)$$

where we drop $n = 0$ since it gives zero contribution to the series. And we arrive, at last, at the problem of choosing the G_n's so that eqn. (5.30) will fit the initial condition:

$$\Theta(\xi, 0) = \sum_{n=1}^{\infty} G_n \sin n\frac{\pi}{2}\xi = 1 \qquad (5.31)$$

---

[2] The word "eigenvalue" is a curious hybrid of the German term "eigenwerte" and its English translation, "characteristic value."

The problem of picking the values of G_n that will make this equation true is called "making a Fourier series expansion" of the function $f(\xi) = 1$. We shall not pursue strategies for making Fourier series expansions in any general way. Instead, we shall merely show how to accomplish the task for the particular problem at hand. We begin with a mathematical trick. We multiply eqn. (5.31) by $\sin(m\pi\xi/2)$, where m may or may not equal n, and we integrate the result between $\xi = 0$ and 2.

$$\int_0^2 \sin \frac{m\pi}{2} \xi \, d\xi = \sum_{n=1}^{\infty} G_n \int_0^2 \sin \frac{m\pi}{2} \xi \left(\sin \frac{n\pi}{2} \xi \right) d\xi \tag{5.32}$$

(The interchange of summation and integration turns out to be legitimate, although we have not proved, here, that it is.) With the help of a table of integrals we find that

$$\int_0^2 \sin \frac{m\pi}{2} \xi \left(\sin \frac{n\pi}{2} \xi \right) d\xi = \begin{cases} 0 & \text{for } n \neq m \\ 1 & \text{for } n = m \end{cases}$$

Thus when we complete the integrations in eqn. (5.32), we get

$$-\frac{2}{m\pi} \cos \frac{m\pi}{2} \xi \Big]_0^2 = \sum_{n=1}^{\infty} G_n \begin{cases} 0 & \text{for } n \neq m \\ 1 & \text{for } n = m \end{cases}$$

This reduces to

$$-\frac{2}{n\pi} \left[(-1)^n - 1 \right] = G_n$$

so

$$G_n = \frac{4}{n\pi} \qquad \text{where } n \text{ is an odd number}$$

Substituting this result into eqn. (5.30), we finally obtain the solution to the problem:

$$\Theta(\xi, \text{Fo}) = \frac{4}{\pi} \sum_{n,\text{odd}}^{\infty} \frac{1}{n} e^{-(n\pi/2)^2 \text{Fo}} \sin \frac{n\pi}{2} \xi \tag{5.33}$$

Such solutions are easy enough to *follow* and to use, but you clearly need more training in mathematics than we have assumed here if you are to be very good at *inventing* them.

Equation (5.33) admits a very nice simplification for larger time (or at larger Fo). Suppose that we wish to evaluate Θ at the center of the slab—at $x = 0$ or $\xi = 1$. Then

$$\Theta = \frac{4}{\pi} \left\{ \underbrace{\exp\left[-\left(\frac{\pi}{2}\right)^2 \text{Fo} \right]}_{\substack{=0.085 \text{ at Fo}=1 \\ =0.781 \text{ at Fo}=0.1 \\ =.0976 \text{ at Fo}=0.01}} - \frac{1}{3} \underbrace{\exp\left[-\left(\frac{3\pi}{2}\right)^2 \text{Fo} \right]}_{\substack{\approx 10^{-10} \text{ at Fo}=1 \\ =0.036 \text{ at Fo}=0.1 \\ =0.267 \text{ at Fo}=0.01}} + \frac{1}{5} \underbrace{\exp\left[-\left(\frac{5\pi}{2}\right)^2 \text{Fo} \right]}_{\substack{\approx 10^{-27} \text{ at Fo}=1 \\ =0.0004 \text{ at Fo}=0.1 \\ =0.108 \text{ at Fo}=0.01}} + \ldots \right\}$$

Thus for values of Fo much greater than 0.1, only the first term in the series need be used in the solution (except at points very close to the boundaries). This makes possible a graphical presentation of such results which we look at in the following section. Before we move to this matter, let us see what happens to the preceding problem if the slab is subjected to b.c.'s of the third kind.

Suppose that the walls of the slab had been cooled by symmetrical convection such that the b.c's were

$$\bar{h}(T_\infty - T)_{x=-L} = -k\frac{\partial T}{\partial X}\bigg|_{x=-L} \qquad \text{and} \qquad \bar{h}(T-T_\infty)_{x=L} = -k\frac{\partial T}{\partial X}\bigg|_{x=L}$$

or in dimensionless form, using $\Theta \equiv (T-T_\infty)/T_i - T_\infty)$ and $\xi = (x/L)+1$,

$$-\Theta_{\xi=0} = -\frac{1}{\text{Bi}}\frac{\partial \Theta}{\partial \xi}\bigg|_{\xi=0} \qquad \text{and} \qquad \frac{\partial \Theta}{\partial \xi}\bigg|_{\xi=1} = 0$$

The solution is somewhat harder to find, but the result is[3]

$$\Theta = \sum_{n=1}^{\infty} \exp\left[-(\lambda L)^2 \text{Fo}\right]\frac{2\sin(\lambda L)\cos[\lambda L(\xi-1)]}{(\lambda L)+\sin(\lambda L)\cos(\lambda L)} \qquad (5.34)$$

where the values of (λL) are given as a function of n and Bi by the transcendental equation

$$\cot(\lambda L) = \frac{\lambda L}{\text{Bi}} \qquad (5.35)$$

The successive positive roots of this equation—$\lambda L = (\lambda L)_1$, $\lambda L = (\lambda L)_2$, etc.—correspond with $n=1$, $n=2$, etc., and they depend upon Bi. Thus $\Theta = \text{fn}(\xi, \text{Fo}, \text{Bi})$, as we would expect. This result, although more complicated than the result for b.c.'s of the first kind, still reduces to a single term for Fo $\gtrsim 0.3$.

5.4 Temperature-response charts

Figure 5.7 is a graphical presentation of eqn. (5.34) for $0 \leqslant \text{Fo} \leqslant 1.5$, and for six x-planes in the slab. (Remember that the x-coordinate goes from zero in the center to L on the boundary, while ξ goes from 0 up to 2 in the preceding solution.)

Notice that, with the exception of points for which $1/\text{Bi} < 0.25$ on the outside boundary, the curves are all straight lines for Fo > 0.3. Since the coordinates are semilogarithmic, this corresponds to the lead term—the only term that retains any importance—in eqn. (5.34). When we take the logarithm of the one-term version of eqn. (5.34), the result is

$$\ell n\,\Theta \cong \ell n\underbrace{\left[\frac{2\sin(\lambda L)\cos(\lambda L)\xi}{(\lambda L)+\sin(\lambda L)\cos(\lambda L)}\right]}_{\substack{\Theta\text{-intercept at Fo}=0\text{ of} \\ \text{the straight portion of} \\ \text{the curve}}} - \underbrace{(\lambda L)^2\text{Fo}}_{\substack{\text{slope of the} \\ \text{straight portion} \\ \text{of the curve}}}$$

[3]See, for example, [1.4, Sec. 4.4] for details of this calculation.

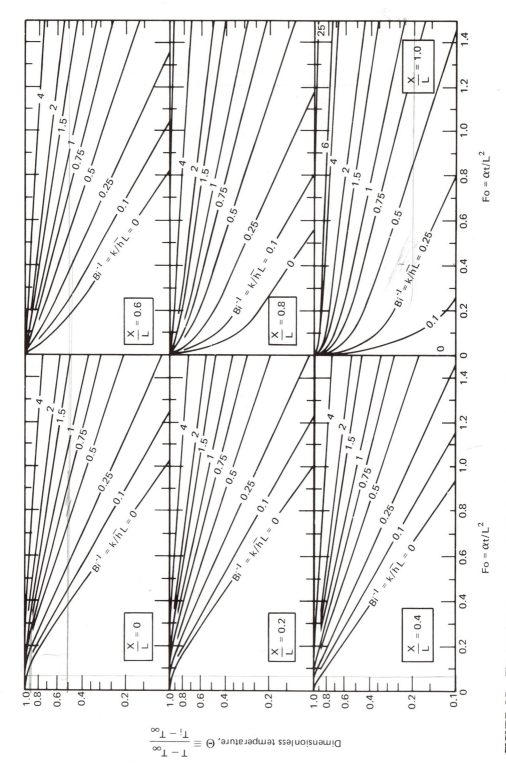

FIGURE 5.7 The transient temperature distribution in a *slab* at six positions. $x/L = 0$ is the center. $x/L = 1$ is one outside boundary.

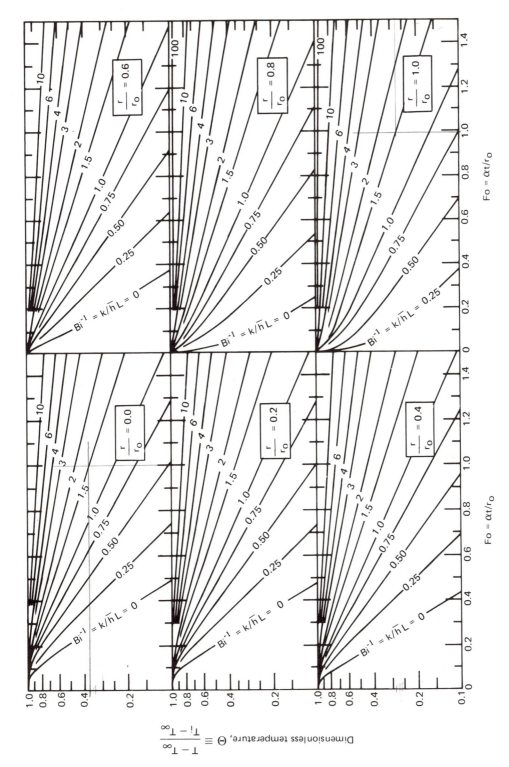

FIGURE 5.8 The transient temperature distribution in a *long cylinder* of radius, r_o, at six positions. $r/r_o = 0$ is the centerline and $r/r_o = 1$ is the outside boundary.

Dimensionless temperature, $\Theta \equiv \dfrac{T - T_\infty}{T_i - T_\infty}$

$Fo = \alpha t/r_o$

$Bi^{-1} = k/\bar{h}L$

157

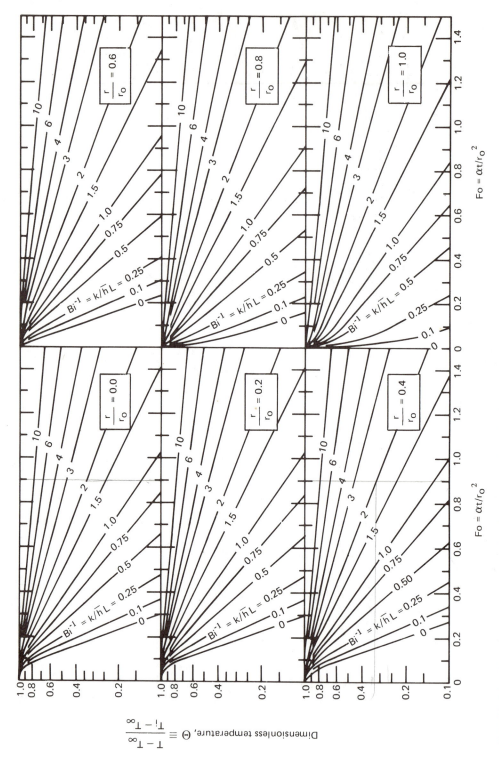

FIGURE 5.9   The transient temperature distribution in a *sphere* of radius, r_o, at six positions. $r/r_o = 0$ is the center, and $r/r_o = 1$ is the outside boundary.

If Fo is greater than 1.5, the following options are then available to us for solving the problem:

- Extrapolate the given curves using a straightedge.
- Evaluate Θ using the first term of eqn. (5.34).
- If Bi is small, use a lumped capacitance result.

Figures 5.8 and 5.9 are similar graphs for cylinders and spheres. Everything that we have said in general about Fig. 5.7 is also true for these graphs. They were simply calculated from different solutions, and the numerical values on them are somewhat different. These charts are from [1.2, Chap. 5], although they are often called Heisler charts, after Heisler's subsequent reorganization of the information contained in them [5.1].

Another useful kind of chart derivable from eqn. (5.34) is one that gives heat removal from a body up to a time of interest:

$$\int_0^t Q\,dt = -\int_0^t kA\,\frac{\partial T}{\partial x}\bigg|_{\text{surface}}\,dt$$

$$= -\int_0^{\text{Fo}} kA\,\frac{T_i - T_\infty}{L}\,\frac{\partial \Theta}{\partial \xi}\bigg|_{\text{surface}}\left(\frac{L^2}{\alpha}\right)d\text{Fo}$$

Dividing this by the total energy of the body above T_∞, we get a quantity, Φ, which approaches unity as $t \to \infty$ and the energy is all transferred to the surroundings:

$$\Phi \equiv \frac{\int_0^t Q\,dt}{\rho c V(T_i - T_\infty)} = -\int_0^{\text{Fo}} \frac{\partial \Theta}{\partial \xi}\bigg|_{\text{surface}}\,d\text{Fo} \tag{5.36}$$

where the volume, $V = AL$. Substituting the appropriate temperature distribution [e.g., eqn. (5.34) for a slab] in eqn. (5.36), we obtain $\Phi(\text{Fo}, \text{Bi})$, which can be plotted once and for all. Such curves are given for the slab, cylinder, and sphere in Fig. 5.10.

Example 5.2

A dozen approximately spherical apples 10 cm in diameter are taken from a 30°C environment and laid out on a rack in a refrigerator at 5°C. Assume that they have approximately the same physical properties as water and that $\bar{h}$ is approximately 6 W/m²-°C as the result of natural convection. What will be the temperature of the centers of the apples after 1 hr? How long will it take to bring the centers to 10°C? How much heat will the refrigerator have to carry away to get the centers to 10°C?

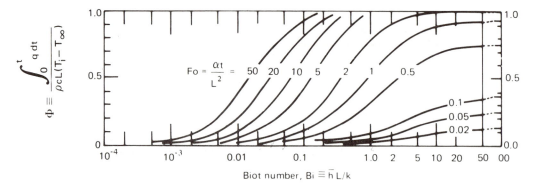

a.) *Slab* of thickness, L, insulated on one side

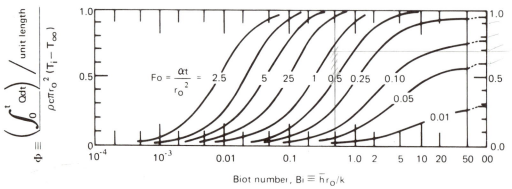

b.) *Cylinder*, of radius, r_O

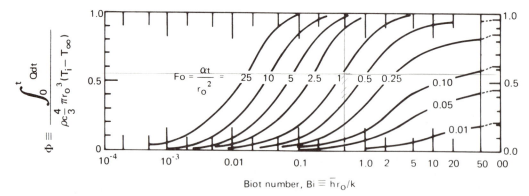

c.) *Sphere*, of radius, r_O

FIGURE 5.10 The heat removal from suddenly cooled bodies as a function of $\bar{h}$ and time.

SOLUTION.   After 1 hr, or 3600 s:

$$\text{Fo} = \frac{\alpha t}{r_0^2} = \left(\frac{k}{\rho c}\right)_{20°C} \frac{3600 \text{ s}}{(0.05 \text{ m})^2}$$

$$= \frac{(0.603 \text{ J/m-s-°C})(3600 \text{ s})}{(997.6 \text{ kg/m}^3)(4180 \text{ J/kg-°C})(0.0025 \text{ m}^2)} = 0.208$$

Furthermore, $\text{Bi}^{-1} = (\bar{h}r_0/k)^{-1} = [6(0.05)/0.603]^{-1} = 2.01$. Therefore, we read from Fig. 5.9 in the upper left-hand corner:

$$\Theta = 0.85$$

Therefore, after 1 hr:

$$T_{center} = 0.85(30 - 5)°C + 5°C = \underline{26.3°C}$$

To find the time required to bring the centers to 10°C, we first calculate

$$\Theta = \frac{10 - 5}{30 - 5} = 0.2$$

and Bi^{-1} is still 2.01. Then from Fig. 5.9 we read

$$\text{Fo} = 1.29 = \frac{\alpha t}{r_0^2}$$

so

$$\text{time} = \frac{1.29(997.6)(4180)(0.0025)}{0.603} = 22,300 \text{ s} = \underline{6 \text{ hr } 12 \text{ min}}$$

Finally, we look up Φ at $\text{Bi} = 1/2.01$ and $\text{Fo} = 1.29$, in Fig. 5.10, for spheres:

$$\Phi = 0.68 = \frac{\int_0^t Q \, dt}{\rho c \frac{4}{3}\pi r_0^3 (T_i - T_\infty)}$$

so

$$\int_0^t Q \, dt = 997.6(4180)\tfrac{4}{3}\pi(0.05)^3(25)(0.68) = 37,121 \text{ J/apple}$$

Therefore, for the 12 apples,

$$\text{Total energy removal} = 12(37.12) = \underline{445 \text{ kJ}}$$

The temperature-response charts in Figs. 5.7 through 5.10 are without doubt among the most useful available since they can be adapted to a host of physical situations. Nevertheless, hundreds of such charts have been formed for other situations. The reader who is faced with a complex temperature-response problem is well advised to scan the literature before he or she attempts to solve it. The two best places to begin are Carslaw and Jaeger's fine treatise on heat conduction [1.14] and a collection of temperature-response charts assembled by Schneider [5.2]. Other charts are scattered about in the technical literature.

Example 5.3

A 1-mm-diameter Nichrome (20% Ni, 80% Cr) wire is simultaneously being used as an electric resistance heater and as a resistance thermometer in a liquid flow. The laboratory

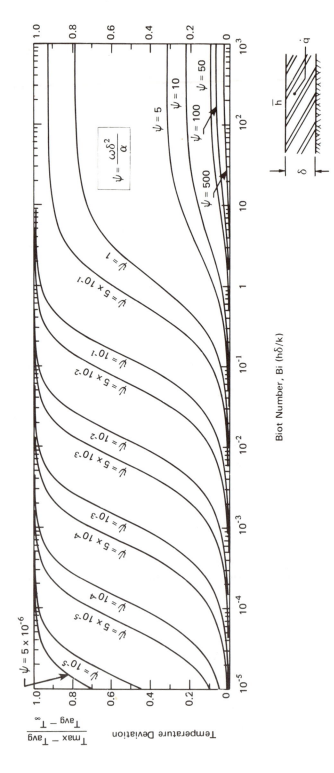

FIGURE 5.11 Temperature deviation at the surface of a *flat plate* heated with alternating current.

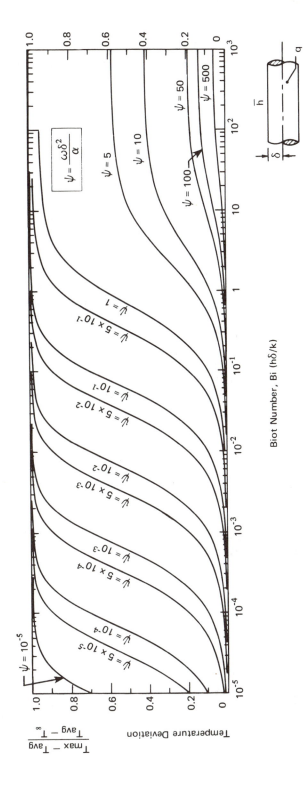

FIGURE 5.12 Temperature deviation at the surface of a *cylinder* heated with alternating current.

workers who operate it are attempting to measure the boiling heat transfer coefficient, $\bar{h}$, by supplying an alternating current and measuring the difference between the average temperature of the heater, T_{av}, and the liquid temperature, T_∞. They get $\bar{h} = 30,000$ W/m^2-°C at a wire temperature of 100°C and are delighted with such a high value. Then a colleague suggests that $\bar{h}$ is so high because the surface temperature is rapidly oscillating as a result of the alternating current. Is this hypothesis correct?

SOLUTION.   Heat is being generated in proportion to EI, or as $\sin^2 \omega t$, where ω is the frequency of the current in rad/s. If the boiling action removes heat rapidly enough in comparison with the heat capacity of the wire, the surface temperature may well vary significantly. The transient conduction problem underlying this situation was solved by Jeglić in 1962 [5.3] and independently in a different form by Switzer and Lienhard [5.4] two years later. The response curves from the latter were of the form

$$\frac{T_{max} - T_{av}}{T_{av} - T_\infty} = fn(Bi, \psi) \tag{5.37}$$

where the left-hand side is the dimensionless range of the temperature oscillation, and $\psi \equiv \omega \delta^2 / \alpha$, where δ is a characteristic length. Because this problem is common and the solution is not widely available, we include the curves for flat plates and cylinders in Figs. 5.11 and 5.12, respectively.

In the present case:

$$Bi = \frac{\bar{h} \text{ radius}}{k} = \frac{30,000(0.0005)}{13.8} = 1.09$$

$$\frac{\omega r^2}{\alpha} = \frac{[2\pi(60)](0.0005)^2}{0.00000343} = 27.5$$

and from the chart for cylinders, Fig. 5.12, we find that

$$\frac{T_{max} - T_{av}}{T_{av} - T_\infty} \cong 0.04$$

A temperature fluctuation of only 4% is probably not serious. It therefore appears that the experiment was valid.

5.5 Transient heat conduction to a semiinfinite region

Introduction

Bronowski's fine television series on the *Ascent of Man* [5.5] includes a brilliant reenactment of the ancient ceremonial procedure by which the Japanese forged Samurai swords (see Fig. 5.13). The metal is heated, folded, beaten, and formed, over and over, to create a blade of remarkable toughness and flexibility. When the blade is formed to its final configuration, a tapered sheath of clay is baked on the outside of it, so the cross section is as shown in Fig. 5.13. The red-hot blade with the clay sheath is then subjected to a rapid quenching which cools the uninsulated cutting edge quickly and the back part of the blade very slowly. The

Clay-coated blade before quench   Case-hardened blade

FIGURE 5.13   The ritualistic case-hardening of a Samurai sword.

result is a layer of case-hardening which is hardest at the edge and less hard at points farther from the edge.

The blade is then tough and ductile, so it will not break, but has a fine hard outer shell that can be honed to sharpness. We need only look a little way up the side of the clay sheath to find a cross section which was thick enough to prevent the blade from experiencing the sudden effects of the cooling quench. The success of the process actually relies on the *failure* of the cooling to penetrate the clay very deeply in a short time.

Now we wish to ask: "How can we say whether or not the influence of a heating or cooling process is restricted to the surface of a body?" Or if we turn the question around: "Under what conditions can we view the depth of a body as *infinite* with respect to the thickness of the region that has felt the heat transfer process?"

Consider next the cooling process within the blade in the absence of the clay retardant and when $\bar{h}$ is very large. Actually, our considerations will apply initially to any finite body whose boundary suddenly changes temperature. The temperature distribution, in this case, is sketched in Fig. 5.14 for four sequential times. Only the fourth curve—that for which $t = t_4$—is noticeably influenced by the opposite wall. Up to that time the wall might as well have infinite depth.

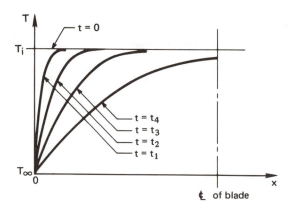

FIGURE 5.14　The initial cooling of a thin sword blade. (Prior to $t = t_4$ the blade might as well be infinitely thick insofar as cooling is concerned.)

Since any body subjected to a sudden change of temperature is infinitely large in comparison with the initial region of temperature change, we must learn how to treat heat transfer in this period.

Solution aided by dimensional analysis

The calculation of the temperature distribution in a semiinfinite region poses a difficulty in that we can impose a definite b.c. only at one position—the exposed boundary. We shall be able to get around that difficulty in a nice way with the help of dimensional analysis.

When a semiinfinite region, initially at $T = T_i$, is suddenly cooled (or heated) at its one boundary to a new temperature, T_∞, as in Fig. 5.14, the dimensional functional equation is

$$T - T_\infty = \text{fn}\left[\, t, x, \alpha, (T_i - T_\infty)\,\right]$$

where there *is no characteristic length*. Since there are five variables in °C, s, and m we should look for two dimensionless groups.

$$\underbrace{\frac{T - T_\infty}{T_i - T_\infty}}_{\Theta} = \text{fn}\left(\underbrace{\frac{x}{\sqrt{\alpha t}}}_{\zeta}\right) \qquad (5.38)$$

The very important thing that we learn from this exercise in dimensional analysis is that position and time collapse into one independent variable. This means that the heat conduction equation must transform from a partial differential equation into a simpler ordinary differential equation in the single variable, $\zeta = x/\sqrt{\alpha t}$. Thus we transform each side of

$$\frac{\partial^2 T}{\partial x^2} = \frac{1}{\alpha}\frac{\partial T}{\partial t}$$

as follows, where we call $T_i - T_\infty \equiv \Delta T$:

$$\frac{\partial T}{\partial t} = (T_i - T_\infty)\frac{\partial \Theta}{\partial t} = \Delta T \frac{\partial \Theta}{\partial \zeta}\frac{\partial \zeta}{\partial t} = \Delta T \frac{\partial \Theta}{\partial \zeta}\left(-\frac{x}{2t\sqrt{\alpha t}}\right)$$

$$\frac{\partial T}{\partial x} = \Delta T \frac{\partial \Theta}{\partial \zeta}\frac{\partial \zeta}{\partial x} = \frac{\Delta T}{\sqrt{\alpha t}}\frac{\partial \Theta}{\partial \zeta}$$

and

$$\frac{\partial^2 T}{\partial x^2} = \frac{\Delta T}{\sqrt{\alpha t}}\frac{\partial^2 \Theta}{\partial \zeta^2}\frac{\partial \zeta}{\partial x} = \frac{\Delta T}{\alpha t}\frac{\partial^2 \Theta}{\partial \zeta^2}$$

Substituting the first and last of these derivatives in the heat conduction equation, we get

$$\frac{d^2\Theta}{d\zeta^2} = -\frac{\zeta}{2}\frac{d\Theta}{d\zeta} \tag{5.39}$$

Notice that we changed from partial to total derivative notation since Θ now depends solely on ζ. The i.c. on eqn. (5.39) is

$$T(t=0) = T_i \qquad \text{or} \qquad \Theta(\zeta=\infty) = 1 \tag{5.40}$$

and the one known b.c. is

$$T(x=0) = T_\infty \qquad \text{or} \qquad \Theta(\zeta=0) = 0 \tag{5.41}$$

If we call $d\Theta/d\zeta \equiv \chi$, then eqn. (5.39) becomes the first-order equation

$$\frac{d\chi}{d\zeta} = -\frac{\zeta}{2}\chi$$

which can be integrated once to get

$$\chi \equiv \frac{d\Theta}{d\zeta} = C_1 e^{-\zeta^2/4} \tag{5.42}$$

and we integrate this a second time to get

$$\Theta = C_1 \int_0^\zeta e^{-\zeta^2/4}d\zeta + \underbrace{\Theta(0)}_{\substack{= 0 \text{ according} \\ \text{to the b.c.}}} \tag{5.43}$$

The b.c. is now satisfied and we need only substitute eqn. (5.43) in the i.c., eqn. (5.40), to solve for C_1:

$$1 = C_1 \int_0^\infty e^{-\zeta^2/4}d\zeta$$

The definite integral is given by integral tables as $\sqrt{\pi}$, so

$$C_1 = \frac{1}{\sqrt{\pi}}$$

Thus the solution to the problem of conduction in a semiinfinite region, subject

Table 5.1   Error function, its complement, and its derivative

$\zeta/2$	$\text{erf}(\zeta/2)$	$\text{erfc}(\zeta/2)$ $\equiv 1 - \text{erf}(\zeta/2)$	$\dfrac{d\,\text{erf}(\zeta/2)}{d(\zeta/2)}$ $= 2e^{-\zeta^2/4}/\sqrt{\pi}$
0	0	1.0	1.1284
0.05	0.05637	0.9436	1.1256
0.1	0.1125	0.8875	1.1172
0.2	0.2227	0.7773	1.0841
0.3	0.3286	0.6714	1.0313
0.4	0.4282	0.5716	0.9615
0.5	0.5205	0.4795	0.8788
0.6	0.6039	0.3961	0.7872
0.7	0.6778	0.3222	0.6913
0.8	0.7421	0.2579	0.5950
0.9	0.7969	0.2031	0.5020
1.0	0.8427	0.1573	0.4151
1.5	0.9661	0.0339	0.3568
2.0	0.9953	0.00468	0.0827
2.5	0.9996	0.00041	0.0109
3.0	0.99998	0.00002	0.0008

to a b.c. of the first kind is

$$\Theta = \frac{1}{\sqrt{\pi}} 2 \int_0^\zeta e^{-\zeta^2/4} d\zeta / 2 \equiv \text{erf}(\zeta/2) \tag{5.44}$$

The middle term of eqn. (5.44) is called *error function*, erf. Its name arises from the importance of the function in certain kinds of statistical problems. We list its value; its complement, $\text{erfc} \equiv 1 - \text{erf}$; and its derivative in Table 5.1. Equation (5.44) is also plotted in Fig. 5.15.

In Fig. 5.15 we see the early-time curves shown in Fig. 5.14 collapse into a single curve. This is accomplished by the *similarity transformation,* as we call it: $\zeta/2 = x/2\sqrt{\alpha t}$. Under this transformation we see immediately that the local

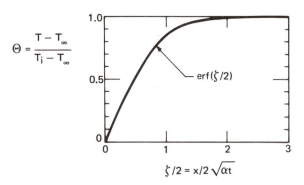

FIGURE 5.15   Temperature distribution in a semiinfinite region.

value of $(T - T_\infty)$ is more than 99% of $(T_i - T_\infty)$ as long as

$$\frac{\zeta}{2} > 1.825 \qquad \text{or} \qquad x > 3.65\sqrt{\alpha t} \tag{5.45}$$

Thus slabs with a half-thickness in excess of $\sqrt{\alpha t}$ are still effectively semiinfinite.

Example 5.4

For what maximum time can a samurai sword be analyzed as a semiinfinite region after it is quenched, if it has no clay coating, and $\bar{h}_{external} \cong \infty$?

SOLUTION.   First we must guess the half-thickness of the sword (say 3 mm), and its material (probably wrought iron with an average α around 1.5×10^{-5} m^2/s). Then we invert eqn. (5.45) and set x equal to the half-thickness, so

$$t \leqslant \frac{x^2}{3.65^2 \alpha} = (0.003 \text{ m})^2 / 13.3(1.5)(10)^{-5} \text{ m}^2/\text{s}$$

$$= \underline{0.045 \text{ s}}$$

Thus the quench would be felt at the centerline of the sword within only $\frac{1}{20}$ s. The thermal diffusivity of clay is smaller than that of steel by a factor of about 30, so the quench time of the coated steel must continue for over 1 s before the temperature of the steel is affected at all, if the clay and sword thicknesses are comparable.

Equation (5.45) provides an interesting foretaste of the notion of a fluid boundary layer. In the context of Figs. 1.9 and 1.10, we observe that free stream flow around an object is disturbed in a thin layer near the object because the fluid adheres to it. It turns out that the thickness of this boundary layer of altered flow velocity increases in the direction of flow. For flow over a flat plate we shall find that this thickness is approximately $4.92\sqrt{\nu t}$, where t is the time required for an element of the stream fluid to move from the leading edge of the plate to a point of interest. This is quite similar to eqn. (5.45) except in that the thermal diffusivity, α, has been replaced by its counterpart, the kinematic viscosity, ν, and the constant is a bit larger. The velocity profile will resemble Fig. 5.15.

If we repeated the problem with a boundary condition of the third kind, we would expect to get $\Theta = \Theta(\text{Bi}, \zeta)$, except in that there is no length, L, upon which to build a Biot number. Therefore, we replace L with $\sqrt{\alpha t}$, which has the dimension of length, so

$$\Theta = \Theta\left(\zeta, \frac{\bar{h}\sqrt{\alpha t}}{k}\right) \equiv \Theta(\zeta, \beta) \tag{5.46}$$

The term $\beta \equiv \bar{h}\sqrt{\alpha t}/k$ is like the product: $\text{Bi}\sqrt{\text{Fo}}$. The solution of this problem (see, e.g., [1.15, Chap. 10]) can be conveniently written in terms of the

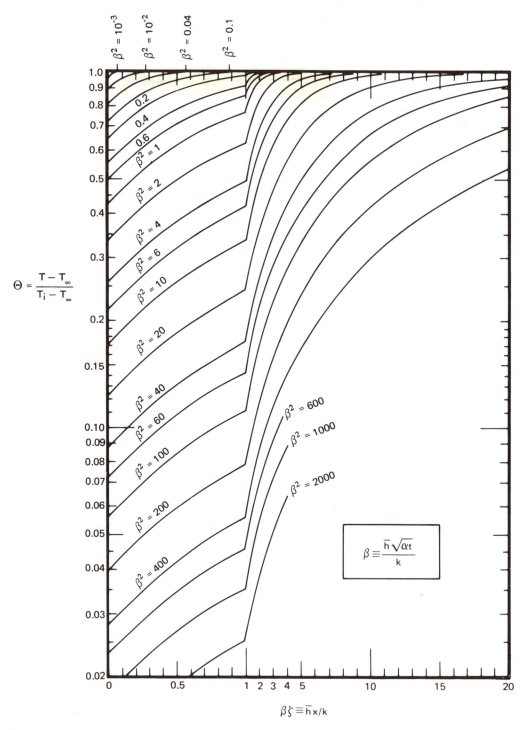

FIGURE 5.16 The cooling of a semiinfinite region by an environment at T_∞, through a heat transfer coefficient, $\bar{h}$.

complementary error function, $\text{erfc} \equiv 1 - \text{erf}$:

$$\Theta = \text{erf}\frac{\zeta}{2} + \exp(\beta\zeta + \beta^2)\left[\text{erfc}\left(\frac{\zeta}{2} + \beta\right)\right] \tag{5.47}$$

This result is plotted in Fig. 5.16.

Example 5.5

Most of us have passed our finger through an 800°C candle flame and know that if we limit exposure to about $\frac{1}{4}$ s we will not be burned. Why not?

SOLUTION. We are discussing a *very* superficial heating, so we consider the finger to be a semiinfinite region and go to eqn. (5.47) to calculate $(T_{\text{burn}} - T_{\text{flame}})/(T_i - T_{\text{flame}})$. It turns out that the burn threshold of human skin, T_{burn}, is about 65°C. (That is why 140°F or 60°C tap water is considered to be "scalding.") Therefore, we shall calculate how long it will take for the surface temperature of the finger to rise from body temperature (37°C) to 65°C, when it is protected by an assumed $\bar{h} \approx 100$ W/m² °C. We shall assume that the thermal conductivity of human flesh equals that of its major component—water—and that the thermal diffusivity is equal to the known value for beef. Then

$$\Theta = \frac{65 - 800}{37 - 800} = 0.963$$

$$\beta\zeta = \frac{\bar{h}x}{k} = 0 \qquad \text{since } x = 0 \text{ at the surface}$$

$$\beta^2 = \frac{\bar{h}^2\alpha t}{k^2} = \frac{100^2(0.135 \times 10^{-6})t}{0.63^2} = 0.0034(t \text{ s})$$

This situation is quite far into the corner of Fig. 5.16. We read $\beta^2 \approx 0.001$, which corresponds with $t \approx 0.3$ s. For greater accuracy we must go to the equation, (5.47):

$$0.963 = \underbrace{\text{erf } 0}_{=0} + \exp(0 + 0.0034t)\left[\text{erfc}(0 + \sqrt{0.0034t}\,)\right]$$

so

$$0.963 = e^{0.0034t}\text{erfc}\sqrt{0.0034t}$$

By trial and error, we get $t \approx 0.36$ s.

Thus it would require about $\frac{1}{3}$s to bring the skin to the burn point.

Experiment 5.1

Immerse your hand in the subfreezing air in the freezer compartment of your refrigerator. Next, immerse your finger in a mixture of ice cubes and water but do not move it. Then, immerse your finger in a mixture of ice cubes and water, swirling it around as you do so. Describe your initial sensation in each case, and explain the differences in terms of Fig. 5.16. What variable has changed from one case to the other?

Heat will be removed from the exposed surface of a semiinfinite region, with a b.c. of either the first or the third kind, in accordance with Fourier's law:

$$q = -k \frac{\partial T}{\partial x}\bigg|_{x=0} = \frac{k(T_\infty - T_i)}{\sqrt{\alpha t}} \frac{d\Theta}{d\zeta}\bigg|_{\zeta=0}$$

Differentiating Θ as given by eqn. (5.44), we obtain for the b.c. of the first kind,

$$q = \frac{k(T_\infty - T_i)}{\sqrt{\alpha t}} \left(\frac{1}{\sqrt{\pi}} e^{-\zeta^2/4} \right)_{\zeta=0} = \frac{k(T_\infty - T_i)}{\sqrt{\pi \alpha t}} \tag{5.48}$$

Thus q decreases with increasing time, as $t^{-1/2}$. When the temperature of the surface is first changed the heat removal rate is enormous. Then it drops off rapidly.

It often occurs that we suddenly apply a specified input heat flux, q_w, at the boundary of a semiinfinite region. In such a case we can differentiate the heat diffusion equation with respect to x, so

$$\alpha \frac{\partial^3 T}{\partial x^3} = \frac{\partial^2 T}{\partial t \partial x}$$

When we substitute $q = -k \partial T / \partial x$ in this we obtain

$$\alpha \frac{\partial^2 q}{\partial x^2} = \frac{\partial q}{\partial t}$$

with the b.c.'s:

$$q(x=0, t>0) = q_w \quad \text{or} \quad \frac{q_w - q}{q_w}\bigg|_{x=0} = 0$$

$$q(x \geqslant 0, t=0) = 0 \quad \text{or} \quad \frac{q_w - q}{q_w}\bigg|_{t=0} = 1$$

What we have done here is quite elegant. We have made the problem of predicting the local heat flux q into exactly the same form as that of predicting the local temperature in a semiinfinite region subjected to a step change of wall temperature. Therefore, the solution must be the same:

$$\frac{q_w - q}{q_w} = \text{erf}\left(\frac{x}{2\sqrt{\alpha t}} \right) \tag{5.49}$$

and the temperature distribution is obtained by integrating Fourier's law. At the wall, for example:

$$\int_{T_\infty}^{T_w} dT = -\int_\infty^0 \frac{q}{k} dx$$

or

$$T_w = T_\infty + \frac{q_w}{k} \int_0^\infty \mathrm{erfc}\left(x/2\sqrt{\alpha t}\right)dx$$

This becomes

$$T_w = T_\infty + \frac{q_w}{k} 2\sqrt{\alpha t} \underbrace{\int_0^\infty \mathrm{erfc}(\zeta/2)d\zeta/2}_{=1/\sqrt{\pi}}$$

so

$$T_w(t) = T_\infty + 2\frac{q_w}{k}\sqrt{\frac{\alpha t}{\pi}} \qquad (5.50)$$

Example 5.6   *Predict the Growth Rate of a Vapor Bubble in an Infinite Superheated Liquid*

This prediction is relevant to a large variety of processes, ranging from nuclear thermohydraulics to direct contact heat exchange. It was originally presented by Jakob and others in the early 1930s (see, e.g., [5.6, Chap. I]). Jakob (pronounced Yah'-kob) was an important figure in heat transfer during the 1920s and 1930s. He left Hitler's Germany in 1936 to come to the United States. We encounter his name again, later.

Figure 5.17 shows how growth occurs. When a liquid is superheated to a temperature somewhat above its boiling point, a small gas or vapor cavity in that liquid will grow. (That is what happens in the superheated water at the bottom of a teakettle.)

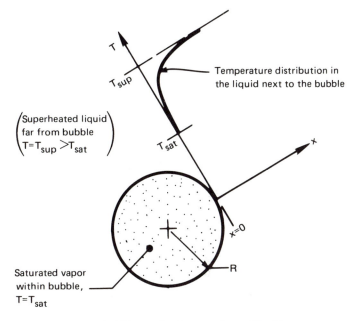

FIGURE 5.17   A bubble growing in a superheated liquid.

This bubble grows into the surrounding liquid because its boundary is kept at the saturation temperature, T_{sat}, by the near-equilibrium coexistence of liquid and vapor. Therefore, heat must flow from the superheated surroundings to the interface, where evaporation occurs. As long as the layer of cooled liquid is thin, we should not suffer too much error by using the one-dimensional semiinfinite region solution to predict the heat flow.

Thus we can write the energy balance at the bubble interface:

$$-q\frac{W}{m^2}(4\pi R^2 m^2) = \left(\rho_g h_{fg}\frac{J}{m^3}\right)\left(\frac{dV}{dt}\frac{m^3}{s}\right)$$

$$\underbrace{\qquad\qquad\qquad}_{Q \text{ into bubble}}\qquad\underbrace{\qquad\qquad\qquad\qquad}_{\substack{\text{rate of energy increase}\\\text{of the bubble}}}$$

and then substitute eqn. (5.48) for q and $\frac{4}{3}\pi R^3$ for the volume, V. This gives

$$\frac{k(T_{sup}-T_{sat})}{\sqrt{\alpha\pi t}} = \rho_g h_{fg}\frac{dR}{dt} \tag{5.51}$$

Integrating eqn. (5.51) from $R=0$ at $t=0$ up to R at t, we obtain Jakob's prediction:

$$R = \frac{2}{\sqrt{\pi}}\frac{k\Delta T}{\rho_g h_{fg}\sqrt{\alpha}}t^{1/2} \tag{5.52}$$

This analysis was done without assuming the curved bubble interface to be plane, 24 years after Jakob's work, by Plesset and Zwick [5.7]. It was verified in a different way after another 5 years by Scriven [5.8]. This exact calculation is

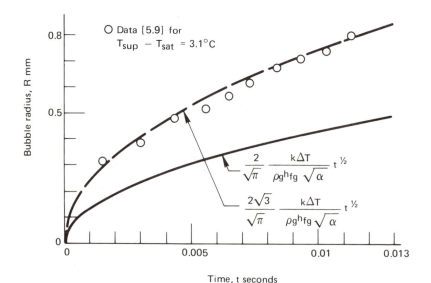

FIGURE 5.18   The growth of a vapor bubble: predictions and measurements.

very complicated and it gives

$$R = \frac{2\sqrt{3}}{\sqrt{\pi}} \frac{k\Delta T}{\rho_g h_{fg} \sqrt{\alpha}} t^{1/2} = \sqrt{3} \, R_{\text{Jakob}} \qquad (5.53)$$

Both predictions are compared with some of the data of Degarabedian [5.9] in Fig. 5.18. The data and the exact theory match almost perfectly. The very simple theory of Jakob et al. shows the correct dependence of R on all its variables, but it shows growth rates that are low by a factor of $\sqrt{3}$. This is because the actual temperature gradient is always being steepened as the region of thermal influence is stretched out circumferentially by bubble growth, and it is made thinner in the radial direction. Therefore, the temperature profile flattens out more slowly than Jakob predicts and the bubble grows more rapidly.

Experiment 5.2

Touch various objects in the room around you: glass, wood, corkboard, paper, steel, and gold or diamond, if available. Rank them in the order of which feels coldest at the first instant of contact.

The more advanced theory of heat conduction (see, e.g., [1.14]) shows that if two semiinfinite regions are placed together suddenly, their interface temperature, T_s, is given by[4]

$$\frac{T_s - T_{\text{obj}}}{T_{\text{body}} - T_{\text{obj}}} = \frac{k_{\text{body}}/\sqrt{\alpha_{\text{body}}}}{k_{\text{obj}}/\sqrt{\alpha_{\text{obj}}} + k_{\text{body}}/\sqrt{\alpha_{\text{body}}}}$$

where we have immediately identified one region with your body ($T_{\text{body}} = 37°C$) and the other with the object being touched ($T_{\text{obj}} \simeq 20°C$). Compare the ranking you obtain experimentally with the ranking given by this equation.

Notice that your bloodstream and capillary system provide a heat source in your finger, so the equation is only valid for a moment. Then you start replacing the heat lost to the objects. If you included a diamond among the objects that you touched, you will notice that it warmed up almost instantly. Most diamonds are quite small but are possessed of the highest known value of α. Therefore, they can only behave as a semiinfinite region for an instant.

Conduction to a semiinfinite region
with a harmonically oscillating temperature at the boundary

Suppose that we approximate the annual variation of the ambient temperature as sinusoidal and then ask what the influence of this variation will be beneath the ground. We want to calculate $T - \overline{T}$ (where $\overline{T}$ is the average surface temperature) as a function of depth, x; α; frequency of oscillation, ω; amplitude

---

[4]For semiinfinite regions, initially at uniform temperatures, T_s will not vary with time. For finite bodies, T_s will eventually change.

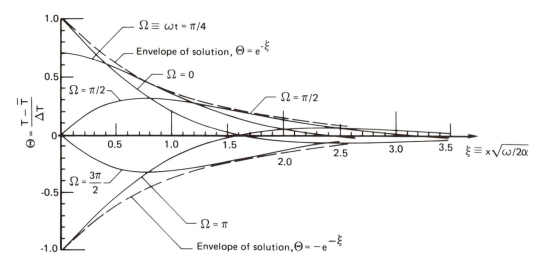

FIGURE 5.19   The temperature variation within a semiinfinite region whose temperature varies harmonically at the boundary.

of oscillation, ΔT; and time, t. There are six variables in °C, m, and s, so the problem can be represented in three dimensionless variables:

$$\Theta \equiv \frac{T - \bar{T}}{\Delta T} \qquad \Omega \equiv \omega t \qquad \xi \equiv x\sqrt{\frac{\omega}{2\alpha}}$$

We pose the problem as follows in these variables: The heat conduction equation is

$$\frac{1}{2}\frac{\partial^2 \Theta}{\partial \xi^2} = \frac{\partial \Theta}{\partial \Omega} \tag{5.54}$$

and the b.c.'s are

$$\Theta_{\xi=0} = \cos\omega t \qquad \text{and} \qquad \Theta_{\xi>0} = \text{finite} \tag{5.55}$$

No i.c. is needed because the steady solution has to be periodic.

The solution is given by Carslaw and Jaeger (see [1.14, Sec. 2.6] or work Problem 5.16). It is

$$\Theta = e^{-\xi}\cos(\Omega - \xi) \tag{5.56}$$

This result is plotted in Fig. 5.19. It shows that the surface temperature variation decays exponentially into the region and it suffers a phase shift as it does so.

Example 5.7

How deep in the earth must we dig to find the temperature wave that was launched by the coldest part of last winter if it is now high summer?

SOLUTION. $\omega = 2\pi$ rad/yr, and $\Omega = \omega t = 0$ at the present. First we must find the depths at which the $\Omega = 0$ curve reaches its local extrema. (We pick the $\Omega = 0$ curve because it gives the highest temperature at $t=0$.)

$$\frac{d\Theta}{d\xi}\bigg|_{\Omega=0} = -e^{-\xi}\cos(0-\xi)+e^{-\xi}\sin(0-\xi)=0$$

This gives

$$\tan(0-\xi)=1 \qquad \text{so} \qquad \xi=\frac{3\pi}{4}, \quad \frac{7\pi}{4}, \quad \text{etc.}$$

and the first minimum occurs where $\xi=3\pi/4=2.356$, as we can see in Fig. 5.19.

$$\xi=x\sqrt{\omega/2\alpha}=2.356$$

or, if we take $\alpha=0.139(10)^{-6}$ m^2/s (given in [1.7] for coarse, gravelly earth),

$$x=2.356\bigg/\sqrt{\frac{2\pi}{2[0.139(10)^{-6}]}\frac{1}{365(24)(3600)}}=2.783 \text{ m}$$

If we dug in the earth we would find it growing colder and colder until it reached a maximum coldness at a depth of about 2.8 m. Farther down it would begin to warm up again, but not much. In midwinter ($\Omega=\pi$) the reverse would be true.

5.6 Steady multidimensional heat conduction

Introduction

The general equation for $T(\vec{r})$ during steady conduction in a region of constant thermal conductivity, without heat sources, is called Laplace's equation:

$$\nabla^2 T=0 \tag{5.57}$$

It looks easier to solve than it is, since [recall eqns. (2.16) to (2.18)] the Laplacian, $\nabla^2 T$, is a sum of several second partial derivatives. We have solved one two-dimensional heat conduction problem in Example 4.1 but were only able to do so because the boundary conditions were made to order.

We shall not undertake any more complicated analytical solutions to multidimensional problems in this course. The reader who is interested in such analysis should go to [1.14], [1.15], or [1.16], where such calculations are done in detail. Faced with a steady multidimensional problem, three routes are open to us:

- Find out whether or not the analytical solution is already available in a heat conduction text or treatise.
- Solve the problem
  - (a) Analytically. (We do not treat such analyses.)
  - (b) Numerically. (Such methods are discussed in Chapter 6.)
- Obtain the solution graphically if the problem is two-dimensional.

It is to the last of these options that we give our attention next.

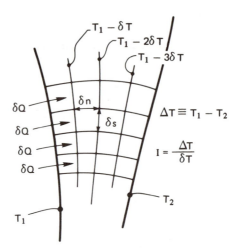

FIGURE 5.20   The two-dimensional flow of heat between two isothermal walls.

The flux plot

The method that we propose here will solve all two-dimensional problems in which all boundaries are held at either of two temperatures or are insulated. With a little skill it will provide accuracies of a few percent. It is quicker and easier than either numerical or analytical methods, its accuracy is almost always greater than the accuracy with which the b.c.'s and k can be specified, and it displays the mathematical sense of the problem very clearly. The method is called *flux plotting*.

Figure 5.20 shows heat flowing from one isothermal wall to another in a regime that does not conform to any convenient coordinate scheme. We identify a series of channels, each of which carries the same heat flow, δQ W/m. We also include a set of equally spaced isotherms, δT apart, between the walls. Since the heat fluxes in all channels are the same,

$$|\delta Q| = k \frac{\delta T}{\delta n} \delta s \tag{5.58}$$

Notice that if we arrange things so that δQ, ΔT, and k are the same for flow through each rectangle in the flow field, then $\delta s/\delta n$ must be the same for each rectangle. We therefore arbitrarily set the ratio equal to unity so all the elements appear as distorted *squares*.

The objective is then to sketch the isothermal lines and the adiabatic,[5] or heat flow, lines perpendicular to them. This sketch is to be done subject to two

---

[5]These are lines *in the direction* of heat flow. It immediately follows that there can be no component of heat flow normal to them; hence, they must be adiabatic.

constraints

- Isothermal and adiabatic lines must intersect at right angles.
- They must subdivide the flow field into elements that are nearly square —"nearly" because they have slightly curved sides.

Once the grid has been sketched, the temperature anywhere in the field can be read directly from the sketch. And the heat flow per unit depth into the paper is

$$Q \text{ W/m} = Nk\delta T \frac{\delta s}{\delta n} = \frac{N}{I} k \Delta T \tag{5.59}$$

where N is the number of heat flow channels and I is the number of temperature increments, $\Delta T / \delta T$.

The first step in constructing a flux plot is to draw the boundaries of the region accurately *in ink* using drafting instruments. The next is to obtain a soft pencil (a no. 2 grade is good) and a soft eraser. We begin with an example which was executed fairly well in the old and influential *Heat Transfer Notes* [1.2]. This example is shown in Fig. 5.21.

The particular example happens to have an axis of symmetry in it. We immediately interpret this as an adiabatic boundary because heat cannot cross it. The problem therefore reduces to the simpler one of sketching lines in only one half of the area. We illustrate this process in four steps. Notice the following steps and features in this plot:

- Begin by dividing the region, by sketching in either a single isothermal or adiabatic line.
- Fill in the lines perpendicular to the original line so as to make squares. Allow the original line to move in such a way as to accommodate squares. This will *always* require some erasing. Therefore:
- *Never* make the original lines dark and firm.
- By successive subdividing of the squares, make the final grid. *Do not make the grid very fine*. If you do, you will lose accuracy because the lack of perpendicularity and squareness will be less evident to the eye. Step IV in Fig. 5.21 is as fine a grid as should ever be made.
- If you have doubts about whether any large ill-shaped regions are correct, fill them in with an extra isotherm and adiabatic line to be sure that they resolve into approximate squares (see the dashed lines in Fig. 5.21).
- Fill in the final grid, when you are sure of it, either in hard pencil or pen, and erase any lingering background sketch lines.
- Your flow channels need not come out even. Notice that there is an extra $\frac{1}{7}$ of a channel in Fig. 5.21. This is simply counted as $\frac{1}{7}$ of a square in eqn. (5.59).
- Never allow isotherms or adiabatic lines to intersect themselves.

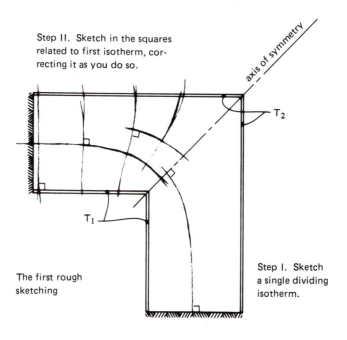

Step II. Sketch in the squares related to first isotherm, correcting it as you do so.

axis of symmetry

T_2

T_1

The first rough sketching

Step I. Sketch a single dividing isotherm.

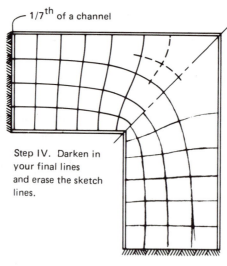

1/7th of a channel

Step III. Sketch and correct until you are reasonably content with the form.

Step IV. Darken in your final lines and erase the sketch lines.

FIGURE 5.21   The evolution of a flux plot.

When the sketch is complete we can return to eqn. (5.59) to compute the heat flux. In this case

$$Q = \frac{N}{I} i \Delta T = \frac{2(6.14)}{4} k \Delta T = 3.07 k \Delta T$$

When the authors of [1.2] did this problem they obtained $N/I = 3.00$—a value only 2% below ours. This kind of agreement is typical when the method is used with care.

One must be careful not to grasp at a false axis of symmetry. Figure 5.22 shows a shape similar to the one that we just treated, but with unequal legs. In this case, no lines must enter (or leave) the corners A and B. The reason is that, since there *is* no symmetry, we have no guidance as to the direction of the lines at these corners. In particular, we know that a line leaving A will no longer arrive at B.

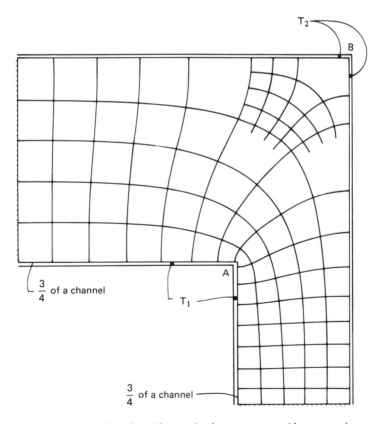

FIGURE 5.22   A flux plot with no axis of symmetry to guide construction.

Example 5.8

A structure consists of metal walls, 8 cm apart with insulating material ($k=0.12$ W/m-°C) between. Ribs 4 cm long protrude from one wall every 14 cm. They can be assumed to stay at the temperature of that wall. Find the heat flux through the wall and the local temperature in the middle if the first wall is at 40°C and the one with the ribs is at 0°C. Find the temperature in the middle of the wall, 2 cm from a rib, as well.

SOLUTION. The flux plot for this configuration is shown in Fig. 5.23. For a typical section there are approximately 5.6 isothermal increments and 6.15 heat flow channels, so

$$Q = \frac{N}{I} k \Delta T = \frac{2(6.15)}{5.6}(0.12)(40-0) = \underline{10.54 \text{ W/m}}$$

where the factor of 2 accounts for the fact that there are two halves in the section. We deduce the temperature for the point, A, of interest by a simple proportionality:

$$T_{\text{point } A} = \frac{2.1}{5.6}(40-0) = \underline{15°C}$$

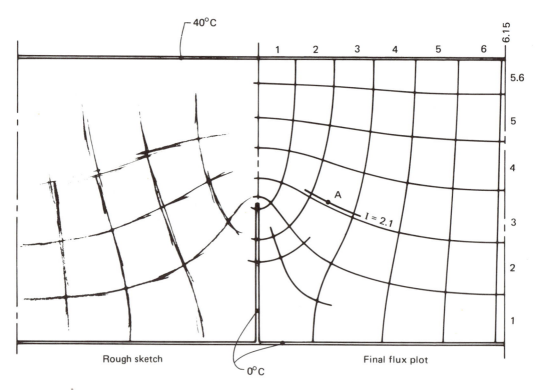

FIGURE 5.23   Heat transfer through a wall with isothermal metal ribs.

The shape factor S is defined such that

$$Q \equiv Sk\,\Delta T \tag{5.60}$$

Thus far, every heat conduction problem we have done has taken this form. The heat flow always equals a function of the geometrical shape of the body multiplied by $k\,\Delta T$. Now let us compare eqns. (5.59) and (5.60):

$$Q\,\frac{\text{W}}{\text{m}} = (S \text{ dimensionless})\left(k\,\Delta T\,\frac{\text{W}}{\text{m}}\right) = \frac{N}{I}k\,\Delta T \tag{5.61}$$

but in three dimensions

$$QW = (S \text{ m})\left(k\,\Delta T\,\frac{\text{W}}{\text{m}}\right) \tag{5.62}$$

It follows that the thermal resistance of a two-dimensional body is

$$R_\text{t} = \frac{1}{kS} \qquad \text{where} \qquad Q = \frac{\Delta T}{R_\text{t}} \tag{5.63}$$

For a three-dimensional body eqn. (5.63) is unchanged except that the dimensions of Q and R_t differ.[6]

The virtue of the shape factor is that it summarizes a heat conduction solution in a given configuration. Once S is known, it can be used again and again. That S is nondimensional in two-dimensional configurations means that Q is independent of the size of the body. Thus, in Fig. 5.21, S is always 3.07—regardless of the size of the figure—and in Example 5.8, S is $2(6.15)/5.6$ or 2.196 whether or not the wall is made larger or smaller. When a body's breadth is increased so as to increase Q, its thickness in the direction of heat flow is also increased so as to decrease Q by the same factor.

Example 5.9

Calculate the shape factor for a one-quarter section of a thick cylinder.

SOLUTION. We already know R_t for a thick cylinder. It is given in eqn. (2.22). From it we compute

$$S_\text{cyl} = \frac{1}{kR_\text{t}} = \frac{2\pi}{\ell\text{n}(r_\text{o}/r_\text{i})}$$

so in the case of a quarter-cylinder,

$$S = \frac{\pi}{2\,\ell\text{n}(r_\text{o}/r_\text{i})}$$

The quarter-cylinder is pictured in Fig. 5.24 for a radius ratio, $r_\text{o}/r_\text{i} = 3$, but for two different sizes. In both cases $S = 1.43$. (Note that the same S is also given by the flux plot shown.)

---

[6]Recall that we noted after eqn. (2.22) how the dimensions of R_t changed, depending on whether or not Q was expressed on a unit-length basis.

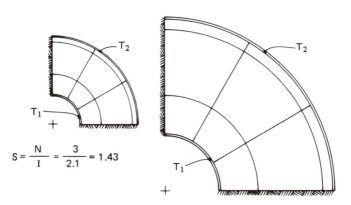

$$S = \frac{N}{I} = \frac{3}{2.1} = 1.43$$

FIGURE 5.24   The shape factor for two similar bodies of different size.

Example 5.10

Calculate S for a thick hollow sphere as shown in Fig. 5.25.

SOLUTION.   The general solution of the heat diffusion equation in spherical coordinates is:

$$T = \frac{C_1}{r} + C_2$$

when $T = \mathrm{fn}(r$ only$)$. The b.c.'s are

$$T(r = r_i) = T_i \qquad \text{and} \qquad T(r = r_o) = T_o$$

substituting the general solution in the b.c.'s, we get

$$\frac{C_1}{r_i} + C_2 = T_i \qquad \text{and} \qquad \frac{C_1}{r_o} + C_1 = T_o$$

Therefore,

$$C_1 = \frac{T_i - T_o}{r_o - r_i} r_i r_o \qquad \text{and} \qquad C_2 = T_i - \frac{T_i - T_o}{r_o - r_i} r_o$$

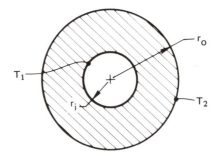

FIGURE 5.25   Heat transfer through a thick hollow sphere.

Putting C_1 and C_2 in the general solution, and calling $T_i - T_o \equiv \Delta T$, we get

$$T = T_i + \Delta T \left[\frac{r_i r_o}{r(r_o - r_i)} - \frac{r_o}{r_o - r_i} \right]$$

Then

$$Q = -kA \frac{dT}{dr} = \frac{4\pi(r_i r_o)}{r_o - r_i} k \Delta T$$

$$\underline{S = \frac{4\pi(r_i r_o)}{r_o - r_i} \text{ m}}$$

where S now has the dimensions of m.

Table 5.2 includes a number of analytically derived shape factors for use in calculating the heat flux in different configurations. Notice that these results will

Table 5.2 Conduction shape factors

Situation	Shape factor, S	Dimen-sions	Source
1. Conduction through a slab	A/L	meter	Example 2.3
2. Conduction through a long thick cylinder	$\dfrac{2\pi}{\ln(r_o/r_i)}$	none	Example 5.9
3. Conduction through a thick-walled hollow sphere	$\dfrac{4\pi(r_i r_o)}{r_o - r_i}$	meter	Example 5.10
4. The boundary of a spherical hole conducting into an infinite medium	$4\pi r$	meter	Problems 5.19 and 2.15

continued

Table 5.2 Continued

Situation	Shape factor, S	Dimensions	Source
5. Submerged pipe of radius, R, and length, L, transferring heat to a parallel isothermal plane; $h \ll L$	$$\dfrac{2\pi L}{\cosh^{-1}(h/R)}$$	meter	[5.10]
6. Same as 5. but $L \to \infty$ (two-dimensional heat conduction)	$$\dfrac{2\pi}{\cosh^{-1}(h/R)}$$	none	[5.10]
7. An isothermal sphere of radius, R, transfers heat to an isothermal plane (see 4.)	$$\dfrac{4\pi R}{1 - R/2h}$$	meter	[5.10]
8. An isothermal sphere of radius, R, near an insulated plane, transfers heat to an infinite medium at T_∞. (see 4. and 7.) insulated	$$\dfrac{4\pi R}{1 + R/2h}$$	meter	[5.11]

continued

Table 5.2   Continued

Situation	Shape factor, S	Dimen-sions	Source
9. Parallel cylinders exchange heat in an infinite conducting medium	$$\dfrac{2\pi}{\cosh^{-1}\dfrac{m_1}{R_1} + \cosh^{-1}\dfrac{m_2}{R_2}}$$ where: $$m_1 = \dfrac{L}{2}\left[1 + \left(\dfrac{R_1}{L}\right)^2 - \left(\dfrac{R_2}{L}\right)^2\right]$$ $$m_2 = \dfrac{L}{2}\left[1 - \left(\dfrac{R_1}{L}\right)^2 + \left(\dfrac{R_2}{L}\right)^2\right]$$	none	[5.10]
10. Same as 9., but with cylinders widely spaced. $(L \gg R_1$ or $R_2)$	$$\dfrac{2\pi}{\cosh^{-1}\dfrac{L}{2R_1} + \cosh^{-1}\dfrac{L}{2R_2}}$$	none	[5.10]
11. Two spheres of radii R_1 and R_2 with centers, a distance, d, apart in a conducting medium. $(d > 5R_{\text{largest}})$	$$\dfrac{4\pi}{\dfrac{R_2}{R_1}\left[1 - \dfrac{(R_1/d)^4}{1-\left(\dfrac{R_2}{d}\right)^2}\right] - \dfrac{2R_2}{d}}$$	none	[5.10]
12. Parallel discs of radius, R, and placed a distance, d, apart in a conducting medium $(d \geqslant 5R)$	$$\dfrac{4\pi}{2\left[\dfrac{\pi}{2} - \tan^{-1}\dfrac{R}{d}\right]}$$	none	[5.10]

not give local temperatures. To obtain that information one must either make a flux plot or do a two- or three-dimensional analysis. Notice, too, that this table is restricted to bodies with isothermal or insulated boundaries.

Example 5.11

A spherical heat source 6 cm in diameter is buried 30 cm below the surface of a very large box of soil and kept at 35°C. The surface of the soil is kept at 21°C. If the steady heat transfer rate is 280 W, what is the thermal conductivity of this sample of soil?

$$Q = Sk\,\Delta T = \left(\frac{4\pi R}{1 - R/2h}\right)k\,\Delta T$$

where S is given by shape factor 7 in Table 5.2. Then

$$k = \frac{280\ \text{W}}{(35-21)°C}\ \frac{1-(6/2)/2(30)}{4\pi(6/2)\ \text{m}} = \underline{0.504\ \text{W}/\text{m-}°C}$$

Readers wanting to look at a broader catalog of shape factors should go to [1.13], [5.10], or [5.11].

The problem of locally vanishing resistance

Suppose that two different temperatures are specified on adjacent sides of a square, as shown in Fig. 5.26. The shape factor in this case is

$$S = \frac{N}{I} = \frac{\infty}{4} = \infty$$

(It is futile to try to count channels beyond $N \simeq 10$, but it is clear that they multiply without limit in the lower left corner.) The problem is that we have violated our rule that isotherms cannot intersect and have created a $1/r$ singularity. If we actually tried to sustain such a situation, the figure would be correct at some distance from the corner. However, where the isotherms are close to one another they will necessarily influence and distort one another in such a way as to avoid intersecting. And S will never really be infinite as it appears to be in the figure.

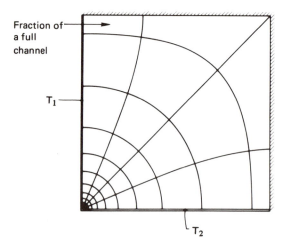

FIGURE 5.26   Resistance vanishes where two isothermal boundaries intersect.

5.7 Transient multidimensional heat transfer—
The tactic of superposition

Consider the cooling of a stubby cylinder such as the one shown in Fig. 5.27a. The cylinder is initially at $T = T_i$, and it is suddenly subjected to a common b.c. on all sides. This problem would be inherently complicated. It requires solving the heat conduction equation for $T = \text{fn}(r, z, t)$, or in dimensionless coordinates, $\Theta = \text{fn}(\rho, \xi, \text{Fo})$, with b.c.'s of the first, second, or third kind.

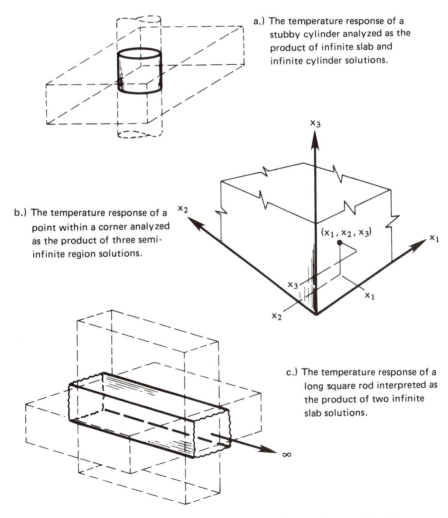

a.) The temperature response of a stubby cylinder analyzed as the product of infinite slab and infinite cylinder solutions.

b.) The temperature response of a point within a corner analyzed as the product of three semi-infinite region solutions.

c.) The temperature response of a long square rod interpreted as the product of two infinite slab solutions.

FIGURE 5.27 Various solid bodies whose transient cooling can be treated as the product of one-dimensional solutions.

However, Fig. 5.27a suggests that this can somehow be viewed as a combination of an infinite cylinder and an infinite slab. It turns out that the problem *can be analyzed* from that point of view.

If the body is subject to uniform b.c.'s of the first, second, or third kind, this problem can be characterized as the product of the component one-dimensional problems with the same b.c.'s and i.c.. In Fig. 5.27a we find

$$\Theta(\rho, \xi, \text{Fo}) = \left[\Theta(\rho, \text{Fo})_{\text{inf cyl}}\right]\left[\Theta(\xi, \text{Fo})_{\text{inf slab}}\right] \tag{5.64}$$

The proof of the legitimacy of such product solutions is given by Carslaw and Jaeger [1.14, Sec. 1.15].

Figure 5.27b shows how to calculate the temperature at a point near the corner, inside a one-eighth-infinite region. Here we would write

$$\Theta(\xi_1, \xi_2, \xi_3, \text{Fo}) = \left[\Theta(\xi_1, \text{Fo})\right]\left[\Theta(\xi_2, \text{Fo})\right]\left[\Theta(\xi_3, \text{Fo})\right] \tag{5.65}$$

Example 5.12

A 4-cm-square iron rod at $T_i = 100°C$ is suddenly immersed in a coolant at $T_\infty = 20°C$ with $\bar{h} = 800 \text{ W/m}^2\text{-}°C$. What is the temperature on a line 1 cm from one side and 2 cm from the adjoining side, after 10 s?

SOLUTION. With reference to Fig. 5.27c, we first evaluate $\text{Fo} = \alpha t / L^2 = (0.0000226 \text{ m}^2/\text{s})(10 \text{ s})/(0.04 \text{ m}/2)^2 = 0.565$, and $\text{Bi} = \bar{h}L/k = 800(0.04/2)/76 = 0.2105$, and then write

$$\Theta\left(\left(\frac{x}{L}\right)_1 = 0, \left(\frac{x}{L}\right)_2 = \frac{1}{2}, \text{Fo} = 0.565, \text{Bi}^{-1} = 4.75\right)$$

$$= \underbrace{\Theta\left(\left(\frac{x}{L}\right)_1 = 0, \text{Fo} = 0.565, \text{Bi}^{-1} = 4.75\right)}_{\substack{= 0.93 \text{ from upper left-hand} \\ \text{side of Fig. 5.7}}} \underbrace{\Theta\left(\left(\frac{x}{L}\right)_2 = \frac{1}{2}, \text{Fo} = 0.565, \text{Bi}^{-1} = 4.75\right)}_{\substack{= 0.91 \text{ from interpolation} \\ \text{between lower left-hand side and} \\ \text{upper right-hand side of Fig. 5.7}}}$$

Thus, at the axial line of interest,

$$\Theta = (0.93)(0.91) = 0.846$$

so

$$\frac{T - 20}{100 - 20} = 0.846 \qquad \underline{T = 87.7°C}$$

PROBLEMS

5.1. Rework Example 5.1, and replot the solution, with one change. This time, insert the thermometer at zero time, at an initial temperature $< (T_i - bT)$.

5.2. A body of known volume and surface area, and temperature T_i, is suddenly immersed in a bath whose temperature is rising as $T_{\text{bath}} = T_i + (T_0 - T_i)e^{t/\tau}$. Let us suppose that $\bar{h}$ is known, that $\tau = 10\rho cV/\bar{h}A$, and that t is measured from the time of immersion. The Biot number of the body is small. Find the temperature response of the body. Plot the response and the bath temperature as a function of time up to $t = 2\tau$. (Do not use Laplace transform methods except, perhaps, as a check.)

5.3. A body of known volume and surface area is immersed in a bath whose temperature is varying sinusoidally with a frequency, ω, about an average value. The heat transfer coefficient is known and the Biot number is small. Find the temperature variation of the body after a long time has passed and plot it along with the bath temperature. Comment on any interesting aspects of the solution.

A suggested program for solving this problem:

- Write the differential equation of response.

- To get the particular integral of the complete equation, guess that $T - T_{\text{mean}} = C_1 \cos \omega t + C_2 \sin \omega t$. Substitute this in the differential equation and find C_1 and C_2 values that will make the resulting equation valid.

- Write the general solution of the complete equation. It will have one unknown constant in it.

- Write any initial condition you wish—the simplest one you can think of—and use it to get rid of the constant.

- Let the time be large and note which terms vanish from the solution. Throw them away.

- Combine two trigonometric terms in the solution into a term involving $\sin(\omega t - \beta)$, where $\beta = \text{fn}(\omega T)$ is the phase lag of the body temperature.

5.4. A block of copper floats within a large region of well-stirred mercury. The system is initially at a uniform temperature, T_i. There is a heat transfer coefficient, $\bar{h}_m$, on the inside of the thin metal container of the mercury, and another one, $\bar{h}_c$, between the copper block and the mercury. The container is then suddenly subjected to a change in ambient temperature from T_i to $T_s < T_i$. Predict the temperature response of the copper block neglecting the internal resistance of both the copper and the mercury. Check your result by seeing that it fits both initial conditions and that it gives the expected behavior at $T \to \infty$.

5.5. Sketch the electrical circuit analogous to the second-order lumped-capacity system treated in the context of Fig. 5.5 and explain it fully.

5.6. A 1-in.-diameter copper sphere with a thermocouple mounted in its center is mounted as shown in Fig. P5.6 and immersed in water that is saturated at 211°F. The figure shows the thermocouple reading as a function of time during the quenching process. If the Biot number is small, the center temperature can be interpreted as the uniform temperature of the sphere during the quench. First draw tangents to the curve, and graphically differentiate it. Then use the resulting values of dT/dt to construct a graph of the heat transfer coefficient as a function of $(T_{\text{sphere}} - T_{\text{sat}})$. The result will give actual values of $\bar{h}$ during boiling over the range of temperature differences. Check to see whether or not the largest value of the Biot number is too great to permit the use of lumped-capacity methods.

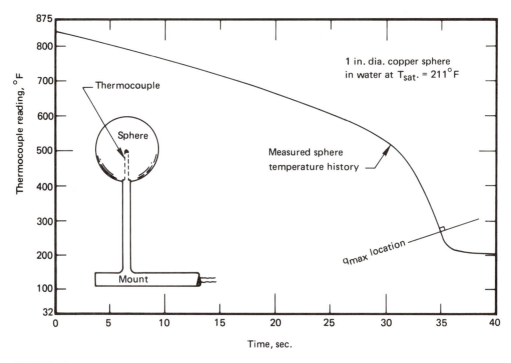

FIGURE P5.6

5.7. A butt-welded 36-gage thermocouple is placed in a gas flow whose temperature rises at the rate 20°C/s. The thermocouple steadily records a temperature 2.4°C below the known gas flow temperature. If ρc is 3800 kJ/m³-°C for the thermocouple material, what is $\bar{h}$ on the thermocouple? [$\bar{h} = 1006$ W/m²-°C.]

5.8. Check the point on Fig. 5.7 at Fo=0.2, Bi=10, and $x/L=0$, analytically.

5.9. Prove that when Bi is large, eqn. (5.34) reduces to eqn. (5.33).

5.10. Check the point at Bi=0.1 and Fo=2.5 on the slab curve in Fig. 5.10 analytically.

5.11. Sketch one of the curves in Figs. 5.7, 5.8, or 5.9 and identify:

- The region in which b.c.'s of the second kind can be replaced with b.c.'s of the first kind.
- The region in which a lumped-capacity response can be assumed.
- The region in which the solid can be viewed as a semiinfinite region.

5.12. Water flows over a flat slab of Nichrome 0.05 mm thick which serves as a resistance heater using ac power. The apparent value of $\bar{h}$ is 2000 W/m²-°C. How much surface temperature fluctuation will there be?

5.13. Put Jakob's bubble growth formula in dimensionless form, identifying a "Jakob number," $\mathrm{Ja} \equiv c_p(T_{\mathrm{sup}} - T_{\mathrm{sat}})/h_{\mathrm{fg}}$ as one of the groups. (Ja is a comparison of sensible heat with latent heat.) Be certain that your nondimensionalization is consistent with the Buckingham pi-theorem.

5.14. A 7-cm-long vertical glass tube is filled with water that is uniformly at a temperature of $T = 102°C$. The top is suddenly opened to the air at 1 atm pressure. Plot the decrease of the height of water in the tube due to evaporation, as a function of time until the bottom of the tube has cooled by 0.05°C.

5.15. A slab is cooled convectively on both sides from a known initial temperature. Compare the variation of surface temperature with time as given by Fig. 5.7 with that given by eqn. (5.47), if Bi = 2. Discuss the meaning of your comparison.

5.16. To obtain eqn. (5.56) assume a complex solution of the type $\Theta = f(\xi)\exp(i\Omega)$, where $i \equiv \sqrt{-1}$. This will assure that the real part of your solution has the required periodicity, and when you substitute it in eqn. (5.54) you will get an easy-to-solve ordinary d.e. in $f(\xi)$.

5.17. A certain steel cylinder wall is subjected to a temperature oscillation which we approximate as $T = 650°C + (300°C)\cos\omega t$, where the piston fires eight times per second. For stress design purposes, plot the amplitude of the temperature variation in the steel as a function of depth. If the cylinder is 1 cm thick, can we view it as having infinite depth?

5.18. A 40-cm-diameter pipe at 75°C is buried in a large block of portland cement. It runs parallel with a 15°C isothermal surface at a depth of 1 m. Plot the temperature distribution along the line normal to the 15°C surface that passes through the center of the pipe. Compute the heat loss from the pipe both graphically and analytically.

5.19. Derive shape factor 4 in Table 5.2.

5.20. Verify shape factor 9 in Table 5.2, with a flux plot. Use $R_1/R_2 = 2$ and $R_1/L = 1/2$. (Be sure to start out with enough blank paper surrounding the cylinders.)

5.21. A copper block 1 in. thick and 3 in. square is maintained at 100°F on one 1-in. by 3-in. surface. The opposing 1-in. by 3-in. surface is adiabatic for 2 in. and 90°F for 1 in. The remaining surfaces are adiabatic. Find the rate of heat transfer. [$Q = 36.8$ W.]

5.22. Obtain the shape factor for any or all of the situations pictured. In each case present a well-drawn flux plot. [$S_b \simeq 1.03$, $S_c \gg S_d$, $S_g = 1$]

5.23. Two copper slabs 3 cm thick and insulated on the outside are suddenly slapped tightly together. The one on the left is initially at 100°C and the one on the right at 0°C. Determine the left-hand adiabatic boundary's temperature after 2.3 s have elapsed. [$T_{wall} \simeq 80.5°C$]

5.24. Estimate the time required to hard-boil an egg if:

- The minor diameter is 3.8 cm.

- k for the egg is about the same as for water.

- $\bar{h}$ between the egg and the water is 140 W/m²-°C.

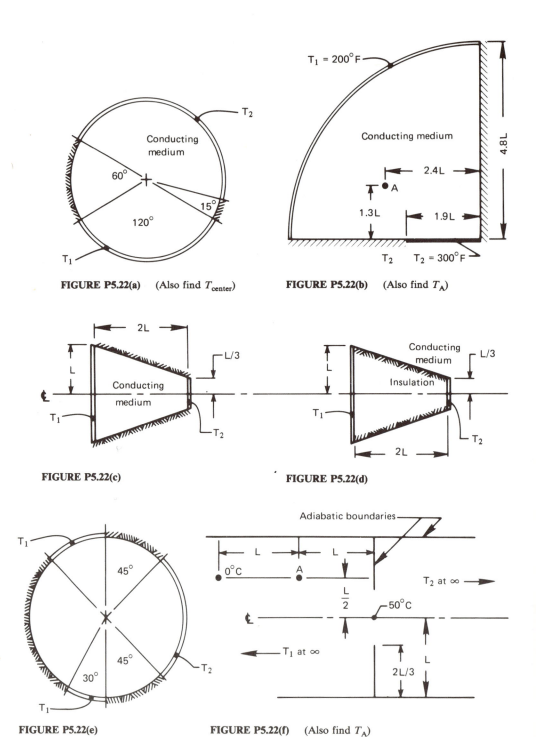

FIGURE P5.22(a)   (Also find T_{center})

FIGURE P5.22(b)   (Also find T_A)

FIGURE P5.22(c)

FIGURE P5.22(d)

FIGURE P5.22(e)

FIGURE P5.22(f)   (Also find T_A)

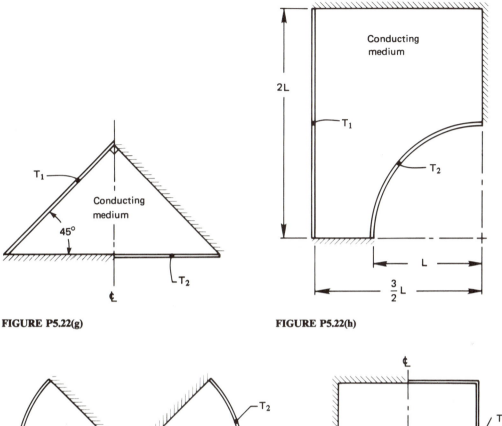

FIGURE P5.22(g)   **FIGURE P5.22(h)**

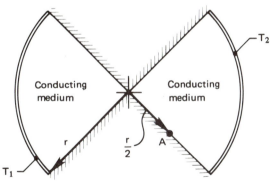   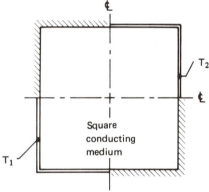

FIGURE P5.22(i)   (Also find T_A)   **FIGURE P5.22(j)**

- The egg is put in boiling water when its temperature is 4°C.
- The interior of the egg coagulates at 96°C with no significant release of heat or change of properties.

5.25. Prove that T_1 in Fig. 5.5 cannot oscillate.

5.26. Show that when isothermal and adiabatic lines are interchanged in a two-dimensional body, the new shape factor is the inverse of the original one.

5.27.   A 0.5 = cm-diameter cylinder at 300°C is suddenly immersed in saturated water at 1 atm. If $\bar{h} = 10,000$ W/m²-°C, find the centerline and surface temperatures after 0.2 s:

   (a)   If the cylinder is copper.

   (b)   If the cylinder is Nichrome V. [$T_{\text{sfc}} \simeq 200$°C.]

   (c)   If the cylinder is Nichrome V, obtain the most accurate value of the temperatures after 0.04 s that you can.

5.28.   A large flat electrical resistance strip heater is fastened to a firebrick wall, uniformly at 15°C. When it is suddenly turned on, it releases heat at the uniform rate of 4000 W/m². Plot the temperature of the brick immediately under the heater as a function of time if the other side of the heater is insulated. What is the heat flux at a depth of 1 cm when the surface reaches 200°C?

5.29.   Do Experiment 5.2 and submit a report on the results.

5.30.   An approximately spherical container, 2 m in diam., containing electronic equipment is placed in wet mineral soil with its center 2 m below the surface. The soil surface is kept at 0°C. What is the maximum rate at which energy can be released by the equipment if the surface of the sphere is not to exceed 30°C.

5.31.   A semi-infinite slab of ice at -10°C is exposed to air at 15°C through a heat transfer coefficient of 10 W/m²-°C. What is the initial rate of melting in kg of water/m²-s? What is the asymptotic rate of melting? Describe the melting process in physical terms. (The latent heat of fusion of ice, $h_{\text{fs}} = 333,300$ W/kg.)

5.32.   One side of a firebrick wall, 10 cm thick, initially at 20°C is exposed to a 1000°C flame through a heat transfer coefficient of 230 W/m²-°C. How long will it be before the other side is too hot to touch? (Estimate properties at 500°C.)

5.33.   A particular lead bullet travels for 0.5 sec within a shock wave that heats the air near the bullet to 300°C. Approximate the bullet as a cylinder 0.8 cm in diameter. What is its surface temperature at impact if $\bar{h} = 600$ W/m²=C, and if the bullet was initially at 200°C. What is its center temperature?

5.34.   A loaf of bread is removed from the oven at 125°C and set on the (insulating) counter to cool. The loaf is 30 cm long, 15 cm high, and 12 cm wide. If $k = 0.05$ W/m-°C and $\alpha = 5 \times 10^{-7}$ m²/s for bread, and $\bar{h} = 10$ W/m²-°C, when will the hottest part of the loaf have cooled to 60°C? [About 1 hr 10 min.]

5.35.   A lead cube, 50 cm on each side, is initially at 20°C. The surroundings are suddenly raised to 200°C and $\bar{h}$ around the cube is 272 W/m²-°C. Plot the cube temperature along a line from the center to the middle of one face, after 20 minutes have elapsed.

REFERENCES

[5.1]   M. P. HEISLER, "Temperature Charts for Induction and Constant Temperature Heating," *Trans. ASME*, vol. 69, 1947, pp. 227–236.

[5.2]   P. J. SCHNEIDER, *Temperature Response Charts*, John Wiley & Sons, Inc., New York, 1963.

[5.3]   F. A. JEGLIĆ, "An Analytical Determination of Temperature Oscillations in Wall Heated by Alternating Current," NASA TN D-1286, July 1962.

[5.4]   K. A. SWITZER and J. H. LIENHARD, "Surface Temperature Variations on Electrical Resistance Elements Supplied with Alternating Current," Wash. State Univ., Inst. of Tech. Bull. 280, 1964.

[5.5]   J. BRONOWSKI, *The Ascent of Man*, Little, Brown and Company, Boston, 1973, Chapter 4.

[5.6]   N. ZUBER, "Hydrodynamic Aspects of Boiling Heat Transfer," AEC Report AECU-4439, Physics and Mathematics, June 1959.

[5.7]   M. S. PLESSET and S. A. ZWICK, "The Growth of Vapor Bubbles in Superheated Liquids," *J. Appl. Phys.*, vol. 25, 1957, pp. 493–500.

[5.8]   L. E. SCRIVEN, "On the Dynamics of Phase Growth," *Chem. Eng. Sci.*, vol. 10, 1959, pp. 1–13.

[5.9]   P. DERGARABEDIAN, "The Rate of Growth of Bubbles in Superheated Water," *J. Appl. Mech., Trans. ASME*, vol. 75, 1953, p. 537.

[5.10]   E. HAHNE and U. GRIGULL, "Formfactor und Formwiderstand der stationären mehrdimensionalen Wärmeleitung," *Int. J. Heat Mass Transfer*, vol. 18, 1975, pp. 751–767.

[5.11]   vR. RÜDENBERG, "Die Ausbreitung der Luft—und Erdfelder um Hochspannungsleitungen besonders bei Erd—und Kurzschlüssen," *Electrotech. Z.*, vol. 36, 1925, pp. 1342–1346.

Chapter 6

Numerical analysis
of heat transfer problems

> *So far I have paid piece rates for the operation
> of about n/18 pence per coordinate point, n
> being the number of digits. The chief trouble to
> the computers has been the intermixture of plus
> and minus signs. ... one of the quickest boys
> averaged 2,000 operations per week for
> numbers of three digits, those done wrong
> being discounted.*
>
> **L. F. Richardson, 1910 [6.1]**

6.1 Introduction

The cost of computing has dropped drastically since 1910. A large-scale contemporary electronic digital computer will typically perform 2 million or more operations per second. At \$750 per hour of CPU time, one can thus perform 1 week's worth of Richardson's hand calculations for a cost of $\frac{1}{15}$ cent, today. The quest for methods to solve problems approximately (now called numerical methods) began much earlier than 1910. Newton recognized the need to resort to calculation, and some of the methods in use today bear his name. Methods are still being developed, but now their focus is directed toward use on the digital computer.

Experiment 6.1

Determine about how long it would require for your relatively slow programmable calculator to do a week's worth of hand calculations.

If it were not for numerical methods and the digital computer, we would simply be unable to solve many of the problems that arise in practice. The techniques of analysis generally become too hard to use—even impossible to use —in many cases. This occurs in problems of heat conduction in complex geometries, problems with temperature-dependent properties, and problems with nonlinear boundary conditions. Even without these complications, large thermal systems with many interacting parts are almost impossible to analyze without numerical methods. We note in Chapter 3, for example, that the selection of an optimum design of a heat exchanger for a given application falls in this category.

Analytical methods do have an important role to play, and we will continue to use them wherever they are feasible. They make it easy to vary the parameters of a problem, as in our treatment of fin lengths in Section 4.5; to explore the limits of a solution, as in the case of large and small Biot numbers in Example 2.5; and to concentrate on the physical nature of a problem and display features often impossible to discern with a few computer simulations. When the problems are complicated, however, they inevitably have to be solved on the computer.

In this chapter we can describe only a few of the many numerical methods suitable for treating heat transfer problems. And we cannot show the full range of computer applications in the heat transfer field. We try to convey the basic ideas and to illustrate them with applications to a few problems. The principal, although by no means the only, use of numerical methods in heat transfer is to solve differential equations. We begin this discussion with a problem that involves a one-dimensional differential equation.

6.2 Illustrative problem in one dimension

Lumped-capacity cooling by combined convection and radiation

Consider the problem described in Fig. 6.1a. A body is subject to heat transfer by both convection and radiation. The ambient gas temperature, T_∞, may be greater than or less than the temperature, T_s, of the distant surroundings with which radiation exchange occurs. The heat transfer coefficient, $\bar{h}$, may be a function of the temperature difference between the body and the ambient gas, $T - T_\infty$. We assume that a lumped-capacity analysis will be adequate[1] so that

---

[1]Two Biot numbers can be defined: $\bar{h}L/k$ and $\bar{h}_r L/k$, where

$L=$ volume/area and h_r is a radiation heat transfer coefficient $\equiv$

$$\frac{q}{\Delta T} = \frac{\mathscr{F}\sigma(T^4 - T_s^4)}{T - T_s} = \mathscr{F}\sigma(T^2 + T_s^2)(T + T_s)$$

Both Biot numbers must be $\ll 1$.

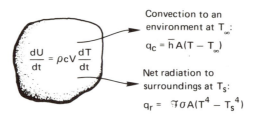

Convection to an
environment at T_∞:

$$q_c = \bar{h}A(T - T_\infty)$$

Net radiation to
surroundings at T_s:

$$q_r = \mathscr{F}\sigma A(T^4 - T_s^4)$$

a) Energy balance terms

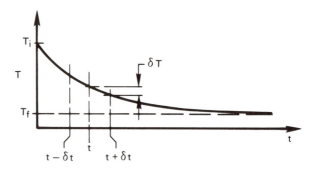

b) Temperature variation with time of the object in (a).

FIGURE 6.1   Cooling of an object by combined convection and radiation.

the equation to be solved is

$$\frac{dT}{dt} = -\frac{\bar{h}A}{\rho c V}(T - T_\infty) - \frac{\mathscr{F}\sigma A}{\rho c V}(T^4 - T_s^4) \tag{6.1}$$

where the i.c. is $T(0) = T_i$. The solution to eqn. (6.1), when we have found it, will look something like the curve in Fig. 6.1b. The final temperature, T_f, at which the temperature of the body stops changing, will be given by the solution of eqn. (6.1) with the derivative set equal to zero:

$$\bar{h}(T_f - T_\infty) + \mathscr{F}\sigma(T_f^4 - T_s^4) = 0 \tag{6.2}$$

Equation (6.1) is nonlinear and has no known analytical solution. Numerical methods must be used. Equation (6.2) is a quartic equation which can be solved by cumbersome analytical methods. Instead, an iterative method is usually used.[2]

---

[2]For this problem, the calculation can easily be done by solving eqn. (6.2) for T_f: $T_f = T_\infty - (\mathscr{F}\sigma/\bar{h})(T_f^4 - T_s^4)$. Insert an estimate for T_f in the right-hand side (T_s or T_∞ will do). The result will be a revised estimate which can be reinserted on the right. Continue the procedure until satisfactory accuracy is obtained. This procedure is called the method of successive approximations.

Two methods can be used to solve eqn. (6.1). One is numerical integration. The other involves directly integrating the differential equation forward in time. In the first case, we will get the solution in the form $t = \text{fn}(T)$ and in the second case, $T = \text{fn}(t)$. Which method to choose depends in part on the form in which the answer is desired. Either of these processes will be much easier if we first simplify eqn. (6.1).

The convection term in eqn. (6.1) gives rise to the usual time constant $T_1 = \rho c V / \bar{h} A$. The radiation term (acting alone) does not lead to an exponential solution, so a time constant, as such, does not exist. However, during the initial period of cooling (or heating) by radiation alone, T will be close to T_i and we can approximate the radiation term by

$$\frac{\mathscr{F}\sigma A}{\rho c V}\left(T^4 - T_s^4\right) \cong \frac{\mathscr{F}\sigma A}{\rho c V} 4 T_i^3 (T - T_i) + \frac{\mathscr{F}\sigma A}{\rho c V}\left(T_i^4 - T_s^4\right)$$

During the brief initial period, radiation alone will cool the body at an exponential rate with a time constant equal to $T_2 = \rho c V / 4 \mathscr{F} T_i^3 \sigma A$. We will not use this time constant explicitly, but it is important to recognize how a numerical solution should behave at small time (see Problem 6.1). Equation (6.1) then takes the form

$$\frac{dT}{d\tau} = -(T - T_\infty) - \frac{\mathscr{F}\sigma}{\bar{h}}\left(T^4 - T_s^4\right) \qquad (6.1a)$$

where we introduce the dimensionless time,[3] $\tau \equiv t / T_1$.

Numerical integration

We can rewrite eqn. (6.1a) as

$$\tau = \int_0^\tau d\tau = -\int_{T_i}^T \frac{dT}{(T - T_\infty) + a\left(T^4 - T_s^4\right)} \qquad (6.3)$$

where $a \equiv \mathscr{F}\sigma / \bar{h}$. If either of the two terms in the denominator of the integrand can be neglected with respect to the other, an analytical solution can be found (see Section 1.3 and Problems 1.12 and 1.18). If this assumption cannot be made, eqn. (6.3) must be integrated numerically. The result will be in the form $\tau = \text{fn}(T)$, so the roles of the dependent and independent variables are reversed.

Most college-level courses in calculus discuss at least the trapezoidal rule and Simpson's rule of numerical integration. These are available as standard computer routines on most computing machines. More sophisticated integration routines such as Gauss quadrature are sometimes used. An elementary discussion of these methods is given by Hornbeck [6.2]. Ralston and Rabinowitz [6.3] give a comprehensive treatment of the general topic of numerical integration.

---

[3]If $T_1 > T_2$, we should probably take $\tau = t / T_2$ instead.

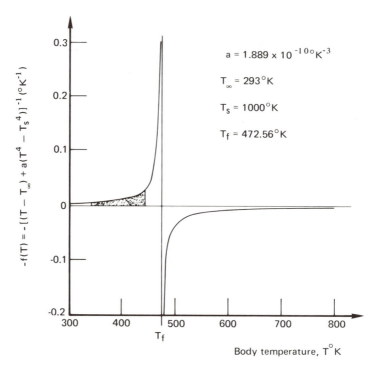

FIGURE 6.2   The integrand of eqn. (6.3). The shaded area is the time required for T to increase from 340° to 440°K.

Figure 6.2 is a graph of the integrand of eqn. (6.3) for the particular case, $T_\infty = 293°$K, $T_s = 1000°$K, and $a = 1.89 \times 10^{-10°}$K^{-3} (see Example 6.1). It shows the final temperature as a singularity at $T_f \cong 472.6°$K. The dimensionless time, τ, from eqn. (6.3), is represented by the area under the curve. For example, the shaded area is equal to the time required for the temperature to increase from 340°K to 440°K.

If we attempt to carry the integration of eqn. (6.3) to $T = T_f$ (from an initial temperature either higher or lower than T_f), the denominator of the integrand will go to zero. We say then that the integrand is singular at $T = T_f$. This fact must be taken into account in any numerical scheme used to evaluate the integrand. The existence of the improper integral is physically significant. It indicates that $T = T_f$ only after the passage of infinite time. The area under a portion of the curve that includes $T = T_f$ is, in fact, infinite.

Figure 6.3 shows a portion of Fig. 6.2 enlarged to illustrate trapezoidal integration. We presume that the numerical integration has been carried out to the point $T - \delta T$, so $\tau(T - \delta T)$ is known. The time required for the temperature to reach T will then be given by

$$\tau = -\int_{T_i}^{T} f(T)\,dT = -\int_{T_i}^{T-\delta T} f(T)\,dT - \int_{T-\delta T}^{T} f(T)\,dT \qquad (6.4)$$

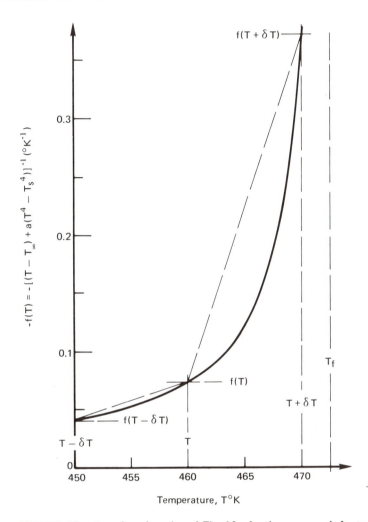

FIGURE 6.3   An enlarged portion of Fig. 6.2a showing area panels for trapezoidal rule integration.

The first term on the right-hand side of this equation is simply $\tau(T - \delta T)$. The last term is the area beneath the curve of the first vertical panel in Fig. 6.3. In trapezoidal integration, we replace the last term by the area of the trapezoid whose corners are $T - \delta T$, T, $f(T)$, and $f(T - \delta T)$. Thus

$$-\int_{T-\delta T}^{T} f(T)\, dT = -\tfrac{1}{2}\big[\, f(T-\delta T) + f(T)\,\big]\delta T \qquad (6.5)$$

If this scheme is used for the entire integral in eqn. (6.5), and δT is taken to be a constant, we have

$$\tau_n = -\frac{\delta T}{2}\left[\, f(T_i) + f(T_n) + 2\sum_{j=1}^{n-1} f(T_j)\,\right] \qquad (6.6)$$

where T_i is the initial temperature, $T_j = T_{j-1} + \delta T$, and τ_n is the total dimensionless time that has elapsed when the temperature has reached T_n.

The type of error associated with the use of eqn. (6.6) is apparent in Fig. 6.3. It can be reduced by reducing the step size, δT, or by using a more sophisticated numerical integration formula [6.2, 6.3]. For the 10°K step size illustrated in Fig. 6.3, a rather bad approximation would be found on the second panel. An attempt to continue the integration (with the same value of δT) beyond $T = 470°K$ would give meaningless results, since τ would begin to decrease with increases in T.

Example 6.1

A black sphere is placed in an enclosure where it is subject to heat transfer by convection with a heat transfer coefficient $\bar{h} = 300$ W/m²-°K. The ambient gas temperature is 293°K and the walls of the enclosure are maintained at 1000°K. What is the temperature of the sphere as a function of time if its initial temperature is (a) 293°K and (b) 1000°K? If the sphere is brass with diameter equal to 2 cm, how long will it take for its temperature to come to within 15°K of T_f in each case?

SOLUTION. The final temperature, T_f, is the same in each case. To find T_f, we have to solve eqn. (6.2), which becomes

$$T_f = 293°K - \frac{5.6697 \times 10^{-8}}{300}(T_f^4 - 10^{12})$$

The result, to five significant figures, is $T_f = 472.56°K$. Equation (6.3) becomes

$$\tau = -\int_{T_i}^{T} \frac{dT}{(T-293) + 1.89 \times 10^{-10}(T^4 - 10^{12})}$$

A block diagram of a computer program to solve this equation by trapezoidal integration is shown in Fig. 6.4.[4] It is designed to be used on a programmable hand calculator. About 50 programming steps are required.

Figure 6.5 shows the dimensionless temperature $(T - T_f)/(T_i - T_f)$ plotted against τ for $T_i = 293°K$ and 1000°K. The step size used was $\Delta T = 1.5625°K$. A curve is also shown for the no-radiation case, eqn. (1.21), $(T - T_f)/(T_i - T_f) = e^{-\tau}$.

The dimensionless temperature for $T = T_f + 15°K = 487.56°K$ is 0.0284 for $T_i = 1000°K$: for $T_i = 293°K$ and $T = T_f - 15°K = 457.56°K$, it is 0.0835. From Fig. 6.5 these values give $\tau = 3.16$ and 2.32, respectively. For the 2-cm-diameter brass sphere, we have

$$T_1 = \frac{\rho c V}{hA}$$

$$= 8522 \frac{kg}{m^3} 385 \frac{W-s}{kg\text{-}°K} \frac{2 \times 10^{-2}\, m}{6} \frac{1}{300} \frac{m^2\text{-}°K}{W}$$

$$= 36.46\ s$$

---

[4]"The best test of whether one understands a (numerical) method—is—to write a computer program, and then to relinquish personal decision making to the impersonal computer. It is remarkable how hazy concepts can become clear under the resulting pressure to be completely precise and unambiguous" (R. W. Hornbeck [6.2]).

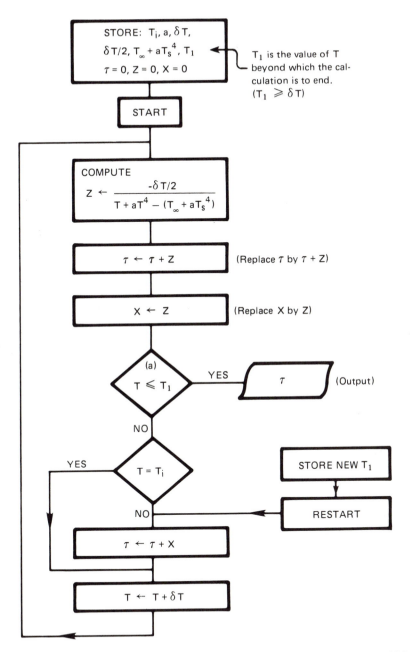

FIGURE 6.4 Computer program for $T_i > T_f$, to solve eqn. (6.2) by trapezoidal integration. Notes: (1) Storage locations must be provided for X, Z, and τ. (2) If $T_i < T_f$, the logical test (a) must be reversed: $T_i \leqslant T$. (3) $\delta T < 0$ for $T_i > T_f$, and $\delta T > 0$ for $T_i < T_f$.

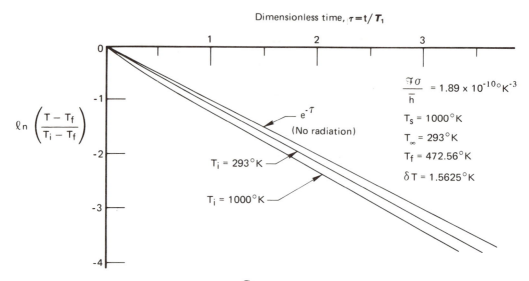

FIGURE 6.5   Solution of $dT/d\tau = -(T - T_\infty) - \dfrac{\mathcal{F}\sigma}{h}(T^4 - T_s^4)$ by numerical integration (Example 6.1).

Thus for $T_i = 1000°K$, 115 s will be required to reach 487.56°K, and for $T_i = 293°K$, 84.6 s will be required to reach 457.56°K.

To work Example 6.1 we had to choose a step size δT. With the trapezoidal integration method, the easiest way to do this is to choose an initial value for δT that will result in 5 to 10 total integration steps. The calculation should be repeated with a step size equal to one-half the previous value. If the result (τ in this case) changes significantly, the calculation is repeated with another reduction in step size. Continued halving of the step size will eventually give accurate results.[5] Integration routines available on larger digital computers often have features that automatically produce a solution to a prescribed degree of accuracy. This will typically be done by halving the step size when an estimate of the error becomes greater than the prescribed value.

In this example, we did not use all the information available to us. From relatively straightforward expansions of eqn. (6.1) (see Problem 6.1), we can show that the limits of the change in temperature with time are given by

$$\lim_{\tau \to 0} \frac{dT}{d\tau} = -(T_i - T_\infty) - \frac{\mathcal{F}\sigma}{h}(T_i^4 - T_s^4)$$

[5]However, the step size cannot be decreased indefinitely. All computers are subject to roundoff error, which could eventually become significant.

and

$$\lim_{\tau \to \infty} \frac{d(T - T_f)}{d\tau} = -\left(1 + 4T_f^3 \frac{\mathscr{F}\sigma}{\bar{h}}\right)(T - T_f)$$

These equations can be used to start the numerical solution and to extrapolate it beyond the point where numerical inaccuracies become significant as $T \to T_f$. Such an approach, which combines the use of numerical methods and analytical methods, should be followed whenever possible. It makes little sense to continue to refine the numerical method when an analytical procedure can be used to extrapolate a numerical result.

Forward numerical integration

Difference representation of the first derivative. Our starting point will be the formulation of a *difference equation* to replace eqn. (6.1a). This equation can be developed either from a mathematical or a physical point of view. Let us apply both approaches.

From a physical point of view, eqn. (6.1) says that the energy of the body in Fig. 6.1 changes as a result of heat transfer with its environment. If the temperature of the body is $T(t)$ at time t, it will lose an amount of heat

$$\left[\bar{h}A(T - T_\infty) + \mathscr{F}\sigma A(T^4 - T_s^4)\right]\delta t$$

over a small period of time, δt. We expect the body to cool by an amount δT at the same time, so the heat that must be removed is given by $\rho c V \delta T$. The result of an energy balance is then

$$T(t + \delta t) = T(t) + \delta T = T(t) - \left[\frac{\bar{h}A}{\rho c V}\left[T(t) - T_\infty\right] - \frac{\mathscr{F}\sigma A}{\rho c V}\left[T(t)^4 - T_s^4\right]\right]\delta t$$

This simple expression can be used to calculate the body's temperature by "marching forward in time." This means that after the temperature at $t + \delta t$ is known, we need only replace t by $t + \delta t$ and repeat the procedure to calculate the temperature at $t + 2\delta t$.

As straightforward as this derivation is, it often requires keen physical insight to develop such a relation for more complicated problems. It also gives scant information on the errors associated with its use and usually permits the derivation of only the simplest and most inefficient numerical methods. A more general result can be developed mathematically in the following way.

The temperature at any time t is $T(t)$. At an earlier time $t - \delta t$ it was $T(t - \delta t)$, while at a later time $t + \delta t$ it will be $T(t + \delta t)$. In general, we denote the size of the time step by $p\,\delta t$, where p is usually a number on the order of unity. (It will be set equal to $+1$ or -1 in certain of the cases we encounter subsequently.) A Taylor's series expansion of $T(t+p\delta t)$ then gives

$$T(t+p\,\delta t)=T(t)+\frac{dT}{dt}\bigg|_t p\,\delta t+\frac{d^2T}{dt^2}\bigg|_t \frac{p^2\,\delta t^2}{2!}+\frac{d^3T}{dt^3}\bigg|_t \frac{p^3\,\delta t^3}{3!}$$

$$+\cdots+\frac{d^nT}{dt^n}\bigg|_t \frac{p^n\,\delta t^n}{n!}+\cdots \qquad (6.7)$$

where the subscript on each derivative implies that it is to be evaluated at time t.
    Equation (6.7) can be solved for the first derivative with the result

$$\frac{dT}{dt}\bigg|_t=\frac{T(t+p\,\delta t)-T(t)}{p\,\delta t}-\frac{d^2T}{dt^2}\bigg|_t \frac{p\,\delta t}{2!}-\frac{d^3T}{dt^3}\bigg|_t \frac{p^2\,\delta t^2}{3!}\cdots \qquad (6.8)$$

Now if we approximate the derivative by retaining only the first term on the right-hand side, eqn. (6.8) reduces to

$$\frac{dT}{dt}\bigg|_t=\frac{T(t+p\,\delta t)-T(t)}{p\,\delta t}+\mathcal{O}(p\,\delta t) \qquad (6.9)$$

where the notation $\mathcal{O}(p\,\delta t)$ means that terms of order of $p\,\delta t$ have been neglected.
    Various common finite-difference formulas which approximate the first derivative can be formed by choosing different values for p. If we take $p=1$, the result is a *forward difference*:

$$\frac{dT}{dt}\bigg|_t=\frac{T(t+\delta t)-T(t)}{\delta t}+\mathcal{O}(\delta t) \qquad (6.10a)$$

It is called a forward difference because it is formed by advancing an increment, δt, forward in time. Likewise, with $p=-1$, eqn. (6.9) gives what is called a *backward difference*:

$$\frac{dT}{dt}\bigg|_t=\frac{T(t)-T(t-\delta t)}{\delta t}-\mathcal{O}(\delta t) \qquad (6.10b)$$

If we take the mean of eqns. (6.10a) and (6.10b), the result is a *central difference*:

$$\frac{dT}{dt}\bigg|_t=\frac{T(t+\delta t)-T(t-\delta t)}{2\delta t}+\mathcal{O}(\delta t^2) \qquad (6.10c)$$

    Notice that the error term in eqn. (6.10c) is *second order*[6] in δt, implying that it is proportional to δt^2. Equation (6.10c) is therefore more accurate than either eqn. (6.10a) or (6.10b). Multipoint formulas, which use information at three or more points, can be derived in a similar way, as is illustrated in Example 6.2 below. Standard handbooks give lists of more elaborate difference formulas (see, e.g., [6.4]).

----

[6]The second term on the right-hand side of eqn. (6.8) carries the sign of $-p$. With $p=\pm 1$, we have $\mp d^2T/dt^2|_t\delta t$, so in forming eqn. (6.10c), the terms multiplied by $d^2T/dt^2|_t$ cancel.

Example 6.2

Derive a second-order three-point forward-difference approximation for the derivative $dT/dt|_t$.

SOLUTION.   Set $p=2$ in eqn. (6.8) to get

$$\frac{dT}{dt}\bigg|_t = \frac{T(t+2\delta t)-T(t)}{2\delta t} - \frac{2\delta t}{2!}\frac{d^2T}{dt^2}\bigg|_t + \mathcal{O}(\delta t^2)$$

Similarly, with $p=1$, eqn. (6.8) becomes

$$\frac{dT}{dt}\bigg|_t = \frac{T(t+\delta t)-T(t)}{\delta t} - \frac{\delta t}{2!}\frac{d^2T}{dt^2}\bigg|_t + \mathcal{O}(\delta t^2)$$

If we multiply the second equation by 2 and subtract it from the first, the result is

$$\frac{dT}{dt}\bigg|_t = \frac{-T(t+2\delta t)+4T(t+\delta t)-3T(t)}{2\delta t} + \mathcal{O}(\delta t^2) \qquad (6.10d)$$

Equation (6.10d) is of second order since the linear term in δt was eliminated by the subtraction.

A three-point backward difference of second order can be developed in the same way as was done in Example 6.2 for the forward difference. The result is the same as we would get by replacing δt by $-\delta t$ in eqn. (6.10d) in Example 6.2:

$$\frac{dT}{dt}\bigg|_t = \frac{3T(t)-4T(t-\delta t)+T(t-2\delta t)}{2\delta t} + \mathcal{O}(\delta t^2) \qquad (6.10e)$$

Example 6.3

In Chapter 9 we find that a physical property of crucial importance in the study of natural convection is the isothermal compressibility, β, for a fluid. Estimate β for water at 300°C.

SOLUTION.

$$\beta \equiv \frac{1}{v}\frac{\partial v}{\partial T}\bigg|_p = -\frac{1}{\rho}\frac{\partial \rho}{\partial T}\bigg|_p$$

where v is the specific volume of the fluid and ρ is its density. Since ρ for a cool liquid is insensitive to pressure, we can write

$$\beta \cong -\frac{1}{\rho}\frac{d\rho}{dT} \qquad \text{where } \rho = \rho(T = T_{\text{sat}})$$

Equation (6.10a) gives the forward-difference estimate of β, with the help of density data from Table A.3, as

$$\beta = -\frac{1}{\rho_{300°C}}\frac{\rho_{320°C}-\rho_{300°C}}{20°C} = -\frac{989.3-996.6}{996.6(20)} = \underline{0.000366°C^{-1}}$$

Equation (6.10b) gives the backward difference as

$$\beta = -\frac{1}{\rho_{300°C}}\frac{\rho_{300°C}-\rho_{280°C}}{20°C} = -\frac{996.6-999.9}{996.6(20)} = \underline{0.000166°C^{-1}}$$

And eqn. (6.10c) gives the central difference as

$$\beta = -\frac{1}{\rho_{300°C}}\frac{\rho_{320°C}-\rho_{280°C}}{40} = -\frac{989.3-999.9}{996.6(40)} = \underline{0.000266°C}$$

In this case, step sizes are rather large, so the forward-difference estimate is 33% above the tabled value of $0.000275°C^{-1}$ and the backward difference is 40% low. The central difference, even with such a crude estimate, is accurate within 3%.

Example 6.4

Evaluate the truncation error incurred by approximating the derivatives of the formulas $T = t^{-(1/4)}$ and $T = t^{(1/4)}$ at $t = 1$, using first-order and second-order formulas.

SOLUTION.   The exact values of the derivatives of $T = t^{\pm(1/4)}$ are

$$\left.\frac{dT}{dt}\right|_{t=1} = \pm\tfrac{1}{4}(t^{-(3/4) \text{ or } -(5/4)})_{t=1} = +\tfrac{1}{4} \quad \text{and} \quad -\tfrac{1}{4}$$

We shall use the first-order forward- and backward-difference formulas, eqns. (6.10a) and (6.10b), and the second-order central-difference formula, eqn. (6.10c), to approximate this known result.

The relative truncation error is given by the expression

$$E = \left(\left.\frac{dT}{dt}\right|_{approx} - \left.\frac{dT}{dt}\right|_{exact}\right)\Big/ \left.\frac{dT}{dt}\right|_{exact}$$

so from eqn. (6.10a), and the Taylor's series expansion, eqn. (6.7), with $p = 1$, for $T = t^{1/4}$, we have

$$E = -\tfrac{3}{8}\delta t + \tfrac{7}{32}\delta t^2 - \cdots$$

Similar formulas can be derived for the other equations. The result is shown in Fig. 6.6, which gives the absolute value of the truncation error for the forward-difference and backward-difference first-order formulas, and the central-difference second-order formula. From Fig. 6.6 we can draw the following conclusions:

- The truncation errors of all the formulas approach zero as δt approaches zero if roundoff errors can be avoided.

- The central-difference formula yields a truncation error which approaches zero much more rapidly than either the forward- or backward-first-order difference formulas. A truncation error of 10^{-5} can be achieved with the central-difference formula with a step size approximately 300 times greater than that required for one of the first-order formulas.

- The forward-difference formula is slightly more accurate than the backward-difference formula, but the improvement is not important for relative errors less than 0.01. This is not a general result but depends on the signs of the higher derivatives of T.

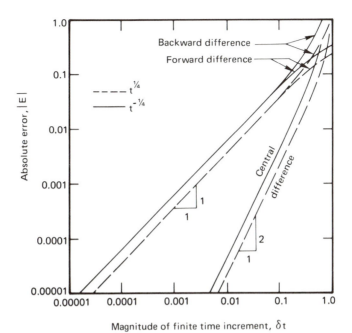

FIGURE 6.6 Error in first- and second-order approximations to the derivative of $T = t^{\pm \frac{1}{4}}$ at $t = 1$ (Example 6.4).

- At a given step size δt, the difference schemes are more accurate when operating on the formula $t^{(1/4)}$ than they are on $t^{-(1/4)}$. This is true because of the relative magnitude of the higher derivatives of the functions. Both the second and third derivatives of $t^{-(1/4)}$ are larger in absolute value than the corresponding derivatives of $t^{(1/4)}$.

The finite-difference equation: Explicit and implicit forms. Let us return to the illustrative one-dimensional problem of Section 6.2. From eqns. (6.1a) and (6.9) we have (in terms of the dimensionless time, $\tau \equiv t/T$)

$$\frac{T(\tau + p\,\delta\tau) - T(\tau)}{p\,\delta\tau} = -\left[T(\tau) - T_\infty \right] - \frac{\mathcal{F}\sigma}{\bar{h}}\left[T(\tau)^4 - T_s^4 \right] \qquad (6.11)$$

With $p = 1$, we can use eqn. (6.11) to calculate the temperature at $\tau + \delta\tau$ from the *known* temperature at time τ. This approach yields what is called an *explicit* method: the temperature at $\tau + \delta\tau$ can be found explicitly once the temperature at τ is known. The use of eqn. (6.11) is called *Euler's method*. Explicit methods (or Euler's methods) usually require a small step size to achieve reasonable accuracy. We discuss the problems of using such methods, subsequently.

As long as $p > 0$, eqn. (6.11) allows us to calculate the temperature at a later time $\tau + p\,\delta\tau$ from information at time τ. When $p < 0$, eqn. (6.11) gives a formula for the temperature at time τ, $T(\tau)$, in terms of the temperature at an earlier time $T(\tau + p\,\delta\tau)$. In this case we refer to the calculation as being *implicit*. It is equivalent to evaluating the function on the right-hand side at the time at which the unknown temperature is desired, instead of at an earlier time as in the explicit method.

Equation (6.11) can be written in an equivalent and somewhat more convenient form by representing the derivative as a forward difference and evaluating the right-hand side at time $\tau + p\,\delta\tau$. The result is

$$\frac{T(\tau + \delta\tau) - T(\tau)}{\delta\tau} = -\left\{\left[\,T(\tau + p\,\delta\tau) - T_\infty\,\right] - \frac{\mathscr{F}\sigma}{\bar{h}}\left[\,T(\tau + p\,\delta\tau)^4 - T_s^4\,\right]\right\}$$

(6.12)

With $p = 0$, this equation can be used as the basis for the explicit method referred to above. It is first order in time. With $p = 1$, eqn. (6.12) is the basis for an implicit first-order method.

An implicit scheme is often chosen because it yields a computational method that is stable. We will learn of the implications of this consideration in Example 6.5. Both the explicit and the implicit first-order methods based on eqn. (6.12) are self-starting: that is, knowledge of the temperature at $\tau = 0$ is sufficient to calculate the temperature at $\tau = \delta\tau$.

Example 6.5

Write a computer program for a programmable calculator to solve eqn. (6.12). Use it to solve the problem stated in Example 6.1.

SOLUTION. We will use $p = 0$ in eqn. (6.12) since it yields an explicit method that requires the smallest number of programming steps. Equation (6.12), with the numerical values from Example 6.1, becomes

$$T(\tau + \delta\tau) = T(\tau) - \delta\tau\left\{\left[\,T(\tau) - 293\right] + 1.89 \times 10^{-10}\left[\,T(\tau)^4 - 10^{12}\right]\right\}$$ (6.13)

A flowchart of a program to solve this equation is shown in Fig. 6.7. The program is designed to calculate and display the temperature at a dimensionless time denoted by τ_{DISPLAY}. When the calculator is restarted, the solution will be advanced to the next value of τ_{DISPLAY}.

The results of calculations made with a program based on the flowchart in Fig. 6.7 are given in Table 6.1 and in Fig. 6.8. The column in the table labeled "accurate solution" was obtained on a large digital computer with a standard "Runge–Kutta" integration routine. It is accurate to the number of significant figures shown in the table.

The accurate solution in Fig. 6.8 cannot be distinguished from the solution of eqn. (6.12) obtained with $\delta\tau = 0.02$. The solution with $\delta\tau = 0.2$ shows the correct trend, but the numerical values are inaccurate.

The solution for $\delta\tau = 2.0$ evidently oscillates and diverges. To see how this comes about, we write eqn. (6.13) for two values of the time and form the ratio of the temperature difference at the successive time steps. For convenience of

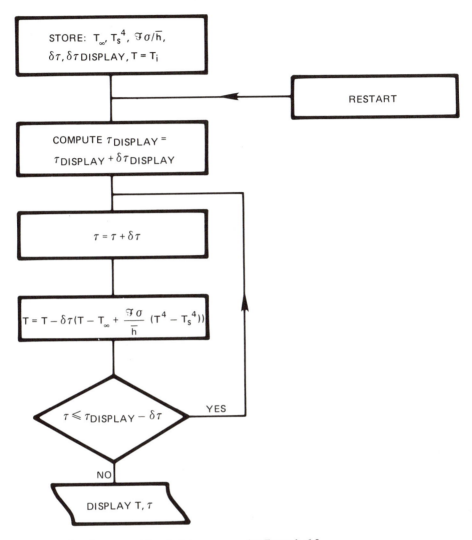

FIGURE 6.7 Programmable calculator program for Example 6.5.

notation, we denote the temperature at time $\tau = k\,\delta\tau$ by T^k and find

$$\frac{T^{k+2} - T^{k+1}}{T^{k+1} - T^k} = 1 - \delta\tau\left\{1 + 1.89 \times 10^{-12}\left[\left[(T^{k+1})^2 + (T^k)^2\right](T^{k+1} + T^k)\right]\right\}$$

(6.14)

According to Fig. 6.8, the correct solution is monotonic in time. This implies that the right-hand side of eqn. (6.14) should be less than unity. Clearly, it always will be so, but for large enough values of $\delta\tau$, the right-hand side can

Table 6.1 Finite-difference solutions for several dimensionless-time step sizes (Example 6.5)

$\delta\tau$	τ					
	0	1.0	2.0	3.0	4.0	∞
2	293.0	[a]	668.21	[a]	220.42	—
0.2	293.0	417.66	456.13	467.68	471.12	472.57
0.02	293.0	410.81	451.65	465.53	470.21	472.57
Accurate solution	293.0	410.12	451.16	465.27	470.09	472.57
2	1000.0	[a]	−414.0	[a]	1366.90	—
0.2	1000.0	604.09	510.34	483.67	475.85	472.57
0.02	1000.0	628.95	523.60	489.56	478.26	472.57
Accurate solution	1000.0	631.43	525.04	490.24	478.56	472.57

[a]When $\delta\tau = 2$, τ can only advance in even increments of 2.

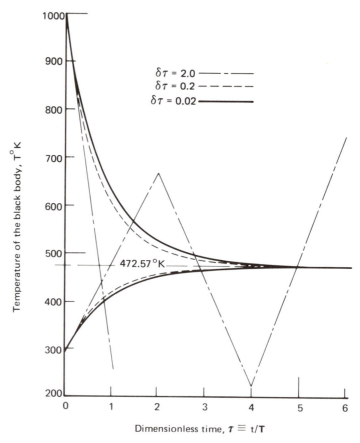

FIGURE 6.8 Temperature of a black body subject to radiation and convection heat transfer (Example 6.5).

become zero or even negative. This explains the oscillation of the solution of eqn. (6.13) for $\delta\tau = 2$.

If the right-hand side of eqn. (6.14) is zero at any time step, the temperature at all succeeding time steps will be the same: $T^{k+2} = T^{k+1}$ and the right-hand side will continue to be zero. The temperature reached in this way will be the final equilibrium temperature. If the right-hand side is < -1, the solution will oscillate with ever-increasing swings. If it is -1, the solution simply oscillates.

Euler's method is simple to program, but it is almost never used in practice because to achieve accuracy and to avoid oscillatory or divergent solutions, small time steps are required. Thus in Example 6.5 a step size smaller than $\delta\tau = 0.001$ is needed to achieve the same accuracy as the "accurate solution" in Table 6.1. This may be feasible on a large digital computer, but a programmable calculator would require several hours to produce the solution.

Fortunately, far more accurate schemes exist and are readily available as standard subroutines on most digital computers. The "accurate solution" in Example 6.5 was obtained using a code based on a Runge–Kutta method. It required a step size of $\delta\tau = 0.05$ and produced the solutions (on an IBM 370) in less than 1 s of computing time.

Most of the available integration routines can be called from the "users' program." It is only necessary to write a few program statements to initialize a problem, to specify the form of the output desired, and to give the formulas that the subroutine needs to calculate the derivatives at each time step. Some subroutines calculate the error at each time step and reduce or increase the size of the time step as needed. Most subroutines are designed to integrate a number (50 or more) of simultaneous first-order differential equations; so very complex problems can be solved conveniently and economically.

So far we have discussed only a simple one-dimensional problem. We turn next to multidimensional problems.

6.3 Numerical methods in more than one dimension

Unsteady problem in one spatial dimension

A familiar problem that can be used to explain numerical methods in more than one dimension[7] is the fin problem introduced in Section 4.5. Equation (4.30) describes the steady temperature of a fin of uniform cross-sectional area. If the fin temperature is allowed to vary with time as well as with axial position, eqn.

---

[7]A "multidimensional problem" is one in which there is more than one independent variable. Thus a problem in which T varies with *time* and one or more spatial variables is multidimensional.

(4.30) becomes

$$\frac{1}{\alpha}\frac{\partial T}{\partial t} = \frac{\partial^2 T}{\partial x^2} - \frac{\bar{h}P}{kA}(T - T_\infty) \qquad (6.15)$$

If the temperature is steady, the left-hand side of eqn. (6.15) is zero and we recover eqn. (4.30). If $\bar{h}$ is zero, eqn. (6.15) reduces to eqn. (5.24), the conventional diffusion equation.

Our plan of attack in this section is to replace eqn. (6.15) with a finite-difference analog whose solution will approximate the true solution to eqn. (6.15). We do this with the help of a particular physical problem: Let us suppose that a fin of length L is initially uniform in temperature at its base temperature, T_0, and that at time $t = 0$, the temperature of the fluid stream is changed from T_0 to T_∞. We assume that the base temperature remains at T_0 but that a variety of end conditions are applied at $x = L$. If the end of the fin is insulated, we expect the temperature distribution along the fin, at various times, t, to resemble the profiles sketched in Fig. 6.9.

Suppose that the temperature distribution at some time $t > 0$ has been found. Then the right-hand side of eqn. (6.15) is known and the time rate of change of the temperature $\partial T/\partial t$ at each position x can be calculated. In this way the solution $T(x, t)$ can be "marched out" in time. A numerical method to solve eqn. (6.15) works just that way. We start from a set of known initial temperatures at positions along the length of the fin and calculate the solution at successive finite intervals of time. This is done by writing the finite-difference

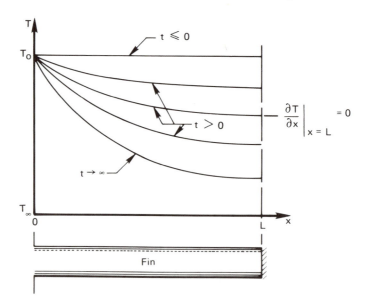

FIGURE 6.9   Fin with an insulated end, cooling from an initially uniform temperature, T_0.

analog for eqn. (6.15), subject to boundary conditions that are also expressed in finite-difference form, and solving it at each time step.

Finite-difference representations

Notation. Equation (6.15) is second order in x and it involves partial derivatives. We must first see how to put it in finite-difference form. A better scheme of notation than we have used above will be helpful. The temperature, T, in eqn. (6.15) depends on two variables, t and x. Let us suppose that x ranges from 0 to L and that the length, L, is subdivided into n intervals of equal length, δx. Then if we denote by x_i the value of x at the junction of two intervals, the entire range of x will be represented by the discrete set of numbers:

$$x_0, x_1, x_2, \ldots, x_{i-1}, x_i, x_{i+1}, \ldots, x_{n-1}, x_n$$

where $x_0 = 0$ and $x_n = L$. Note that $x_{i+1} = x_i + \delta x$.

We can represent t by a similar set of numbers. Thus

$$t_0, t_1, t_2, \ldots, t_{k-1}, t_k, t_{k+1}, \ldots$$

will be our discrete representation of time, where $t_{k+1} = t_k + \delta t$.

The temperature, $T(x, t)$, will be defined only at the discrete points x_i and t_k. We will write

$$T(x_i, t_k) \equiv T_i^k$$

Figure 6.10 shows a grid that represents the x, t domain. Time is shown as the vertical axis and distance as the horizontal axis. The approximate solution, $T(x_i, t_k) = T_i^k$, is known up to time $t = k \, \delta t$ and that information is used to find the solution, $T(x_i, t_{k+1}) = T_i^{k+1}$, at $t = (k+1)\delta t$. Note the contrast between a solution that advances in time just as the physical process does and an analytical solution that is valid for all times. In the numerical method for unsteady problems, *all* of the nodal solutions at earlier times must be calculated before the solution at a specific time and location can be obtained.

Time derivative. We have already considered the finite-difference representation of the derivative dT/dt. The same formulas hold for the partial derivative $\partial T/\partial t$. For example, eqn. (6.10a), a forward-difference representation of the derivative at $x = x_i$, gives

$$\frac{\partial T}{\partial t} = \frac{T(x, t + \delta t) - T(x, t)}{\delta t} + \mathcal{O}(\delta t)$$

or

$$\frac{\partial T}{\partial t} = \frac{T_i^{k+1} - T_i^k}{\delta t} + \mathcal{O}(\delta t) \tag{6.16}$$

Similar formulas are obtained for the other time derivatives given earlier. All are intended to represent the change in temperature with time at a fixed position, $x = x_i$, and all can be identified as differences between values of the solution $T(x, t)$ at discrete points on Fig. 6.10.

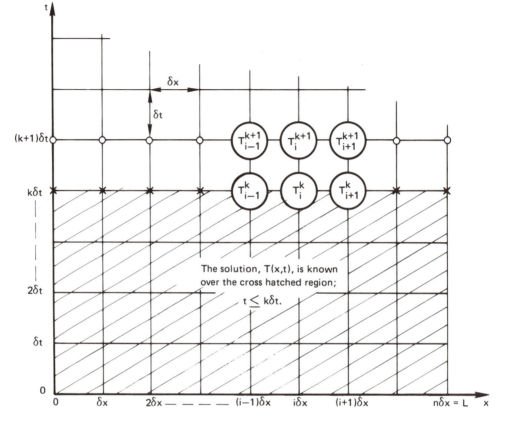

FIGURE 6.10   The x,t solution domain. Knowledge of T at the points marked x allows T to be calculated at the points marked 0.

Spatial derivative.   A finite-difference expression for the second derivative can be found from the expansion of eqn. (6.7). Writing eqn. (6.7) for a change δx instead of δt, and noting that the derivatives should be partial since T depends on both x and t, we get

$$T(x+p\,\delta x,t) = T(x,t) + \frac{\partial T}{\partial x}\bigg|_{x,t} p\,\delta x + \frac{\partial^2 T}{\partial x^2}\bigg|_{x,t} \frac{p^2 \delta x^2}{2!}$$

$$+ \frac{\partial^3 T}{\partial x^3}\bigg|_{x,t} \frac{p^3 \delta x^3}{3!} + \cdots \qquad (6.17)$$

Now, if we write eqn. (6.17) for $p=1$ and $p=-1$ and add these two equations, all odd-order derivatives cancel, and we obtain

$$T(x+\delta x,t)+T(x-\delta x,t)=2T(x,t)+2\left.\frac{\partial^2 T}{\partial x^2}\right|_{x,t}\frac{\delta x^2}{2!}$$

$$+2\left.\frac{\partial^4 T}{\partial x^4}\right|_{x,t}\frac{\delta x^4}{4!}+\dots \tag{6.18}$$

Equation (6.18) can be solved for the second derivative with the result

$$\left.\frac{\partial^2 T}{\partial x^2}\right|_{x,t}=\frac{T(x+\delta x,t)+T(x-\delta x,t)-2T(x,t)}{\delta x^2}-\frac{1}{12}\left.\frac{\partial^4 T}{\partial x^4}\right|_{x,t}\delta x^2\dots$$

or

$$\left.\frac{\partial^2 T}{\partial x^2}\right|_{x,t}=\frac{T_{i+1}^k+T_{i-1}^k-2T_i^k}{\delta x^2}+\mathcal{O}(\delta x^2) \tag{6.19}$$

Equation (6.19) is "second order in δx" and is the most widely used difference form of the second derivative. Because eqn. (6.19) is based on values of T that are equally spaced a distance δx from x_i, the location at which the derivative is approximated, it is often referred to as a *central difference*. The solution values at time t_k, needed to evaluate $\partial^2 T/\partial x^2$, can be identified in Fig. 6.10.

Finite-difference equation.   We now have the elements we need to derive finite-difference analogs to eqn. (6.15). The resulting formulas make it possible to calculate the temperatures T_i^{k+1} from the known temperatures T_i^k. As we found with the unsteady cooling problem in Section 6.2, the calculations may be made explicitly or implicitly. The implicit formulas are harder to solve, but they turn out to be stable and one can use them to make calculations with large time steps, δt.

Equation (6.19) can be used directly in eqn. (6.15) to replace the term $\partial^2 T/\partial x^2$. A formula that is the basis for an *explicit* calculation of the temperature results when we introduce eqns. (6.19) and (6.16) into eqn. (6.15) is

$$\frac{T_i^{k+1}-T_i^k}{\alpha\delta t}=\frac{T_{i+1}^k+T_{i-1}^k-2T_i^k}{\delta x^2}-\frac{\bar{h}P}{kA}\left(T_i^k-T_\infty\right) \tag{6.20}$$

or

$$T_i^{k+1}=\frac{\alpha\,\delta t}{\delta x^2}\left(T_{i+1}^k+T_{i-1}^k\right)+\left(1-\frac{2\alpha\,\delta t}{\delta x^2}-\frac{\bar{h}P}{kA}\alpha\,\delta t\right)T_i^k+\frac{\bar{h}P}{kA}\alpha\,\delta t\,T_\infty \tag{6.21}$$

There are $n-2$ such equations, each of which must be solved to predict the temperature distribution at t_{k+1}. We shall derive two additional equations, one each for $i=0$ and $i=n$, from the boundary conditions for eqn. (6.15).

Since the right-hand side of eqn. (6.20) or (6.21) can be calculated as soon as the temperature distribution is known at time step $k(t=k\,\delta t)$, the temperature distribution at time step $k+1$ [where $t=(k+1)\delta t$] can be found directly.

An alternative *implicit* equation for T_i^{k+1} can be developed easily by evaluating the right-hand side of eqn. (6.15) at time step $k+1$ and using eqn. (6.16) for the left-hand side. The result is

$$\frac{T_i^{k+1}-T_i^k}{\alpha\,\delta t}=\frac{T_{i+1}^{k+1}+T_{i-1}^{k+1}-2T_i^{k+1}}{\delta x^2}-\frac{\bar{h}P}{kA}\left(T_i^{k+1}-T_\infty\right) \qquad (6.22)$$

This formula is much harder to solve than eqn. (6.20). We consider one solution method later in the section "Steady-State Temperature Distribution in a Fin."

Equation (6.21) is a representation of Euler's method. We encountered Euler's method earlier in eqn. (6.11). Equation (6.22) allows an implicit determination of the temperature distribution. The difference between the two methods is illustrated in Fig. 6.11, which shows a portion of the solution domain of Fig. 6.10. The connected circles are called "calculation molecules." From Fig. 6.11a we see that the temperature T_i^{k+1} at the isolated point i can be calculated from eqn. (6.20) once the solution at t_k is known. Figure 6.11b shows how, with the implicit method, eqn. (6.22) gives a value of T_i^{k+1} which also depends on the solution at the neighboring points at time step t_{k+1}. All of the eqns. (6.22) for different values of i are interrelated—they must be solved simultaneously.

The explicit method uses information at the previous time step, t_k, to calculate the second derivative $\partial^2 T/\partial x^2$. The implicit method uses information at the current time step, t_{k+1}, for the same purpose. There are also implicit calculation methods that use information at both time steps. One such, the Crank–Nicolson method [6.5], uses the average value of the right-hand side of eqn. (6.15) evaluated at the previous and the current time step. The Crank–Nicolson formula is

$$\frac{T_i^{k+1}-T_i^k}{\alpha\,\delta t}=\frac{1}{2}\left[\frac{T_{i+1}^k+T_{i-1}^k-2T_i^k}{\delta x^2}-\frac{\bar{h}P}{kA}\left(T_i^k-T_\infty\right)\right.$$

$$\left.+\frac{T_{i+1}^{k+1}+T_{i-1}^{k+1}-2T_i^{k+1}}{\delta x^2}-\frac{\bar{h}P}{kA}\left(T_i^{k+1}-T_\infty\right)\right] \qquad (6.23)$$

This relationship is illustrated in the partial solution domain in Fig. 6.11c.

The explicit and implicit methods are of the same order of accuracy. The explicit method, however, can become unstable for large time steps while the implicit method never does. We encountered this effect in the one-dimensional problem of Example 6.5. In multidimensional problems, as we will see below, the maximum size of the time step is linked to the size of the spatial division, δx. Both must be decreased to avoid instability—often to sizes so small that the solution cannot be calculated in a practical length of time and to sizes smaller than the limit required to achieve a reasonable level of accuracy. The Crank–Nicolson method is slightly harder to program than the implicit method,

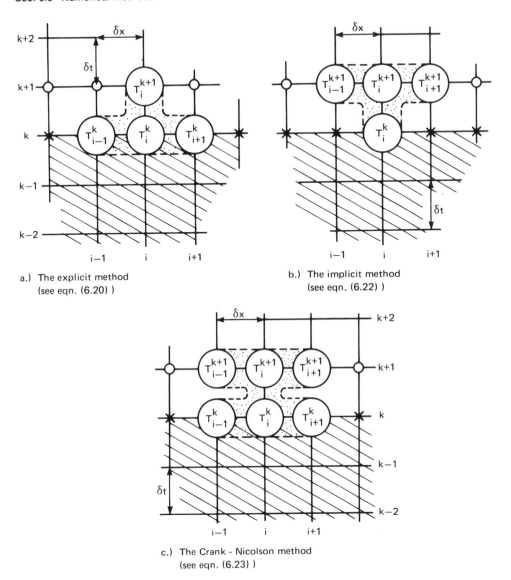

a.) The explicit method
(see eqn. (6.20))

b.) The implicit method
(see eqn. (6.22))

c.) The Crank - Nicolson method
(see eqn. (6.23))

FIGURE 6.11   Illustrations of the ways in which known information at the kth time step is used to obtain information at the $(k+1)$th time step.

but it gives more accurate results than either the explicit or implicit method. It is subject to oscillations, but they never become unstable [6.2, 1.17].

Example 6.6

A 1-cm-diameter 316 stainless steel rod connects two surfaces that are separated by 8 cm and are maintained at different temperatures, 100°C and 0°C, respectively. Initially, the fluid occupying the space between the surfaces is at rest and the heat transfer coefficient between the rod and the fluid is effectively zero. At $t=0$, the fluid is set in motion, the heat transfer coefficient is 1000 W/m²-°K, and the ambient temperature is 20°C. Use eqn. (6.21) to find the temperature distribution as a function of time. Use $\delta x = 2$ cm and study the effect of varying δt.

SOLUTION. Because eqn. (6.15) is linear in temperature, we can simplify the problem by introducing the variable $\theta = T - T_\infty$. Then the term including T_∞ disappears from eqn. (6.12), which becomes

$$\theta_i^{k+1} = \frac{\alpha\,\delta t}{\delta x^2}\left(\theta_{i+1}^k + \theta_{i-1}^k\right) + \left(1 - \frac{2\alpha\,\delta t}{\delta x^2} - \frac{\bar{h}P}{kA}\alpha\,\delta t\right)\theta_i^k$$

The initial temperature distribution is linear in x, so we have

$$\theta(x,0) = 100 - 20 + \frac{(0-100)x}{8} = (80 - 12.5x)°C$$

From Appendix A, we find $k = 13.5$ W/m-°K and $\alpha = 0.37 \times 10^{-5}$ m²/s. Then

$$\theta_i^{k+1} = 9.25 \times 10^{-3}\,\delta t\left(\theta_{i+1}^k + \theta_{i-1}^k\right) + (1 - 0.128\,\delta t)\theta_i^k$$

Since the rod is 8 cm long, three of these equations must be solved, one each for $i=1$ or $x=2$ cm, $i=2$ or $x=4$ cm, and $i=3$ or $x=6$ cm. The values of θ at the ends are fixed at $\theta_0 = 80°C$ and $\theta_4 = -20°C$, respectively.

The equations for θ_i^{k+1} are particularly easy to solve if we choose $\delta t = 1/0.128 = 7.81$ s. Results for this value and for[8] $\delta t = 0$, 3.91 and 15.62 s are shown in Fig. 6.12a, for the central node of the rod, $i=2$ or $x=4$ cm.

The numerical solution for $\delta t = 15.62$ s is oscillatory and it diverges for large values of t. The solution for $\delta t = 7.81$ s reaches the final value of the solution for small δt within a few steps. The curve for $\delta t = 3.91$ s reaches the final value for $t \to \infty$ more rapidly than it should but is a fair representation of that solution.

Figure 6.12b shows the temperature distribution along the rod for $t=0$, 3.91, 7.82, 11.73, and 39.10 s and the exact result for $t=\infty$, the steady-state value. The latter was obtained by the methods developed in Chapter 4. Since mL for this problem is 13.77, a large value, the two ends of the rod behave almost as separate infinite fins. The solution for the temperature at $t=\infty$ can then be written as

$$\theta = \theta_0 e^{-mx} \qquad 0 \leqslant x < 4$$

and

$$\theta = \theta_4 e^{-m(L-x)} \qquad 4 < x \leqslant 8$$

These formulas give the curve labeled $t=\infty$ (exact solution) on Fig. 6.12b.

Clearly, the value of δx chosen for the numerical solution is too large to give a close representation of the physical phenomenon. The curve in Fig. 6.12b labeled $t=39.10$ s is of the same general shape as the exact solution, but it comes far from the mark in representing the temperature distribution near the wall.

---

[8]The solution for $\delta t = 0$ can be found by reducing δt in steps until θ_i becomes independent of the value of δt. But it can also be found analytically as follows: In the limit as $\delta t \to 0$, eqn. (6.20) or (6.22) becomes

$$\frac{d\theta_i}{dt} = -A\theta_i + B(\theta_{i-1} + \theta_{i+1})$$

where $A \equiv 2\alpha/\delta x^2 + \bar{h}P\alpha/\delta x^2$ and $B \equiv \alpha/\delta x^2$. One such equation must be written for each node and they all must be solved simultaneously. (See Problem 6.24.)

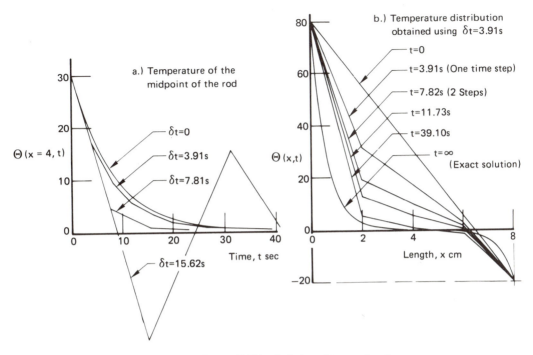

FIGURE 6.12 Results of Euler's method [eqn. (6.21)] calculation of temperature in a rod with fixed end temperatures, cooling with time (Example 6.6).

Physical derivation. We must still consider boundary conditions and solution methods, but first it will be useful to study the physical implication of eqns. (6.20), (6.22), and (6.23). We can easily derive these equations on physical grounds as we did in Section 6.2. Here, too, we will find that the accuracy of the equations is not obtained in the derivation.

Consider the sketch of a fin in Fig. 6.13. An energy balance on the mass within the control volume, centered about x_i and spanning the distance $(x_i - \delta x/2) < x < (x_i + \delta x/2)$, is

$$\dot{q}_s A \, \delta x \, \delta t = (q_L + q_R) A \, \delta t - q_c P \, \delta x \, \delta t$$

where $\dot{q}_s$ = rate of heat stored/unit volume

q_L, q_R = rates of heat transfer by conduction/unit cross-sectional area

q_c = rate of heat transfer by convection/unit surface area

This energy balance is already in finite difference form. It can be transformed into eqn. (6.20) in the following way:

- Heat flow by conduction (q_L and q_R) and surface convection (q_c) over the time period t_k to t_{k+1} will be assumed constant and equal to the values prevailing at time t_k.

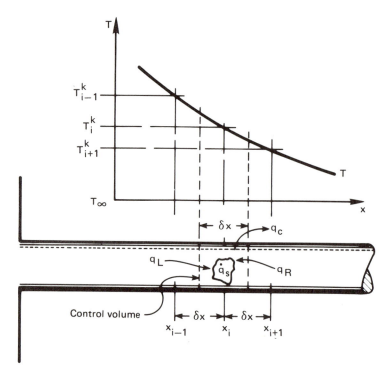

FIGURE 6.13   Energy balance on a finite control element in a fin of uniform area, A, and perimeter, P.

- q_L and q_R will be calculated with Fourier's law by assuming that the temperature varies linearly between the nodes x_{i-1} and x_i, and x_i and x_{i+1}.
- The net quantity of heat thus determined will cause a change in the fin temperature at x_i from T_i^k to T_i^{k+1}.

Then we have

$$\dot{q}_s = \frac{\rho c \left(T_i^{k+1} - T_i^k \right)}{\delta t}$$

$$q_L = \frac{k \left(T_{i-1}^k - T_i^k \right)}{\delta x}$$

$$q_R = \frac{k \left(T_{i+1}^k - T_i^k \right)}{\delta x}$$

$$q_c = \bar{h} \left(T_i^k - T_\infty \right)$$

By substituting these equations in the energy balance, we obtain the finite-difference formula, eqn. (6.20). The implicit formula, eqn. (6.22), can be derived in the same way by evaluating q_L, q_R, and q_c at time t_{k+1}. If q_L, q_R, and q_c are taken as averages of the values prevailing at times t_k and t_{k+1}, the procedure results in the Crank–Nicolson formula, eqn. (6.23).

Boundary conditions.   Figure 6.14 is a sketch of the last few spatial elements of the fin and it shows the energy balance on the last element. This energy balance can be written

$$\dot{q}_s A\left(\frac{\delta x}{2}\right)\delta t = (q_L + q_R)A\,\delta t - q_c P\left(\frac{\delta x}{2}\right)\delta t$$

This equation is the same as the one developed above for a centrally located control volume except that the element length is now $\delta x/2$ instead of δx.

The heat transfer terms in this equation can be approximated as:

$$\dot{q}_s = \frac{\rho c\left(T_n^{k+1} - T_n^k\right)}{\delta t}$$

$$q_L = \frac{k\left(T_{n-1}^k - T_n^k\right)}{\delta x}$$

$$q_R = \begin{cases} \bar{h}_L\left(T_\infty - T_n^k\right) & \text{(convective end condition with heat} \\ & \quad\text{transfer coefficient } \bar{h}_{\text{end}} = \bar{h}_L) \\[2mm] 0 & \text{(insulated end)} \end{cases}$$

and

$$q_c = \bar{h}\left(T_n^k - T_\infty\right)$$

Notice that two possible end conditions are listed and that the formulas are appropriate to an explicit calculation method. For an implicit calculation method we need only change the index k to $k+1$ in the formulas for q_L, q_R, and q_c. We can also handle a prescribed end temperature, T_L, if we replace the expression for q_R with $T_n^k = T_L$

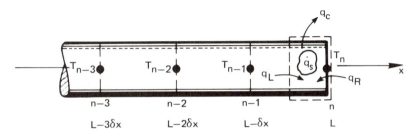

FIGURE 6.14   Boundary conditions on the free end of the fin.

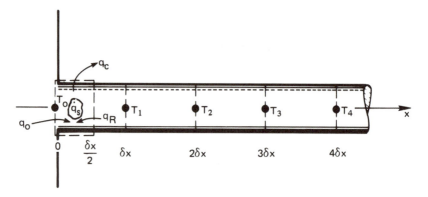

FIGURE 6.15   Boundary conditions on the root end of the fin.

The formulas for the end node that result from the relations above are similar to eqn. (6.20). For a convective end condition,

$$\frac{T_n^{k+1} - T_n^k}{\alpha\,\delta t} = \frac{2\left(T_{n-1}^k - T_n^k\right)}{\delta x^2} + \frac{2\bar{h}_L}{k\,\delta x}\left(T_\infty - T_n^k\right) - \frac{\bar{h}P}{kA}\left(T_n^k - T_\infty\right) \quad (6.24a)$$

For an insulated end,

$$\frac{T_n^{k+1} - T_n^k}{\alpha\,\delta t} = \frac{2\left(T_{n-1}^k - T_n^k\right)}{\delta x^2} - \frac{\bar{h}P}{kA}\left(T_n^k - T_\infty\right) \quad (6.24b)$$

and for a fixed end temperature,

$$T_n^{k+1} = T_n^k = T_L \quad (6.24c)$$

Figure 6.15 shows the root end of the fin. If the temperature, T_0, is prescribed as it was for most of the cases we considered in Chapter 4, we need not write an energy balance at all. The first equation in the set that describes the evolution of the fin temperature will be the one for $i = 1$. If the root heat transfer rate is prescribed,[9] an energy balance can be written in the same way as was done above. The result is an equation which determines the root temperature:

$$\frac{T_0^{k+1} - T_0^k}{\alpha\,\delta t} = \frac{2q_0^k}{k\,\delta x} + 2\frac{T_1^k - T_0^k}{\delta x^2} - \frac{\bar{h}P}{kA}\left(T_0^k - T_\infty\right) \quad (6.25)$$

No mention of root-temperature depression has been made in the preceding discussion. If the fin is attached to a large thermal mass, and if conditions are

---

[9]A constant-root heat flux is an uncommon boundary condition for a fin, although it can occur in some electronic cooling applications. However, the present results are adaptable to a more general class of one-dimensional transient heat conduction problems in which a prescribed wall heat flux might be imposed.

such that Fig. 4.10b suggests significant temperature depression, then the transient fin problem must be coupled with a matching transient problem in the two- or three-dimensional region at the root. Such a procedure would be fairly complicated and we do not treat it here.

Equations (6.24) and (6.25) correctly model the energy balance on the ends of the fin but they lead to an error of the order $\mathcal{O}(\delta x)$, instead of $\mathcal{O}(\delta x^2)$, as in the equations for the internal nodes. It turns out that the global error for the solution of the differential equation is not changed by this fact but remains $\mathcal{O}(\delta x^2)$.

Nondimensionalization

The finite-difference equations can be expressed in nondimensional form just as the analytical equations can. Of course, we are sometimes forced to use numerical methods because the physical properties or $\bar{h}$ depend on T, making the problems nonlinear. In such cases dimensionless groups, such as Bi, Fo, and so on, would all vary, and their use would be most impractical. In those cases one is normally better off simply to solve the problem in the physical coordinates.

When the system parameters can generally be treated as constants, the dimensionless groups for the fin problem represented by eqn. (6.15) are the dimensionless temperature,

$$\Theta \equiv \frac{T - T_\infty}{T_0 - T_\infty}$$

the dimensionless time (a Fourier number),

$$\text{Fo} \quad \text{or} \quad \tau \equiv \frac{\alpha t}{L^2}$$

the dimensionless fin length,

$$mL \equiv \sqrt{\frac{\bar{h}P}{kA}} \; L$$

and the dimensionless axial coordinate,

$$\xi \equiv \frac{x}{L}$$

With these definitions, eqn. (6.15) becomes

$$\frac{\partial \Theta}{\partial \tau} = \frac{\partial^2 \Theta}{\partial \xi^2} - m^2 L^2 \Theta \tag{6.26}$$

The boundary conditions for eqn. (6.26) are

$$\left.\begin{array}{l} \Theta(0, \tau) = 1.0 \\[2mm] \dfrac{\partial \Theta}{\partial \xi}\bigg|_{\xi=1} = -\text{Bi}_{\text{axial}}\Theta(1, \tau) \end{array}\right\} \tag{6.27a}$$

where $\mathrm{Bi}_{\mathrm{axial}} = h_L L / k$ is another dimensionless group. The initial temperature distribution (the i.c.) must also be given:

$$\Theta(\xi, 0) = f_1(\xi) \tag{6.27b}$$

Now we can introduce the discrete variables Θ_i^k, ξ_i, and τ_k. Increments in ξ and τ are $\delta\xi$ and $\delta\tau$, respectively. The difference equations for the central nodes become

$$\frac{\Theta_i^{k+1} - \Theta_i^k}{\delta\tau} = \frac{\left(\Theta_{i+1}^k + \Theta_{i-1}^k - 2\Theta_i^k\right)}{\delta\xi^2} - m^2 L^2 \Theta_i^k \tag{6.28a}$$

as a replacement for eqn. (6.20). There are $n-2$ such equations to be solved for the fin temperature. Two more equations are developed from the boundary conditions at $\xi = 0$ $(i = 0)$ and $\xi = 1$ $(i = n)$. For a fixed root temperature and a convective end condition they are

$$\Theta_0^k = 1.0 \tag{6.28b}$$

and

$$\frac{\Theta_n^{k+1} - \Theta_n^k}{\delta\tau} = \frac{2\left(\Theta_{n-1}^k - \Theta_n^k\right)}{\delta\xi^2} - \left(m^2 L^2 - 2\mathrm{Bi}_{\mathrm{axial}}\right)\Theta_n^k \tag{6.28c}$$

This is the dimensionless form of eqn. (6.24).

Similar dimensionless formulas can be developed for the other types of boundary conditions. We will not consider them here.

Steady-state temperature distribution in a fin

In Example 6.6 we were able to evaluate the steady-state temperature distribution that is approached at large values of the time, t. If we are only interested in the steady-state temperature distribution, it can be found directly without solving an unsteady problem. The finite-difference equations for the steady-state temperature distribution are easily obtained from eqns. (6.20), (6.24), and (6.25) by setting the left-hand sides equal to zero and dropping the superscript, k. Thus

$$T_{i+1} + T_{i-1} - \left(2 + \frac{h_i P}{kA} \delta x^2\right) T_i + \frac{h_i P}{kA} \delta x^2 T_\infty = 0 \qquad 1 \leqslant i \leqslant n - 1 \tag{6.29}$$

at any node, and

$$2T_{n-1} - \left(2 + 2\frac{\bar{h}_L \delta x}{k} + \frac{h_n P}{kA} \delta x^2\right) T_n + \left(\frac{2\bar{h}_L \delta x}{k} + \frac{h_n P}{kA} \delta x^2\right) T_\infty = 0 \tag{6.30}$$

at the free end of the fin. A prescribed root end temperature is $T_0 = $ constant, as before. The possibility of a variable heat transfer coefficient has been included in eqns. (6.29) and (6.30) by denoting the surface heat transfer coefficient by h_i so that it can be prescribed at each node point. Equation (6.29) is not correct for either a variable thermal conductivity or a variable area, however.

Method of solution. Equations (6.29) and (6.30) represent a set of n simultaneous equations for the temperatures, T_i. It is useful to write them in matrix form. For $n=4$, we have

$$
\begin{bmatrix}
1 & 0 & 0 & 0 & 0 \\
1 & -\left(2+\dfrac{h_1 P}{kA}\delta x^2\right) & 1 & 0 & 0 \\
0 & 1 & -\left(2+\dfrac{h_2 P}{kA}\delta x^2\right) & 1 & 0 \\
0 & 0 & 1 & -\left(2+\dfrac{h_3 P}{kA}\delta x^2\right) & 1 \\
0 & 0 & 0 & 2 & -\left(2+\dfrac{2\bar{h}_L \delta x}{k}+\dfrac{h_4 P}{kA}\delta x^2\right)
\end{bmatrix}
\cdot
\begin{bmatrix}
T_0 \\ T_1 \\ T_2 \\ T_3 \\ T_4
\end{bmatrix}
$$

$$
= -
\begin{bmatrix}
T_0 \\
\dfrac{h_1 P}{kA}\delta x^2 T_\infty \\
\dfrac{h_2 P}{kA}\delta x^2 T_\infty \\
\dfrac{h_3 P}{kA}\delta x^2 T_\infty \\
\dfrac{2\bar{h}_L \delta x}{k}+\dfrac{h_4 P}{kA}\delta x^2 T_\infty
\end{bmatrix}
\tag{6.31}
$$

where each row represents the equation that determines the temperature of a single node. The first row is usually not included because it merely states the fact that $T_0 = T_0$. We include it for completeness and to illustrate the fact that the first row must be included for certain types of boundary conditions.

The square matrix in eqn. (6.31) is called tridiagonal, because there are nonzero values only along the main diagonal and the diagonals immediately above and below it. This fact makes the set of equations particularly easy to solve using the procedure called Gauss elimination. To apply it, we convert the square matrix into one with nonzero values only above the main diagonal. Thus the last equation in the set can be solved for T_4, the one preceding it for T_3, and so forth. Let us apply the procedure to eqn. (6.31). The process will be simpler if we replace eqn. (6.31) with

$$
\begin{bmatrix}
A_0 & 0 & 0 & 0 & 0 \\
1 & A_1 & 1 & 0 & 0 \\
0 & 1 & A_2 & 1 & 0 \\
0 & 0 & 1 & A_3 & 1 \\
0 & 0 & 0 & 2 & A_4
\end{bmatrix}
\cdot
\begin{bmatrix}
T_0 \\ T_1 \\ T_2 \\ T_3 \\ T_4
\end{bmatrix}
=
\begin{bmatrix}
B_0 \\ B_1 \\ B_2 \\ B_3 \\ B_4
\end{bmatrix}
\tag{6.32}
$$

where the A_i's and B_i's are the coefficients in the equations

$$A_0 T_0 = B_0$$
$$T_{i+1} + T_{i-1} + A_i T_i = B_i \qquad 1 \leqslant i \leqslant 3 \tag{6.33}$$
$$2T_3 + A_4 T_4 = B_4$$

and are defined in eqn. (6.31).

To change the elements to the left and below the main diagonal to zero, we proceed as follows:

1. Multiply the first row by $1/A_0$ and subtract it from the second row. The result is

$$
\begin{bmatrix}
A_0 & 0 & 0 & 0 & 0 \\
0 & A_1 & 1 & 0 & 0 \\
0 & 1 & A_2 & 1 & 0 \\
0 & 0 & 1 & A_3 & 1 \\
0 & 0 & 0 & 2 & A_4
\end{bmatrix}
\cdot
\begin{bmatrix}
T_0 \\ T_1 \\ T_2 \\ T_3 \\ T_4
\end{bmatrix}
=
\begin{bmatrix}
B_0 \\ B_1 - \dfrac{B_0}{A_0} \\ B_2 \\ B_3 \\ B_4
\end{bmatrix}
$$

2. Multiply the second row by $1/A_1$ and subtract it from the third row. We get

$$
\begin{bmatrix}
A_0 & 0 & 0 & 0 & 0 \\
0 & A_1 & 1 & 0 & 0 \\
0 & 0 & A_2 - \dfrac{1}{A_1} & 1 & 0 \\
0 & 0 & 1 & A_3 & 1 \\
0 & 0 & 0 & 2 & A_4
\end{bmatrix}
\cdot
\begin{bmatrix}
T_0 \\ T_1 \\ T_2 \\ T_3 \\ T_4
\end{bmatrix}
=
\begin{bmatrix}
B_0 \\ B_1 - \dfrac{B_0}{A_0} \\ B_2 - \left(B_1 - \dfrac{B_0}{A_0} \right) \dfrac{1}{A_1} \\ B_3 \\ B_4
\end{bmatrix}
$$

3. Continue the procedure, row by row, until all the terms to the left and below the main diagonal are zero. The result is

$$
\begin{bmatrix}
A_0 & 0 & 0 & 0 & 0 \\
0 & A_1 & 1 & 0 & 0 \\
0 & 0 & A_2 - \dfrac{1}{A_1} & 1 & 0 \\
0 & 0 & 0 & A_3 - \dfrac{1}{A_2 - \dfrac{1}{A_1}} & 1 \\
0 & 0 & 0 & 0 & A_4 - \dfrac{2}{A_3 - \dfrac{1}{A_2 - \dfrac{1}{A_1}}}
\end{bmatrix}
\begin{bmatrix}
T_0 \\ T_1 \\ T_2 \\ T_3 \\ T_4
\end{bmatrix}
$$

$$
=
\begin{bmatrix}
B_0 \\[4pt]
B_1 - \dfrac{B_0}{A_0} \\[8pt]
B_2 - \left(B_1 - \dfrac{B_0}{A_0}\right)\dfrac{1}{A_1} \\[8pt]
B_3 - \left[B_2 - \left(B_1 - \dfrac{B_0}{A_0}\right)\dfrac{1}{A_1}\right]\left[\dfrac{1}{A_2 - \dfrac{1}{A_1}}\right] \\[16pt]
B_4 - \left\{ B_3 - \left[B_2 - \left(B_1 - \dfrac{B_0}{A_0}\right)\dfrac{1}{A_1}\right]\left[\dfrac{1}{A_2 - \dfrac{1}{A_1}}\right]\right\}\left\{\dfrac{2}{A_3 - \dfrac{1}{A_2 - \dfrac{1}{A_1}}}\right\}
\end{bmatrix}
$$

The last equation in the set can be solved for T_4. We get

$$
T_4 = \frac{B_4 - \left\{ B_3 - \left[B_2 - \left(B_1 - \dfrac{B_0}{A_0}\right)\dfrac{1}{A_1}\right]\left[\dfrac{1}{A_2 - \dfrac{1}{A_1}}\right]\right\}\left\{\dfrac{2}{A_3 - \dfrac{1}{A_2 - \dfrac{1}{A_1}}}\right\}}{A_4 - \dfrac{2}{A_3 - \dfrac{1}{A_2 - \dfrac{1}{A_1}}}}
$$

Since T_4 is now known, the next-to-last equation can be solved for T_3. The temperature T_2 is calculated next and then T_1.

This procedure may seem hopelessly cumbersome, but it is really very simple when the calculations are made on a digital computer. Standard subroutines, using the "tridiagonal algorithm," are available on most digital computers. They are somewhat different from the procedure illustrated above, which can break down in certain problems. The user has only to set up the arrays representing the square matrix on the left-hand side and the vector on the right-hand side of eqn. (6.32), and then call the subroutine and print out the solution vector after the subroutine has returned it to the users' program. It is also possible to write a simple program for use on a hand-held calculator, but approximately $5n$ storage locations are required.

Example 6.7

Find the steady (final) temperature distribution for the rod in Example 6.6.

SOLUTION.   Equation (6.29) becomes

$$\theta_0 = 80°C$$

$$\theta_{i+1} + \theta_{i-1} - \left(2 + \frac{\bar{h}P}{kA}\delta x^2\right)\theta_i = 0 \qquad i = 1,2,3$$

$$\theta_4 = -20°C$$

We have $\bar{h}P\delta x^2/kA = 11.85$. Thus eqn. (6.31) becomes

$$
\begin{bmatrix}
1 & 0 & 0 & 0 & 0 \\
1 & -13.850 & 1 & 0 & 0 \\
0 & 1 & -13.850 & 1 & 0 \\
0 & 0 & 1 & -13.850 & 1 \\
0 & 0 & 0 & 0 & 1
\end{bmatrix}
\cdot
\begin{bmatrix}
\theta_0 \\ \theta_1 \\ \theta_2 \\ \theta_3 \\ \theta_4
\end{bmatrix}
=
\begin{bmatrix}
80 \\ 0 \\ 0 \\ 0 \\ -20
\end{bmatrix}
$$

Note that the solutions of the first and last equations in the set are known: $\theta_0 = 80$ and $\theta_4 = -20$. This is because of the type of boundary conditions in the problem. Our procedure yields the following progression of equations:

$$
\begin{bmatrix}
1 & 0 & 0 & 0 & 0 \\
0 & -13.850 & 1 & 0 & 0 \\
0 & 1 & -13.850 & 1 & 0 \\
0 & 0 & 1 & -13.850 & 1 \\
0 & 0 & 0 & 0 & 1
\end{bmatrix}
\cdot
\begin{bmatrix}
\theta_0 \\ \theta_1 \\ \theta_2 \\ \theta_3 \\ \theta_4
\end{bmatrix}
=
\begin{bmatrix}
80 \\ -80 \\ 0 \\ 0 \\ -20
\end{bmatrix}
$$

$$
\begin{bmatrix}
1 & 0 & 0 & 0 & 0 \\
0 & -13.850 & 1 & 0 & 0 \\
0 & 0 & -13.778 & 1 & 0 \\
0 & 0 & 1 & -13.850 & 1 \\
0 & 0 & 0 & 0 & 1
\end{bmatrix}
\cdot
\begin{bmatrix}
\theta_0 \\ \theta_1 \\ \theta_2 \\ \theta_3 \\ \theta_4
\end{bmatrix}
=
\begin{bmatrix}
80 \\ -80 \\ -5.776 \\ 0 \\ -20
\end{bmatrix}
$$

$$\begin{bmatrix} 1 & 0 & 0 & 0 & 0 \\ 0 & -13.850 & 1 & 0 & 0 \\ 0 & 0 & -13.778 & 1 & 0 \\ 0 & 0 & 0 & -13.774 & 1 \\ 0 & 0 & 0 & 0 & 1 \end{bmatrix} \cdot \begin{bmatrix} \theta_0 \\ \theta_1 \\ \theta_2 \\ \theta_3 \\ \theta_4 \end{bmatrix} = \begin{bmatrix} 80 \\ -80 \\ -5.776 \\ -0.4192 \\ -20 \end{bmatrix}$$

The solution "vector" is found by solving the last set of equations:

$$\theta_4 = -20$$
$$-13.774\theta_3 + \theta_4 = -0.4192 \rightarrow \theta_3 = -1.422$$
$$-13.778\theta_2 + \theta_3 = -5.776 \rightarrow \theta_2 = 0.316$$
$$-13.850\theta_1 + \theta_2 = -80 \rightarrow \theta_1 = 5.779$$
$$\theta_0 = 80$$

These numerical values cannot be distinguished, on Fig. 6.12, from those plotted for $t = 39.1$ s.

One-dimensional solid

The equations derived above can be used to solve one-dimensional unsteady heat conduction problems. We need only eliminate heat convection by setting $\bar{h} = 0$ in eqn. (6.15) and wherever it appears in the finite-difference equations. For example, the explicit form of the finite-difference equations for a slab with a prescribed heat flux at $x = 0$, and convective cooling at $x = L$, can be found from eqns. (6.25), (6.22), and (6.24a). We have

$$\frac{T_0^{k+1} - T_0^k}{\alpha \delta t} = \frac{2q_0^k}{k \delta x} + 2\left(\frac{T_1^k - T_0^k}{\delta x^2}\right)$$

$$\frac{T_i^{k+1} - T_i^k}{\alpha \delta t} = \frac{T_{i+1}^k + T_{i-1}^k - 2T_i^k}{\delta x^2} \qquad 1 \leqslant i \leqslant n-1$$

$$\frac{T_n^{k+1} - T_n^k}{\alpha \delta t} = \frac{2(T_{n-1}^k - T_n^k)}{\delta x^2} + \frac{2\bar{h}_L}{k \delta x}(T_\infty - T_n^k)$$

The implicit form of these equations is obtained simply by changing the superscript k to $k+1$ on the right-hand sides. The result is a set of n equations similar to eqns. (6.29) and (6.30):

$$\left.\begin{aligned} -T_0^{k+1}\left(1 + \frac{1}{2}\frac{\delta x^2}{\alpha \delta t}\right) + T_1^{k+1} + \frac{1}{2}\frac{\delta x^2}{\alpha \delta t}T_0^k + \frac{q_0^{k+1}\delta x}{k} &= 0 \\ T_{i-1}^{k+1} - \left(2 + \frac{\delta x^2}{\alpha \delta t}\right)T_i^{k+1} + T_{i+1}^{k+1} + \frac{\delta x^2}{\alpha \delta t}T_i^k &= 0 \\ T_{n-1}^{k+1} - \left(1 + \frac{1}{2}\frac{\delta x^2}{\alpha \delta t} + \frac{\bar{h}_L \delta x}{k}\right)T_n^{k+1} + \frac{1}{2}\frac{\delta x^2}{\alpha \delta t}T_n^k + \frac{\bar{h}_L \delta x}{k}T_\infty &= 0 \end{aligned}\right\} \quad (6.34)$$

Numerical solutions. At any time step, the underlined quantities in eqn. (6.34) are known. We can thus write the equations in matrix form as follows:

$$
\begin{bmatrix}
-\left(1+\dfrac{1}{2}\dfrac{\delta x^2}{\alpha\,\delta t}\right) & 1 & 0 & & \text{-----} \\[2ex]
1 & -\left(2+\dfrac{\delta x^2}{\alpha\,\delta t}\right) & 1 & & \text{-----} \\[2ex]
0 & 1 & -\left(2+\dfrac{\delta x^2}{\alpha\,\delta t}\right) & & \text{-----} \\[2ex]
& & & -\left(2+\dfrac{\delta x^2}{\alpha\,\delta t}\right) & 1 & 0 \\[2ex]
& & & 1 & -\left(2+\dfrac{\delta x^2}{\alpha\,\delta t}\right) & 1 \\[2ex]
& & & 0 & 1 & -\left(1+\dfrac{1}{2}\dfrac{\delta x^2}{\alpha\,\delta t}+\dfrac{\bar{h}_L\,\delta x}{k}\right)
\end{bmatrix}
$$

$$
\times
\begin{bmatrix}
T_0^{k+1} \\[1ex]
T_1^{k+1} \\[1ex]
T_2^{k+1} \\[1ex]
\vdots \\[1ex]
T_{n-2}^{k+1} \\[1ex]
T_{n-1}^{k+1} \\[1ex]
T_n^{k+1}
\end{bmatrix}
= -
\begin{bmatrix}
\dfrac{1}{2}\dfrac{\delta x^2}{\alpha\,\delta t}T_0^k+\dfrac{q_0^{k+1}\delta x}{k} \\[2ex]
\dfrac{\delta x^2}{\alpha\,\delta t}T_1^k \\[2ex]
\dfrac{\delta x^2}{\alpha\,\delta t}T_2^k \\[2ex]
\vdots \\[2ex]
\dfrac{\delta x^2}{\alpha\,\delta t}T_{n-2}^k \\[2ex]
\dfrac{\delta x^2}{\alpha\,\delta t}T_{n-1}^k \\[2ex]
\dfrac{\delta x^2}{2\alpha\,\delta t}T_n^k+\dfrac{\bar{h}_L\,\delta x}{k}T_\infty
\end{bmatrix}
\qquad (6.35)
$$

Equation (6.35) is similar to eqn. (6.32) and can be solved the same way. However, the right-hand side of the set of equations changes with time. A computer program to march the solution forward in time can be constructed on the following outline:

1. Set up formulas to compute the values of the diagonals of the square array, and the right-hand-side vector.

2. Initialize a time step counter: $n=0$.

3. Advance the time step counter one step: $n=n+1$.

4. Find the solution vector for $t=n\,\delta t$ from the tridiagonal algorithm (Gauss elimination).

5. Store or/and print the solution vector.

6. Test for the final time step. If the test is satisfied, stop or proceed to the next part of the program: step 9.

7. Alter the elements of the square array, if necessary, and the right-hand-side vector as required.

8. Return to step 3.

9. Stop or continue to the next part of the program.

We illustrate the method in the following example:

Example 6.8

A granite wall 9 cm thick initially at a uniform temperature of 20°C is subjected to a surface heat flux of 100 W/m² on one side and is cooled on the other by a fluid at 20°C with a heat transfer coefficient of 100 W/m²-°C. What is the distribution of temperature within the slab and on its surface as a function of time? Use the implicit method with $\delta x = 3$ cm, $\delta t = 1$ hr.

SOLUTION. From Table A.2, we obtain $\alpha = 7.4 \times 10^{-7}$ m²/s, $k = 1.6$ W/m-°C. Thus

$$\frac{\delta x^2}{\alpha\,\delta t} = 0.33784$$

$$\frac{\bar{h}_L\,\delta x}{k} = 1.87500$$

$$\frac{q_0\,\delta x}{k} = 1.87500 \;(\degree C)$$

and eqn. (6.35) becomes

$$
\begin{bmatrix}
-1.16892 & 1 & 0 & 0 \\
1 & -2.33784 & 1 & 0 \\
0 & 1 & -2.33784 & 1 \\
0 & 0 & 1 & -3.04392
\end{bmatrix}
\cdot
\begin{bmatrix}
T_0^{k+1} \\
T_1^{k+1} \\
T_2^{k+1} \\
T_3^{k+1}
\end{bmatrix}
$$

$$
=
\begin{bmatrix}
-0.16892\,T_0^k - 1.875 \\
-0.33784\,T_1^k \\
-0.33784\,T_2^k \\
-0.16892\,T_3^k - 37.50
\end{bmatrix}
$$

To solve these equations we will use the method employed in Example 6.7. With the same set of operations, the set of equations above becomes

$$
\begin{bmatrix}
-1.16892 & 1 & 0 & 0 \\
0 & -1.48235 & 1 & 0 \\
0 & 0 & -1.66364 & 1 \\
0 & 0 & 0 & -2.44268
\end{bmatrix}
\cdot
\begin{bmatrix}
T_0^{k+1} \\
T_1^{k+1} \\
T_2^{k+1} \\
T_3^{k+1}
\end{bmatrix}
$$

$$
=
\begin{bmatrix}
-0.16892\,T_0^k - 1.875 \\
-0.33784\,T_1^k - 0.14451\,T_0^k - 1.60404 \\
-0.33784\,T_2^k - 0.22791\,T_1^k - 0.09749\,T_0^k - 1.08209 \\
-0.16892\,T_3^k - 0.20307\,T_2^k - 0.13699\,T_1^k - 0.05860\,T_0^k - 38.1504
\end{bmatrix}
$$

Now we solve the equations in reverse order to get

$$T_3^{k+1} = 0.06915\,T_3^k + 0.08313\,T_2^k + 0.5608\,T_1^k + 0.02399\,T_0^k + 15.6183$$

$$T_2^{k+1} = 0.60109\,T_3^{k+1} + 0.20307\,T_2^k + 0.13699\,T_1^k + 0.05860\,T_0^k + 0.65044$$

$$T_1^{k+1} = 0.67460\,T_2^{k+1} + 0.22791\,T_1^k + 0.09749\,T_0^k + 1.08209$$

$$T_0^{k+1} = 0.85549\,T_1^{k+1} + 0.14451\,T_0^k + 1.60404$$

To find the temperature distribution after one time step, we insert $T_i^0 = 20°C$ on the right-hand side of these equations and solve for T_i^1. Successive temperature distributions are found in the same way. The result is shown in Fig. 6.16, which also shows the exact solution.[10] After a long time, a steady state is reached. Then the temperature distribution is linear and the surface temperatures can be calculated easily:

$$T_0 - T_3 = \frac{q_0 x}{k} = 5.625°C \quad \text{and} \quad T_3 - T_\infty = \frac{q_0}{\bar{h}_L} = 1.00°C$$

These values agree well with the temperature given by the implicit method at $t = 10$ hr.

Neither the spatial step size, δx, nor the time step, δt, are small enough to give a solution independent of δx and δt, in this example. If a computer program is written to do the calculations, it is easy to reduce δx and δt until the solution is independent of either step size, and in agreement with the exact solution. The calculation depends more strongly on the magnitude of δt than it does on δx. To obtain results within $0.05°C$ of the true temperature at $t = 1$ hr, one must pick $\delta t < 3$ min and $\delta x < 1$ cm.

We found in Chapter 5 that the temperature within the wall does not begin to change immediately when an external change is imposed. Therefore, the

---

[10] The exact solution is given by the formula

$$
T - T_{\text{initial}} = \frac{q_0}{\bar{h}_L}\left[1 + \text{Bi}\left(1 - \frac{x}{L}\right) - \sum_{n=1}^{\infty} \frac{2\,\text{Bi}(\kappa_n^2 + \text{Bi}^2)\cos[\kappa_n(x/L)]}{\kappa_n^2(\text{Bi} + \text{Bi}^2 + \kappa_n^2)}\,e^{-\kappa_n^2(\alpha t/L^2)}\right]
$$

where the κ_n are the roots of the equation $\kappa_n \tan \kappa_n = \text{Bi}$ [1.14].

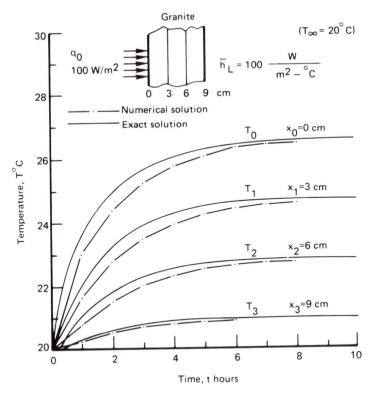

FIGURE 6.16   Temperature distribution in a 9 cm thick granite slab heated on one surface ($T = T_0$) and cooled on the other ($T = T_3$) (Example 6.8).

temperatures T_2, T_3, and T_4 in Fig. 6.16 should have an initial slope of zero. The present implicit numerical solution does not display this feature because the time and step size are too coarse. We shall not describe more efficient methods here, but several do exist. The Crank–Nicolson method is an example of one such method.

Graphical solutions.   If we set $\bar{h} = 0$ in eqn. (6.21), we obtain the following formula for the explicit calculation of the one-dimensional unsteady temperature distribution:

$$T_i^{k+1} = \frac{T_{i-1}^k + T_{i+1}^k}{\delta x^2 / \alpha \, \delta t} + \left(1 - \frac{2}{\delta x^2 / \alpha \, \delta t}\right) T_i^k \qquad (6.36)$$

A particularly simple formula is obtained for this equation if we set $\delta x^2 / \alpha \, \delta t = 2$:

$$T_i^{k+1} = \frac{T_{i-1}^k + T_{i+1}^k}{2} \qquad (6.37)$$

Thus the temperature at position i, at time step t_{k+1}, is the mean of the

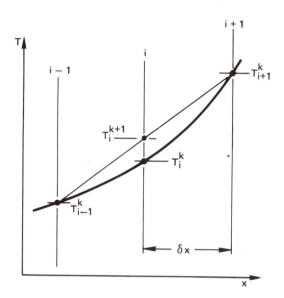

FIGURE 6.17   Illustration of Schmidt's graphical method to solve the unsteady heat conduction equation.

temperatures on either side of position i, at the *previous time step*. A well-known calculation method for unsteady heat conduction problems, which uses this fact, is known as a Schmidt plot (see, e.g., [1.7, pp. 201–206]).

A segment of a Schmidt plot is shown in Fig. 6.17. These solutions can be constructed simply with a straightedge and a sharp pencil with hard lead. They suffer from the same drawback that affects any explicit method, but here matters are somewhat worse: because $\alpha\,\delta t / \delta x^2$ can only be set equal to $\frac{1}{2}$, the only way to set smaller increments of x is to greatly reduce the time step, δt.

Graphical solution methods have been superseded largely by numerical solution methods on the digital computer. However, because the Schmidt plot method offers a unique physical insight into the process of heat conduction, we illustrate it in the following example.

Example 6.9

Solve Example 6.8 with a Schmidt plot. Begin the solution with $\delta x = 3$ cm and choose δt to give $\delta x^2 / \alpha\,\delta t = 2.0$.

SOLUTION.   The initial time step is $\delta t = \delta x^2 / 2\alpha = 0.169$ hr. Adjacent to the constant heat flux boundary, the slope of the temperature profile in the granite is constant and has the value $-q_0 / k$. Therefore, after each time step the surface temperature increases in such a way that $q_0 = k(T_0^{k+1} - T_1^{k+1}) / \delta x$, or $T_0^{k+1} = q_0 \delta x / k + T_1^{k+1}$. On the convectively cooled surface, the boundary condition at each time step is

$$\frac{k(T_{n-1} - T_n)}{\delta x} = \bar{h}(T_n - T_\infty)$$

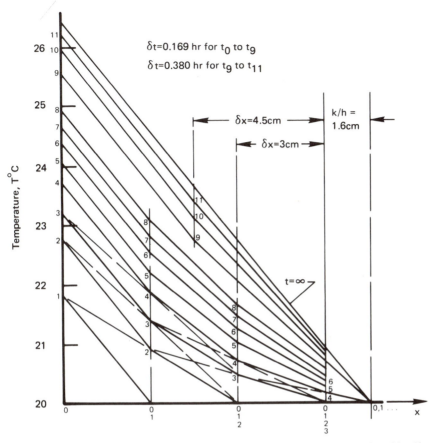

a.) Schmidt-plot construction. Dashed lines are construction lines; small numbers identify time steps. Solid lines are solutions connecting points at each time step.

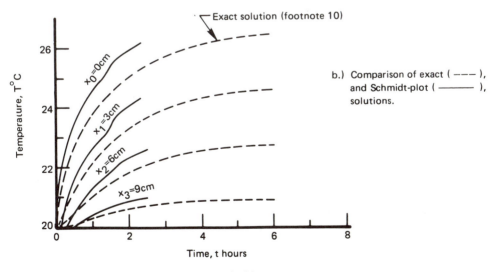

b.) Comparison of exact (−−−), and Schmidt-plot (———), solutions.

FIGURE 6.18 Schmidt-plot solution of Example 6.8.

or

$$\frac{T_{n-1}-T_n}{\delta x}=\frac{T_n-T_\infty}{k/\bar{h}}$$

Therefore, as we saw in Figs. 1.10 and 1.11, the temperature profile within the solid near the convective boundary can be extended a distance $k/\bar{h}$ to intersect a point on the line $T=T_\infty$.

We begin the Schmidt plot by laying out a vertical temperature scale and a horizontal distance scale, as shown in Fig. 6.18a.

At the end of the first time step, $t=\delta t$, the surface temperature on the left-hand boundary is 21.875°C, but each of the other nodal temperatures remains at 20°C. The solution is advanced in time by applying eqn. (6.37) graphically. For example, after the second time step, $t=2\delta t$, the temperature at the second node is found by connecting the point representing the surface temperature with the point representing the temperature of the third node. The surface temperature at time $t=2\delta t$ is found by drawing a line of slope $-q_0/k$ through the point representing the temperature of the second node.

Figure 6.18a shows the temperature distribution within the granite slab (Example 6.8 and Fig. 6.16) for the first eight time steps, using $\delta x = 3$ cm. The number of internal nodes was reduced to one after time step eight. Three additional temperature profiles are shown, each with the new time step, $\delta t = \delta x^2/\alpha\,\delta t = 0.380$ hr.

Figure 6.18b shows the temperature history of the slab at the four positions shown in Fig. 6.16. Each temperature profile shows a glitch at $t=1.35$ hr caused by changing the time step after $t=8\,\delta t$. The graphical Schmidt method, like any other explicit calculation scheme, leads to a temperature distribution that develops more quickly in time than the implicit method illustrated in Example 6.9. Although the method exhibits the correct initial zero slope of temperature with time, the effect is exaggerated when δx is large. The approximate solution would approach the exact solution only when δx (and therefore δt) are kept small. However, when many time steps must be taken with the graphical method, very large drawings are required.

Example 6.9 shows how to treat two different kinds of boundary conditions: constant heat flux and convective heat transfer to isothermal surroundings. A nonuniform initial temperature, unsteady surface or ambient temperatures, and time-varying surface heat transfer coefficients can easily be treated as well.

Finite-difference formulas
in two spatial dimensions

Equation 6.15 was written for a fin. We could just as well have begun our discussion with the unsteady two-dimensional heat conduction equation, eqn. (2.15):

$$\frac{1}{\alpha}\frac{\partial T}{\partial t}=\frac{\partial^2 T}{\partial x^2}+\frac{\partial^2 T}{\partial y^2}+\frac{\dot{q}}{k} \tag{6.38}$$

The methods we used above to find the finite-difference form of eqn. (6.15) works as well with eqn. (6.38). Before we apply them, however, we must expand our notational scheme to include discretization in y.

Notation.   The temperature $T(x,y,t)$ can be written in discrete form as

$$T(x_i, y_j, t_k) = T_{i,j}^k$$

Thus eqn. (6.19), the second-order, central-difference approximation for $\partial^2 T / \partial x^2$ becomes

$$\frac{\partial^2 T}{\partial x^2} = \frac{T_{i+1,j}^k + T_{i-1,j}^k - 2T_{i,j}^k}{\delta x^2} + \mathcal{O}(\delta x^2) \tag{6.39a}$$

Similarly, we can write an approximate expression for the other spatial derivative $\partial^2 T / \partial y^2$:

$$\frac{\partial^2 T}{\partial y^2} = \frac{T_{i,j+1}^k + T_{i,j-1}^k - 2T_{i,j}^k}{\delta y^2} + \mathcal{O}(\delta y^2) \tag{6.39b}$$

where δy is the mesh size in the y direction. The time derivative in eqn. (6.38) is easily found from eqn. (6.16) simply by adding the additional subscript, j. Thus a first-order (in δt) forward difference for $\partial T / \partial t$ is

$$\frac{\partial T}{\partial t} = \frac{T_{i,j}^{k+1} - T_{i,j}^k}{\delta t} + \mathcal{O}(\delta t) \tag{6.39c}$$

Equations (6.38) and (6.39) can be combined to yield an explicit formula for the temperature field at time t^{k+1}:

$$\frac{T_{i,j}^{k+1} - T_{i,j}^k}{\alpha \, \delta t} = \frac{T_{i+1,j}^k + T_{i-1,j}^k - 2T_{i,j}^k}{\delta x^2}$$
$$+ \frac{T_{i,j+1}^k + T_{i,j-1}^k - 2T_{i,j}^k}{\delta y^2} + \frac{\dot{q}_{i,j}^k}{k} \tag{6.40}$$

where $\dot{q}_{i,j}^k$ is the discrete representation of the internal heat generation rate. Equation (6.40) can be written in implicit form by changing the index k to $k+1$ on the right-hand side; a Crank–Nicolson formula results, where we replace the right-hand side of eqn. (6.40) with an average, as we did to obtain eqn. (6.23).

Figure 6.19 shows a portion of the grid or mesh representing the x,y domain, and shows how the four temperatures surrounding the point i,j are related to the temperature $T_{i,j}$ through eqn. (6.40). If an energy balance is written for the square centered about i,j, the physical basis for eqn. (6.40) can be easily shown (see Problem 6.23).

If we choose $\delta x = \delta y$, eqn. (6.40) becomes

$$T_{i,j}^{k+1} = \frac{\alpha \, \delta t}{\delta x^2} \left[T_{i+1,j}^k + T_{i-1,j}^k + T_{i,j+1}^k + T_{i,j-1}^k \right.$$
$$\left. + \left(\frac{\delta x^2}{\alpha \, \delta t} - 4 \right) T_{i,j}^k \right] + q_{i,j}^k \frac{\alpha \, \delta t}{k} \tag{6.41}$$

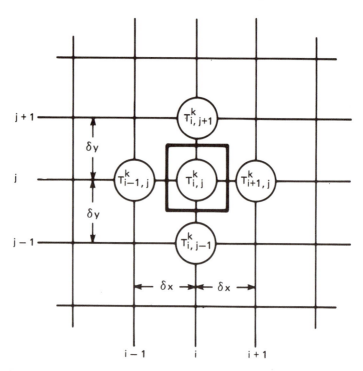

FIGURE 6.19 Grid system for two dimensional heat conduction showing the temperatures involved in a finite difference representation [eqn. (6.40)] of the diffusion equation.

This formula is convenient to use because it permits a marching type of solution. Unfortunately, the solutions diverge if the Fourier number based on the mesh size and the time step, $\alpha\,\delta t/\delta x^2$, is too large.

Stability. A discussion of the stability of numerical calculation methods is beyond the scope of this book, but a few remarks are in order. Equation (6.41) will be stable and will give convergent solutions if

$$\frac{\alpha\,\delta t}{\delta x^2} \leqslant \frac{1}{4}$$

A one-dimensional version of eqn. (6.41) [eqn. (6.20) with $\bar{h}=0$] will be stable and will give convergent solutions if

$$\frac{\alpha\,\delta t}{\delta x^2} \leqslant \frac{1}{2}$$

Similarly, the three-dimensional equations will require

$$\frac{\alpha\,\delta t}{\delta x^2} \leqslant \frac{1}{8}$$

The implicit method and the Crank–Nicolson method are always stable and convergent.

In Example 6.6 we found that the solution diverged for $\delta t = 15.6$ s. A stability criterion can be derived for the problem of Example 6.6 (see [1.17], for example). The criterion is

$$\delta t < \frac{2}{(1 + \sqrt{2}\,)(\alpha/\delta x^2) + (\bar{h}P/kA)\alpha}$$

for the number of mesh points used in Example 6.6, and gives $\delta t < 15.11$ s. The five mesh points chosen for Example 6.6 did not result in an accurate solution of the differential equations. To improve the accuracy would require more mesh points (smaller δx), which, in turn, will require smaller values of δt. The unfortunate consequence of this requirement is that more computing time is required and that values of δt much smaller than are needed for the sake of accuracy must be used just to achieve convergence. For these reasons, fully implicit (or Crank–Nicolson) methods are usually used.

Boundary conditions. The boundary conditions for two-dimensional problems can be developed by an extension of the methods we used above for one-dimensional problems. Figure 6.20 illustrates the procedure for an arbitrary point on the left-hand boundary of a rectangular body exchanging heat with an ambient fluid at temperature T_∞. We have merely to sum the heat flows into the rectangular control volume adjacent to the boundary and equate them to the energy stored within the control volume in time step δt. The result for an explicit

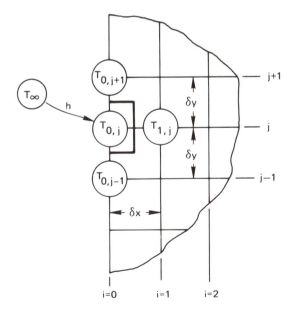

FIGURE 6.20   Grid system for the point $x_i = x_0, y_j$ on the left-hand boundary of a solid. The temperatures indicated are used in a finite difference representation [eqn. (6.42)] of the diffusion equation.

calculation method, in terms of rates of heat flow, is

$$\underbrace{\frac{k\left(T_{0,j-1}^k - T_{0,j}^k\right)}{\delta y} \frac{\delta x}{2}}_{\substack{\text{conduction from}\\\text{below}}} + \underbrace{\frac{k\left(T_{1,j}^k - T_{0,j}^k\right)}{\delta x} \delta y}_{\substack{\text{conduction from}\\\text{the right}}} + \underbrace{\frac{k\left(T_{0,j+1}^k - T_{0,j}^k\right)}{\delta y} \frac{\delta x}{2}}_{\substack{\text{conduction from}\\\text{above}}}$$

$$+ \underbrace{\bar{h}\left(T_\infty^k - T_{0,j}^k\right)\delta y}_{\substack{\text{convection from}\\\text{the left}}} + \underbrace{\dot{q}_{0,j}^k \frac{\delta x}{2} \delta y}_{\substack{\text{heat}\\\text{generation}}} = \underbrace{\frac{\rho c\left(T_{0,j}^{k+1} - T_{0,j}^k\right)}{\delta t} \frac{\delta x}{2} \delta y}_{\text{heat stored}}$$

We can rewrite this equation in a form similar to eqn. (6.41) if δx is chosen to be equal to δy.

$$T_{0,j}^{k+1} = \frac{\alpha\,\delta t}{\delta x^2}\left[2T_{1,j}^k + T_{0,j+1}^k + T_{0,j-1}^k + \left(\frac{1}{2}\frac{\delta x^2}{\alpha\,\delta t} - 2\frac{\bar{h}\,\delta x}{k} - 4\right)T_{0,j}^k\right]$$

$$+ \dot{q}_{0,j}^k \frac{\alpha\,\delta t}{k} + 2\left(\frac{\bar{h}\,\delta x}{k}\right)\left(\frac{\alpha\,\delta t}{\delta x^2}\right)T_\infty^k \qquad (6.42)$$

Expressions similar to eqn. (6.42) can be derived for other kinds of boundary conditions and for irregular boundaries [1.17, 1.3], but we will not consider them here.

Both eqns. (6.41) and (6.42) are explicit in form and are first order in time. They can be replaced by implicit or Crank–Nicolson forms in the usual way. Equation (6.41) is second order in δx, while eqn. (6.42) is first order in δx. However, the resulting temperature field is second order in δx. Usually, δx is chosen small enough to achieve reasonable accuracy.

We shall do no more with unsteady two-dimensional problems. One can march calculations out in time using the explicit forms of the difference equation, described in this section. Calculations with the implicit forms become harder to carry out, but they can be done by the same methods that we use next to solve steady two-dimensional problems.

Steady two-dimensional problems

The temperature distribution in steady two-dimensional conduction is given by eqn. (6.38) with the time derivative, $\partial T/\partial t$, set equal to zero. Difference representations of the spatial derivatives, $\partial^2 T/\partial x^2$ and $\partial^2 T/\partial y^2$, are given by eqns. (6.39a) and (6.39b) with the superscript k deleted. Thus for a node that does not lie on a boundary,

$$\frac{T_{i+1,j} + T_{i-1,j} - 2T_{i,j}}{\delta x^2} + \frac{T_{i,j+1} + T_{i,j-1} - 2T_{i,j}}{\delta y^2} + \frac{\dot{q}_{i,j}}{k} = 0$$

or if δx is set equal to δy,

$$\left(T_{i+1,j} + T_{i-1,j} + T_{i,j+1} + T_{i,j-1} - 4T_{i,j}\right) + \frac{\dot{q}_{i,j}}{k}\delta x^2 = 0 \qquad (6.43)$$

Equation (6.43) has the same physical interpretation as eqn. (6.41) (as illustrated in Fig. 6.19), although it now represents a steady-state energy balance. The convective b.c. given in eqn. (6.42) and illustrated in Fig. 6.20 for unsteady conduction becomes

$$\left(T_{0,j+1} + T_{0,j-1} + 2T_{i,j} - 4T_{0,j}\right) + \frac{\dot{q}_0 \delta x^2}{k} + \frac{2h\,\delta x}{k}\left(T_\infty - T_{0,j}\right) = 0 \quad (6.44)$$

When $\delta x = \delta y$, similar equations result for other types of boundary conditions.

For any given problem, one equation of the type illustrated by eqn. (6.43) is needed for each interior node and one equation like (6.44) is needed for each node on a boundary. Thus when a rectangular region is divided into m by n nodes, there will be a total of mn unknown temperatures and mn equations. These equations have to be solved simultaneously. For those points on a boundary whose temperature is prescribed, it is not necessary to write an equation and the number of unknown temperatures is therefore reduced.

The problem of determining the steady temperature distribution in a two-dimensional region thus reduces to the solution of a set of simultaneous, linear algebraic equations. We consider a few of the many ways to solve such a set of equations in the next section.

Solution methods for two-dimensional problems.   Figure 6.21 shows a heat conduction configuration that can be used to illustrate some solution methods

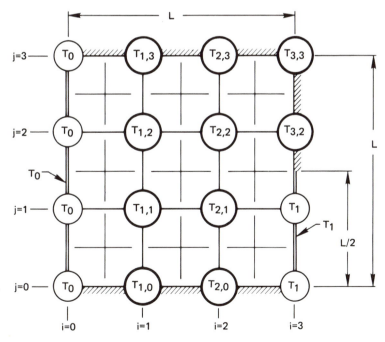

FIGURE 6.21   A two-dimensional heat conduction problem arranged for numerical solution.

for two-dimensional problems. The magnitude of heat flow from the higher temperature on the left to the lower temperature on the right can most easily be evaluated on the left, using

$$Q = k\frac{\delta y}{2}\frac{T_0-T_{1,0}}{\delta x} + k\,\delta y\frac{T_0-T_{1,1}}{\delta x} + k\,\delta y\frac{T_0-T_{1,2}}{\delta x} + k\frac{\delta y}{2}\frac{T_0-T_{1,3}}{\delta x} \quad (6.45)$$

once the temperature distribution throughout the medium is known.

The 10 unknown temperatures (identified in the larger circles in Fig. 6.21) can be written as the vector

$$\begin{bmatrix} T_{1,0} \\ T_{1,1} \\ T_{1,2} \\ T_{1,3} \\ T_{2,0} \\ T_{2,1} \\ T_{2,2} \\ T_{2,3} \\ T_{3,2} \\ T_{3,3} \end{bmatrix}$$

where we have chosen to order the elements of the vector $T_{i,j}$ such that the j values are taken in sequence for each value of i beginning with $i=1$. All of the temperatures are known at $i=0$, in this case.

Each of the temperatures in this vector is related to the temperatures of the nodes surrounding it, by an energy balance. For the node $i=1, j=0$, we have

$$k\frac{\delta y}{2}\frac{T_0-T_{1,0}}{\delta x} + k\frac{\delta y}{2}\frac{T_{2,0}-T_{1,0}}{\delta x} + k\,\delta x\frac{T_{1,1}-T_{1,0}}{\delta y} = 0$$

If we set $\delta x = \delta y$, then

$$-4T_{1,0}+2T_{1,1}+T_{2,0}= -T_0 \quad \left.\rule{0pt}{1em}\right\}$$

where the primary unknown is $T_{1,0}$. Similarly, for the other nodes in Fig. 6.21,

$$\left.\begin{aligned}
-4T_{1,1}+T_{1,0}+T_{1,2}+T_{2,1}&= -T_0 \\
-4T_{1,2}+T_{1,1}+T_{1,3}+T_{2,2}&= -T_0 \\
-4T_{1,3}+2T_{1,2}+T_{2,3}&= -T_0 \\
-4T_{2,0}+T_{1,0}+2T_{2,1}&= -T_1 \\
-4T_{2,1}+T_{2,0}+T_{2,2}+T_{1,1}&= -T_1 \\
-4T_{2,2}+T_{2,1}+T_{2,3}+T_{1,2}+T_{3,2}&=0 \\
-4T_{2,3}+2T_{2,2}+T_{1,3}+T_{3,3}&=0 \\
-4T_{3,2}+2T_{2,2}+T_{3,3}&= -T_1 \\
-4T_{3,3}+2T_{3,2}+2T_{2,3}&=0
\end{aligned}\right\} \quad (6.46)$$

Coefficient matrices for systems of equations such as this—the kind of systems that arise in two- and three-dimensional conduction problems—are more complex than the tridiagonal matrix that arose previously in one-dimensional problems. The determination of the solution vector, $T_{i,j}$, is accordingly much more difficult. Two classes of methods are commonly used to solve such problems as represented by eqn. (6.46). The first class consists of "direct methods," the second consists of "iterative methods."

Direct methods are normally variants of the Gaussian elimination procedure which exploit the special structure of the coefficient matrix. In recent years much effort has been devoted to the construction of economical, stable direct methods. Subroutines for solving systems such as eqn. (6.46) using reliable direct methods are beginning to appear in the software libraries of large digital computers. (A discussion of direct methods is beyond the scope of this work.)

The software of most large digital computers includes subroutines that will determine the solution vector, $T_{i,j}$, by iterative methods. We shall illustrate how such routines are constructed, using the Gauss–Seidel iterative method. Solving each equation in (6.46) for the primary unknown we get

$$
\begin{aligned}
T_{1,0} &= \frac{T_0 + 2T_{1,1} + T_{2,0}}{4} \\[4pt]
T_{1,1} &= \frac{T_0 + T_{1,0} + T_{1,2} + T_{2,3}}{4} \\[4pt]
T_{1,2} &= \frac{T_0 + T_{1,1} + T_{1,3} + T_{2,2}}{4} \\[4pt]
T_{1,3} &= \frac{T_0 + 2T_{1,2} + T_{2,3}}{4} \\[4pt]
T_{2,0} &= \frac{T_1 + T_{1,0} + 2T_{2,1}}{4} \\[4pt]
T_{2,1} &= \frac{T_1 + T_{2,0} + T_{2,2} + T_{1,1}}{4} \\[4pt]
T_{2,2} &= \frac{T_{2,1} + T_{2,3} + T_{1,2} + T_{3,2}}{4} \\[4pt]
T_{2,3} &= \frac{2T_{2,2} + T_{1,3} + T_{3,3}}{4} \\[4pt]
T_{3,2} &= \frac{T_1 + 2T_{2,2} + T_{3,3}}{4} \\[4pt]
T_{3,3} &= \frac{2T_{3,2} + 2T_{2,3}}{4}
\end{aligned}
\qquad (6.47)
$$

The solution vector, $T_{i,j}$, is still unknown in eqn. (6.47).

One step of the Gauss–Seidel iteration involves guessing values for the $T_{i,j}$, inserting them in the right-hand sides of the formulas in eqn. (6.47), and calculating revised values. These values are then reinserted in eqn. (6.47) and new values are calculated. The procedure is continued until the values of the $T_{i,j}$ no longer change, or until the change between iterations is arbitrarily small. A common criterion used to determine the adequacy of convergence is that the maximum deviation between successive iterations must be less than some prescribed value. We can state this criterion as $|T_{i,j}^{(k+1)} - T_{i,j}^{(k)}|_{max} \leqslant \varepsilon$, where (k) denotes the iteration number and ε is the desired maximum deviation.

In Gauss–Seidel iteration, new values are substituted in the formulas as soon as they are calculated. For example, an initial assumed value of $T_{i,j}$ produces a new value of $T_{1,0}$ from the first formula of eqn. (6.47). $T_{1,0}$ appears on the right-hand side of the formulas for $T_{1,1}$ and $T_{2,0}$, and the new value of $T_{1,0}$ is used in these formulas.

While the Gauss–Seidel process will always converge in heat conduction problems, it might converge slowly. The rate of convergence can be accelerated by a process called "overrelaxation." In a general relaxation process, the new value of each temperature is modified before it is stored or used in subsequent equations of the set being solved. The modified value is a weighted mean of the initial and final value of the temperature. For example, to calculate the temperatures, $T_{i,j}^{(k+1)}$ at level $(k+1)$ from the temperatures, $T_{i,j}^{(k)}$ at level (k), we first estimate $T_{i,j}^{(k+1)'}$ using eqn. (6.47). The relaxed value of $T_{i,j}^{(k+1)}$ is calculated from a weighted mean of this temperature and the preceding one, $T_{i,j}^{(k)}$:

$$T_{i,j}^{(k+1)} = \beta T_{i,j}^{(k+1)'} + (1-\beta)T_{i,j}^{(k)} \tag{6.48}$$

The weighting factor, β, is called a "relaxation factor." If $0 < \beta < 1$, the process is called *under*relaxation, and if $1 < \beta < 2$ it is called *over*relaxation. The iteration process will diverge if $\beta > 2$. In the linear problems that we are treating, overrelaxation can be used to speed the convergence process, but in nonlinear problems, underrelaxation sometimes *has* to be used just to achieve convergence. For some types of problems, it is possible to determine the optimum value of β analytically [6.6].

Example 6.10

Use the Gauss–Seidel iteration method to find the shape factor for the problem depicted in Fig. 6.21.

SOLUTION. We can set $T_0 = 1$ and $T_1 = 0$ without loss of generality. Then from eqn. (6.45) we can write the shape factor as

$$S = \frac{Q}{k(T_0 - T_1)} = 3 - \frac{T_{1,0}}{2} - T_{1,1} - T_{1,2} - \frac{T_{1,3}}{2}$$

A flowchart of a computer program to find the temperature distribution is presented in Fig. 6.22. The computer program itself is shown in Fig. 6.23, a portion of the output in Fig. 6.24, and the results in Fig. 6.25. The shape factor to two significant figures is 0.79 as determined by Gauss–Seidel iteration. The value is not accurate to more than two significant figures because of truncation errors associated with the large value of δx.

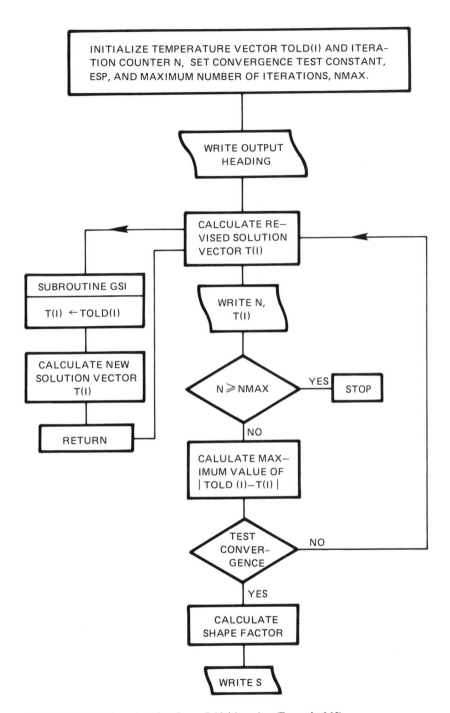

FIGURE 6.22   Flow chart for Gauss-Seidel iteration (Example 6.10).

```
      C
      C
      C     ***************************************************
      C     ***************************************************
      C     **                                             **
      C     **                                             **
      C     **   THIS PROGRAM CALCULATES THE TEMPERATURE DISTRIBUTION   **
      C     **   FOR EXAMPLE 6.10.  GAUSS-SEIDEL ITERATION IS USED.     **
      C     **   TOLD(I)= SOLUTION VECTOR(TEMPERATURE) AT PREVIOUS      **
      C     **   ITERATION STEP.  T(I)= SOLUTION VECTOR AT CURRENT      **
      C     **   ITERATION STEP.  N= ITERATION COUNTER.  NMAX= MAXIMUM  **
      C     **   NUMBER OF ITERATIONS TO BE ALLOWED.  EPS= TERMINATION  **
      C     **   CRITERION.                                **
      C     **                                             **
      C     **                                             **
      C     **                                             **
      C     ***************************************************
      C     ***************************************************
      C
      C
      C-----MAIN PROGRAM
      C
   1          DIMENSION TOLD(10),T(10),R(10)
   2          N=0
   3          NMAX=50
   4          EPS=0.0001
   5          WRITE(6,100) EPS
   6      100 FORMAT(31H1GAUSS-SEIDEL ITERATION   EPS=,F10.6///2H N,
         13X,6HT(1,0),4X,6HT(1,1),4X,6HT(1,2),4X,6HT(1,3),4X,6HT(2,0),4X,
         26HT(2,1),4X,6HT(2,2),4X,6HT(2,3),4X,6HT(3,2),4X,6HT(3,3)//)
   7          DO 110 I=1,10
   8          TOLD(I)=1.0
   9          T(I)=1.0
  10      110 CONTINUE
      C-----START OF MAIN DO LOOP
  11      115 N=N+1
      C
  12          CALL GSI(TOLD,T)
      C
  13          WRITE(6,120) N,(T(I),I=1,10)
  14      120 FORMAT(I3,10F10.6)
      C-----TEST FOR TERMINATION CRITERION
  15          IF (N.GE.NMAX) STOP
  16          DO 140 I=1,10
  17          R(I)=ABS(TOLD(I)-T(I))
  18      140 CONTINUE
  19          DO 150 I=2,10
  20          R(I)=AMAX1(R(I),R(I-1))
  21      150 CONTINUE
      C-----READY TO COMPUTE SHAPE FACTOR?
  22          IF (R(10).LE.EPS) GO TO 160
  23          DO 130 I=1,10
  24          TOLD(I)=T(I)
  25      130 CONTINUE
      C-----START DO LOOP AGAIN
  26          GO TO 115
  27      160 S=3.0-T(1)/2.0-T(2)-T(3)-T(4)/2.0
      C-----COMPUTE SHAPE FACTOR
  28          WRITE(6,170) S
  29      170 FORMAT(///14H SHAPE FACTOR=,F10.6////)
  30          STOP
  31          END
      C
      C
  32          SUBROUTINE GSI(TOLD,T)
  33          DIMENSION TOLD(10),T(10)
  34          DO 100 I=1,10
  35          T(I)=TOLD(I)
  36      100 CONTINUE
  37          T(1)=(1.0+T(5)+2.0*T(2))/4.0
  38          T(2)=(1.0+T(6)+T(1)+T(3))/4.0
  39          T(3)=(1.0+T(7)+T(2)+T(4))/4.0
  40          T(4)=(1.0+T(8)+2.0*T(3))/4.0
  41          T(5)=(T(1)+2.0*T(6))/4.0
  42          T(6)=(T(2)+T(5)+T(7))/4.0
  43          T(7)=(T(3)+T(9)+T(6)+T(8))/4.0
  44          T(8)=(T(4)+T(10)+2.0*T(7))/4.0
  45          T(9)=(T(10)+2.0*T(7))/4.0
  46          T(10)=(T(8)+T(9))/2.0
  47          RETURN
  48          END
```

FIGURE 6.23 Computer program used to solve Example 6.10 by Gauss-Seidel iteration.

N	T(1,0)	T(1,1)	T(1,2)	T(1,3)	T(2,0)	T(2,1)	T(2,2)	T(2,3)	T(3,2)	T(3,3)
1	1.000000	1.000000	1.000000	1.000000	0.750000	0.687500	0.921875	0.960938	0.710938	0.835938
2	0.937500	0.906250	0.957031	0.968750	0.578125	0.601563	0.807617	0.854980	0.612793	0.733887
3	0.847656	0.851563	0.906982	0.917236	0.512695	0.542969	0.729431	0.777496	0.548187	0.662842
4	0.803955	0.813477	0.865036	0.876892	0.472473	0.503845	0.673641	0.681748	0.469874	0.575810
5	0.774857	0.785934	0.834117	0.847497	0.445637	0.476303	0.633676	0.653057	0.446467	0.549762
6	0.754376	0.766199	0.811843	0.826358	0.426746	0.456655	0.605030	0.632487	0.429688	0.531087
7	0.739786	0.752071	0.795864	0.811196	0.413274	0.442593	0.584495	0.617742	0.417660	0.517701
8	0.729354	0.741952	0.784410	0.800327	0.403635	0.432520	0.569776	0.607172	0.409038	0.508105
9	0.721885	0.734704	0.776201	0.792536	0.396731	0.425303	0.559226	0.599596	0.402858	0.501227
10	0.716535	0.729510	0.770318	0.786952	0.391785	0.420130	0.551664	0.594166	0.398429	0.496297
11	0.712701	0.725787	0.766100	0.782949	0.388240	0.416423	0.546244	0.590274	0.395254	0.492764
12	0.709953	0.723119	0.763078	0.780080	0.385700	0.413765	0.542359	0.587484	0.392978	0.490231
13	0.707984	0.721207	0.760911	0.778024	0.383879	0.411861	0.539575	0.585485	0.391347	0.488416
14	0.706573	0.719836	0.759358	0.776550	0.382574	0.410496	0.537579	0.584051	0.390178	0.487115
15	0.705561	0.718854	0.758245	0.775494	0.381638	0.409518	0.536148	0.583024	0.389340	0.486182
16	0.704836	0.718149	0.757448	0.774737	0.380968	0.408816	0.535123	0.582288	0.388739	0.485514
17	0.704316	0.717645	0.756876	0.774194	0.380487	0.408314	0.534388	0.582288	0.388739	0.485514
18	0.703944	0.717283	0.756466	0.773805	0.380143	0.407953	0.533861	0.581760	0.388309	0.485035
19	0.703677	0.717024	0.756172	0.773526	0.379896	0.407695	0.533484	0.581382	0.388000	0.484691
20	0.703486	0.716838	0.755962	0.773326	0.379719	0.407510	0.533213	0.581111	0.387779	0.484445
21	0.703349	0.716705	0.755811	0.773183	0.379592	0.407377	0.533019	0.580916	0.387621	0.484269
22	0.703250	0.716609	0.755703	0.773080	0.379501	0.407282	0.532880	0.580777	0.387507	0.484142
23	0.703180	0.716541	0.755625	0.773007	0.379436	0.407214	0.532781	0.580678	0.387426	0.484052

SHAPE FACTOR= 0.789739

FIGURE 6.24   Twenty-three successive Gauss-Seidel iterations $(N=1,2,\ldots 23)$ for the 10 unknown $T_{i,j}$'s in Example 6.10. This is the output of the program in Fig. 6.23.

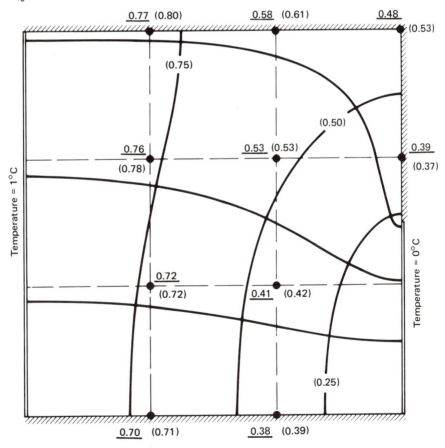

FIGURE 6.25   Flux plot for Example 6.10. The underlined numbers are from Fig. 6.24. Those enclosed in parentheses were interpolated from the flux plot.

A flux plot for this example is shown in Fig. 6.25. It includes numerical values of the temperature distribution both from Fig. 6.24 and from the flux plot itself. The shape factor computed from the flux plot is 0.78, and it is probably just as accurate as the one we have calculated by finite differences. Problem 6.14 affords an opportunity for the reader to produce an improved value by numerical methods.

The computer program listing in Fig. 6.23 was written specifically for this problem. It can be modified to handle more general problems by developing algorithms to compute the matrix coefficients and the right-hand vector and incorporating them in the main program. Gauss–Seidel iteration subroutines are available in most computer libraries to compute a revised estimate of the solution vector each time they are called.

In this particular example, 23 iterations were needed to produce a solution to the difference equations to the tolerance implied by $\varepsilon = 0.0001$. The rate of convergence can be increased significantly by using overrelaxation (see Problem 6.21), a technique that is important for problems with many nodes or for problems that must be solved repetitively.

More powerful and more general calculation methods than we have described are needed to handle problems with temperature-dependent thermal properties or nonlinear boundary conditions, as might occur if a body for which $\mathrm{Bi} \geqslant \mathcal{O}(1)$ should experience heat exchange by radiation. Descriptions of such methods can be found in the literature and in more advanced textbooks.

PROBLEMS

6.1. Show, by linearizing eqn. (6.1), that initially the temperature of the body will change exponentially with a time constant given by

$$T = \left(\frac{\bar{h}A}{\rho c V} + 4T_i^3 \frac{\mathcal{F}\sigma A}{\rho c V} \right)^{-1}$$

Express this result in terms of the two time constants T_1 and T_2 given in the text. [*Hint:* If $T = T_i + \varepsilon$, $T^4 = (T_i + \varepsilon)^4 \cong T_i^4 + 4\varepsilon T_i^3$, by the binomial theorem.]

6.2. Equation (6.2) can be written in the form

$$T_f = \frac{T_\infty + (\mathcal{F}\sigma/\bar{h}) T_s^4}{1 + T_f^3(\mathcal{F}\sigma/\bar{h})}$$

Hence an estimate of T_f inserted on the right-hand side will produce a revised estimate of T_f. Compare this successive approximation scheme with the one suggested in footnote 2. Plot T_f vs. number of trials for both methods and indicate which one gives the more rapid convergence.

6.3. The heat transfer coefficient for a horizontal cylinder in air can be represented by the relation

$$\bar{h} = 0.18(T_w - T_\infty)^{1/3} \text{ Btu/hr-ft}^2\text{-°F}$$

for a certain range of cylinder diameters and temperature differences. Devise a numerical calculation scheme to find the equilibrium temperature of a cylinder exchanging heat by convection with air at a temperature T_∞ and by radiation with surroundings at T_s. Assume that the cylinder and its surroundings are both black.

6.4. A hollow copper sphere which is thermally black and has a 5-cm diameter and a 3-cm inner diameter is heated by induction at the rate of 100 W. The heat transfer coefficient between the sphere and the ambient fluid is 200 W/m²-°K; and the initial temperature of the sphere, the ambient fluid temperature, and the temperature of the surroundings are all 300°K. Find the final equilibrium temperature of the sphere and the time it takes for 95% of the overall temperature change to take place. To find the time, use an integration scheme similar to that developed in Section 6.2. [$T_f = 361.136°$K]

6.5. Calculate the specific heat at constant pressure, $c_p = (\partial h / \partial T)_p$, for superheated steam (where h is the enthalpy) at 1 atm and several temperatures. Do this by applying (a) a forward-difference, (b) a backward-difference, and (c) a central-difference approximation to data from your thermodynamics textbook or other steam tables. Plot the results as a function of temperature, and compare the results to predicted values of $c_{p_{\text{steam}}}$ at low pressure (which should also be available in your thermodynamics text).

6.6. Calculate the coefficient of thermal expansion, β, for superheated steam at 500 psia and several temperatures by applying (a) a forward-difference, (b) a backward-difference, and (c) a central-difference approximation to data from your thermodynamics text or other steam tables. Plot the results as a function of temperature and compare them with the ideal gas result:

$$\beta = -\frac{1}{\rho} \frac{\partial \rho}{\partial T}\Big|_p = \frac{1}{T}$$

Comment on this result.

6.7. Solve Example 6.5 with an implicit formula obtained by setting $p = 1$ in eqn. (6.12). Write a digital computer program or a program for a hand-held programmable calculator. Compare your results with those in Table 6.1. You will find that the instability for $\delta\tau = 2.0$ does not occur with the implicit method.

6.8. Devise a Crank–Nicolson scheme [Section 6.3, eqn. (6.23)] to solve eqn. (6.1). Write a program for a digital computer to solve Example 6.5. Compare your results with those in Table 6.1. You should find that your solution is more accurate for a given $\delta\tau$ than the results of either Example 6.5 or Problem 6.6.

6.9. Develop a computer program that uses Simpson's rule to evaluate eqn. (6.3). Use it to solve the problem of Example 6.1. Vary the step size, δT, and estimate the largest value that will permit an accurate solution to three significant figures.

6.10. Develop an implicit scheme to solve Example 6.6 [see eqn. (6.22)]. Write a computer program using Gauss–Seidel iteration at each time step. Compare the results of your method with Fig. 6.12.

6.11. Extend the computer program from Problem 6.10 to permit a Crank–Nicolson solution of Example 6.6. Compare your result with that in Example 6.6 (and that in Problem 6.10, if you were asked to work it).

6.12. Use Gauss–Seidel iteration to find the shape factor for the solid shown in Fig. P6.12. Prepare a flux plot for the same problem and compare the results. Explain why this shape factor differs from that found in Example 6.10. [$S = 0.83$]

6.13. Repeat Problem 6.12 for the configuration in Fig. P6.13. [$S = 1.17$]

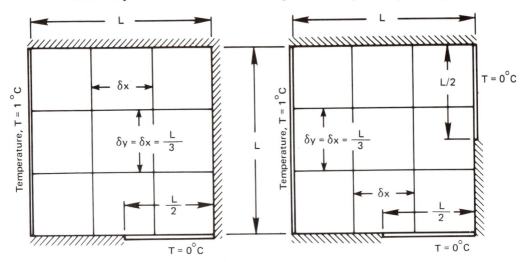

FIGURE P6.12   **FIGURE P6.13**

6.14. The computer program for Example 6.10 was written for $\delta x = \delta y = L/3$, where L is the length of the side of the square. Write a computer program that will solve Example 6.10 if $\delta x = L/n$, where n is an odd number. Find the shape factor for $n = 5$ and 7. Compare your results with those given in the text.

6.15. The top surface of a 10-cm 1%-carbon-steel cube is maintained at 30°C and the lower quarter of one side at 10°C, as shown in Fig. P6.15. The rest of the cube is

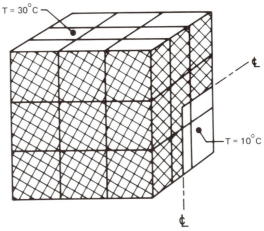

FIGURE P6.15

insulated. Write a computer program to find the temperature distribution in the cube, compute the shape factor, and report the heat flux through it. Use Gauss–Seidel iteration, using $4 \times 4 \times 4$ node points. Note that $64 - 4 \times 4 - 4 \times 2 = 40$ equations must be solved. (This is a rather lengthy undertaking.) $[S = 0.063 \text{ m}]$

6.16. The heat loss from a fin can be calculated either from the heat flow into the fin from its ends or from the heat flow to the fluid from the fin surface. Evaluate the heat loss for the rod in Example 6.7 both ways and compare them.

6.17. Determine and plot, as a function of time, the net heat flow from the walls to the rod and from the rod to the fluid in Example 6.6. Determine the net amount of heat stored within the rod by integrating the temperature distribution with x as a function of time. How does it compare with the difference between the heat flow into the rod from the walls, and the heat from the rod to the fluid?

6.18. Devise a formula to calculate the shape factor for the problem illustrated in Fig. 6.21, using the heat flow out of the T_1 boundary. Apply it to the results of Example 6.10 and compare your value with the one obtained there.

6.19. A commonly used method for deriving the finite-difference equation for the end node on a fin (Fig. 6.14) is to employ a fictitious node as shown in Fig. P6.19. An energy balance is written on the volume centered about $i = n$, and a second-order central difference is used to evaluate dT/dx in the end boundary condition. Perform this derivation and show that it leads to eqn. (6.24a).

6.20. Derive a finite-difference equation for an inside corner as shown in Fig. P6.20.

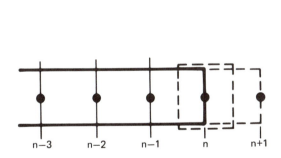

FIGURE P6.19

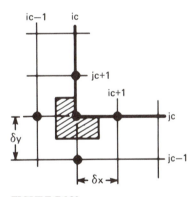

FIGURE P6.20

6.21. Modify the program in Example 6.10 to permit calculations with a relaxation factor, β [eqn. (6.48)]. Run the program for several values of β between 0.5 and 1.75. Plot the number of iterations needed to achieve a value of $\varepsilon = 0.0001$. Estimate the value of β needed to reach the solution in a minimum number of iterations. $[\beta \cong 1.4]$

6.22. Find the temperature distribution for the solid in Example 6.10 if the left hand side communicates with a fluid at temperature, T_0, through a heat transfer coefficient $\bar{h}$ and the right hand side is maintained at temperature T_1. Define a dimensionless temperature, $\Theta = (T - T_0)/(T_1 - T_0)$ and a Biot number, $\mathrm{Bi} = hL/k$, where L is the length of a side. Set up the equations for an arbitrary value of Bi. Solve them for $\mathrm{Bi} = 3.0$.

6.23. Use the energy balance method to derive the finite-difference equation for Fig. 6.19.

6.24. Find the solution for the midpoint temperature in Example 6.6, for $\delta t = 0$ by solving the three simultaneous differential equations

$$\frac{d\theta_i}{dt} = -A\theta_i + B(\theta_{i-1} + \theta_{i+1}); \qquad i = 1, 2, 3$$

where $A = 0.1281$ s^{-1}, $B = 9.25 \times 10^{-3}$ s^{-1}, $\theta_0 = 80°C$ and $\theta_4 = -20°C$. Compare your solution with the results of Example 6.6, Problem 6.10, and Problem 6.11. (*Note:* The Laplace transform solution method can be applied simply and directly to this problem.)

6.25. Write a computer program or use a hand calculator to evaluate eqn. (5.44). Choose one or two step sizes and find a step size that will allow you to compute results that agree with those in Table 5.1 to within two significant figures.

REFERENCES

[6.1] L. F. RICHARDSON, "The Approximate Arithmetical Solution by Finite Differences of Physical Problems Involving Differential Equations, with an Application to the Stresses in a Masonry Dam," *Trans. R. Soc. Lond., Ser. A*, vol. 210, 1910, pp. 307–357.

[6.2] R. W. HORNBECK, *Numerical Methods*, Quantum Publishers, Inc., New York, 1975.

[6.3] A. RALSTON and P. RABINOWITZ, *A First Course in Numerical Analysis*, McGraw-Hill Book Company, New York, 1978.

[6.4] R. C. WEAST and S. M. SELBY, *Handbook of Tables for Mathematics*, 3rd ed., The Chemical Rubber Co., Cleveland, Ohio, 1967.

[6.5] J. CRANK and P. NICOLSON, "A Practical Method for Numerical Evaluation of Solutions of Partial Differential Equations of the Heat-Conduction Type," *Proc. Camb. Phil. Soc.*, vol. 43, no. 50, 1947, pp. 50–67.

[6.6] R. S. VARGA, *Matrix Iterative Analysis*, Prentice-Hall, Inc., Englewood Cliffs, N.J., 1962.

Part III

Convective
Heat Transfer

Chapter 7

Laminar and turbulent boundary layers

In cold weather, if the air is calm, we are not so much chilled as when there is wind along with the cold; for in calm weather, our clothes and the air entangled in them receive heat from our bodies; this heat... brings them nearer than the surrounding air to the temperature of our skin. But in windy weather, this heat is prevented... from accumulating; the cold air, by its impulse ... both cools our clothes faster and carries away the warm air that was entangled in them

Joseph Black's notes on "The General Effects of Heat," ca. 1790s

7.1 Some introductory ideas

Joseph Black's perception about forced convection (above) represents a very correct understanding of the way forced convective cooling works. When cold air moves past a warm body, it constantly sweeps away warm air which has become, as Black put it, "entangled" with the body and replaces it with cold air. In this chapter we learn to form analytical descriptions of these convective heating (or cooling) processes.

Our aim is to predict h and $\bar{h}$, and it is clear that such predictions must begin in the motion of fluid around the bodies that they heat or cool. It is by predicting such motion that we will be able to find out how much heat is removed during the replacement of hot fluid with cold, and vice versa.

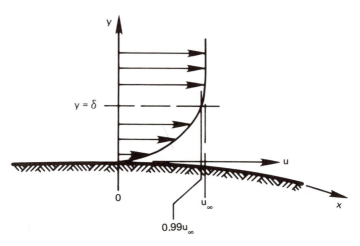

FIGURE 7.1   A boundary layer of thickness, δ.

Flow boundary layer

Fluids flowing past solid bodies adhere to them, so a region of variable velocity must be built up between the body and the free fluid stream as indicated in Fig. 7.1. This region is called a *boundary layer*, which we will often abbreviate as "b.l." The b.l. has a thickness, δ. The boundary layer thickness is arbitrarily defined as the distance from the wall at which the flow velocity approaches to within 1% of u_∞. The boundary layer is normally very thin in comparison with the dimensions of the body immersed in the flow.[1]

The first step that has to be taken before h can be predicted is the mathematical description of the boundary layer. This description was first made by Prandtl[2] (see Fig. 7.2) and his students, starting in 1904, and it depended upon simplifications that followed after he recognized how thin the layer must be.

The dimensional functional equation for the boundary layer thickness on a flat surface is

$$\delta = \text{fn}(u_\infty, \rho, \mu, x)$$

where x is the length along the surface and ρ and μ are the fluid density in kg/m^3 and viscosity in kg/m$-$s. We have five variables in kg, m, and s, so we

---

[1]We qualify this remark when we treat the b.l. quantitatively.

[2]Prandtl was educated at the Technical University in München and finished his doctorate there in 1900. He was given a chair in a new fluid mechanics institute at Göttingen University in 1904—the same year that he presented his historic paper explaining the boundary layer. His work at Göttingen, during the period up to Hitler's regime, set the course of modern fluid mechanics and aerodynamics and laid the foundations for the analysis of heat convection.

FIGURE 7.2   Ludwig Prandtl (1875 to 1953). (Photo courtesy *Appl. Mech. Revs.*, vol. 26, Feb. 1973.)

anticipate two pi-groups:

$$\frac{\delta}{x} = \text{fn}(\text{Re}_x) \qquad \text{Re}_x \equiv \frac{\rho u_\infty x}{\mu} = \frac{u_\infty x}{\nu} \tag{7.1}$$

where ν is the kinematic viscosity μ/ρ and Re_x is called the *Reynolds number.* It characterizes the relative influences of inertial and viscous forces in a fluid problem. The subscript on Re—x in this case—tells what length it is based upon.

We discover shortly that the actual form of eqn. (7.1) for a flat surface, where u_∞ remains constant, is

$$\frac{\delta}{x} = \frac{4.92}{\sqrt{\text{Re}_x}} \tag{7.2}$$

which means that if the velocity is great or the viscosity is low, δ/x will be relatively small. Heat transfer will be relatively high in such cases. If the velocity is low, the b.l. will be relatively thick. A good deal of nearly stagnant fluid will accumulate near the surface and be "entangled" with the body, although in a different way than Black envisioned it to be.

Osborne Reynolds (1842 to 1912)

> Reynolds was born in Ireland but he taught at the University of
> Manchester. He was a significant contributor to the subject of
> fluid mechanics in the late 19th C. His original laminar-to-tur-
> bulent flow transition experiment, pictured below, is still in
> use as a student experiment at the University of Manchester.

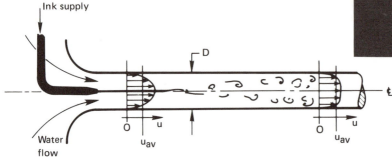

FIGURE 7.3   Osborne Reynolds and his laminar-turbulent flow transition experiment. (Photo courtesy *Appl. Mech. Revs.,* vol. 26, Feb. 1973.)

The Reynolds number is named after Osborne Reynolds (see Fig. 7.3), who
discovered the laminar–turbulent transition during fluid flow in a tube. He
injected ink into a steady and undisturbed flow of water and found that, beyond
a certain average velocity, u_{av}, the liquid streamline marked with ink would
become wobbly and then break up into increasingly disorderly eddies, and it
would finally be completely mixed into the water, as is suggested in the sketch.

To define the transition we first note that $(u_{av})_{crit}$, the transitional value of
the average velocity, must depend on the pipe diameter, D, on μ, and on ρ—four
variables in kg, m, and s. There is therefore only one pi-group:

$$\mathrm{Re}_{critical} \equiv \frac{\rho D (u_{av})_{crit}}{\mu} \tag{7.3}$$

The maximum Reynolds number for which fully developed laminar flow in a
pipe will always be stable, regardless of the level of background noise, is 2100. In

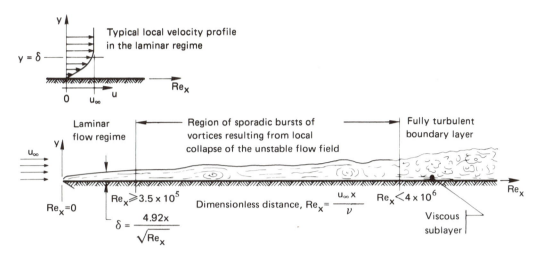

FIGURE 7.4   Boundary layer on a long flat surface with a sharp leading edge.

a reasonably careful experiment, laminar flow can be made to persist up to Re = 10,000. With enormous care it can be increased still another order of magnitude. But the constant in eqn. (7.3)—the critical value of Re—is 2100.

Much the same sort of thing happens in a boundary layer. Figure 7.4 shows fluid flowing over a plate with a sharp leading edge. The flow is laminar up to a transitional Reynolds number based on x:

$$\text{Re}_{x_{\text{critical}}} = \frac{u_\infty x_{\text{crit}}}{\nu} = (3 \text{ to } 5) \times 10^5 \qquad (7.4)$$

At larger values of x the b.l. exhibits sporadic vortexlike instabilities over a fairly long range, and it finally settles into a fully turbulent b.l. The actual onset of turbulent behavior depends strongly on background noise and on the precise shape of the leading edge. It can rather easily be delayed until $\text{Re}_x = 2.8 \times 10^6$, but not much further. The b.l. is always turbulent for $\text{Re}_x \geqslant 4 \times 10^6$.

These specifications of the critical Re are restricted to flat surfaces. Furthermore, if the surface is curved into the flow as shown in Fig. 7.1, turbulence might be triggered at greatly lowered values of Re_x.

Thermal boundary layer

If the wall is at a temperature, T_w, different from that of the free stream, T_∞, there is a *thermal boundary layer* thickness, δ_t—different from the *flow* or *momentum* b.l. thickness, δ. A thermal b.l. is pictured in Fig. 7.5. Now with reference to this picture, we equate the heat conducted away from the wall by the fluid, to the same heat transfer expressed in terms of a convective heat

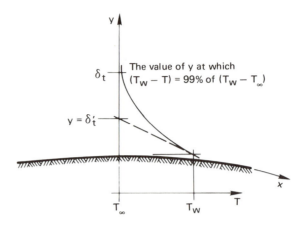

FIGURE 7.5   The thermal boundary layer during the flow of cool fluid over a warm plate.

transfer coefficient:

$$-k_f \frac{\partial T}{\partial y}\bigg|_{y=0} = h(T_w - T_\infty) \tag{7.5}$$

$$\underbrace{\phantom{-k_f \frac{\partial T}{\partial y}\bigg|_{y=0}}}_{\substack{\text{conduction} \\ \text{into the fluid}}}$$

where k_f is the conductivity of the fluid. Notice two things about this result. In the first place it is correct to express heat removal *at the wall* using Fourier's law of conduction because there is no fluid motion in the direction of q. The other point is that while eqn. (7.5) looks like a b.c. of the third kind, it is not. This condition *defines h within* the fluid instead of specifying it as known information on the boundary. Equation (7.5) can be arranged in the form

$$\frac{\partial\left(\dfrac{T_w - T}{T_w - T_\infty}\right)}{\partial(y/L)}\Bigg|_{y/L=0} = \frac{hL}{k_f} = \text{Nusselt number, Nu}_L \tag{7.5a}$$

where L is a characteristic dimension of the body under consideration—the length of a plate, the diameter of a cylinder, or [if we write eqn. (7.5) at a point of interest along a flat surface] $\text{Nu}_x \equiv hx/k_f$. From eqn. (7.5) we see immediately that the physical significance of Nu is given by

$$\text{Nu}_L = \frac{L}{\delta_t'} \tag{7.6}$$

FIGURE 7.6   Ernst Kraft Wilhelm Nusselt (1882 to 1957). This photograph, provided by his student, G. Lück, shows Nusselt at the Kesselberg waterfall in 1912. He was an avid mountain climber.

In other words, the Nusselt number is inversely proportional to the thickness of the thermal b.l.

The Nusselt number is named after Wilhelm Nusselt,[3] whose work on convective heat transfer was as basic as Prandtl's was in analyzing the related fluid dynamics (see Fig. 7.6).

We now turn to the detailed evaluation of h. And, as the preceding remarks make very clear, this evaluation will have to start with a development of the flow field in the boundary layer.

[3]Nusselt finished his doctorate in mechanical engineering at the Technical University in Münich in 1907. During an indefinite teaching appointment at Dresden (1913 to 1917) he made two of his most important contributions: He did the dimensional analysis of heat convection before he had access to Buckingham and Rayleigh's work. In so doing, he showed how to generalize limited data, and he set the pattern of subsequent analysis. He also showed how to predict convective heat transfer during film condensation. After moving about Germany and Switzerland from 1907 until 1925, he was named to the important Chair of Theoretical Mechanics at Münich. During his early years in this post he made basic contributions to heat exchanger design methodology. He held this position until 1952 during which time his, and Germany's, great influence in heat transfer and fluid mechanics waned. He was succeeded in the Chair by another of Germany's heat transfer luminaries, Ernst Schmidt.

We predict the boundary layer flow field by solving the equations that express conservation of mass and momentum in the b.l. Thus the first order of business is to develop these equations.

Conservation of mass—The continuity equation

A two- or three-dimensional velocity field can be expressed in vectorial form:

$$\vec{u} = \vec{i}\,u + \vec{j}\,v + \vec{k}\,w$$

where u, v, and w are the x, y, and z components of velocity. Figure 7.7 shows a two-dimensional velocity flow field. If the flow is steady, the paths of individual particles appear as steady *streamlines*. The streamlines can be expressed in terms of a mathematical function, $\psi(x,y) = $ constant, where each value of the constant identifies a separate streamline as shown in the figure.

The velocity, $\vec{u}$, is directed along the streamlines, so that no flow can cross them. Any pair of adjacent streamlines thus resembles a heat flow channel in a flux plot (Section 5.6); such channels were adiabatic—no heat flow could cross them. Therefore, we write the equation for the conservation of mass by summing the inflow and outflow of mass on two faces of a triangular element of unit depth as shown in Fig. 7.7:

$$\rho v\,dx - \rho u\,dy = 0 \qquad (7.7)$$

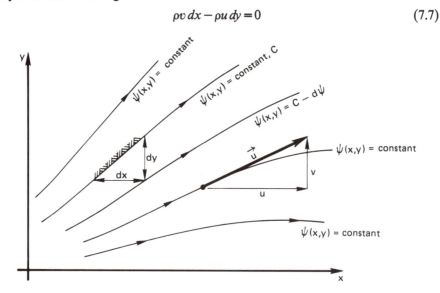

FIGURE 7.7 A steady, incompressible, two-dimensional flow field represented by streamlines, or lines of constant ψ.

or, as long as ρ = constant,

$$- v\, dx + u\, dy = 0 \tag{7.8}$$

But we can also differentiate the stream function along any streamline, $\psi(x,y) =$ constant, in Fig. 7.7:

$$d\psi = \left.\frac{\partial \psi}{\partial x}\right|_y dx + \left.\frac{\partial \psi}{\partial y}\right|_x dy = 0 \tag{7.9}$$

If we compare eqns. (7.8) and (7.9), we immediately see that the coefficients of dx and dy must be the same, so

$$v = -\left.\frac{\partial \psi}{\partial x}\right|_y \qquad \text{and} \qquad u = \left.\frac{\partial \psi}{\partial y}\right|_x \tag{7.10}$$

Furthermore

$$\frac{\partial^2 \psi}{\partial y\, \partial x} = \frac{\partial^2 \psi}{\partial x\, \partial y}$$

so it follows that

$$\boxed{\frac{\partial u}{\partial x} + \frac{\partial v}{\partial y} = 0} \tag{7.11}$$

This is called the two-dimensional *continuity equation* for steady incompressible flow, because it expresses mathematically the fact that the flow is *continuous*. It has no breaks in it. In three dimensions the continuity equation is

$$\boxed{\frac{\partial u}{\partial x} + \frac{\partial v}{\partial y} + \frac{\partial w}{\partial z} = 0}$$

Example 7.1

Fluid moves with a uniform velocity, u_∞, in the x-direction. Find the stream function and see if it gives plausible behavior (see Fig. 7.8).

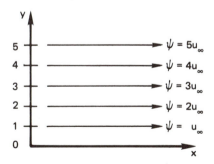

FIGURE 7.8　　Streamlines in a uniform horizontal flow field, $\psi = u_\infty y$.

SOLUTION. $u = u_\infty$ and $v = 0$. Therefore, from eqns. (7.10)

$$u_\infty = \frac{\partial \psi}{\partial y}\bigg|_x \qquad \text{and} \qquad 0 = \frac{\partial \psi}{\partial x}\bigg|_y$$

Integrating these equations, we get

$$\psi = u_\infty y + \text{fn}(x) \qquad \text{and} \qquad \psi = 0 + \text{fn}(y)$$

Comparing these equations, we get $\text{fn}(x) = \text{constant}$ and $\text{fn}(y) = u_\infty y + \text{constant}$, so

$$\psi = u_\infty y + \text{constant}$$

This gives a series of equally spaced, horizontal streamlines, as we would expect (see Fig. 7.8). We set the arbitrary constant equal to zero in the figure.

Conservation of momentum

The momentum equation in a viscous flow is a complicated vectorial equation called the Navier–Stokes equation. Its derivation is carried out in any advanced fluid mechanics text (see, e.g., [7.1, Chapt. III]). We shall offer a very restrictive derivation of the equation—one that applies only to a two-dimensional incompressible b.l. flow as shown in Fig. 7.9.

Here we see that shear stresses act upon any element such as to continuously distort and rotate it. In the lower part of the figure one such element is enlarged, so we can see the horizontal shear stresses[4] and the pressure forces that act upon it. They are shown as heavy arrows. We also display, as lighter arrows, the momentum fluxes entering and leaving the element.

Notice that both x and y directed momentum enters and leaves the element. To understand this one can envision a boxcar moving down the railroad track with a man standing, facing its open door. A child standing at a crossing throws him a baseball as the car passes. When he catches the ball its momentum will push him back, but a component of momentum will also jar him toward the rear of the train, owing to the relative motion. Particles of fluid entering element A will likewise influence its motion with their x components of momentum carried in by both components of flow.

The velocities must adjust themselves to satisfy the principle of conservation of linear momentum. Thus we require that the sum of the external forces in the x-direction, which act on the control volume, A, must be balanced by the rate at which the control volume, A, forces x-directed momentum out. The external forces, shown in Fig. 7.9, are

$$\left(\tau_{yx} + \frac{\partial \tau_{yx}}{\partial y}\, dy\right) dx - \tau_{yx}\, dx + p\, dy - \left(p + \frac{\partial p}{\partial x}\, dx\right) dy = \left(\frac{\partial \tau_{yx}}{\partial y} - \frac{\partial p}{\partial x}\right) dx\, dy$$

---

[4]The stress, τ, is often given two subscripts. The first one identifies the direction normal to the plane on which it acts, and the second one identifies the line along which it acts. Thus if both subscripts are the same, the stress must act normal to a surface—it must be a pressure or tension instead of a shear stress.

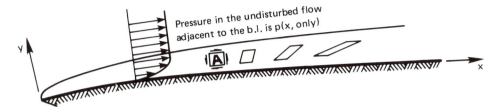

Pressure in the undisturbed flow adjacent to the b.l. is p(x, only)

a b.l., Showing how a fluid element, A, distorts, subject to flow and to shear stress, τ_{xy}.

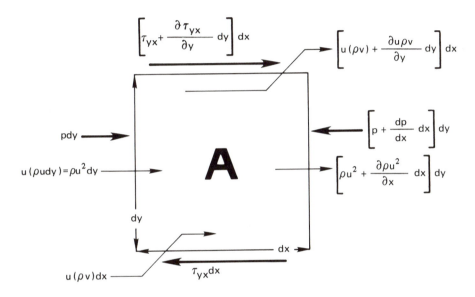

The x-direction forces and momentum fluxes acting upon element A in the b.l. shown above.

FIGURE 7.9   Forces acting in a two-dimensional incompressible boundary layer (b.l.).

The rate at which A loses x-directed momentum to its surroundings is

$$\left(\rho u^2 + \frac{\partial \rho u^2}{\partial x}\, dx\right) dy - \rho u^2\, dy + \left[u(\rho v) + \frac{\partial \rho uv}{\partial y}\, dy\right] dx - \rho uv\, dx$$

$$= \left(\frac{\partial \rho u^2}{\partial x} + \frac{\partial \rho uv}{\partial y}\right) dx\, dy$$

We equate these results and obtain the basic statement of conservation of x-directed momentum for the b.l.

$$\frac{\partial \tau_{yx}}{\partial y}\, dy\, dx - \frac{dp}{dx}\, dx\, dy = \left(\frac{\partial \rho u^2}{\partial x} + \frac{\partial \rho uv}{\partial y}\right) dx\, dy$$

The shear stress in this result can be eliminated with the help of Newton's law of viscous shear:

$$\tau_{yx} = \mu \frac{\partial u}{\partial y}$$

so the momentum equation becomes

$$\frac{\partial}{\partial y}\left(\mu \frac{\partial u}{\partial y}\right) - \frac{dp}{dx} = \left(\frac{\partial \rho u^2}{\partial x} + \frac{\partial \rho u v}{\partial y}\right)$$

Finally, we remember that the analysis is limited to $\rho \simeq$ constant, and we limit use of the equation to temperature ranges in which $\mu \cong$ constant. Then

$$\frac{\partial u^2}{\partial x} + \frac{\partial uv}{\partial y} = -\frac{1}{\rho}\frac{dp}{dx} + \nu \frac{\partial^2 u}{\partial y^2} \qquad (7.12)$$

This is one form of the steady, two-dimensional, incompressible, boundary layer equation on a flat surface. It we multiply eqn. (7.11) by u and subtract the result from the left-hand side of eqn. (7.12), we obtain a second form of the momentum equation:

$$u \frac{\partial u}{\partial x} + v \frac{\partial u}{\partial y} = -\frac{1}{\rho}\frac{dp}{dx} + \nu \frac{\partial^2 u}{\partial y^2} \qquad (7.13)$$

Equation (7.13) has a number of so-called "boundary layer assumptions" built into it. They are:

- $|\partial u/\partial x|$ is generally $\ll |\partial u/\partial y|$.
- v is generally $\ll u$.
- $p \neq \mathrm{fn}(y)$.

The Bernoulli equation for the free stream flow just above the boundary layer where there is no viscous shear,

$$\frac{p}{\rho} + \frac{u_\infty^2}{2} = \text{constant}$$

can be differentiated and used to eliminate the pressure gradient,

$$\frac{1}{\rho}\frac{dp}{dx} = -u_\infty \frac{du_\infty}{dx}$$

so from eqn. (7.12):

$$\frac{\partial u^2}{\partial x} + \frac{\partial(uv)}{\partial y} = u_\infty \frac{du_\infty}{dx} + \nu \frac{\partial^2 u}{\partial y^2} \qquad (7.14)$$

And if there is no pressure gradient in the flow—if p and u_∞ are constant—then eqns. (7.12), (7.13), and (7.14) become

$$\frac{\partial u^2}{\partial x} + \frac{\partial (uv)}{\partial y} = u\frac{\partial u}{\partial x} + v\frac{\partial u}{\partial y} = \nu\frac{\partial^2 u}{\partial y^2} \qquad (7.15)$$

Predicting the velocity profile
in the laminar boundary layer without a pressure gradient

Exact solution. There are two strategies for solving eqn. (7.15) for the velocity profile. The first was developed by Prandtl's student Blasius[5] before World War I. It is exact, and we shall only sketch it briefly. First we introduce the stream function, ψ, into eqn. (7.15). This reduces the number of dependent variables from two (u and v) to just one, namely ψ. We do this by substituting eqns. (7.10) in (7.15):

$$\frac{\partial \psi}{\partial y}\frac{\partial^2 \psi}{\partial y\partial x} - \frac{\partial \psi}{\partial x}\frac{\partial^2 \psi}{\partial y^2} = \nu\frac{\partial^3 \psi}{\partial y^3} \qquad (7.16)$$

It turns out that eqn. (7.16) can be converted into an ordinary d.e. with the following change of variables:

$$\psi(x,y) \equiv \sqrt{u_\infty \nu x}\, f(\eta) \qquad \text{where } \eta \equiv \sqrt{\frac{u_\infty}{\nu x}}\, y \qquad (7.17)$$

where $f(\eta)$ is an as-yet-undetermined function. [This transformation is rather similar to the one that we used to make an ordinary d.e. of the heat conduction equation, between eqn. (5.38) and (5.39).] After some manipulation of partial derivatives, this substitution gives (Problem 7.2)

$$f\frac{d^2 f}{d\eta^2} + 2\frac{d^3 f}{d\eta^3} = 0 \qquad (7.18)$$

and

$$\frac{u}{u_\infty} = \frac{df}{d\eta} \qquad \frac{v}{\sqrt{u_\infty \nu / x}} = \frac{1}{2}\left(\eta\frac{df}{d\eta} - f\right) \qquad (7.19)$$

The boundary conditions for this flow are

$$\left.\begin{aligned}
u(y=0) &= 0 &\quad \text{or} \quad& \left.\frac{df}{d\eta}\right|_{\eta=0} = 0 \\[2mm]
u(y=\infty) &= u_\infty &\quad \text{or} \quad& \left.\frac{df}{d\eta}\right|_{\eta=\infty} = 1 \\[2mm]
v(y=0) &= 0 &\quad \text{or} \quad& f(\eta=0) = 0
\end{aligned}\right\} \qquad (7.20)$$

[5]Blasius achieved great fame for many accomplishments in fluid mechanics and then gave it up. He is quoted as saying: "I decided that I had no gift for it; all of my ideas came from Prandtl."

Table 7.1 Exact velocity profile in the boundary layer on a flat surface with no pressure gradient

η $= y\sqrt{\dfrac{u_\infty}{\nu x}}$	$f(\eta)$	u/u_∞ $= df/d\eta$	$\nu\sqrt{\dfrac{x}{\nu u_\infty}}$ $= [(\eta df/d\eta) - f]/2$	$\dfrac{d^2f}{d\eta^2}$
0	0	0	0	0.33206
0.2	0.00664	0.06641	0.00332	0.33199
0.4	0.02656	0.13277	0.01322	0.33147
0.6	0.05974	0.19894	0.02981	0.33008
0.8	0.10611	0.26471	0.05283	0.32739
1.0	0.16557	0.32979	0.08211	0.32301
2.0	0.65003	0.62977	0.30476	0.26675
3.0	1.39682	0.84605	0.57067	0.16136
4.0	2.30576	0.95552	0.75816	0.06424
4.918	3.20169	0.99000	0.83344	0.01837
6.0	4.27964	0.99898	0.85712	0.00240
8.0	6.27923	1.00000^-	0.86039	0.00001

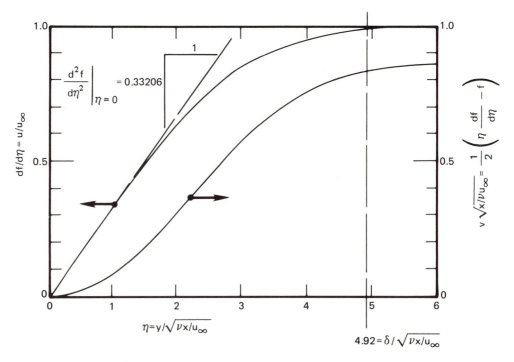

FIGURE 7.10 The dimensionless velocity components in a laminar boundary layer.

The solution of eqn. (7.18) subject to these b.c.'s must be done numerically. (The student who wants to practice programing should try it—see Problem 7.3.)

The solution of the Blasius problem is listed in Table 7.1, and the dimensionless velocity components are plotted in Fig. 7.10. The u-component increases from zero at the wall ($\eta = 0$) to 99% of u_∞ at $\eta = 4.92$. Thus the b.l. thickness is given by

$$4.92 = \frac{\delta}{\sqrt{\nu x / u_\infty}}$$

or, as we anticipated earlier,

$$\frac{\delta}{x} = \frac{4.92}{\sqrt{u_\infty x / \nu}} = \frac{4.92}{\sqrt{Re_x}} \qquad (7.2)$$

Concept of similarity. The exact solution for $u(x,y)$ reveals a most useful fact, namely that u can be expressed as a function of a single variable, η.

$$\frac{u}{u_\infty} = f'(\eta) = f'\left(y \sqrt{\frac{u_\infty}{\nu x}} \right)$$

This is called a *similarity solution*. To see why, we solve eqn. (7.2) for

$$\sqrt{\frac{u_\infty}{\nu x}} = \frac{4.92}{\delta(x)}$$

and substitute this in $f'\left(y / \sqrt{u_\infty / \nu x} \right)$. The result is

$$f' = \frac{u}{u_\infty} = \text{fn}\left[\frac{y}{\delta(x)} \right] \qquad (7.21)$$

The velocity profile thus has the same shape with respect to the b.l. thickness at each x-station. We say, in other words, that the profile is *similar* at each station. This is what we found to be true for heat transfer into a semiinfinite region. In that case [recall eqn. (5.45)], $x/\sqrt{t}$ always had the same value at the outer limit of the thermally disturbed region.

Boundary layer similarity makes it especially easy to use a simple method for solving other b.l. problems. This method, called the momentum integral method, is the subject of the next subsection.

Example 7.2

Air at 27°C blows over a flat surface with a sharp leading edge at 1.5 m/s. Find the b.l. thickness, $\frac{1}{2}$ m from the leading edge. Check the b.l. assumption that $u \gg v$, at the trailing edge.

SOLUTION.

$$\mu = 1.853 \times 10^{-5} \text{ kg/m-s}; \quad \nu = 1.566 \times 10^{-5} \text{ m}^2/\text{s}$$

$$Re_x = \frac{u_\infty x}{\nu} = \frac{1.5(0.5)}{1.566 \times 10^{-5}} = 47,893$$

The Reynolds number is low enough to permit the use of a laminar flow analysis. Then

$$\delta = \frac{4.92x}{\sqrt{\text{Re}_x}} = \frac{4.92(0.5)}{\sqrt{47,893}} = 0.01124 = \underline{1.124 \text{ cm}}$$

(Remember that the b.l. analysis is only valid of $\delta/x \ll 1$. In this case, $\delta/x = 1.124/50 = 0.0225$.) Finally, according to Fig. 7.10 or Table 7.1, v at $x = 0.5$ m is

$$v = \frac{0.8604}{\sqrt{x/\nu u_\infty}} = 0.8604\sqrt{\frac{(1.566)(10^{-5})(1.5)}{(0.5)}}$$

$$= 0.00590 \text{ m/s}$$

or

$$\frac{v}{u_\infty} = \frac{0.00590}{1.5} = 0.00393$$

Therefore, v is always $\ll u$, at least as long as we are not near the leading edge, where the b.l. assumptions themselves break down. We say more about this breakdown after eqn. (7.34).

Momentum integral method.[6] The second method for solving the b.l. momentum equation is approximate and much easier to apply to a wide range of problems than is any exact method of solution. The idea is this: we are not really interested in the details of the velocity or temperature profiles in the b.l., beyond learning their slopes at the wall. (These slopes give us the shear stress at the wall, $\tau_w = \mu(\partial u/\partial y)_{y=0}$, and the heat flux at the wall, $q_w = -k(\partial T/\partial y)_{y=0}$.) Therefore, we integrate the b.l. equations from the wall, $y = 0$, to the b.l. thickness, $y = \delta$, to make ordinary d.e.'s of them. It turns out that these much simpler equations do not reveal anything new about the temperature and velocity profiles, but they do give accurate explicit equations for τ_w and q_w.

Let us see how this procedure works with the b.l. momentum equation. We integrate eqn. (7.15), as follows, for the case in which there is no pressure gradient ($dp/dx = 0$):

$$\int_0^\delta \frac{\partial u^2}{\partial x}\, dy + \int_0^\delta \frac{\partial(uv)}{\partial y}\, dy = \nu \int_0^\delta \frac{\partial^2 u}{\partial y^2}\, dy$$

At $y = \delta$, u is within 1% of the free stream value, u_∞, and other quantities can also be evaluated at $y = \delta$ just as though y were infinite:

$$\int_0^\delta \frac{\partial u^2}{\partial x}\, dy + \left[\underbrace{(uv)_{y=\delta}}_{= u_\infty v_\infty} - \underbrace{(uv)_{y=0}}_{=0} \right] = \nu\left[\left(\frac{\partial u}{\partial y}\right)_{y=\delta} - \left(\frac{\partial u}{\partial y}\right)_{y=0} \right] \qquad (7.22)$$

---

[6]This method was developed by Pohlhausen, von Kármán, and others. See the discussion in [7.1, Chap. XII].

The continuity equation (7.11) can be integrated thus:

$$v_\infty - \underbrace{v_{y=0}}_{=0} = -\int_0^\delta \frac{\partial u}{\partial x}\, dy \tag{7.23}$$

Multiplying this by u_∞ gives

$$u_\infty v_\infty = -\int_0^\delta \frac{\partial u u_\infty}{\partial x}\, dy$$

Using this result in eqn. (7.22), we obtain

$$\int_0^\delta \frac{\partial}{\partial x}\left[u(u - u_\infty) \right] dy = -\left. v\frac{\partial u}{\partial y} \right|_{y=0}$$

Finally, we note that $\mu \left. \dfrac{\partial u}{\partial y} \right|_{y=0}$ is the shear stress exerted on the wall, $\tau_w = \tau_w(x$ only), so this becomes[7]

$$\boxed{\frac{d}{dx} \int_0^{\delta(x)} u(u - u_\infty)\, dy = -\frac{\tau_w}{\rho}} \tag{7.24}$$

Equation (7.24) expresses the conservation of linear momentum in integrated form. It shows that the rate of momentum loss caused by the b.l. is balanced by the shear force on the wall. When we use it in place of eqn. (7.15), we are said to be *using an integral method*. To make use of eqn. (7.24), we first nondimensionalize it as follows:

$$\frac{d}{dx}\left[\delta \int_0^1 \frac{u}{u_\infty}\left(\frac{u}{u_\infty} - 1\right) d\left(\frac{y}{\delta}\right) \right] = -\frac{v}{u_\infty \delta} \left. \frac{\partial(u/u_\infty)}{\partial(y/\delta)} \right|_{y=0}$$

$$= -\frac{\tau_w(x)}{\rho u_\infty^2} \equiv -\frac{1}{2} C_f(x) \tag{7.25}$$

where $\tau_w/(\rho u_\infty^2/2)$ is defined as the *skin friction coefficient*, C_f.

Equation (7.25) will be satisfied precisely by the exact solution (Problem 7.4) for u/u_∞. However, the point is to use eqn. (7.25) to determine u/u_∞ when we do not already have an exact solution. To do this we recall that the exact solution exhibits *similarity*. First, we guess the solution in the form of eqn. (7.21): $u/u_\infty = \text{fn}(y/\delta)$. This guess is made in such a way that it will fit the

---

[7] The interchange of integration and differentiation is consistent with Leibnitz's rule for differentiation of an integral (Problem 7.14).

following four things which are true of the velocity profile:

- $u/u_\infty = 0$ at $y/\delta = 0$
- $u/u_\infty \cong 1$ at $y/\delta = 1$
- $d\left(\dfrac{u}{u_\infty}\right)\Big/d\left(\dfrac{y}{\delta}\right) \cong 0$ at $y/\delta = 1$
- and from eqn. (7.15) we know that at $y/\delta = 0$:

$$\underset{=0}{\underbrace{u}}\frac{\partial u}{\partial x} + \underset{=0}{\underbrace{v}}\frac{\partial u}{\partial y} = \nu \frac{\partial^2 u}{\partial y^2}\Big|_{y=0}$$

(7.26)

so

$$\frac{\partial^2 (u/u_\infty)}{\partial (y/\delta)^2}\Big|_{y/\delta=0} = 0 \tag{7.27}$$

If the function $fn(y/\delta)$ is written as a polynomial with four constants, a, b, c, and d, in it,

$$\frac{u}{u_\infty} = a + b\frac{y}{\delta} + c\left(\frac{y}{\delta}\right)^2 + d\left(\frac{y}{\delta}\right)^3 \tag{7.28}$$

the four things that are known about the profile give

- $0 = a$, which eliminates a immediately
- $1 = 0 + b + \quad c + d$
- $0 = \qquad b + 2c + 3d$
- $0 = \qquad\quad 2c$, which eliminates c as well

Solving the middle two equations (above) for b and d, we obtain $d = -\frac{1}{2}$ and $b = +\frac{3}{2}$, so

$$\frac{u}{u_\infty} = \frac{3}{2}\frac{y}{\delta} - \frac{1}{2}\left(\frac{y}{\delta}\right)^3 \tag{7.29}$$

This approximate velocity profile is compared with the exact Blasius profile in Fig. 7.11, and they prove to be equal within a maximum error of 8%. The only remaining problem is then that of calculating $\delta(x)$. To do this we substitute eqn. (7.29) in eqn. (7.25) and get, after integration (see Problem 7.5):

$$-\frac{d}{dx}\left[\delta\left(\frac{39}{280}\right)\right] = -\frac{\nu}{u_\infty \delta}\left(\frac{3}{2}\right) \tag{7.30}$$

or

$$-\frac{39}{280}\left(\frac{2}{3}\right)\left(\frac{1}{2}\right)\frac{d\delta^2}{dx} = -\frac{\nu}{u_\infty}$$

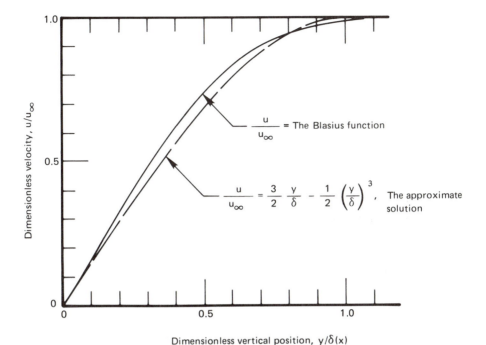

FIGURE 7.11   Comparison of the third-degree polynomial fit with the exact b.l. velocity profile. (Notice that the approximate result has been forced to $u/u_\infty=1$ instead of 0.99 at $y=\delta$.)

We integrate this using the b.c. $\delta^2=0$ at $x=0$:

$$\delta^2=\frac{280}{13}\frac{\nu x}{u_\infty}$$

or

$$\frac{\delta}{x}=\frac{4.64}{\sqrt{Re_x}} \tag{7.31}$$

This b.l. thickness is of the correct functional form and the constant is low by only 5.6%.

The friction coefficient

The fact that the function $u/u_\infty=f'(\eta)$ or $fn(y/\delta)$ gives all information about flow in the b.l. must be stressed. For example, the shear stress can be obtained from it by using Newton's law of viscous shear. Thus

$$\tau_w=\mu\frac{\partial u}{\partial y}\bigg|_{y=0}=\mu u_\infty\left[\frac{\partial f'}{\partial \eta}\frac{\partial \eta}{\partial y}\right]_{\eta=0}=\mu u_\infty\frac{\sqrt{u_\infty}}{\sqrt{\nu x}}\frac{d^2f}{d\eta^2}\bigg|_{\eta=0}$$

But from Fig. 7.10 and Table 7.1, we see that $(d^2f/d\eta^2)_{\eta=0}=0.33206$, so

$$\tau_w = 0.332 \frac{\mu u_\infty}{x} \sqrt{\mathrm{Re}_x} \tag{7.32}$$

The integral method that we just outlined would have given 0.323 for the constant in eqn. (7.32) instead of 0.332 (Problem 7.6).

The skin friction coefficient or *local drag coefficient* is defined as

$$\boxed{C_f \equiv \frac{\tau_w}{\rho u_\infty^2/2} = \frac{0.664}{\sqrt{\mathrm{Re}_x}}} \tag{7.33}$$

The overall drag coefficient, $\overline{C}_f$, is based on the average of the shear stress, τ_w, over the length, L, of the plate

$$\overline{\tau_w} = \frac{1}{L}\int_0^L \tau_w\, dx = \frac{\rho u_\infty^2}{2L}\int_0^L \frac{0.664}{\sqrt{\dfrac{u_\infty x}{\nu}}}\, dx = 1.328\frac{\rho u_\infty^2}{2}\sqrt{\frac{\nu}{u_\infty L}}$$

so

$$\boxed{\overline{C}_f = \frac{1.328}{\sqrt{\mathrm{Re}_L}}} \tag{7.34}$$

As a matter of interest we note that $C_f(x)$ approaches infinity at the leading edge of the flat surface. This means that to stop the fluid which first touches the front of the plate—dead in its tracks—would require infinite shear stress right at that point. Nature, of course, will not allow such a thing to happen; and it turns out that the boundary layer analysis is not really valid right at the leading edge. Actually, we must declare that the range $x \lesssim 5\delta$ (where the b.l. is relatively thick) is too close to the edge to use this analysis with accuracy. This converts to

$$x > 600\, \nu/u_\infty \qquad \text{for a boundary layer to exist}$$

In Example 7.2 this condition is satisfied for all x's greater than about 6 mm. This region is usually very small.

Example 7.3

Calculate the average shear stress and the overall drag coefficient for the surface in Example 7.2 if its total length is $L=0.5$ m. Compare $\overline{\tau_w}$ with τ_w at the trailing edge. At what point on the surface does $\tau_w = \overline{\tau_w}$? Finally, estimate what fraction of the surface can legitimately be analyzed using boundary layer theory.

SOLUTION.

$$\overline{C}_f = \frac{1.328}{\sqrt{\mathrm{Re}_{0.5}}} = \frac{1.328}{\sqrt{47{,}893}} = \underline{0.00607}$$

and

$$\overline{\tau_w} = \frac{\rho u_\infty^2}{2}\,\overline{C_f} = \frac{1.183(1.5)^2}{2}\,0.00607 = 0.00808\,\underbrace{\text{kg/m-s}^2}_{\text{N/m}^2}$$

(This is very little drag. It amounts only to about $\frac{1}{50}$ ounce/m^2.)

At $x = L$,

$$\left.\frac{\tau_w(x)}{\overline{\tau_w}}\right|_{x=L} = \frac{\rho u_\infty^2/2}{\rho u_\infty^2/2}\left[\frac{0.664/\sqrt{\text{Re}_L}}{1.328/\sqrt{\text{Re}_L}}\right] = \frac{1}{2}$$

and

$$\tau_w(x) = \overline{\tau_w} \qquad \text{where} \quad \frac{0.664}{\sqrt{x}} = \frac{1.328}{\sqrt{0.5}}$$

so the local shear stress equals the average value, where

$$x = \tfrac{1}{8}\text{m} \qquad \text{or} \qquad \frac{x}{L} = \frac{1}{4}$$

Thus the shear stress, which is initially infinite, plummets to $\overline{\tau_w}$ one-fourth of the way from the leading edge, and drops only to one-half of $\overline{\tau_w}$ in the remaining 75% of the plate.

The boundary layer assumptions fail when

$$x < 600\,\frac{\nu}{u_\infty} = 600\,\frac{1.566\times 10^{-5}}{1.5} = 0.0063\,\text{m}$$

Thus the preceding analysis should be good over almost 99% of the $\frac{1}{2}$-m length of the surface.

7.3 The energy equation

Derivation

We now know how the fluid moves in the b.l. Next we must extend the heat conduction equation to allow for the motion of the fluid. This equation can be solved for the temperature field in the b.l., and its solution can be used to calculate h, using Fourier's law:

$$h = \frac{q}{T_w - T_\infty} = -\frac{k}{T_w - T_\infty}\left.\frac{\partial T}{\partial y}\right|_{y=0} \tag{7.35}$$

To predict T we extend the analysis done in Section 2.1. Figure 2.5 shows an element of a solid body subjected to a temperature field. We now allow this volume to contain fluid with a velocity field $\vec{u}(x,y,z)$ in it, as shown in Fig. 7.12.

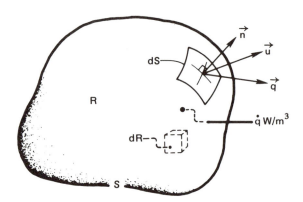

FIGURE 7.12   Control volume in a heat-flow and fluid-flow field.

We make the following restrictive assumptions:

- The fluid is incompressible.
- The fluid shear does not dissipate enough energy to heat the liquid up significantly.
- The temperature dependence of properties is weak.
- Kinetic and potential energy changes are negligible.

Just as we wrote eqn. (2.11) in Section 2.1, we now write conservation of energy in the form

$$-\underbrace{\int_S (-k\vec{\nabla}T)\cdot\vec{n}\,dS}_{\substack{\text{net heat flow}\\\text{rate from }R}} - \underbrace{\int_S [\rho\vec{u}(\text{enthalpy})]\cdot\vec{n}\,dS}_{\substack{\text{rate of energy and}\\\text{flow work out of }R}} + \underbrace{\int_R \dot{q}\,dR}_{\substack{\text{rate of}\\\text{heat gen-}\\\text{eration}\\\text{in }R}} = \underbrace{\int_R \rho c\frac{\partial T}{\partial t}\,dR}_{\substack{\text{rate of}\\\text{energy in-}\\\text{crease in }R}}$$

$$(7.36)$$

But the enthalpy of an incompressible fluid is $c_p(T - T_{\text{ref}})$, so after we call in Gauss's theorem [eqn. (2.12)] to make volume integrals of the surface integrals, eqn. (7.36) becomes

$$\int_R \left(\vec{\nabla}\cdot k\vec{\nabla}T - \rho c_p\vec{\nabla}\cdot\vec{u}T - \rho c_p\frac{\partial T}{\partial t} + \dot{q}\right)dR = 0$$

Because the integrand must vanish identically (recall footnote 1, Chapter 1) and because k depends weakly on T,

$$k\nabla^2 T - \rho c_p\left(\underbrace{\vec{\nabla}\cdot\vec{u}T}_{} + \frac{\partial T}{\partial t}\right) + \dot{q} = 0$$

$$= \vec{u}\cdot\vec{\nabla}T + T\underbrace{\vec{\nabla}\cdot\vec{u}}_{}$$

$$= \frac{\partial u}{\partial x} + \frac{\partial v}{\partial y} + \frac{\partial w}{\partial z} = 0,\text{ by continuity}$$

It follows that

$$\boxed{\nabla^2 T \ + \ \frac{\dot{q}}{k} \ = \ \frac{1}{\alpha}\left(\frac{\partial T}{\partial t} \ + \ \vec{u}\cdot\vec{\nabla}T\right)} \tag{7.37}$$

heat	heat	energy	enthalpy
con-	genera-	increase	transport,
duction	tion	term	or convec-
term	term		tion term

This is the energy equation for an incompressible flow field. It is the same as the corresponding equation (2.15) for a solid body, except for the added "convection" term, $\vec{u}\cdot\vec{\nabla}T/\alpha$.

Consider the term in parentheses in eqn. (7.37):

$$\frac{\partial T}{\partial t} + \vec{u}\cdot\vec{\nabla}T = \frac{\partial T}{\partial t} + u\frac{\partial T}{\partial x} + v\frac{\partial T}{\partial y} + w\frac{\partial T}{\partial z} \equiv \frac{DT}{Dt} \tag{7.38}$$

D/Dt is exactly the so-called "substantial derivative" which is treated in some detail in every fluid mechanics course. DT/Dt is the rate of change of the temperature of a fluid particle as it moves in a flow field.

In a steady two-dimensional flow field without heat sources, eqn. (7.37) takes the form

$$\frac{\partial^2 T}{\partial x^2} + \frac{\partial^2 T}{\partial y^2} = \frac{1}{\alpha}\left(u\frac{\partial T}{\partial x} + v\frac{\partial T}{\partial y}\right) \tag{7.39}$$

Furthermore, in a b.l. $\partial^2 T/\partial x^2 \ll \partial^2 T/\partial y^2$, so the b.l. energy equation is

$$\boxed{u\frac{\partial T}{\partial x} + v\frac{\partial T}{\partial y} = \alpha\frac{\partial^2 T}{\partial y^2}} \tag{7.40}$$

Heat and momentum transfer analogy

Consider a b.l. in a fluid of bulk temperature, T_∞, flowing over a flat surface at temperature, T_w. The momentum equation and its b.c.'s can be written as

$$u\frac{\partial(u/u_\infty)}{\partial x} + v\frac{\partial(u/u_\infty)}{\partial y} = \nu\frac{\partial^2(u/u_\infty)}{\partial y^2} \qquad \left.\begin{array}{l} \left.\dfrac{u}{u_\infty}\right|_{y=0} = 0 \\[2mm] \left.\dfrac{u}{u_\infty}\right|_{y=\infty} = 1 \\[2mm] \left.\dfrac{\partial(u/u_\infty)}{\partial y}\right|_{y=\infty} = 0 \end{array}\right\} \tag{7.41}$$

And the energy equation (7.40) can be written in terms of a dimensionless temperature, $\Theta = (T - T_w)/(T_\infty - T_w)$ as

$$u\frac{\partial \Theta}{\partial x} + v\frac{\partial \Theta}{\partial y} = \alpha\frac{\partial^2 \Theta}{\partial y^2} \qquad \left.\begin{array}{l} \Theta(y=0) = 0 \\[2mm] \Theta(y=\infty) = 1 \\[2mm] \left.\dfrac{\partial \Theta}{\partial y}\right|_{y=\infty} = 0 \end{array}\right\} \qquad (7.42)$$

Notice that the problems of predicting u/u_∞ and Θ are *identical* with one exception: eqn. (7.41) has ν in it whereas eqn. (7.42) has α. If ν and α should happen to be equal, the temperature distribution in the b.l. is

$$\text{for } \nu = \alpha: \qquad \frac{T - T_w}{T_\infty - T_w} = f'(\eta), \qquad \text{derivative of the Blasius function}$$

since each problem must have the same solution.

In this case we can immediately calculate the heat transfer coefficient using eqn. (7.5):

$$h = \frac{k}{T_\infty - T_w}\left.\frac{\partial(T - T_w)}{\partial y}\right|_{y=0} = k\left[\frac{\partial f'}{\partial \eta}\frac{\partial \eta}{\partial y}\right]_{\eta=0}$$

but $(\partial^2 f/\partial \eta^2)_{\eta=0} = 0.33206$ (see Fig. 7.10) and $\partial \eta/\partial y = \sqrt{u_\infty/\nu x}$, so

$$\frac{hx}{k} = \text{Nu}_x = 0.33206\sqrt{\text{Re}_x} \qquad \text{for } \nu = \alpha \qquad (7.43)$$

Normally in using eqn. (7.43) or any other forced convection equation, properties should be evaluated at $T_{av} = (T_w + T_\infty)/2$.

Example 7.4

Water flows over a flat heater, 0.06 m in length, under high pressure at 300°C. The free stream velocity is 2 m/s and the heater is held at 315°C. What is the average heat flux?

SOLUTION.

$$\text{At } T_{av} = (315 + 300)/2 = 307°C: \qquad \nu = 0.124 \times 10^{-6} \text{ m}^2/\text{s}$$
$$\alpha = 0.124 \times 10^{-6} \text{ m}^2/\text{s}$$

Therefore, $\nu = \alpha$ and we can use eqn. (7.43). First we must calculate the average heat flux, $\bar{q}$. To do this we call $T_w - T_\infty \equiv \Delta T$ and write

$$\bar{q} = \frac{1}{L}\int_0^L h\Delta T\, dx = \frac{k\Delta T}{L}\int_0^L \frac{1}{x}\text{Nu}_x\, dx = 0.332\frac{k\Delta T}{L}\int_0^L \sqrt{\frac{u_\infty}{\nu x}}\, dx$$

so

$$\bar{q} = 2\Delta T\left(0.332\frac{k}{L}\sqrt{\text{Re}_L}\right) = 2q_{x=L}$$

Thus

$$\bar{h}=2h_{x=L}=0.664\frac{0.520}{0.06}\sqrt{\frac{2(0.06)}{0.124\times10^{-6}}}\ =5661\text{ W/m}^2\text{-}°\text{C}$$

and

$$q=\bar{h}\Delta T=5661(315-300)=84{,}915\text{ W/m}^2=\underline{84.9\text{ kW/m}^2}$$

Clearly, eqn. (7.43) is a very restrictive heat transfer solution. We now must try to find how to evaluate q when ν does not equal α.

7.4 The Prandtl number and the boundary layer thicknesses

Dimensional analysis

We must now look more closely at the implications of the similarity between the velocity and thermal boundary layers. We first ask what dimensional analysis reveals about heat transfer in the laminar b.l. We know by now that the dimensional functional equation for the heat transfer coefficient, h, should be

$$h=\text{fn}(k,x,\rho,c_{\text{p}},\mu,u_\infty)$$

We have excluded T_w-T_∞ on the basis of Newton's original hypothesis, borne out in eqn. (7.43), that $h\neq\text{fn}(\Delta T)$ during forced convection. This gives seven variables in J/°C, m, kg, and s, or $7-4=3$ pi-groups. They are

$$\Pi_1=\frac{hx}{k}\equiv\text{Nu}_x\qquad\Pi_2=\frac{\rho u_\infty x}{\mu}\equiv\text{Re}_x$$

and a new group:

$$\Pi_3=\frac{\mu c_{\text{p}}}{k}\equiv\frac{\nu}{\alpha}\equiv\text{Pr, Prandtl number}$$

Thus

$$\text{Nu}_x=\text{fn}(\text{Re}_x,\text{Pr})\tag{7.44}$$

in forced convection flow situations. Equation (7.43) was developed for the case in which $\nu=\alpha$ or $\text{Pr}=1$; therefore, it is of the same form as eqn. (7.44), although it does not display the Pr dependence of Nu_x.

To better understand the physical meaning of the Prandtl number, let us briefly consider how to predict its value in a gas.

Kinetic theory of μ and k

Figure 7.13 shows a small neighborhood of a point of interest in a gas in which there exists a velocity or temperature gradient. We identify the *mean free path* of

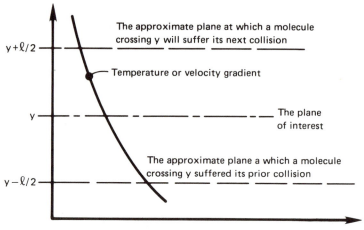

FIGURE 7.13   Momentum and energy transfer in a gas with a velocity or temperature gradient.

molecules between collisions as l and indicate planes at $y \pm l/2$ which bracket the average travel of molecules to be found at point y. (Actually, these planes should be located closer to $y \pm l$ for a variety of subtle reasons. This and other fine points of these arguments are explained in detail in [7.2].)

The shear stress, τ_{yx}, can be expressed as the change of momentum of all molecules that pass through the y-plane of interest, per unit area:

$$\tau_{yx} = \left(\begin{array}{c} \text{net mass flux of molecules} \\ \text{from } y - l/2 \text{ to } y + l/2 \end{array} \right) \cdot \left(\begin{array}{c} \text{change in fluid} \\ \text{velocity} \end{array} \right)$$

The net mass flux from top to bottom is proportional to $\rho\bar{c}$, where $\bar{c}$, the mean molecular speed, is $\gg u$ or v in incompressible flow. Thus

$$\tau_{yx} = C_1(\rho\bar{c})\left(l\frac{du}{dy} \right)\frac{\text{N}}{\text{m}^2} \quad \text{and this also equals} \quad \mu\frac{du}{dy} \quad (7.45)$$

By the same token

$$q_y = C_2(\rho c_p \bar{c})\left(l\frac{dT}{dy} \right) \quad \text{and this also equals} \quad -k\frac{dT}{dy}$$

The constants, C_1 and C_2, are on the order of unity. It follows immediately that

$$\mu = C_1(\rho\bar{c}l) \qquad \text{so} \qquad \nu = C_1(\bar{c}l)$$

and

$$k = C_2(\rho c_p \bar{c}l) \qquad \text{so} \qquad \alpha = C_2(\bar{c}l)$$

Thus, for a gas,

$$\text{Pr} \equiv \frac{\nu}{\alpha} = \text{a constant on the order of unity}$$

More detailed use of the kinetic theory of gases reveals more specific information as to the value of the Prandtl number, and these points are borne out reasonably well experimentally, as you can determine from Appendix A:

- For simple monatomic gases, $Pr = \frac{2}{3}$.

- For diatomic gases in which vibration is unexcited (such as N_2 and O_2 at room temperature) $Pr = \frac{5}{7}$.

- As the complexity of gas molecules increases, Pr approaches an upper value of unity.

- Pr is most insensitive to temperature in the simplest molecules because their structure is least responsive to temperature changes.

In a liquid the physical mechanisms of molecular momentum and energy transport are much more complicated and Pr can be far from unity. For example (cf. Table A.3):

- For liquids composed of fairly simple molecules, excluding metals, Pr is of the order of magnitude of 1 to 10.

- For liquid metals Pr is of the order of magnitude of 10^{-2} or less.

- If the molecular structure of a liquid is very complex, Pr might reach values on the order of 10^5. This is true of oils made of long-chain hydrocarbons, for example.

Thus, while Pr can vary over almost eight orders of magnitude in common fluids, it is still the result of analogous mechanisms of heat and momentum transfer. Both the numerical values of Pr, and the analogy itself, have their origins in the same basic process of molecular transport.

Boundary layer thicknesses, δ and δ_t

We have seen that the exact solution of the b.l. equations gives $\delta = \delta_t$ for $Pr = 1$, and it gives dimensionless velocity and temperature profiles that are identical on a flat surface. Two other things should be easy to see:

- When $Pr > 1$, $\delta > \delta_t$, and when $Pr < 1$, $\delta < \delta_t$. This is true because high viscosity leads to a thick velocity b.l., and a high thermal diffusity should give a thick thermal b.l.

- Since the exact governing equations (7.41) and (7.42) are identical for either b.l., except for the appearance of α in one and ν in the other, we expect that

$$\frac{\delta_t}{\delta} = fn\left(\frac{\nu}{\alpha} \text{ only}\right)$$

Therefore, we can combine these two observations, defining $\delta_t/\delta \equiv \phi$, and get

$$\phi = \text{monotonically decreasing function of Pr only} \tag{7.46}$$

The exact solution of the thermal b.l. equations proves this to be precisely true.

The fact that ϕ is independent of x will greatly simplify the use of the integral method. We shall establish the correct form of eqn. (7.46) in the following section.

7.5 Heat transfer coefficient for laminar, incompressible flow over a flat surface

The integral method for solving the energy equation

Integrating the b.l. energy equation in the same way as the momentum equation gives

$$\int_0^{\delta_t} u \frac{\partial T}{\partial x}\,dy + \int_0^{\delta_t} v \frac{\partial T}{\partial y}\,dy = \alpha \int_0^{\delta_t} \frac{\partial^2 T}{\partial y^2}\,dy$$

And the chain rule of differentiation in the form $x\,dy \equiv dxy - y\,dx$, reduces this to

$$\int_0^{\delta_t} \frac{\partial u T}{\partial x}\,dy - \int_0^{\delta_t} T \frac{\partial u}{\partial x}\,dy + \int_0^{\delta_t} \frac{\partial v T}{\partial y}\,dy - \int_0^{\delta_t} T \frac{\partial v}{\partial y}\,dy = \alpha \frac{\partial T}{\partial y} \Big]_0^{\delta_t}$$

or

$$\int_0^{\delta_t} \frac{\partial u T}{\partial x}\,dy \; + \underbrace{vT\,]_0^{\delta_t}}_{= \, v_\infty T_\infty - 0} \; - \int_0^{\delta_t} T \underbrace{\left(\frac{\partial u}{\partial x} + \frac{\partial v}{\partial y} \right)}_{\substack{=0, \\ \text{eqn. (7.11)}}}\,dy = \alpha \left[\underbrace{\frac{\partial T}{\partial y}\Big|_{\delta_t} }_{=0} - \frac{\partial T}{\partial y}\Big|_0 \right]$$

We evaluate v_∞ at $y = \delta$, using the continuity equation in the form of eqn. (7.23), in the preceding expression:

$$\int_0^{\delta_t} \frac{\partial}{\partial x} u(T - T_\infty)\,dy = \frac{1}{\rho c_{\mathrm{p}}} \left(-k \frac{\partial T}{\partial y}\Big|_0 \right) = \text{fn}(x \text{ only})$$

or

$$\boxed{\frac{d}{dx} \int_0^{\delta_t} u(T - T_\infty)\,dy = \frac{q_w}{\rho c_{\mathrm{p}}}} \tag{7.47}$$

Equation (7.47) expresses the conservation of thermal energy, in integrated form. It shows that the rate thermal energy is carried away by the b.l. flow, is matched by the rate heat is transferred in at the wall.

We can continue to paraphrase the development of the velocity profile in the laminar b.l., from the preceding section. We previously guessed the velocity profile in such a way as to make it match what we know to be true. We also know certain things to be true of the temperature profile. The temperatures at the wall and at the outer edge of the b.l. are known. Furthermore, the temperature distribution should be smooth as it blends into T_∞ for $y > \delta_t$. This condition is imposed by setting dT/dy equal to zero at $y = \delta_t$. A fourth condition is obtained by writing eqn. (7.40) at the wall where $u = v = 0$. This gives $(\partial^2 T/\partial y^2)_{y=0} = 0$. These four conditions take the following dimensionless form:

$$\left. \begin{array}{ll} \dfrac{T - T_\infty}{T_w - T_\infty} = 1 & \text{at } y/\delta_t = 0 \\[2mm] \dfrac{T - T_\infty}{T_w - T_\infty} = 0 & \text{at } y/\delta_t = 1 \\[2mm] \dfrac{d[(T - T_\infty)/(T_w - T_\infty)]}{d(y/\delta_t)} = 0 & \text{at } y/\delta_t = 1 \\[2mm] \dfrac{\partial^2[(T - T_\infty)/(T_w - T_\infty)]}{\partial(y/\delta_t)^2} = 0 & \text{at } y/\delta_t = 0 \end{array} \right\} \tag{7.48}$$

Equations (7.48) provide enough information to approximate the temperature profile with a cubic function.

$$\frac{T - T_\infty}{T_w - T_\infty} = a + b\frac{y}{\delta_t} + c\left(\frac{y}{\delta_t}\right)^2 + d\left(\frac{y}{\delta_t}\right)^3 \tag{7.49}$$

Substituting eqn. (7.49) into eqns. (7.48), we get

$$a = 1 \qquad -1 = b + c + d \qquad 0 = b + 2c + 3d \qquad 0 = 2c$$

which gives

$$a = 1 \qquad b = -\tfrac{3}{2} \qquad c = 0 \qquad d = \tfrac{1}{2}$$

so the temperature profile is

$$\boxed{\frac{T - T_\infty}{T_w - T_\infty} = 1 - \frac{3}{2}\frac{y}{\delta_t} + \frac{1}{2}\left(\frac{y}{\delta_t}\right)^3} \tag{7.50}$$

Equation (7.47) contains an as-yet-unknown quantity—the thermal b.l. thickness, δ_t. To calculate δ_t, we substitute the temperature profile, eqn. (7.50), and the velocity profile, eqn. (7.29), in the integral form of the energy equation, (7.47), which we first express as

$$u_\infty(T_w - T_\infty)\frac{d}{dx}\left[\delta_t \int_0^1 \frac{u}{u_\infty}\left(\frac{T - T_\infty}{T_w - T_\infty}\right)d\left(\frac{y}{\delta_t}\right)\right]$$

$$= -\frac{\alpha(T_w - T_\infty)}{\delta_t}\left.\frac{d\left(\dfrac{T - T_\infty}{T_w - T_\infty}\right)}{d(y/\delta_t)}\right|_{y/\delta_t = 0} \qquad (7.51)$$

There will be no problem in completing this integration if $\delta_t < \delta$. However, if $\delta_t > \delta$, there will be a problem because the equation, $u/u_\infty = 1$, instead of eqn. (7.29), defines the velocity beyond $y = \delta$. Let us proceed in the hope that the requirement that $\delta_t < \delta$ will be satisfied. Introducing $\phi \equiv \delta_t/\delta$ in eqn. (7.51) and calling $y/\delta_t \equiv \eta$, we get

$$\delta_t \frac{d\delta_t}{dx}\left[\underbrace{\int_0^1 \left(\frac{3}{2}\eta\phi - \frac{1}{2}\eta^3\phi^3\right)\left(1 - \frac{3}{2}\eta + \frac{1}{2}\eta^3\right)d\eta}_{= \frac{3}{20}\phi - \frac{3}{280}\phi^3}\right] = \frac{3\alpha}{2u_\infty} \qquad (7.52)$$

Since ϕ is a constant for any Pr [recall eqn. (7.46)], we separate variables

$$\frac{d\delta_t^2}{dx} = \frac{3\alpha/u_\infty}{\dfrac{3}{20}\phi - \dfrac{3}{280}\phi^3}$$

Integrating this result with respect to x and taking $\delta_t = 0$ at $x = 0$, we get

$$\delta_t = \sqrt{\frac{3\alpha x}{u_\infty}}\left/\sqrt{\frac{3}{20}\phi - \frac{3}{280}\phi^3}\right. \qquad (7.53)$$

But $\delta = 4.64x/\sqrt{\mathrm{Re}_x}$ in the integral formulation [eqn. (7.31)]. We divide by this value of δ to be consistent, and obtain

$$\frac{\delta_t}{\delta} \equiv \phi = 0.9638\left/\sqrt{\mathrm{Pr}\phi(1 - \phi^2/14)}\right.$$

Rearranging this gives

$$\frac{\delta_t}{\delta} = \frac{1}{1.025\mathrm{Pr}^{1/3}\left[1 - (\delta_t^2/14\delta^2)\right]^{1/3}} \simeq \frac{1}{1.025\mathrm{Pr}^{1/3}} \qquad (7.54)$$

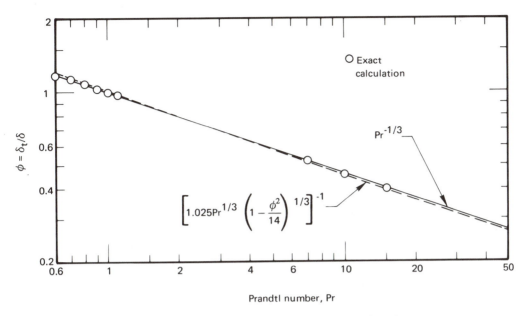

FIGURE 7.14   The exact and approximate Prandtl number dependence of the ratio of b.l. thicknesses.

The unapproximated result above is shown in Fig. 7.14 along with the results of Pohlhausen's precise calculation (see Schlichting [7.1, Chap. 14]). It turns out that the exact ratio, δ/δ_t, is represented with great accuracy by

$$\boxed{\frac{\delta_t}{\delta} = Pr^{-1/3}} \qquad 0.6 \leqslant Pr \leqslant 50 \tag{7.55}$$

So the integral method is accurate within 2.5% in the Prandtl number range indicated.

Notice that Fig. 7.14 is terminated for Pr less than 0.6. The reason for doing this is that the lowest Pr for gases is 0.67, and the next lower values of Pr are on the order of 10^{-2} for liquid metals. For Pr = 0.67, $\delta_t/\delta = 1.143$, which violates the assumption that $\delta_t < \delta$, but only by a small margin. For, say, mercury at 100°C, Pr = 0.0162 and $\delta_t/\delta = 3.952$, which violates the condition by an intolerable margin. We therefore have a theory that is acceptable for gases and all liquids except the metallic ones.

The final step in predicting the heat flux is to write Fourier's law:

$$q = -k\frac{\partial T}{\partial y}\bigg|_{y=0} = -k\frac{T_w - T_\infty}{\delta_t}\frac{\partial\left(\dfrac{T-T_\infty}{T_w-T_\infty}\right)}{\partial(y/\delta_t)}\bigg|_{y/\delta_t=0} \tag{7.56}$$

Using the dimensionless temperature distribution given by eqn. (7.50), we get

$$q = + k \frac{T_w - T_\infty}{\delta_t} \frac{3}{2}$$

or

$$h \equiv \frac{q}{\Delta T} = \frac{3k}{2\delta_t} = \frac{3}{2} \frac{k}{\delta} \frac{\delta}{\delta_t} \tag{7.57}$$

and substituting eqns. (7.54) and (7.31) for δ/δ_t and δ, we obtain

$$Nu_x \equiv \frac{hx}{k} = \frac{3}{2} \frac{\sqrt{Re_x}}{4.64} 1.025 Pr^{1/3} = 0.3314 Re_x^{1/2} Pr^{1/3}$$

Considering the various approximations, this is very close to the result of the exact calculation, which turns out to be

$$\boxed{Nu_x = 0.332 \, Re_x^{1/2} Pr^{1/3}} \qquad 0.6 \leqslant Pr \leqslant 50 \tag{7.58}$$

This expression gives very accurate results under the assumptions it is based on: a laminar two-dimensional b.l. on a flat surface with $T_w = $ constant, and $0.6 \leqslant Pr \leqslant 50$.

Some other laminar boundary layer heat transfer equations

High Pr.   At high Pr, eqn. (7.58) is still close to correct. The solution of the exact equation gives

$$Nu_x \to 0.339 Re_x^{1/2} Pr^{1/3} \qquad Pr \to \infty \tag{7.59}$$

Low Pr.   Figure 7.15 shows a low-Pr liquid flowing over a flat plate. In this case $\delta_t \gg \delta$, and for all practical purposes $u = u_\infty$ everywhere within the thermal

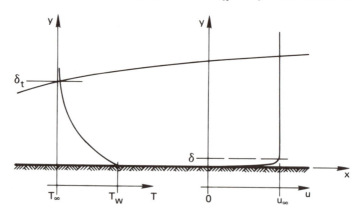

FIGURE 7.15   A laminar b.l. in a low Pr liquid. The velocity b.l. is so thin that $u \simeq u_\infty$ in the thermal b.l.

b.l. It is as though the no-slip condition $[u(y=0)=0]$ and the influence of viscosity were removed from the problem. Thus the dimensional functional equation for h becomes

$$h = \mathrm{fn}(x, k, \rho c_p, u_\infty) \tag{7.60}$$

There are five variables in J/°C, m, and s, so there are only two pi-groups. They are

$$\mathrm{Nu}_x = \frac{hx}{k} \quad \text{and} \quad \Pi_2 \equiv \mathrm{Re}_x\mathrm{Pr} = \frac{u_\infty x}{\alpha}$$

The new group, Π_2, is called a *Péclét number*, Pe_x, where the subscript identifies the length upon which it is based. It can be interpreted as follows:

$$\mathrm{Pe}_x \equiv \frac{u_\infty x}{\alpha} = \frac{\rho c_p u_\infty \delta_t}{k\delta_t/x} = \frac{\begin{array}{c}\text{heat capacity rate of}\\ \text{fluid in the b.l.}\end{array}}{\begin{array}{c}\text{axial heat conduction}\\ \text{in the b.l.}\end{array}} \tag{7.61}$$

As long as Pe_x is large, the b.l. assumption that $\partial^2 T/\partial x^2 \ll \partial^2 T/\partial y^2$ will be valid, but for small Pe_x (i.e., $\mathrm{Pe}_x \ll 100$) it will be violated and a boundary layer solution cannot be used.

The exact solution of the b.l. equations gives, in this case:

$$\left. \begin{array}{l} \mathrm{Nu}_x = 0.565\mathrm{Pe}_x^{1/2} \quad \mathrm{Pe}_x \geqslant 100 \quad \text{and} \\[4pt] \qquad\qquad\qquad \mathrm{Pr} \lesssim \frac{1}{100} \quad \text{or} \\[4pt] \qquad\qquad\qquad \mathrm{Re}_x \geqslant 10^4 \end{array} \right\} \tag{7.62}$$

General relationship.   Churchill and Ozoe [7.3] recommend the following empirical correlation for laminar flow on a constant-temperature flat surface for the entire range of Pr.

$$\boxed{\mathrm{Nu}_x = \frac{0.3387\mathrm{Pr}^{1/3}\mathrm{Re}_x^{1/2}}{\left[1+(0.0468/\mathrm{Pr})^{2/3}\right]^{1/4}}} \qquad \mathrm{Pe}_x > 100 \tag{7.63}$$

This relationship proves to be quite accurate, and it goes approximately to eqns. (7.59) and (7.62) in the high- and low-Pr limits. The calculations of an average Nusselt number for the general case is left as a homework exercise (Problem 7.10).

Boundary layer with an unheated starting length.   Figure 7.16 shows a b.l. with a heated region which starts at a distance, x_0, from the leading edge. The heat transfer in this instance is easily obtained using integral methods (see, e.g.,

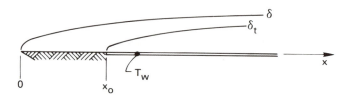

FIGURE 7.16   A b.l. with an unheated region at the leading edge.

[1.3, Chap. 5]).

$$\mathrm{Nu}_x = 0.332 \mathrm{Pr}^{1/3} \mathrm{Re}_x^{1/2} \Big/ \left[1 - \left(\frac{x_0}{x} \right)^{3/4} \right]^{1/3} \qquad x > x_0 \qquad (7.64)$$

Average heat transfer coefficient, $\bar{h}$.   The heat transfer coefficient, h, is the ratio of two quantities, q and ΔT, either of which might vary with x. So far we have only dealt with the *constant wall temperature problem*. Equations (7.58), (7.59), (7.62), and (7.63), for example, can all be used to calculate $q(x)$ when $(T_w - T_\infty) \equiv \Delta T$ is a specified constant. In the next subsection we discuss the problem of predicting $[T(x) - T_\infty]$ when q is a specified constant. This is called the *constant wall heat flux problem*.

The term $\bar{h}$ is used to designate either $\bar{q}/\Delta T$ in the constant wall temperature problem, or $q/\overline{\Delta T}$ in the constant wall heat flux problem. Thus

$$\text{constant wall temp.:} \quad \bar{h} \equiv \frac{\bar{q}}{\Delta T} = \frac{1}{\Delta T} \left[\frac{1}{L} \int_0^L q \, dx \right] = \frac{1}{L} \int_0^L h(x) \, dx \quad (7.65)$$

$$\text{constant heat flux:} \quad \bar{h} \equiv \frac{q}{\overline{\Delta T}} = \frac{q}{\dfrac{1}{L} \int_0^L \Delta T(x) \, dx} \qquad\qquad (7.66)$$

The Nusselt number based on $\bar{h}$ and a characteristic length, L, is designated $\overline{\mathrm{Nu}_L}$. This is not to be construed as an average of Nu_x, which would be meaningless in either of these cases.

Thus for a flat surface (with $x_0 = 0$), we use eqn. (7.58) in eqn. (7.65) to get

$$\bar{h} = \frac{1}{L} \int_0^L \underbrace{h(x)}_{\frac{k}{x}\mathrm{Nu}_x} dx = \frac{0.332 k \mathrm{Pr}^{1/3}}{L} \sqrt{\frac{u_\infty}{\nu}} \int_0^L \frac{\sqrt{x}\,dx}{x} = 0.664 \mathrm{Pr}^{1/3} \mathrm{Re}_L^{1/2} \left(\frac{k}{L} \right)$$

$$(7.67)$$

Thus $\bar{h} = 2h(x = L)$ in a laminar flow, and

$$\boxed{\overline{\mathrm{Nu}_L} = \frac{\bar{h}L}{k} = 0.664 \mathrm{Pr}^{1/3} \mathrm{Re}_L^{1/2}} \qquad\qquad (7.68)$$

Likewise for liquid metal flows:

$$\overline{\text{Nu}_L} = 1.13\text{Pe}_L^{1/2} \tag{7.69}$$

Some final observations. The preceding results are restricted to the two-dimensional, incompressible, laminar b.l. on a flat isothermal wall, at velocities that are not too high. These conditions are met if:

- Re_x or Re_L is not much in excess of 3.5×10^5.
- The Mach number of the flow is less than about $\frac{1}{4}$. (Even gaseous flows behave incompressibly at velocities well below sonic.) A related condition is:
- The "Eckert number," $\text{Ec} \equiv u_\infty^2 / c_p(T_w - T_\infty)$, is substantially less than unity. (This means that viscous dissipation—which we have neglected—does not play any role in the problem.)

It is worthwhile to notice how h and Nu depend on their independent variables:

$$h \text{ or } \bar{h} \propto \frac{1}{\sqrt{x}} \text{ or } \frac{1}{\sqrt{L}}, \ \sqrt{u_\infty}, \ \frac{1}{\nu^{1/6}}, k^{2/3}, (\rho c_p)^{1/3}$$
$$\text{Nu}_x \text{ or } \overline{\text{Nu}_L} \propto \sqrt{x} \text{ or } L, \ \sqrt{u_\infty}, \ \frac{1}{\nu^{1/6}}, \frac{1}{k^{1/3}}, (\rho c_p)^{1/3} \tag{7.70}$$

Thus $h \to \infty$ and Nu_x vanishes, at the leading edge, $x = 0$. Of course, an infinite value of h, like infinite shear stress, will not really occur at the leading edge because the b.l. description will actually break down in a small neighborhood of $x = 0$.

In all of the preceding considerations the fluid properties have been assumed constant. Actually, k, ρc_p, and especially ν might all vary noticeably with T, within the b.l. It turns out that if properties are all evaluated at the average temperature of the b.l., $(T_w + T_\infty)/2$, the results will normally be quite accurate.

Example 7.5

Air at 20°C and moving at 15 m/s is warmed by an isothermal steam-heated plate at 110°C, $\frac{1}{2}$ m in length and $\frac{1}{2}$ m in width. Find the average heat transfer coefficient and the total heat transferred. What are h, δ_t, and δ at the trailing edge?

SOLUTION. We evaluate properties at $T = (110 + 20)/2 = 65$°C. Then

$$\text{Pr} = 0.707 \quad \text{and} \quad \text{Re}_L = \frac{u_\infty L}{\nu} = \frac{15(0.5)}{0.0000194} = 386,600$$

so the flow ought to be laminar up to the trailing edge. The Nusselt number is then

$$\overline{\text{Nu}_L} = 0.664\text{Re}_L^{1/2}\text{Pr}^{1/3} = 367.8$$

and

$$\bar{h} = 367.8 \frac{k}{L} = \frac{367.8(0.02885)}{0.5} = \underline{21.2 \text{ W/m}^2\text{-}°C}$$

The value is quite low, owing to the low conductivity of air. The total heat flux is then

$$Q = \bar{h} A \,\Delta T = 21.2(0.5)^2(110-20) = \underline{477 \text{ W}}$$

By comparing eqns. (7.58) and (7.68), we see that $h(x=L) = \frac{1}{2}\bar{h}$, so

$$h \text{ (trailing edge)} = \frac{1}{2}(21.2) = \underline{10.6 \text{ W/m}^2\text{-}°C}$$

And finally,

$$\delta(x=L) = 4.92 L \Big/ \sqrt{\text{Re}_L} = \frac{4.92(0.5)}{\sqrt{386,600}} = 0.00396 \text{ m}$$

$$= \underline{3.96 \text{ mm}}$$

and

$$\delta_t = \frac{\delta}{\sqrt[3]{\text{Pr}}} = \frac{3.96}{\sqrt[3]{0.707}} = \underline{4.44 \text{ mm}}$$

The problem of constant wall heat flux

When the heat flux at the heater wall, q_w, is specified instead of the temperature, it is T_w that we need to know. We leave the problem of finding Nu_x for $q_w = $ constant as an exercise (Problem 7.11). The exact solution is

$$\text{Nu}_x = 0.453 \text{Pr}^{1/3}\text{Re}_x^{1/2} \tag{7.71}$$

where $\text{Nu}_x = hx/k = q_w x/k(T_w - T_\infty)$. The integral method gives the same result with a slightly lower constant (0.417).

We must be very careful in discussing *average* results in the constant heat flux case. The problem now might be that of finding an average temperature difference:

$$\overline{T_w - T_\infty} = \frac{1}{L}\int_0^L (T_w - T_\infty)\,dx = \frac{1}{L}\int_0^L \frac{q_w x}{k\big(0.453\text{Pr}^{1/3}\sqrt{u_\infty/\nu}\,\big)} \frac{dx}{\sqrt{x}}$$

or

$$\overline{T_w - T_\infty} = \frac{q_w L/k}{0.6795\text{Pr}^{1/3}\text{Re}_L^{1/2}} \tag{7.72}$$

which can be put into the form $\overline{\text{Nu}_L} = 0.6795\text{Pr}^{1/3}\text{Re}_L^{1/2}$ [although the Nusselt number is an awkward kind of nondimensionalization for $\overline{T_w - T_\infty}$]. Churchill and Ozoe [7.3] have pointed out that their eqn. (7.63) will describe $(T_w - T_\infty)$ with high accuracy over the full range of Pr if the constants are changed as follows:

- 0.3387 is changed to 0.4637.

- 0.0468 is changed to 0.02052.

Example 7.6

Air at 15°C flows at 1.8 m/s over a 0.6-m-long heating panel. The panel is intended to supply 420 W/m^2 to the air, but the surface can only sustain about 105°C without being damaged. Is it safe? What is the average temperature of the plate?

SOLUTION.   In accordance with eqn. (7.71),

$$\Delta T_{max} = \Delta T_{x=L} = \frac{qL}{k\mathrm{Nu}_{x=L}} = \frac{qL/k}{0.453\mathrm{Pr}^{1/3}\mathrm{Re}_x^{1/2}}$$

or if we evaluate properties at $(85+15)/2 = 50°C$, for the moment,

$$\Delta T_{max} = \frac{420(0.6)/0.0278}{0.453(0.709)^{1/3}[0.6(1.8)/1.794\times 10^{-5}]^{1/2}} = 91.5°C$$

This will give $T_{w_{max}} = 15 + 91.5 = 106.5°C$. This is very close to 105°C. If 105°C is at all conservative, $q = 420$ W/m^2 should be safe—particularly since it only occurs over a very small distance at the end of the plate.

From eqn. (7.72) we find that

$$\overline{\Delta T} = \frac{0.453}{0.6795}\Delta T_{max} = 61.0°C$$

so

$$\overline{T_w} = 15 + 61.0 = 76.0°C$$

7.6 Reynolds's analogy

The analogy between heat and momentum transfer can now be generalized to provide a very useful result. We begin by recalling eqn. (7.25), which is restricted to a flat surface with no pressure gradient:

$$\frac{d}{dx}\left[\delta\int_0^1 \frac{u}{u_\infty}\left(\frac{u}{u_\infty} - 1\right)d\left(\frac{y}{\delta}\right)\right] = -\frac{C_f}{2} \tag{7.25}$$

and rewriting eqns. (7.47) and (7.51), we obtain for the constant wall temperature case:

$$\frac{d}{dx}\left[\phi\delta\int_0^1 \frac{u}{u_\infty}\left(\frac{T-T_\infty}{T_w-T_\infty}\right)d\left(\frac{y}{\delta_t}\right)\right] = \frac{q_w}{\rho c_p u_\infty(T_w-T_\infty)} \tag{7.73}$$

But the similarity of temperature and flow boundary layers to one another [see, e.g., eqns. (7.29) and (7.50)], requires that

$$\frac{T-T_\infty}{T_w-T_\infty}\delta=\left(1-\frac{u}{u_\infty}\right)\delta_t$$

Substituting this result in eqn. (7.73) and comparing it to eqn. (7.25), we get

$$-\frac{d}{dx}\left[\delta\int_0^1\frac{u}{u_\infty}\left(\frac{u}{u_\infty}-1\right)d\left(\frac{y}{\delta}\right)\right]=-\frac{C_f}{2}=-\frac{q_w}{\rho c_p u_\infty(T_w-T_\infty)\phi^2} \quad (7.74)$$

Finally, we substitute eqn. (7.55) to eliminate ϕ from eqn. (7.74). This result is one instance of the *Reynolds–Colburn analogy*:[8]

$$\frac{h}{\rho c_p u_\infty}\mathrm{Pr}^{2/3}=\frac{C_f}{2} \quad (7.75)$$

The dimensionless group $h/\rho c_p u_\infty$ is called the Stanton number and defined as follows:

$$\text{St, Stanton number}\equiv\frac{h}{\rho c_p u_\infty}\equiv\frac{\mathrm{Nu}_x}{\mathrm{Re}_x\mathrm{Pr}}$$

The physical significance of the Stanton number is:

$$\text{St}=\frac{h\Delta T}{\rho c_p u_\infty\Delta T}=\frac{\text{actual heat flux to the fluid}}{\text{heat flux capacity of the fluid flow}} \quad (7.76)$$

The group $\mathrm{StPr}^{2/3}$ was dealt with by the chemical engineer Colburn, who gave it a special symbol:

$$j\equiv\text{Colburn }j\text{-factor}=\mathrm{StPr}^{2/3}=\frac{\mathrm{Nu}_x}{\mathrm{Re}_x\mathrm{Pr}^{1/3}} \quad (7.77)$$

Example 7.7

Does the equation for the Nusselt number on an isothermal flat surface in laminar flow satisfy Reynolds's analogy?

SOLUTION. If we rewrite eqn. (7.58), we obtain

$$\frac{\mathrm{Nu}_x}{\mathrm{Re}_x\mathrm{Pr}^{1/3}}=\mathrm{StPr}^{2/3}=\frac{0.332}{\sqrt{\mathrm{Re}_x}} \quad (7.78)$$

But comparison with eqn. (7.33) reveals that the left-hand side of eqn. (7.78) is precisely $C_f/2$, so the analogy is satisfied perfectly. Likewise, from eqns. (7.68) and (7.34) we get

$$\frac{\overline{\mathrm{Nu}_L}}{\mathrm{Re}_L\mathrm{Pr}^{1/3}}\equiv\overline{\mathrm{St}}\mathrm{Pr}^{2/3}=\frac{0.664}{\sqrt{\mathrm{Re}_L}}=\frac{\overline{C_f}}{2} \quad (7.79)$$

[8]Reynolds developed the analogy in 1874 [7.4]. Colburn made important use of it in this century.

There is a great deal of value to the Reynolds–Colburn analogy. It can be used directly to infer heat transfer data from measurements of the shear stress, or vice versa. It can also be extended to turbulent flow, which is much harder to predict analytically. We shall undertake that problem next.

Example 7.8

How much drag force does the air flow in Example 7.5 exert on the heat transfer surface?

SOLUTION.   From eqn. (7.79) in Example 7.5 we obtain

$$\overline{C_f} = \frac{2\overline{Nu_L}}{Re_L Pr^{1/3}}$$

From Example 7.5 we obtain $\overline{Nu_L}$, Re_L, and $Pr^{1/3}$:

$$\overline{C_f} = \frac{2(367.8)}{(386,600)(0.707)^{1/3}} = 0.002135$$

so

$$\overline{\tau_{yx}} = (0.002135)\frac{1}{2}\rho u_\infty^2 = \frac{(0.002135)(1.05)(15)^2}{2} = 0.2522 \text{ kg/m-s}^2$$

and the force is

$$\overline{\tau_{yx}}A = 0.2522(0.5)^2 = 0.06305 \text{ kg-m/s}^2 = 0.06305 \text{ N}$$

$$= 0.23 \text{ oz}$$

7.7 Turbulent boundary layers

Turbulence

> Big whirls have little whirls,
> That feed on their velocity.
> Little whirls have littler whirls,
> And so on, to viscosity.

This bit of doggerel by the English fluid mechanic, L. F. Richardson, tells us a great deal about the nature of turbulence. Turbulence in a fluid can be viewed as a spectrum of coexisting vortices of different sizes that dissipate energy from the larger ones to the smaller ones until we no longer see macroscopic vortices (or "whirls"). Then we identify the process as viscous dissipation.

The next time the weatherman shows a satellite photograph of North America on the 11:00 P.M. news, notice the cloud patterns. There will be one or two enormous vortices of continental proportions. These huge vortices in turn feed smaller "weather-making" vortices on the order of hundreds of miles in

diameter. These further dissipate into vortices of cyclone and tornado proportions—sometimes with that level of violence but more often not. These
dissipate into still smaller whirls as they interact with the ground and its various
protrusions. The next time the wind blows, stand behind any tree and *feel* the
vortices. In the great plains, where there are not a lot of ground vortex
generators, you will see small cyclonic eddies called "dust devils." The process
continues right on down to molecular dimensions. There, momentum exchange
is no longer identifiable as turbulence, but appears as viscosity.

The same kind of process exists within, say, a turbulent pipe flow at high
Reynolds number. Such a flow is shown in Fig. 7.17. Turbulence in such a case
consists of coexisting vortices which vary in size from a substantial fraction of
the pipe radius, down to molecular dimensions. The spectrum of sizes varies
with location in the pipe. The size and intensity of vortices at the wall must
clearly approach zero, since the fluid velocity approaches zero at the wall.

Figure 7.17 shows the fluctuation of a typical flow variable—namely
velocity—both with location in the pipe, and with time. This fluctuation arises
because of the turbulent motions that are superposed on the average local
flow. Other flow variables, such as T or ρ, can also vary in flows in the same
manner. For any variable we can write a local time-average value as

$$\bar{u} \equiv \frac{1}{\mathbf{T}} \int_0^{\mathbf{T}} u \, dt \qquad (7.80)$$

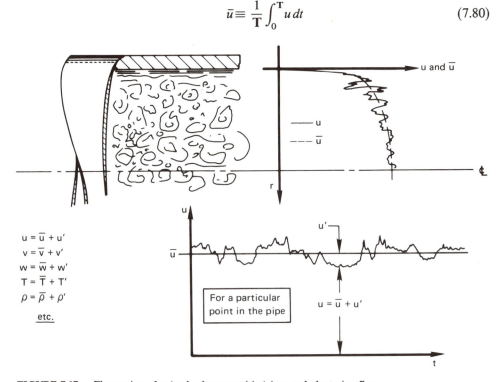

FIGURE 7.17   Fluctuation of u (and other quantities) in a turbulent pipe flow.

where **T** is a time that is much longer than the period of typical fluctuations.[9] Equation (7.80) can only be written for so-called "stationary" processes—ones for which $\bar{u}$ is nearly time-independent.

If we substitute $u = \bar{u} + u'$ in eqn. (7.80), where u is the actual local velocity and u' is the magnitude of the fluctuation, we obtain

$$\bar{u} = \underbrace{\frac{1}{\mathbf{T}} \int_0^{\mathbf{T}} \bar{u}\, dt}_{\bar{u}} + \underbrace{\frac{1}{\mathbf{T}} \int_0^{\mathbf{T}} u'\, dt}_{\overline{u'}} \qquad (7.81)$$

This is consistent with the fact that

$$\overline{u'} \text{ or any other average fluctuation} = 0 \qquad (7.82)$$

We now want to create a measure of the size, or *scale*, of turbulent fluctuations. This might be done experimentally by placing two velocity-measuring devices very close to one another in a turbulent flow field. There will then be a very high correlation between the two measurements. Then, suppose that the two velocity probes are moved apart until the measurements first become unrelated to one another. This spacing gives an indication of the average size of the turbulent motions.

Prandtl invented a slightly different (although related) measure of the scale of turbulence called the *mixing length*, *l*. He saw *l* as an average distance that a lump of fluid moved between interactions. It has a physical significance similar to that of the molecular mean free path. It would be harder to get a clean experimental measure of *l* than it is the scale of turbulence. But we can use *l* to fix the notion of turbulent shear stress.

The turbulent shear stress arises from the same kind of momentum exchange process that gives rise to the molecular viscosity. Recall that in that case, eqn. (7.45),

$$\tau_{yx} = C_1(\rho\bar{c})\left(l\frac{du}{dy}\right) = \mu\frac{du}{dy} \qquad (7.45)$$

where *l* was the molecular mean free path. In the turbulent flow case, pictured in Fig. 7.18, we rewrite eqn. (7.45) in the following way:

- *l* changes from the mean free path to the mixing length.
- $\bar{c}$ is replaced with $v = \bar{v} + v'$.
- The derivative du/dy is approximated as u'/l.

Then

$$\tau'_{yx} = C_1\rho(\bar{v} + v')u' \qquad (7.83)$$

Equation (7.83) can also be derived formally and precisely with the help of the Navier–Stokes equation. When this is done, C_1 comes out equal to -1.

---

[9]Take care not to interpret this **T** as a time constant; time constants are denoted as *T*.

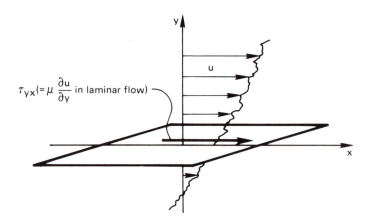

FIGURE 7.18   The shear stress, τ_{yx}, in a laminar or turbulent flow.

Then

$$\overline{\tau_{yx}} = -\frac{\rho}{\mathbf{T}} \int_0^{\mathbf{T}} (\bar{v}u' + v'u')\, dt = -\rho\bar{v}\underbrace{\overline{u'}}_{=0} - \rho\overline{v'u'} \qquad (7.84)$$

Notice that, while $\overline{u'} = \overline{v'} = 0$, averages of cross products of fluctuations (such as $\overline{u'v'}$ or $\overline{u'^2}$) do not generally vanish. Thus the turbulent shear stress is

$$\overline{\tau_{yx}} = -\rho\overline{v'u'} = \overline{\tau_{xy}} \qquad (7.85)$$

It is not easy to know how to write $\overline{v'u'}$ a priori, so we shall not make direct use of eqn. (7.85). Still, the essential similarity of the mechanisms giving rise to laminar and turbulent shear stresses suggests that the shear stress might be expressed as a combination of similar laminar and turbulent contributions.

$$\tau = \mu\frac{\partial\bar{u}}{\partial y} + \left(\begin{array}{c}\text{some other constant which}\\ \text{reflects turbulent mixing}\end{array}\right)\frac{\partial\bar{u}}{\partial y} \qquad (7.86)$$

or

$$\tau_{yx} = \rho(\nu + \varepsilon_{\mathrm{m}})\frac{\partial\bar{u}}{\partial y} \qquad (7.87)$$

where ε_{m} is called the momentum, or "eddy," diffusivity. We now use this characterization in calculating the heat transfer.

Reynolds–Colburn analogy for turbulent flow

The eddy diffusivity was actually introduced by Boussinesq [7.5] in 1877. It was subsequently proposed that Fourier's law might likewise be modified to

$$q = -k\frac{\partial\overline{T}}{\partial y} - \left(\begin{array}{c}\text{another constant which}\\ \text{reflects turbulent mixing}\end{array}\right)\frac{\partial\overline{T}}{\partial y}$$

where $\overline{T}$ is the average of the fluctuating temperature. Therefore,

$$q = -\rho c_p (\alpha + \varepsilon_h) \frac{\partial \overline{T}}{\partial y} \qquad (7.88)$$

where ε_h is called the "eddy diffusivity of heat." This immediately suggests yet another definition:

$$\text{turbulent Prandtl number, } \mathrm{Pr}_t \equiv \frac{\varepsilon_m}{\varepsilon_h} \qquad (7.89)$$

Equation (7.88) can be written in terms of ν and ε_m by introducing Pr and Pr_t into it. Thus

$$\frac{q}{\rho c_p} = -\left(\frac{\nu}{\mathrm{Pr}} + \frac{\varepsilon_m}{\mathrm{Pr}_t} \right) \frac{d\overline{T}}{dy} \qquad (7.90)$$

which looks a little like eqn. (7.87) when it is written in the form

$$\frac{\tau_{yx}}{\rho} = (\nu + \varepsilon_m) \frac{d\bar{u}}{dy} \qquad (7.91)$$

Notice that the derivatives have been changed from partial to total. This restricts the use of eqns. (7.90) and (7.91), in which $\bar{u}$ and $\overline{T}$ are predominantly y-dependent. This is strictly true only in the so-called "parallel flows"—ones in which all streamlines and isotherms are parallel. Parallel flow exists in pipes, but it is only an approximation in boundary layers.

Before trying to build a form of Reynolds's analogy for turbulent flow, we must note the behavior of Pr and Pr_t:

- Pr is a physical property of the fluid. It is both theoretically and actually near unity for ideal gases, but it differs radically from unity for liquids.
- Pr_t is a property of the flow field more than of the fluid. The numerical value of Pr_t is normally well within a factor of 2 of unity.

Let us first consider what will happen if $\mathrm{Pr} = \mathrm{Pr}_t = 1$. Then

$$\frac{q}{\rho c} = -(\nu + \varepsilon_m) \frac{d\overline{T}}{dy} = -\frac{\tau_{yx}}{\rho} \frac{dy}{d\bar{u}} \frac{d\overline{T}}{dy} = -\frac{\tau_{yx}}{\rho} \frac{d\overline{T}}{d\bar{u}}$$

So at the wall,

$$q(x) = -c_p \tau_w(x) \frac{d(\overline{T} - T_w)}{d\bar{u}} \qquad (7.92)$$

For $\mathrm{Pr} = 1$, $(T - T_w)/(T_\infty - T_w) = u/u_\infty$ in laminar flow. Therefore, we presume this same fact to be true for turbulent flow when $\mathrm{Pr} = \mathrm{Pr}_t = 1$. Equation (7.92) then becomes

$$q(x) = -c_p \tau_w(x) \frac{d}{d\bar{u}} \left[(T_\infty - T_w) \frac{\bar{u}}{u_\infty} \right]$$

or

$$q(x) = \frac{k}{\mu} \frac{T_w - T_\infty}{u_\infty} \tau_w(x) \tag{7.93}$$

since $\text{Pr} = \mu c_p / k = 1$. We call $(T_w - T_\infty) \equiv \Delta T$ and rearrange eqn. (7.93) to obtain

$$\frac{q(x)}{k \Delta T} x = \frac{1}{2} \frac{u_\infty x}{\nu} \frac{\tau_w(x)}{\rho u_\infty^2 / 2}$$

or

$$\text{Nu}_x = \tfrac{1}{2} \text{Re}_x C_f(x) \tag{7.94}$$

Equation (7.94) is based upon the assumption that $\text{Pr} = \text{Pr}_t = 1$ and upon the notion that the flow is parallel. It is also identical with the corresponding laminar flow equation for heat transfer in a b.l. with $\text{Pr} = 1$. Recall eqns. (7.75) and (7.77) which can be written as

$$\boxed{j = \text{St}_x \text{Pr}^{2/3} = \frac{C_f}{2}} \tag{7.95}$$

This suggests that the same result might also apply to the turbulent b.l. on an isothermal plate when $\text{Pr} \neq 1$. It turns out that we can do so with reasonable accuracy. We have noted already that eqn. (7.95) is called the Reynolds–Colburn analogy.

Predictions of heat transfer in the turbulent boundary layer

The frictional drag coefficient, C_f, in this case is no longer the laminar value, $0.664 / \sqrt{\text{Re}_x}$. It is, instead, the value appropriate to the turbulent flow in question. For example, Schlichting ([7.1, Chap. XXI]) shows that on a smooth flat plate in the low-Re turbulent b.l. range:

$$C_f = \frac{0.0592}{\text{Re}_x^{1/5}} \qquad 5 \times 10^5 \leqslant \text{Re}_x \leqslant 10^7 \tag{7.96}$$

In this case eqn. (7.95) becomes

$$\text{St}_x \text{Pr}^{2/3} = \frac{0.0296}{\text{Re}_x^{1/5}}$$

or

$$\text{Nu}_x = 0.0296 \text{Pr}^{1/3} \text{Re}_x^{0.8} \tag{7.97}$$

The Nusselt number based on $\bar{h}$ is obtained from eqn. (7.97) as follows:

$$\overline{\text{Nu}_L} = \frac{L}{k} \bar{h} = \frac{0.0296 \text{Pr}^{1/3} L}{k} \left[\frac{1}{L} \int_0^L \left(\frac{1}{x} \text{Re}_x^{0.8} \right) dx \right]$$

where we ignore the fact that there is a laminar region at the front of the plate. Thus

$$\overline{Nu_L} = 0.0370Pr^{1/3}Re_L^{0.8} \tag{7.98}$$

A flat heater with a turbulent b.l. on it actually has a laminar b.l. between $x=0$ and $x=x_{transition}$, as is indicated in Fig. 7.4. The obvious way to calculate $\bar{h}$ in this case is to write

$$\bar{h} = \frac{1}{L\Delta T}\int_0^L q\,dx = \frac{1}{L}\left[\int_0^{x_{transition}} h_{laminar}\,dx + \int_{x_{transition}}^L h_{turbulent}\,dx\right] \tag{7.99}$$

where $x_{transition} = (\nu/u_\infty)Re_{transition}$. Thus we substitute eqns. (7.58) and (7.97) in eqn. (7.99) and obtain, for $0.6 \leqslant Pr \leqslant 50$,

$$\boxed{\overline{Nu_L} = 0.037Pr^{1/3}\left\{Re_L^{0.8} - \left[Re_{transition}^{0.8} - 17.95(Re_{transition})^{1/2}\right]\right\}} \tag{7.100}$$

If $Re_L \gg Re_{transition}$, this result reduces to eqn. (7.98).

Whitaker [7.6] offers the following correlation for $\overline{Nu_L}$, which is similar in form to eqn. (7.100):

$$\boxed{\overline{Nu_L} = 0.036Pr^{0.43}(Re_L^{0.8} - 9200)\left(\frac{\mu_\infty}{\mu_w}\right)^{1/4}} \qquad Pr > 0.7 \tag{7.101}$$

This expression has been corrected to account for the variability of liquid viscosity with the factor $(\mu_\infty/\mu_w)^{1/4}$, where μ_∞ is evaluated at the free stream temperature, T_∞, and μ_w is evaluated at the wall temperature, T_w. If eqn. (7.101) is used to predict heat transfer to a gaseous flow, the viscosity-ratio correction term should not be used. This is because the viscosity of a gas rises with temperature instead of dropping, and the correction will be incorrect. Notice too that eqn. (7.101) compares very well with eqn. (7.100) when Pr is on the order of unity, if $Re_{transition}$ is only about 200,000.

Finally, it is important to remember that eqns. (7.100) and (7.101) should only be used when Re_L is substantially above the transitional value.

A problem with the preceding relations is that they do not really deal with the question of heat transfer in the rather lengthy transition region. Both eqns. (7.100) and (7.101) are based on the assumption that flow abruptly passes from laminar to turbulent at a critical value of x, and we have noted in the context of Fig. 7.4 that this is not what occurs. The location of the transition depends upon such variables as surface roughness and the turbulence, or lack of it, in the stream approaching the heater. Churchill [7.7] suggests correlating any *particular set* of data with

$$\frac{\overline{Nu_L} - 0.45}{0.6774\phi^{1/2}} = \left\{1 + \frac{(\phi/12,500)^{3/5}}{\left[1 + (\phi_{um}/\phi)^{7/2}\right]^{2/5}}\right\}^{1/2} \tag{7.102}$$

where

$$\phi \equiv \mathrm{Re}_L \mathrm{Pr}^{2/3} \left[1 + \left(\frac{0.0468}{\mathrm{Pr}} \right)^{2/3} \right]^{1/2}$$

and ϕ_{um} is a number between about 10^5 and 10^7. The actual value of ϕ_{um} must be fit to the particular set of data. In a very "clean" system ϕ_{um} will be larger and in a very "noisy" one it will be smaller.

The advantage of eqn. (7.102) is that, once ϕ_{um} is known, it will predict q *through* the transition regime.

Example 7.9

Ammonia at 100°C flows at 15 m/s over a flat surface 1.6 m in length at 200°C. Evaluate $\bar{h}$.

SOLUTION.   The properties of NH_3 at $(100+200)/2=150°C$ are $\nu=2.97\times10^{-5}$ m²/s, $k=0.0391$ W/m-°C, and $\mathrm{Pr}=0.87$. $\mathrm{Re}_L=1.6(15)/297(10)^{-5}=808{,}000$, so the flow is turbulent over a part of the surface. Then if we take $\mathrm{Re}_{transition}$ as 400,000 in eqn. (7.100), we get

$$\overline{\mathrm{Nu}_L}=0.037(0.87)^{1/3}\left\{(808{,}000)^{0.8}-\left[400{,}000^{0.8}-17.95(400{,}000)^{1/2}\right]\right\}=1209$$

so

$$\bar{h}=\frac{1209k}{L}=\frac{1209(0.0391)}{1.6}=29.5\ \mathrm{W/m^2\text{-}°C}$$

Whitaker's eqn. (7.101), on the other hand, gives

$$\overline{\mathrm{Nu}_L}=0.036(0.87)^{0.43}\left[(808{,}000)^{0.8}-9200\right]=1492$$

where we have deleted the viscosity correction, since the NH_3 is gaseous. This gives a 19% higher value of $\bar{h}$.

$$\bar{h}=\frac{1492(0.0391)}{1.6}=36.5\ \mathrm{W/m^2\text{-}°C}$$

Finally, using Churchill's formulation, we get $\phi=7.87\times10^5$, so eqn. (7.102) gives $\overline{\mathrm{Nu}_L}=697$ at $\phi_{um}=10^7$ and 2168 at $\phi_{um}=10^5$. These values spread over a factor of three and they embrace the values above. This serves to show how minor system variations can introduce a great deal of uncertainty into a combined laminar–turbulent system.

Example 7.10

Compare eqns. (7.100) and (7.101) at high Re_L, say $\mathrm{Re}_L \geqslant 10^7$.

SOLUTION.   Neglecting the viscosity ratio,

$$\frac{\overline{\mathrm{Nu}}_{L7.100}}{\overline{\mathrm{Nu}}_{L7.101}}=\frac{1.03}{\mathrm{Pr}^{0.13}}\left[1-\frac{\dfrac{\mathrm{Re}_{trans}^{0.8}-17.95\sqrt{\mathrm{Re}_{trans}}}{\mathrm{Re}_L^{0.8}}}{1-9200/\mathrm{Re}_L^{0.8}}\right]$$

In the worst case, $Re_{transition} = 500,000$ and $Re_L = 10^7$, this reduces to

$$\frac{\overline{Nu_{L_{7.100}}}}{\overline{Nu_{L_{7.101}}}} = \frac{1.066}{Pr^{0.13}}$$

up to $Pr \simeq 3$ the disagreement is within $\pm 7\%$. For higher Pr we should use Whitaker's relation, eqn. (7.101), with its stronger Pr dependence.

Example 7.11

What is $\overline{\tau_w}$ in Example 7.9?

SOLUTION. From Reynolds's analogy we obtain

$$\overline{C_f} = 2\overline{St_L}Pr^{2/3} = \frac{2\overline{Nu_L}}{Re_L Pr^{1/3}} = \frac{2(1492)}{808,000(0.87)^{1/3}} = 0.00387$$

Therefore,

$$\overline{\tau_w} = \frac{1}{2}\rho u_\infty^2 \overline{C_f} = \frac{0.4934(15)^2}{2}(0.00387) = \underline{0.215 \text{ N/m}^2}$$

(If the plate were 1 m wide, this would be a drag force of 0.344 N or 1.2 oz.)

A word about the analysis
of turbulent boundary layers

The preceding discussion has circumvented serious *analysis* of heat transfer in turbulent flows. Current methods of analysis are simply beyond the scope of this book. In the past, boundary layer heat transfer has been analyzed in many flows (with and without pressure gradients, dp/dx) using integral methods. However, in recent years, new computer-based methods have largely supplanted these techniques. These approaches are still described in the technical literature, more than by textbooks. The *Handbook of Heat Transfer* describes the state of recent work up to about 1970 (see [1.13, Chaps. 7 and 8]).

We have gotten around analysis by presenting some very recent correlations which are generally more accurate than analyses in the restrictive situations that they describe. In the next chapter we deal with more complicated configurations than the simple plane surface. A few of these configurations will be amenable to a level of analysis appropriate to a first course, but for others we shall only be able to present the best data correlations available.

PROBLEMS

7.1. Verify that eqn. (7.13) follows from eqns. (7.11) (7.12).

7.2. The student with some analytical ability (or some assistance from the instructor) should complete the algebra between eqns. (7.16) and (7.20).

7.3. Use a digital computer to solve eqn. (7.18) subject to b.c.'s (7.20). To do this you need all three b.c.'s at $\eta = 0$, but one is presently at $\eta = \infty$. There are three ways to get around this:

- Start out by guessing a value of $df/d\eta$ at $\eta = 0$—say $df/d\eta = 1$. When η is large—say 6 or 10—$df/d\eta$ will asymptotically approach a constant. If the constant > 1 go back and guess a lower value of $df/d\eta$, or vice versa, until the constant converges on unity.

- The correct value of $df'/d\eta$ is approximately 0.33206 at $\eta = 0$. You might cheat and begin with it.

- There exists a clever way to map $df/d\eta = 1$ at $\eta = \infty$ back into the origin. (Consult your instructor.)

7.4. Verify that the Blasius solution (Table 7.1) satisfies eqn. (7.25). To do this, carry out the required integration graphically.

7.5. Verify eqn. (7.30).

7.6. Obtain the counterpart of eqn. (7.32) based on the velocity profile given by the integral method.

7.7. Assume a laminar b.l. velocity profile of the form $u/u_\infty = y/\delta$ and calculate δ and C_f on the basis of this very rough estimate, using the momentum integral method. How accurate is each? [C_f is about 13% low.]

7.8. In a certain flow of water at 40°C over a flat plate δ m $= 0.005\sqrt{x}$ m . Plot *to scale* on a common graph (with an appropriately expanded y-scale):

- δ and δ_t for the water.
- δ and δ_t for air at the same temperature and velocity.

7.9. A thin film of liquid with a constant thickness, δ_0, falls down a vertical plate. It has reached its terminal velocity so that viscous shear and weight are in balance, and the flow is steady. The b.l. equation for such a flow is the same as eqn. (7.13) except that it has a gravity force in it. Thus

$$u\frac{\partial u}{\partial x} + v\frac{\partial u}{\partial y} = -\frac{1}{\rho}\frac{dp}{dx} - g + v\frac{\partial^2 u}{\partial y^2}$$

Assume that the surrounding air density $\simeq 0$, so there is no hydrostatic pressure gradient in the surrounding air. Then:

- Simplify the equation to describe this situation.
- Write the b.c.'s for the equation neglecting any air drag on the film.
- Solve for the velocity distribution in the film, assuming that you known δ_0 (cf. Chap. 9).

(This solution is the starting point in the study of many process heat and mass transfer problems.)

7.10. Develop an equation for $\overline{Nu_L}$ that is valid over the entire range of Pr, for a laminar b.l. over a flat isothermal surface.

7.11. Use an integral method to develop a prediction of Nu_x for a laminar b.l. over a constant heat flux surface. Compare your result with eqn. (7.71). What is the temperature difference at the leading edge of the surface?

7.12. Verify eqn. (7.100).

7.13. It is known from flow measurements that transition to turbulence occurs when the Reynolds number based on mean velocity and diameter exceeds 4000 in a certain pipe. Use the fact that the laminar boundary layer on a flat plate grows according to the relation

$$\frac{\delta}{x} = 4.92\sqrt{\frac{\nu}{u_{max}x}}$$

to find an equivalent value for the Reynolds number of transition based on distance from the leading edge of the plate and u_{max}. (Note that $u_{max} = 2\bar{u}_{av}$ during laminar flow in a pipe.)

7.14. Execute the differentiation in eqn. (7.24) with the help of Leibnitz's rule for the differentiation of an integral and show that the preceding equation results.

7.15. Liquid at 23°C flows at 2 m/s over a smooth, sharp-edged, flat surface 12 cm in length which is kept at 57°C. Calculate h at the trailing edge: (a) if the fluid is water; (b) if the fluid is glycerin ($h = 346$ W/m²-°C). (c) Compare the drag forces in the two cases. [There is 23.4 times as much drag in the glycerin.]

7.16. Air at -10°C flows over a smooth, sharp-edged, almost-flat, aerodynamic surface at 240 km/hr. The surface is at 10°C. Find: (a) the approximate location of the laminar turbulent transition; (b) the overall $\bar{h}$ for a 2-m chord; (c) h at the trailing edge for a 2-m chord; (d) δ and h at the transition point. [$\delta_{x_t} = 0.54$ mm.]

7.17. Find $\bar{h}$ in Example 7.9 using eqn. (7.102) with $\phi_{um} = 10^7$ and 10^6. Discuss the result.

7.18. Suppose that you had one data point with which to fix ϕ_{um} in Churchill's equation for $\overline{Nu_L}$ on a flat plate. This value is $\bar{h} = 32$ W/m²-°C in the system in Example 7.9. Evaluate ϕ_{um} and then use eqn. (7.102) to predict $\bar{h}$ if u_∞ is increased to 21 m/s.

7.19. Mercury at 25°C flows at 0.7 m/s over a 4-cm-long flat heater at 60°C. Find $\bar{h}$, τ_w, $h(x = 0.04$ m), and $\delta(x = 0.04$ m).

7.20. A large plate is at rest in water at 15°C. The plate is suddenly translated parallel with itself, at 1.5 m/s. The resulting fluid movement is not exactly like that in a b.l. because the velocity profile builds up uniformly, all over, instead of from an edge. The governing transient momentum equation, $Du/Dt = \nu(\partial^2 u/\partial y^2)$, takes the form

$$\frac{1}{\nu}\frac{\partial u}{\partial t} = \frac{\partial^2 u}{\partial y^2}$$

Determine u, 0.015 m from the plate at $t = 1$, 10, and 1000 s. Do this by first posing the problem fully and then comparing it with the solution in Section 5.5. [$u \simeq 0.003$ m/s after 10 s.]

7.21. Notice that when Pr is large, the velocity b.l. on an isothermal flat heater is much larger than δ_t. The small part of the velocity b.l. inside the thermal b.l. is approximately $u/u_\infty = \frac{3}{2}y/\delta = \frac{3}{2}\phi(y/\delta_t)$. Derive Nu_x for this case based on this velocity profile.

REFERENCES

[7.1]   H. Schlichting, *Boundary-Layer Theory*, 6th ed. (trans. J. Kestin), McGraw-Hill Book Company, New York, 1968.

[7.2]   C. L. Tien and J. H. Lienhard, *Statistical Thermodynamics*, rev. ed., Hemisphere Publishing Corp.: McGraw-Hill Book Company, Washington D.C., 1978.

[7.3]   S. W. Churchill and H. Ozoe, "Correlations for Laminar Forced Convection in Flow over an Isothermal Flat Plate and in Developing and Fully Developed Flow in an Isothermal Tube," *J. Heat Trans., Trans. ASME, Ser. C*, vol. 95, 1973, p. 78.

[7.4]   O. Reynolds, "On the Extent and Action of the Heating Surface for Steam Boilers," *Proc. Manchester Lit. Phil. Soc.*, vol. 14, 1874, pp. 7–12.

[7.5]   J. Boussinesq, "Théorie de l'écoulement tourbillant," *Mem. Pres. Acad. Sci.*, vol. 23, Paris, 1877, p. 46.

[7.6]   S. Whitaker, "Forced Convection Heat Transfer Correlation for Flow in Pipes Past Flat Plates, Single Cylinders, Single Spheres, and for Flow in Packed Beds and Tube Bundles," *AIChE J.*, vol. 18, 1972, p. 361.

[7.7]   S. W. Churchill, "A Comprehensive Correlating Equation for Forced Convection from Flat Plates," *AIChE J.*, vol. 22, 1976, pp. 264–268.

Chapter 8

Forced convection in a variety of configurations

8.1 Introduction

Consider for a moment the fluid flow pattern within a shell-and-tube heat exchanger such as that shown in Fig. 3.5. The shell-pass flow moves up and down across the tube bundle from one baffle to the next. The flow around each pipe is determined by the complexities of the one before it, and the direction of the mean flow relative to each pipe can vary. Yet the problem of determining the heat transfer in this situation, however difficult it appears to be, is a task that must be undertaken.

The flow within the tubes of the exchanger is somewhat more tractable, but it too brings with it several problems that do not arise in the flow of fluids over a flat surface. Heat exchangers thus present a kind of microcosm of internal and external forced convection problems. Other such problems arise everywhere that energy is delivered, controlled, utilized, or produced. They arise in the complex flow of water through nuclear heating elements, or in the liquid heating tubes of a solar collector—in the flow of a cryogenic liquid coolant in certain digital computers, or in the circulation of refrigerant in the spacesuit of a lunar astronaut.

We dealt with the simple configuration of flow over a flat surface in Chapter 7. This situation has considerable importance in its own right, and it also reveals a number of analytical methods that apply to other configurations. Now we wish to undertake a sequence of progressively harder problems of forced convection heat transfer in more complicated flow configurations.

Incompressible forced convection heat transfer problems normally admit an extremely important simplification: the fluid flow problem can be solved without reference to the temperature distribution in the fluid. Thus we can first find the velocity distribution, and then put it in the energy equation as known information and solve for the temperature distribution. Two things can impede this procedure, however:

- If the fluid properties (especially μ and ρ) vary significantly with temperature, we cannot predict the velocity without knowing the temperature, and vice versa. The problems of predicting velocity and temperature become intertwined and harder to solve. We encounter such a situation later in the study of natural convection, where the fluid is driven by thermally induced density changes.

- Either the fluid flow solution or the temperature solution can, itself, become prohibitively hard to find. When that happens we resort to the correlation of experimental data with the help of dimensional analysis.

Our aim in this chapter is to present the analysis of a few simple problems and to show the progression toward increasingly empirical solutions as the problems become progressively more unwieldy. We begin this undertaking with one of the simplest problems: that of predicting laminar convection in a pipe.

8.2 Heat transfer
to and from laminar flows in pipes

Not many industrial pipe flows are laminar, but laminar heating and cooling does occur in an increasing variety of modern instruments and equipment: laser coolant lines, certain cryogenic coolant systems, and many compact heat exchangers, for example. As in any forced convection problem, we first describe the flow field. This description will include a number of ideas that apply to turbulent as well as laminar flow.

Development of a laminar flow

Figure 8.1 shows the evolution of fully developed laminar flow from the entrance of a pipe. A b.l. builds up from the front, generally accelerating the otherwise undisturbed core. The b.l. eventually occupies the entire flow area and

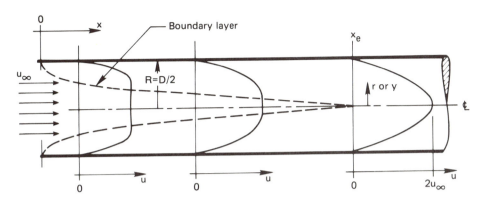

FIGURE 8.1 The development of a laminar velocity profile in a pipe.

defines a velocity profile that changes very little thereafter. We call such a flow *fully developed*. A flow is fully developed from the dynamical standpoint, when

$$\frac{\partial u}{\partial x} = 0 \quad \text{or} \quad v = 0 \tag{8.1}$$

at each radial location in the cross section. An attribute of a dynamically fully developed flow is that the streamlines are all parallel to one another.

The concept of a fully developed flow, from the thermal standpoint, is a little more complicated. We must first understand the notion of a mixing-cup, or bulk, temperature, T_b:

$$T_b \equiv \frac{\text{rate of flow of thermal}}{\text{rate of flow of heat capacity}}$$
$$\quad\quad\quad \frac{\text{energy through a cross section}}{\text{through a cross section}}$$

If we identify $\rho u(2\pi r)\,dr$ and $c_p T[\rho u(2\pi r)\,dr]$ as the mass and energy flows through an annular element of the cross section, respectively, this becomes

$$T_b = \frac{\int_0^R \rho u c_p T(2\pi r)\,dr}{c_p \int_0^R \rho u(2\pi r)\,dr} \tag{8.2}$$

Thus if the pipe were broken at any x-station and allowed to discharge into a mixing cup, the enthalpy of the mixed fluid in the cup would equal the average enthalpy of the fluid flowing through the cross section, and the temperature of the fluid in the cup would be T_b. This definition of T_b is perfectly general and applies to either laminar or turbulent flow.

A fully developed flow, from the thermal standpoint, is one for which

$$\frac{\partial}{\partial x}\left(\frac{T_w - T}{T_w - T_b}\right) = 0 \tag{8.3}$$

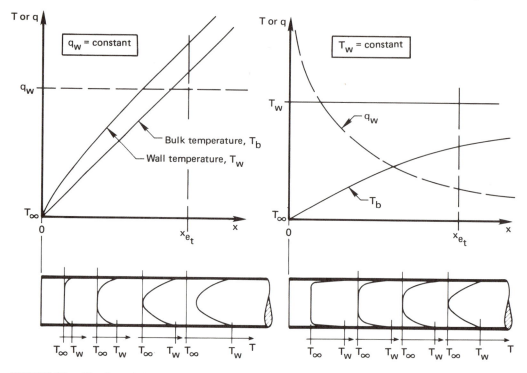

FIGURE 8.2   The thermal development of flows in tubes with a constant wall heat flux and with a constant wall temperature. (The *entrance region*.)

where T generally depends on x and r. That is, the relative shape of the temperature profile does not change with x, but the profile can be scaled up or down with $T_w - T_b$. Of course, a flow must be hydrodynamically developed if it is to be thermally developed.

Figures 8.2 and 8.3 show the development of two flows and their subsequent behavior, respectively. The two flows are subjected to either a constant wall heat flux or a constant wall temperature. In Fig. 8.2 we see each flow develop until its temperature profile achieves a shape which, except for a linear stretching, it will retain thereafter.

In Fig. 8.3 we see the fully developed variation of the temperature profile. When q_w is constant, the profile retains the same shape while the temperature rises at a constant rate at all values of r. Thus we can write an energy balance

$$2\pi R q_w \, dx = u_{av} \rho c_p (\pi R^2) \, dT_b$$

at any radial position, and get

$$\frac{dT_b}{dx} = \frac{2 q_w \alpha}{u_{av} R k} \tag{8.4}$$

This result is also valid for the average temperature in a turbulent flow.

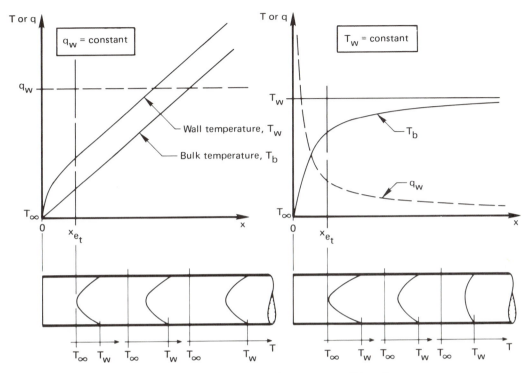

FIGURE 8.3   The thermal behavior of flows in tubes with a constant wall heat flux and with a constant temperature. (The *thermally developed regime.*)

In the constant wall temperature case, the temperature profile keeps the same shape, but its magnitude decreases with x. The lower right-hand corner of Fig. 8.3 has been drawn to conform with this requirement, as expressed in eqn. (8.3).

The velocity profile in laminar tube flows

The Buckingham pi-theorem tells us that if the hydrodynamic *entry length*, x_e, required to establish a fully developed velocity profile depends on u_{av}, μ, ρ, and D in three dimensions (kg, m, and s), then we expect to find two pi-groups:

$$\frac{x_e}{D} = \text{fn}(\text{Re}_D)$$

where $\text{Re}_D \equiv u_{av} D / \nu$. The matter of entry length is discussed by White ([8.1, Chap. 4]), who quotes

$$\frac{x_e}{D} \simeq 0.05 \text{Re}_D \qquad (8.5)$$

The constant, 0.05, guarantees that the laminar shear stress on the pipe wall will be within 1.4% of the value for fully developed flow when $x > x_e$. The number

0.03 can be used instead, if a deviation of 5% is tolerable. The thermal entry length, x_{e_t}, turns out to be different from x_e. We deal with it shortly.

The hydrodynamic entry length for a pipe carrying fluid at speeds near the transitional Reynolds number (2100) will extend beyond 100 diameters. Since heat transfer in pipes shorter than this is very often important, we will eventually have to deal with the entry region.

The velocity profile for a fully developed laminar incompressible pipe flow can be derived from the momentum equation for an axisymmetric flow. It turns out that the b.l. assumptions all happen to be valid for a developed pipe flow:

- The pressure is constant across any section.
- $\partial^2 u/\partial x^2$ is exactly zero.
- The radial velocity is not just small, but it is zero.
- The term $\partial u/\partial x$ is not just small, but it is zero.

The boundary layer equation for cylindrically symmetrical flows is quite similar to that for a flat surface, eqn. (7.13):

$$u\frac{\partial u}{\partial x}+v\frac{\partial u}{\partial r}=-\frac{1}{\rho}\frac{dp}{dx}+\frac{\nu}{r}\frac{\partial}{\partial r}\left(r\frac{\partial u}{\partial r}\right) \qquad (8.6)$$

For fully developed flows we go beyond the b.l. assumptions and set v and $\partial u/\partial x$ equal to zero as well, so eqn. (8.6) becomes

$$\frac{1}{r}\frac{d}{dr}\left(r\frac{du}{dr}\right)=\frac{1}{\mu}\frac{dp}{dx}$$

We integrate this twice and get

$$u=\left(\frac{1}{4\mu}\frac{dp}{dx}\right)r^2+C_1\ln r+C_2$$

The two b.c.'s on u express the no-slip (or zero-velocity) condition at the wall, and the fact that u must be symmetrical in r:

$$u(r=R)=0 \qquad \text{and} \qquad \frac{du}{dr}\bigg|_{r=0}=0$$

They give $C_1=0$ and $C_2=(-dp/dx)R^2/4\mu$, so

$$u=\frac{R^2}{4\mu}\left(-\frac{dp}{dx}\right)\left[1-\left(\frac{r}{R}\right)^2\right] \qquad (8.7)$$

This is the familiar Hagen-Poiseuille[1] parabolic velocity profile. We can identify the lead constant $(-dp/dx)R^2/4\mu$ as the maximum centerline velocity, $u_{\max}$. In

---

[1] The German scientist G. Hagen showed experimentally how u varied with r, dp/dx, μ, and R, in 1839. J. Poiseuille (pronounced Pwa-zói or more precisely Pwä-zə̄e) did the same thing, almost simultaneously (1840), in France. Poiseuille was a physician interested in blood flow, and we find today that if medical students know nothing else about fluid flow, they know "Poiseuille's law."

accordance with the conservation of mass (see Problem 8.1) $2u_{av} = u_{max}$, so

$$\frac{u}{u_{av}} = 2\left[1 - \left(\frac{r}{R}\right)^2\right] \tag{8.8}$$

Thermal behavior of a flow with a constant heat flux at the wall

The b.l. energy equation for a fully developed laminar incompressible flow, eqn. (7.40), takes the following simple form in a pipe flow where the radial velocity is equal to zero:

$$u\frac{\partial T}{\partial x} = \alpha \frac{1}{r}\frac{\partial}{\partial r}\left(r\frac{\partial T}{\partial r}\right) \tag{8.9}$$

For a fully developed flow with $q_w = $ constant, T_w and T_b are constants, also. Then, using eqns. (8.3) and (8.4), we get

$$\frac{\partial T_w}{\partial x} = \frac{\partial T_b}{\partial x} = \frac{dT_b}{dx} = \frac{\partial T}{\partial x} = \frac{2q_w\alpha}{u_{av}Rk}$$

Using this result and eqn. (8.8) in eqn. (8.9), we obtain

$$4\left(1 - \left[\frac{r}{R}\right]^2\right)\frac{q_w}{Rk} = \frac{1}{r}\frac{d}{dr}\left(r\frac{dT}{dr}\right) \tag{8.10}$$

This ordinary d.e. in r can be integrated twice to obtain

$$T = \frac{4q_w}{Rk}\left(\frac{r^2}{4} - \frac{r^4}{16R^2}\right) + C_1\ell n\, r + C_2 \tag{8.11}$$

The first b.c. on this equation is the symmetry condition, $\partial T/\partial r = 0$ at $r = 0$ and it gives $C_1 = 0$. The second b.c. is the definition of the mixing-cup temperature, eqn. (8.2). Substituting eqn. (8.11) with $C_1 = 0$ into eqn. (8.2) and carrying out the indicated integrations, we get

$$C_2 = T_b - \frac{7}{24}\frac{q_w R}{k}$$

so

$$T - T_b = \frac{q_w R}{k}\left[\left(\frac{r}{R}\right)^2 - \frac{1}{4}\left(\frac{r}{R}\right)^4 - \frac{7}{24}\right] \tag{8.12}$$

and at $r = R$ eqn. (8.12) gives

$$T_w - T_b = \frac{11}{24}\frac{q_w R}{k} = \frac{11}{48}\frac{q_w D}{k} \tag{8.13}$$

so the local Nu_D for fully developed flow, based on $h(x) = q_w/[T_w(x) - T_b]$, is

$$Nu_D \equiv \frac{q_w D}{(T_w - T_b)k} = \frac{48}{11} = 4.364 \tag{8.14}$$

Equation (8.14) is surprisingly simple. Indeed, the fact that there is only one dimensionless group in it is predictable by dimensional analysis. In this case the dimensional functional equation is merely

$$h = \mathrm{fn}(D, k)$$

We exclude ΔT, because h should be independent of ΔT in forced convection; μ, because the velocity profile is the same regardless of the viscosity; and ρu_{av}^2, because there is no influence of momentum in a laminar incompressible flow that never changes direction. This gives three variables in W/°C and m, or one dimensionless group, Nu_D, which must be a constant.

Example 8.1

Water at 20°C flows through a small bore tube 1 mm in diameter at an average speed of 0.2 m/s. The flow is fully developed at a point beyond which a constant heat flux of 6000 W/m² is imposed. How much farther down the tube will the water reach 74°C at its hottest point?

SOLUTION.   As a fairly rough approximation, we evaluate properties at $(74+20)/2 = $ 47°C: $k = 0.6367$ W/m-°C, $\alpha = 1.541 \times 10^{-7}$, and $\nu = 0.556 \times 10^{-6}$ m²/s. Therefore, $\mathrm{Re}_D = (0.001 \text{ m})(0.2 \text{ m/s})/0.556 \times 10^{-6}$ m²/s $= 360$, and the flow is laminar. Then, noting that T is greatest at the wall, we integrate eqn. (8.4) from $x = 0$ to $x = L$, where $T_{wall} = 74$°C:

$$T_b(x = L) = 20 + \frac{4 q_w \alpha}{u_{av} D k} L$$

And eqn. (8.13) gives

$$74 = T_b(x = L) + \frac{11}{48} \frac{q_w D}{k} = 20 + \frac{4 q_w \alpha}{u_{av} D k} L + \frac{11}{48} \frac{q_w D}{k}$$

so

$$\frac{L}{D} = \left(54 - \frac{11}{48} \frac{q_w D}{k} \right) \frac{u_{av} k}{4 q_w \alpha}$$

or

$$\frac{L}{D} = \left[54 - \frac{11}{48} \frac{6000(0.001)}{0.6367} \right] \frac{0.2(0.6367)}{4(6000)1.541(10)^{-7}} = 1785$$

so the wall temperature reaches the limiting temperature of 74°C at

$$L = 1785(0.001 \text{ m}) = \underline{1.785 \text{ m}}$$

Notice that we showed no concern over the thermal entry length. While we do not know what it would be, it should be much less than 1785 diameters.

Thermal behavior of the flow in an isothermal pipe

The dimensional analysis that showed $\mathrm{Nu}_D = $ constant for flow with a constant heat flux at the wall is unchanged when the pipe wall is isothermal. Thus Nu_D should still be constant. But this time (see, e.g., [1.18, Chap. 8]) the constant changes to

$$\mathrm{Nu}_D = 3.658 \qquad (T_w = \text{constant}) \tag{8.15}$$

for fully developed flow.

The thermal entrance region is of great importance in laminar flow because the thermally undeveloped region becomes extremely long for higher-Pr fluids. The entry-length equation (8.5) takes the following form for the thermal entry region,[2] where the velocity profile is fully developed at $x = 0$:

$$\frac{x_{e_t}}{D} \simeq 0.05 \mathrm{Re}_D \mathrm{Pr} \tag{8.16}$$

Thus the thermal entry length for the flow of cold water (Pr $\simeq$ 10) can be over 600 diameters in length near the transitional Reynolds number, and oil flows (Pr on the order of 10^4) practically never achieve fully developed temperature profiles.

A complete analysis of the heat transfer rate in the thermal entry region becomes quite complicated. The reader interested in details should look at [1.18, Chap. 8.] Dimensional analysis of the entry problem shows that the local value of h depends on u_{av}, μ, ρ, D, c_p, k, and x—eight variables in m, s, kg, and J/s. This means that we should anticipate four pi-groups:

$$\mathrm{Nu}_D = \mathrm{fn}(\mathrm{Re}_D, \mathrm{Pr}, x/D) \tag{8.17}$$

In other words, to the already familiar Nu, Re_D, and Pr, we add a new length parameter, x/D. The solution of the constant wall temperature problem, originally formulated by Graetz in 1885 [8.2] and solved in convenient form by Sellars, Tribus, and Klein in 1956 [8.3], includes an arrangement of these dimensionless groups, called the Graetz number:

$$\text{Graetz number, } \mathrm{Gz} \equiv \frac{\mathrm{Re}_D \mathrm{Pr} D}{x} \tag{8.18}$$

Figure 8.4 shows values of $\mathrm{Nu}_D \equiv hD/k$ for both the constant wall temperature and constant wall heat flux cases. The independent variable in the figure is a dimensionless length equal to 2/Gz. The figure also presents an average Nusselt number, $\overline{\mathrm{Nu}_D}$ for the isothermal wall case:

$$\overline{\mathrm{Nu}_D} \equiv \frac{\bar{h}D}{k} = \frac{D}{k}\left(\frac{1}{L}\int_0^L h\,dx\right) = \frac{1}{L}\int_0^L \mathrm{Nu}_D\,dx$$

where, since $h = q(x)/[T_w - T_b(x)]$, it is not possible to average just q or ΔT.

---

[2]The Nusselt number will be within 1.4% of the fully developed value if $x_{e_t} > 0.05 \mathrm{Re}_D \mathrm{Pr} D$. The constant can be reduced from 0.05 to 0.03 if we can tolerate 5% error.

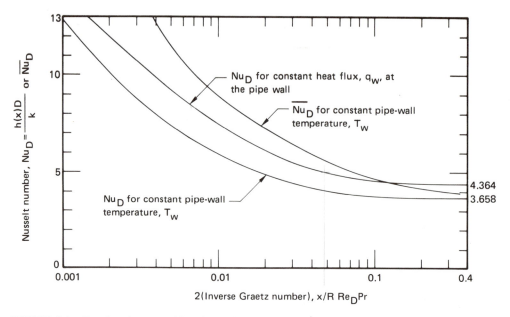

FIGURE 8.4   Local and average Nusselt numbers for the entry region of a pipe during laminar flow.

Example 8.2

A fully developed flow of air at 27°C moves at 2 m/s in a 1-cm-I.D. pipe. An electric resistance heater surrounds the last 20 cm of the pipe and supplies a constant heat flux to bring the air out at $T_b = 40$°C. What power input is needed to do this? What will be the wall temperature at the exit?

SOLUTION.   This is a case in which the wall heat flux is constant along the pipe. We first must compute $(Gz_{20 \text{ cm}})^{-1}$, evaluating properties at $(27+40)/2 \simeq 34$°C.

$$2Gz_{20 \text{ cm}}^{-1} = \frac{x}{R \, Re_D Pr}$$

$$= \frac{0.2 \text{ m}}{(0.005 \text{ m}) \dfrac{(2 \text{ m/s})(0.01 \text{ m})}{16.4 \times 10^{-6} \text{ m}^2/\text{s}} (0.711)} = 0.0461$$

From Fig. 8.4 we read $Nu_x = 5.0$, so

$$T_{w_{\text{exit}}} - T_b = \frac{q_w D}{5.0 k}$$

Notice that we still have two unknowns, q_w and T_w. The bulk temperature is specified as 40°C and q_w is obtained from this number by a simple energy balance:

$$q_w(2\pi Rx) = \rho c u_{\text{av}} (T_b - T_{\text{entry}})\pi R^2$$

so

$$q_w = 1.159 \frac{\text{kg}}{\text{m}^3} 1004 \frac{\text{J}}{\text{kg-}°\text{C}} 2 \frac{\text{m}}{\text{s}} (40-27)°\text{C} \left(\underbrace{\frac{R}{2x}}_{\frac{1}{80}} \right) = 378 \text{ W/m}^2$$

Then

$$T_{w_{\text{exit}}} = 40 + \frac{(378 \text{ W/m}^2)(0.01 \text{ m})}{5.0(0.0266 \text{ W/m-}°\text{C})} = \underline{68.4°\text{C}}$$

8.3 Turbulent pipe flow

Turbulent entry length

The entry lengths x_e and x_{e_t} are generally shorter in turbulent flow than they are in laminar flow. However, x_{e_t} depends on both Re and Pr in a complicated way. Table 8.1 gives the thermal entry length for various values of Pr and Re_D, based on a maximum of 10% error in Nu_D.

Here we see that x_{e_t} is very strongly dependent on Pr and influenced rather less by Re_D. Notice too that x_{e_t} decreases with Pr in turbulent flow while it increases in laminar flow.

Only liquid metal flows give fairly long thermal entry regimes, and they require a separate discussion, owing to certain problems that emerge at low Pr's.

The discussion that follows will deal almost entirely with fully developed turbulent pipe flows.

Illustrative experiment

Figure 8.5 shows average heat transfer data given by Kreith ([1.4, Chap. 8]) for air flowing in a 1-in.-I.D. isothermal pipe 60 in. in length. Let us see how these data compare with what we know about pipe flows thus far.

Table 8.1 Thermal entry lengths for which Nu_D will be no more than 10% above its fully developed value in turbulent flow

Pr	Re_D	x_{e_t}/D
0.01	200,000	28
0.01	100,000	20
0.01	50,000	12
0.7	200,000	7
0.7	100,000	7
0.7	50,000	7
10.0	100,000	$\mathcal{O}(1)$

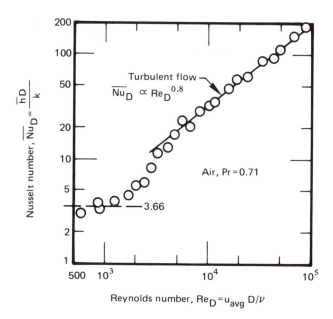

FIGURE 8.5   Heat transfer to air flowing in a 1 in. I.D., 60 in. long pipe (after Kreith [1.4]).

The data are plotted for a single Prandtl number on $\overline{Nu}_D$ vs. Re_D coordinates. This format is consistent with eqn. (8.17) in the fully developed range, but the actual pipe incorporates a significant entry region. Therefore, the data will reflect entry behavior.

The laminar data equal $\overline{Nu}_D \simeq 3.66$—the correct value for an isothermal pipe—at $Re_D = 750$. However, the pipe is too short for flow to be fully developed over much, if any, of its length. Therefore, $\overline{Nu}_D$ is not constant in the laminar range. The rate of rise of $\overline{Nu}_D$ with Re_D becomes very great in the transitional range, which lies between $Re_D = 2100$ and about 5000 in this case. Above $Re_D \simeq 5000$ the flow is turbulent and it turns out that $\overline{Nu}_D \simeq Re_D^{0.8}$.

Reynolds's analogy and heat transfer

The form of Reynolds's analogy appropriate to fully developed turbulent flow in a pipe can be derived from eqn. (7.95), in the form[3]

$$\frac{hx}{k} = \frac{1}{2}\frac{u_{av}x}{\nu}Pr^{1/3}C_f(x)$$

---

[3]C_f must be a *skin* friction coefficient. The "profile drag" that results from the variation of pressure around the body is unrelated to heat transfer. Reynolds analogy does not apply when it is included in C_f.

where h, in this case, is defined as $q/(T_w - T_b)$. We merely replace $C_f(x)$ with a constant value of the friction coefficient, C_f, and we multiply the equation by D/x to get

$$\mathrm{St}_D \mathrm{Pr}^{2/3} = \frac{C_f}{2} \tag{8.19}$$

This should not be used at very low or very high Pr's, but it can be used in either constant q_w or constant T_w situations.

The frictional resistance to flow in a pipe is normally expressed in terms of the Darcy–Weisbach friction factor, f [recall eqn. (3.23)]:

$$f \equiv \frac{\text{head loss}}{\dfrac{\text{pipe length}}{D}\dfrac{u_{av}^2}{2}} = \frac{\Delta p}{\dfrac{L}{D}\dfrac{\rho u_{av}^2}{2}} \tag{8.20}$$

where Δp is the pressure drop in a pipe of length, L. However,

$$\tau_w = \frac{\text{frictional force on liquid}}{\text{surface area of pipe}} = \frac{\Delta p\left[(\pi/4)D^2\right]}{\pi D L} = \frac{\Delta p D}{4L}$$

so

$$f = \frac{\tau_w}{\rho u_{av}^2/8} = 4C_f \tag{8.21}$$

Substituting eqn. (8.21) in eqn. (8.19) and rearranging the result, we obtain for fully developed flow,

$$\mathrm{Nu}_D = \mathrm{Re}_D \mathrm{Pr}^{1/3}(f/8) \tag{8.22}$$

The friction factor is given graphically in Fig. 8.6 as a function of Re_D and the relative roughness, ε/D, where ε is the root-mean-square roughness of the pipe wall. Equation (8.22) can be used directly along with Fig. 8.6 to calculate the Nusselt number.

For smooth pipes, the line $\varepsilon/D = 0$ is approximately tangent to the following line:

$$\frac{f}{4} = C_f = \frac{0.046}{\mathrm{Re}_D^{0.2}} \tag{8.23}$$

in the range $20{,}000 < \mathrm{Re}_D < 300{,}000$, so eqn. (8.22) becomes

$$\mathrm{Nu}_D = 0.023 \mathrm{Pr}^{1/3} \mathrm{Re}_D^{0.8}$$

for smooth pipes. This result was given by Colburn [8.4] in 1933. Actually, it is quite similar to an earlier result developed by Dittus and Boelter in 1930 (see [1.2, p. 552]) for smooth pipes.

$$\mathrm{Nu}_D = 0.0243 \mathrm{Pr}^{0.4} \mathrm{Re}_D^{0.8} \tag{8.24}$$

These equations are intended for reasonably low temperature differences under which properties can be evaluated at a mean temperature $(T_b + T_w)/2$. In

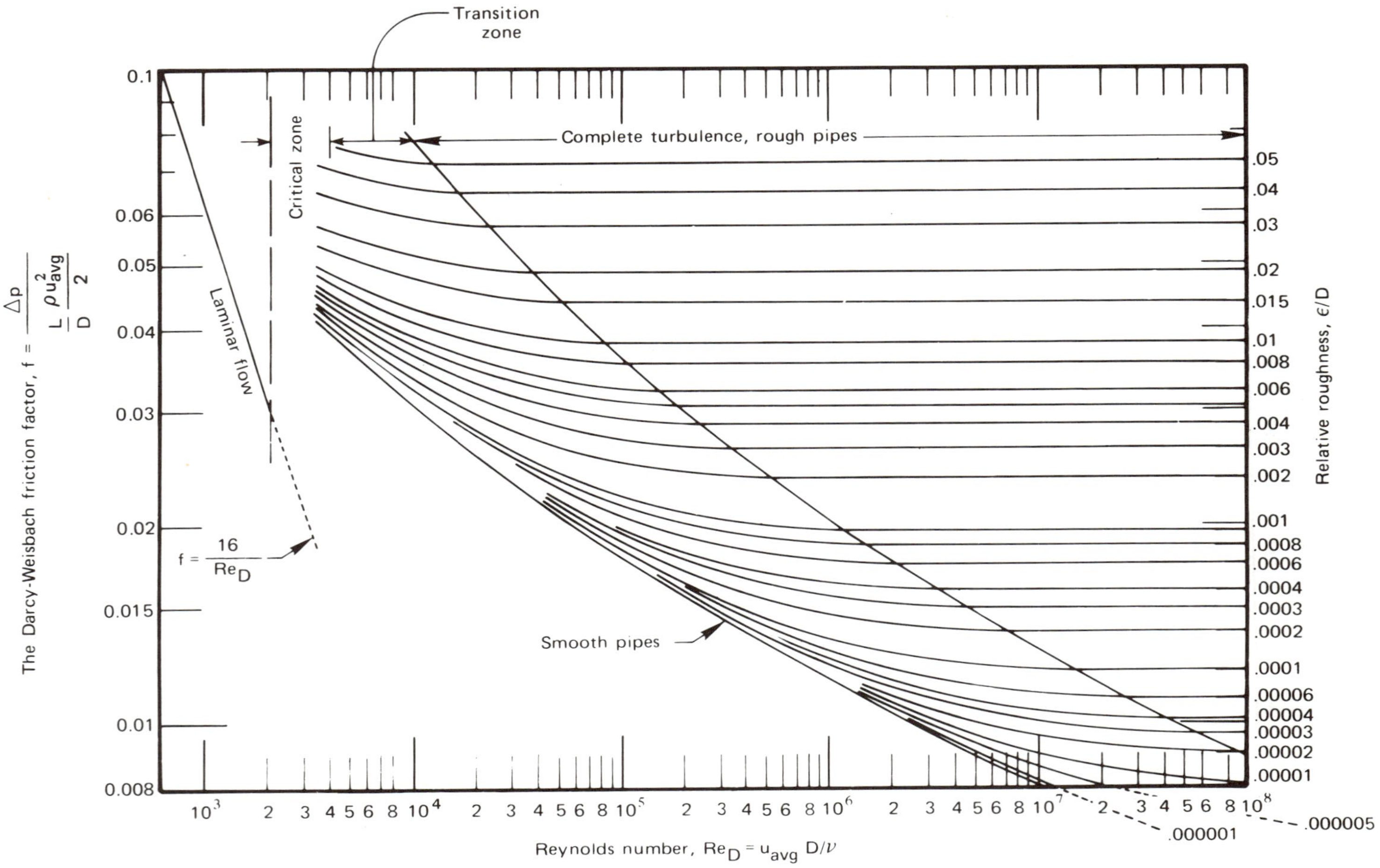

FIGURE 8.6 Pipe friction factors.

1936, Sieder and Tate [8.5] suggested that when $|T_w - T_b|$ is large enough to cause serious changes of μ, the Colburn equation can be modified in the following way for liquids:

$$\mathrm{Nu}_D = 0.023(\mathrm{Re}_D)^{0.8}(\mathrm{Pr})^{1/3}\left(\frac{\mu_b}{\mu_w}\right)^{0.14} \tag{8.25}$$

where all properties are evaluated at the local bulk temperature except μ_w, which is the viscosity evaluated at the wall temperature.

These early relations proved to be reasonably accurate. They gave maximum errors of $+25\%$ and -40% in the range $0.67 < \mathrm{Pr} < 100$ and usually were considerably more accurate than this. However, recent years have provided a great many more data, far more powerful analytical methods, and clever strategies for correlating data. During the 1950s and 1960s, B. S. Petukhov and his coworkers at the Moscow Institute for High Temperature developed a vastly improved description of forced convection heat transfer in pipes. Much of this work is described in a 1970 survey article by Petukhov [8.6].

Petukhov recommends the following equation for the local Nusselt number in fully developed flow, where all properties but μ_b and μ_w are evaluated at $(T_b + T_w)/2$.

$$\boxed{\mathrm{Nu}_D = \frac{(f/8)\mathrm{Re}_D\mathrm{Pr}}{1.07 + 12.7\sqrt{f/8}\ (\mathrm{Pr}^{2/3}-1)}\left(\frac{\mu_b}{\mu_w}\right)^n} \tag{8.26}$$

where

$$0 < \frac{\mu_b}{\mu_w} < 40$$

$$10^4 < \mathrm{Re}_D < 5\times10^6$$

$$0.5 < \mathrm{Pr} < 200 \qquad \text{for 6\% accuracy}$$

$$200 < \mathrm{Pr} < 2000 \qquad \text{for 10\% accuracy}$$

$$n = 0.11 \qquad \text{for uniform } T_w > T_b$$

$$n = 0.25 \qquad \text{for uniform } T_w < T_b$$

$$n = 0 \qquad \text{for } q_w = \text{uniform and/or for gases}$$

and where

$$f = \frac{1}{(1.82\log_{10}\mathrm{Re}_D - 1.64)^2} \tag{8.27}$$

for smooth pipes. For rough *or* smooth pipes, the Moody diagram, Fig. 8.6, can be used to obtain f.

Example 8.3

A 21.5-kg/s flow of water is dynamically and thermally developed in a 12-cm-I.D. pipe. The pipe is held at 90°C and ε is 0.00015. Find h where the bulk temperature of the fluid has reached 50°C.

SOLUTION.

$$u_{av} = \frac{\dot{m}}{\rho A} = \frac{21.5}{977\pi(0.06)^2} = 1.946 \text{ m/s}$$

so

$$\text{Re}_D = \frac{u_{av}D}{\nu} = \frac{1.946(0.12)}{4.07 \times 10^{-7}} = 573,700$$

and

$$\text{Pr} = 2.47 \qquad \frac{\mu_b}{\mu_w} = \frac{5.38 \times 10^{-4}}{3.10 \times 10^{-4}} = 1.74$$

From Fig. 8.6 we read $f = 0.0148$, and since $T_w > T_b$, $n = 0.11$. Thus eqn. (8.26) gives

$$\text{Nu}_D = \frac{(0.0148/8)(5.74 \times 10^5)(2.47)}{1.07 + 12.7\sqrt{0.0148/8}\ (2.47^{2/3} - 1)}(1.74)^{0.11} = 1831$$

or

$$h = 1831\frac{k}{D} = 1831\frac{0.661}{0.12} = \underline{10,080 \text{ W/m}^2\text{-}°C}$$

Heat transfer to fully developed
liquid-metal flows in tubes

A dimensional analysis of the forced convection flow of a liquid metal over a flat surface [recall eqn. (7.60) et seq.] showed that

$$\text{Nu} = \text{fn(Pe)} \tag{8.28}$$

because viscous influences were confined to a region very close to the wall. Thus the thermal b.l., which extends far beyond δ, is hardly influenced by the dynamic b.l. or by viscosity. During heat transfer to liquid metals in pipes the same thing occurs as is illustrated in Fig. 8.7. The region of thermal influence extends far beyond the laminar sublayer, when $\text{Pr} \ll 1$, and the temperature profile is not influenced by the sublayer. Conversely, if $\text{Pr} \gg 1$, the temperature profile is largely shaped within the laminar sublayer. At high or even moderate Pr's, ν is therefore very important, but at low Pr's it vanishes from the functional equation. Equation (8.28) thus applies to pipe flows as well as to flow over a flat surface.

Numerous measured values of Nu_D for liquid metals flowing in pipes with a constant wall heat flux, q_w, were assembled by Lubarsky and Kaufman in [8.7]. They are included in Fig. 8.8. It is clear that while most of the data correlate fairly well on Nu_D vs. Pe coordinates, certain sets of data are badly scattered. This occurs in part because liquid metal experiments are hard to carry out.

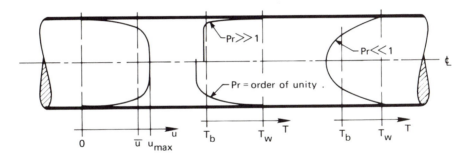

FIGURE 8.7   Velocity and temperature profiles during fully developed turbulent flow in a pipe.

Temperature differences are small and must often be measured at high temperatures. Some of the very low data might possibly be the result of the failure of metals to wet the inner surface of the pipe.

Another problem that besets liquid metal heat transfer measurements is the very great difficulty involved in keeping such liquids pure. Most impurities tend to result in lower values of h. Thus most of the Nusselt numbers in Fig. 8.8 have probably been lowered by impurities in the liquids, while the few high values are probably the more accurate ones.

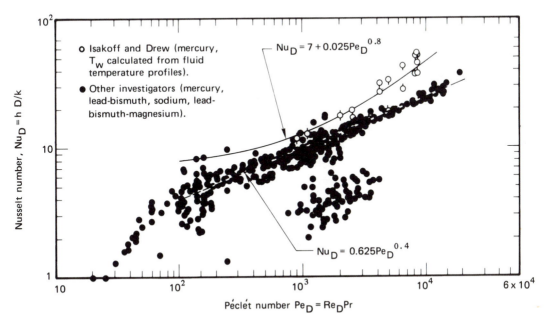

FIGURE 8.8   Comparison of measured and predicted Nusselt numbers for liquid metals heated in long tubes with wall heat flux, q_w. (See NACA TN 336, 1955 for details and data source references.)

There is a body of theory for turbulent liquid metal heat transfer that yields a prediction of the form

$$\mathrm{Nu}_D = C_1 + C_2 \mathrm{Pe}_D^{0.8} \tag{8.29}$$

The constants are normally in the ranges $2 \leqslant C_1 \leqslant 7$ and $0.0185 \leqslant C_2 \leqslant 0.386$ according to the test circumstances. Figure 8.8 includes Seban and Shimazaki's [8.8] relation for uniform wall temperature,

$$\mathrm{Nu}_D = 4.8 + 0.025 \mathrm{Pe}_D^{0.8} \tag{8.30}$$

For uniform wall heat flux, Lyon [8.9] recommends

$$\mathrm{Nu}_D = 7 + 0.025 \mathrm{Pe}_D^{0.8} \tag{8.31}$$

However we cannot overlook the fact that, although eqns. (8.30) and (8.31) are probably correct for pure liquids, the liquid metals in actual use are seldom pure. Lubarsky and Kaufman put the following line through the bulk of the data in Fig. 8.8:

$$\mathrm{Nu}_D = 0.625 \mathrm{Pe}_D^{0.4} \tag{8.32}$$

The use of eqn. (8.32) for $q_w = $ constant is far less optimistic than the use of eqns. (8.30) or (8.31). It should probably be used if it is safer to err on the low side.

8.4 Heat transfer during cross flow over cylinders

Fluid flow pattern

It will help us to understand the complexity of heat transfer from bodies in a cross flow if we first look in detail at the fluid flow patterns that occur in one cross-flow configuration—a cylinder with fluid flowing normal to it. Figure 8.9 shows how the flow develops as $\mathrm{Re} \equiv u_\infty D / \nu$ is increased from below 5 to near 10^7. An interesting feature of this evolving flow pattern is the fairly continuous way in which one flow transition follows another. The flow field degenerates to greater and greater degrees of disorder with each successive transition until, rather strangely, it regains order at the highest values of Re_D.

An important reflection of the complexity of the flow field is the vortex-shedding frequency, f_v. Dimensional analysis shows that a dimensionless frequency called the Strouhal number, Str, depends on the Reynolds number of the flow:

$$\mathrm{Str} \equiv \frac{f_v D}{u_\infty} = \mathrm{fn}(\mathrm{Re}_D) \tag{8.33}$$

Figure 8.10 defines this relationship experimentally on the basis of about 550 of the best data available (see [8.10]). The Strouhal number stays a little over 0.2

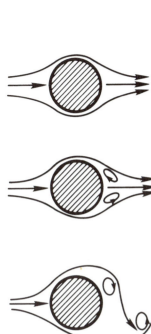

$Re_D < 5$ Regime of unseparated flow.

5 to $15 \leqslant Re_D < 40$ A fixed pair of Föppl
vortices in the wake

$40 \leqslant Re_D < 90$ and $90 \leqslant Re_D < 150$

Two regimes in which vortex street
is laminar:
   Periodicity governed in low Re_D
   range by wake instability

   Periodicity governed in high Re_D
   range by vortex shedding.

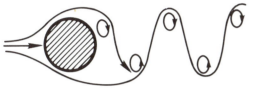

$150 \leqslant Re_D < 300$ Transition range to turbulence
in vortex.

$300 \leqslant Re_D \gtrsim 3 \times 10^5$ Vortex street is fully
turbulent, and the flow
field is increasingly
3-dimensional.

$3 \times 10^5 \gtrsim Re_D < 3.5 \times 10^6$

Laminar boundary layer has undergone
turbulent transition. The wake is
narrower and disorganized. No vortex
street is apparent.

$3.5 \times 10^6 \leqslant Re_D < \infty$ (?)

Re-establishment of the turbulent
vortex street that was evident in
$300 \leqslant Re_D \gtrsim 3 \times 10^5$. This time
the boundary layer is turbulent
and the wake is thinner.

FIGURE 8.9   Regimes of fluid flow across circular cylinders [8.10].

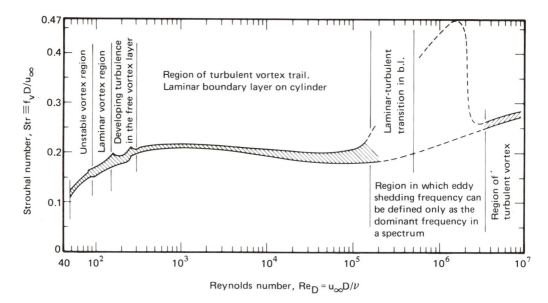

FIGURE 8.10   The Strouhal-Reynolds number relationship for circular cylinders, as defined by existing data [8.10].

over most of the range of Re_D. This means that behind a given object the vortex shedding frequency rises almost linearly with velocity.

Experiment 8.1

When there is a gentle breeze blowing outdoors, go out and locate a large tree with a straight trunk, or the shaft of a water tower. Wet your finger and place it in the wake a couple of diameters downstream and about one radius off center. Estimate the vortex shedding frequency and use $Str \simeq 0.21$ to estimate u_∞. Is your value of u_∞ reasonable?

Heat transfer

The action of vortex shedding greatly complicates the heat removal process. Giedt's data [8.11] in Fig. 8.11 show how the heat removal changes as the constantly fluctuating motion of the fluid to the rear of the cylinder changes with Re_D. Notice, for example, that Nu_D is near its minimum at 110° when $Re_D = 71,000$, but it maximizes at the same place when $Re_D = 140,000$. Direct prediction by the sort of b.l. methods that we discussed in Chapter 7 is out of the question. However, a great deal can be done with the data using relations of the form

$$\overline{Nu_D} = fn(Re_D, Pr)$$

The broad study of Churchill and Bernstein [8.12] probably brings the correlation of heat transfer data from cylinders about as far as it is possible. For

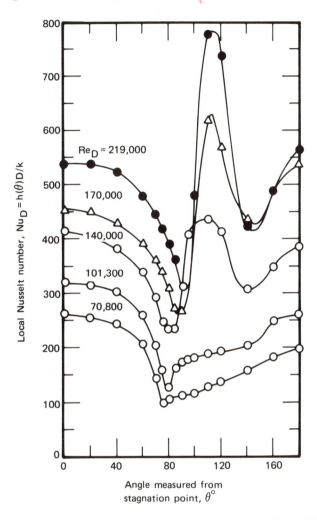

FIGURE 8.11 Giedt's local measurements of heat transfer around a cylinder in a normal cross flow of air.

the entire range of the available data, they offer

$$\overline{Nu_D} = 0.3 + \frac{0.62 Re_D^{1/2} Pr^{1/3}}{[1+(0.4/Pr)^{2/3}]^{1/4}} \left[1 + \left(\frac{Re_D}{282,000} \right)^{5/8} \right]^{4/5} \qquad (8.34)$$

This expression underpredicts most of the data by about 20% in the range $20,000 < Re_D < 400,000$ but is quite good at other Reynolds numbers, above $Pe_D \equiv Re_D Pr = 0.2$. This can be seen in Fig. 8.12, where eqn. (8.34) is compared with data.

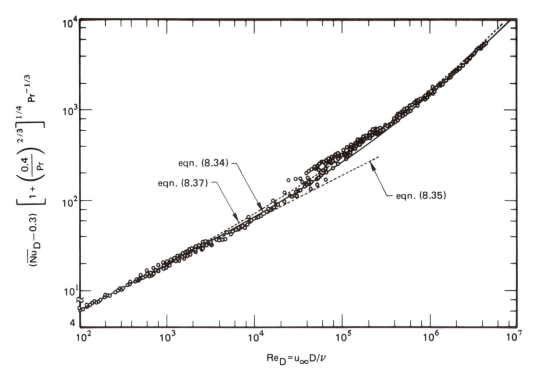

FIGURE 8.12 Comparison of Churchill and Bernstein's correlation with data by many workers from several countries, for heat transfer during cross flow over a cylinder. See [8.12] for data sources. Fluids include air, water, and *sodium*, with both q_w and T_w constant.

Greater accuracy and, in most cases, greater convenience results from breaking the correlation into component equations:

- Below $Re_D = 4000$ the bracketed term $[1 + (Re_D/282,000)^{5/8}]^{4/5}$ can be neglected, so

$$\overline{Nu_D} = 0.3 + \frac{0.62 Re_D^{1/2} Pr^{1/3}}{\left[1 + (0.4/Pr)^{2/3}\right]^{1/4}} \qquad (8.35)$$

- Below $Pe = 0.2$, the Nakai–Okazaki relation [8.13]

$$\overline{Nu_D} = \frac{1}{0.8237 - \ell n(Pe^{1/2})} \qquad (8.36)$$

should be used.

- In the range $20,000 < Re_D < 400,000$, somewhat better results are given by

$$\overline{Nu_D} = 0.3 + \frac{0.62 Re_D^{1/2} Pr^{1/3}}{\left[1 + (0.4/Pr)^{2/3}\right]^{1/4}} \left[1 + \left(\frac{Re_D}{282,000}\right)^{1/2}\right] \quad (8.37)$$

than by eqn. (8.34).

All properties in eqns. (8.34) to (8.37) are to be evaluated at $T = (T_w + T_\infty)/2$.

Example 8.4

An electric resistance wire heater 0.0001 m in diameter is placed perpendicular to an air flow. It holds a temperature of 40°C in a 20°C air flow while it dissipates 17.8 W/m of heat to the flow. How fast is the air flowing?

SOLUTION. $\bar{h} = (17.8 \text{ W/m})/[\pi(0.0001 \text{ m})(40-20)°C] = 2833 \text{ W/m}^2\text{-}°C$. Therefore, $\overline{Nu_D} = 2833(0.0001)/0.0264 = 10.75$, where we have evaluated $k = 0.0264$ at $T = 30°C$. We now want to find the Re_D for which $\overline{Nu_D}$ is 10.75. From Fig. 8.12 we see that Re_D is around 300 when the ordinate is on the order of 10. This means that we can solve eqn. (8.35) to get an accurate value of Re_D:

$$Re_D = \left\{ (\overline{Nu_D} - 0.3) \left[1 + \left(\frac{0.4}{Pr}\right)^{2/3}\right]^{1/4} \middle/ 0.62 Pr^{1/3} \right\}^2$$

but Pr = 0.71, so

$$Re_D = \left\{ (10.75 - 0.3) \left[1 + \left(\frac{0.40}{0.71}\right)^{2/3}\right]^{1/4} \middle/ 0.62(0.71)^{1/3} \right\}^2 = 463$$

Then

$$u_\infty = \frac{\nu}{D} Re_D = \left(\frac{1.596 \times 10^{-5}}{10^{-4}}\right) 463 = \underline{73.9 \text{ m/s}}$$

The data scatter in Re_D is quite small—less than 10%, it would appear—in Fig. 8.12. Therefore, this method can be used to measure local velocities with good accuracy. If the device is calibrated, its accuracy can be improved further. Such an air speed indicator is called a *hot wire anemometer*.

Heat transfer during flow across tube bundles

A rod or tube bundle is an arrangement of parallel cylinders which heat, or are being heated by, a fluid that might flow normal to them, parallel with them, or at some angle in between. The flow of coolant through the fuel elements of all nuclear reactors being used in this country is parallel to the heating rods. The flow on the shell side of most shell-and-tube heat exchangers is generally normal to the tube bundles.

Figure 8.13 shows the two basic configurations of a tube bundle in a cross flow. In one, the tubes are in a line with the flow and in the other, the tubes are staggered in alternating rows. For either of these configurations heat transfer

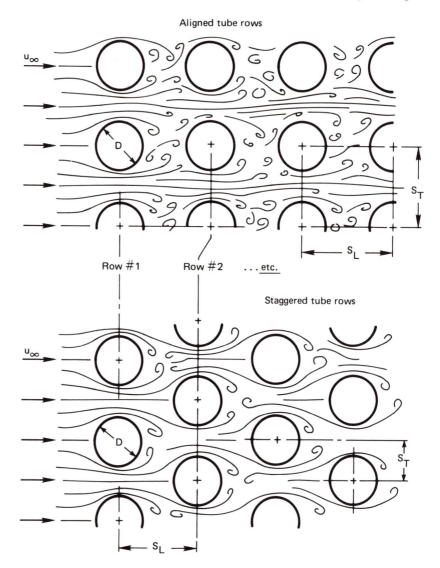

FIGURE 8.13   Aligned and staggered tube rows in tube bundles.

data can be correlated reasonably well with power-law relations of the form

$$\overline{\mathrm{Nu}_D} = C\,\mathrm{Re}_D^n\,\mathrm{Pr}^{1/3} \tag{8.38}$$

but in which the Reynolds number is based on the maximum velocity,

$$u_{max} = \bar{u}_{av} \text{ in the narrowest transverse area of the passage}$$

Thus the Nusselt number based on the average heat transfer coefficient over any particular isothermal tube is

$$\overline{Nu_D} = \frac{\bar{h}D}{k} \quad \text{and} \quad Re_D = \frac{u_{max}D}{\nu}$$

Žukauskas at the Lithuanian Academy of Sciences Institute in Vilnius has written a comprehensive review article on tube-bundle heat transfer [8.14]. In it he summarizes his work and that of other Soviet workers together with earlier work from the West. He is able to correlate data over very large ranges of Pr, Re_D, S_T/D, and S_L/D (see Fig. 8.13) with an expression of the form

$$\overline{Nu_D} = Pr^{0.36}(Pr/Pr_w)^n fn(Re_D) \begin{cases} n = 0 \text{ for gases} \\ n = \dfrac{1}{4} \text{ for liquids} \end{cases} \quad (8.39)$$

where properties are to be evaluated at the local fluid bulk temperature, except for Pr_w, which is evaluated at the constant tube wall temperature, T_w.

The function $fn(Re_D)$ takes the following form for the various circumstances of flow and tube configuration:

$$10 \leqslant Re_D \leqslant 100: \quad fn(Re_D) = 0.8 Re_D^{0.4}, \text{ aligned rows} \quad (8.40)$$

$$fn(Re_D) = 0.9 Re_D^{0.4}, \text{ staggered rows} \quad (8.41)$$

$$100 < Re_D < 10^3: \quad \text{treat tubes as though they were isolated}$$

$$10^3 \leqslant Re_D \leqslant 2 \times 10^5: \quad fn(Re_D) = 0.27 Re_D^{0.63}, \text{ aligned rows} \quad (8.42)$$
$$S_T/S_L < 0.7$$

For $S_T/S_L \geqslant 0.7$, heat exchange is much less effective. Therefore, tube bundles are not designed in this range and no correlation is given

$$fn(Re_D) = 0.35(S_T/S_L)^{0.2} Re_D^{0.6}, \text{ staggered rows} \quad (8.43)$$
$$S_T/S_L \leqslant 2$$

$$fn(Re_D) = 0.40 Re_D^{0.6}, \text{ aligned rows} \quad (8.44)$$
$$S_T/S_L > 2$$

$$Re_D > 2 \times 10^5: \quad fn(Re_D) = 0.021 Re_D^{0.84}, \text{ aligned rows} \quad (8.45)$$

$$fn(Re_D) = 0.022 Re_D^{0.84}, \text{ staggered rows} \quad (8.46)$$
$$Pr > 1$$

$$\overline{Nu_D} = 0.019 Re_D^{0.84}, \text{ staggered rows} \quad (8.47)$$
$$Pr = 0.7$$

All of *the foregoing relations apply to the inner rows* of tube bundles. The heat transfer coefficient is smaller in the rows at the front of a bundle, facing the oncoming flow. The heat transfer coefficient can be corrected so that it will apply to any of the front rows using Fig. 8.14.

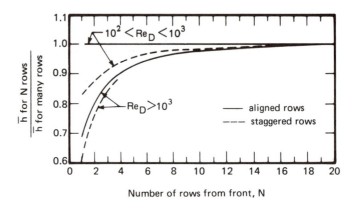

FIGURE 8.14 Correction for the heat transfer coefficients in the front rows of a tube bundle [8.14].

Early in this chapter we alluded to the problem of predicting the heat transfer coefficient during the flow of a fluid at an angle other than 90° to the axes of the tubes in a bundle. Žukauskas provides the empirical corrections in Fig. 8.15 to account for this problem.

The work of Žukauskas does not extend to liquid metals. However, Kalish and Dwyer [8.15] present the results of an experimental study of heat transfer to the liquid eutectic mixture of 77.2% potassium and 22.8% sodium (called NAK). NAK is a fairly popular low-melting point metallic coolant which has received a good deal of attention for its potential use in certain kinds of nuclear reactors. Kalish and Dwyer give for isothermal tubes in an equilateral triangular array as shown in Fig. 8.16:

$$\mathrm{Nu}_D = (5.44 + 0.228\mathrm{Pe}^{0.614})\sqrt{C\frac{P-D}{P}\left(\frac{\sin\phi + \sin^2\phi}{1+\sin^2\phi}\right)} \qquad (8.48)$$

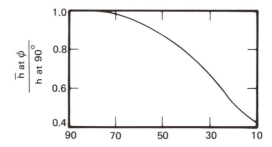

FIGURE 8.15 Correction for the heat transfer coefficient in flows that are not perfectly perpendicular to heat exchanger tubes [8.14].

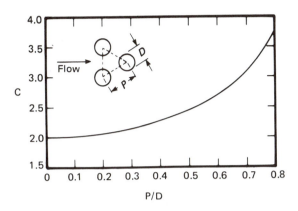

FIGURE 8.16   Geometric correction for the Kalish-Dwyer equation (8.48).

where

- ϕ is the angle between the flow direction and the rod axis.
- P is the "pitch" of the tube array as shown in Fig. 8.16 and D is the tube diameter.
- C is the constant given in Fig. 8.16.
- Pe_D is the Péclét number based on the mean flow velocity through the narrowest opening between the tubes.
- For the same uniform heat flux around each tube, the constants in eqn. (8.48) change as follows:

$$5.44 \text{ becomes } 4.60$$
$$0.228 \text{ becomes } 0.193$$

8.5 Other configurations

At the outset we noted that this chapter would move further and further beyond the reach of analysis in the heat convection problems that it dealt with. However, we must not forget that even the most completely empirical relations in Section 8.4 were devised by people who were keenly aware of the theoretical framework into which these relations had to fit. Notice, for example, that eqn. (8.35) reduces to $\mathrm{Nu}_D \sim \sqrt{\mathrm{Pe}_D}$ as Pr becomes small. That sort of theoretical requirement did not just pop out of a data plot. Instead, it was a consideration that led the authors to select an empirical equation that agreed with theory at low Pr.

Thus the theoretical considerations in Chapter 7 guide us in correlating limited data in situations that cannot be analyzed. Such correlations can be found for all kinds of situations, but all must be viewed critically. Many are based on limited data, and many incorporate systematic errors of one kind or another.

In the face of a heat transfer situation that has to be predicted, one can often find a correlation of data from similar systems. This might involve flow in or across noncircular ducts; axial flow through tube or rod bundles; flow over such bluff bodies as spheres, cubes, or cones; or flow in circular and noncircular annuli. The *Handbook of Heat Transfer* [1.13], the shelf of heat transfer texts in your library, or the journals referred to by the *Engineering Index* are among the first places to look for a correlation curve or equation. When you find a correlation, there are many questions that you should ask yourself:

- Is my case included within the range of dimensionless parameters upon which the correlation is based, or must I extrapolate to reach my case?

- What geometrical differences exist between the situation represented in the correlation and the one I am dealing with? (Such elements as these might differ:

    (a)   inlet flow conditions.

    (b)   small but important differences in hardware, mounting brackets, and so on.

    (c)   minor aspect ratio, or other geometric, nonsimilarities.)

- Does the form of the correlating equation that represents the data, if there is one, have any basis in theory? (If it is only a curve fit to the existing data, one might be unjustified in using it for more than interpolation of those data.)

- What nuisance variables might make our systems different? (For example:

    (a)   surface roughness.

    (b)   fluid purity.

    (c)   problems of surface wetting.)

- To what extent do the data scatter around the correlation line? Can I actually see the data points? (In this regard you must notice whether you are looking at a correlation on linear or logarithmic coordinates. Errors usually appear smaller than they really are on logarithmic coordinates. Compare, for example, the data of Figs. 9.3 and 9.10.)

- Are the ranges of physical variables large enough to guarantee that I can rely on the correlation for the full range of dimensionless groups that it purports to embrace?

- Am I looking at a primary or secondary source (i.e., is this the author's original presentation or someone's report of the original)? If it is a secondary source, have I been given enough information to question it?

- Has the correlation been signed by the persons who formulated it? (When technical work is issued without clear identification of the *individuals* who are responsible for it, one must be sure that there is a person who is *willing* to assume responsibility for it. And there is no way for a company or institution to shoulder responsibility.)

8.6 Heat transfer surface viewed as a heat exchanger

Let us reconsider the problem of a fluid flowing through a pipe with a constant wall temperature. By now we can predict $\bar{h}$ for a pretty wide range of conditions. Suppose that we need to know the net heat transfer to a pipe of known length once $\bar{h}$ is known. This problem is complicated by the fact that the bulk temperature, T_b, is varying along its length.

However, we need only recognize that such a section of pipe is a heat exchanger whose overall heat transfer coefficient, U (between the wall and the bulk), is just $\bar{h}$. Thus if we wish to know how much pipe surface area is needed to raise the bulk temperature from $T_{b_{in}}$ to $T_{b_{out}}$, we can calculate it as follows:

$$Q = (\dot{m}c_p)_b (T_{b_{out}} - T_{b_{in}}) = \bar{h} A \, \text{LMTD}$$

or

$$A = \frac{(\dot{m}c_p)_b (T_{b_{out}} - T_{b_{in}})}{\bar{h}} \frac{\ln \dfrac{T_{b_{out}} - T_w}{T_{b_{in}} - T_w}}{(T_{b_{out}} - T_w) - (T_{b_{in}} - T_w)} \tag{8.49}$$

By the same token, heat transfer in a duct can be analyzed with the effectiveness method (Section 3.3) if the exiting fluid temperature is unknown. Suppose that we do not know $T_{b_{out}}$ in the example above. Then we can write an energy balance at any cross section.

$$dQ = \bar{h}(T_w - T_b) \underbrace{dA}_{\pi D \, dx} = \rho c_p u_{av} \left(\frac{\pi}{4} D^2 \right) dT_b$$

Integrating this between $T_b(x=0) = T_{b_{in}}$ and $T_b(x=L) = T_{b_{out}}$ yields

$$\frac{4L\bar{h}}{\rho c_p u_{av} D} = -\ln \frac{T_w - T_{b_{out}}}{T_w - T_{b_{in}}}$$

which can be rearranged as

$$\frac{T_{b_{out}} - T_{b_{in}}}{T_w - T_{b_{in}}} = 1 - \exp\left(-\frac{\bar{h}}{\rho c_p u_{av}} \frac{4L}{D} \right) \tag{8.50}$$

This can also be expressed in terms of dimensionless groups that are familiar:

$$\varepsilon = 1 - \exp(-\text{NTU}) = 1 - \exp\left(-\frac{4L}{D}\text{St}\right) \qquad (8.51)$$

Thus the NTU and Stanton number, St, are virtually the same thing in this case. And eqn. (8.51) is the same sort of effectiveness equation as given in Chapter 3. If we set $C_{min}/C_{max} = 0$ in eqn. (3.20) or (3.21), we obtain eqn. (8.51), directly.

Example 8.5

Air at 20°C is fully thermally developed as it flows in a 1-cm-I.D. pipe. The average velocity is 0.7 m/s. If the pipe wall is at 60°C, what is the temperature 0.25 m farther downstream?

SOLUTION.

$$\text{Re}_D = \frac{u_{av}D}{\nu} = \frac{(0.7)(0.01)}{1.70 \times 10^{-5}} = 412$$

The flow is therefore laminar, so

$$\text{Nu}_D = \frac{\bar{h}D}{k} = 3.658$$

Thus

$$\bar{h} = \frac{3.658(0.0271)}{0.01} = 9.91$$

Then

$$\varepsilon = 1 - \exp\left(-\frac{\bar{h}}{\rho c_p u_{av}}\frac{4L}{D}\right) = 1 - \exp\left[-\frac{9.91}{1.14(1004)(0.7)}\frac{4(0.25)}{0.01}\right]$$

or

$$\frac{T_b - 20}{60 - 20} = 0.698$$

so

$$T_b = 47.9°C$$

PROBLEMS

8.1. Prove that in fully developed laminar pipe flow, $(-dp/dx)R^2/4\mu$ is twice the average velocity in the pipe. To do this set the mass flow rate through the pipe equal to (ρu_{av})(area).

8.2. Air at 27°C and 1 atm flows fully developed in a 1-cm-I.D. pipe with $u_{av} = 2$ m/s. Plot (to scale) T_w, q_w, and T_b as a function of the distance x after T_w is changed or q_w is imposed:

(a) In the case for which $T_w = 68.4°C = $ constant.

(b)   In the case for which $q_w = 378$ W/m^2 = constant.
Indicate x_{e_t} on your graphs.

8.3.   Prove that C_f is $16/\text{Re}_D$ in fully developed laminar flow.

8.4.   Air at 200°C flows at 4 m/s over a 3-cm-O.D. pipe which is kept at 240°C. (a) Find $\bar{h}$. (b) If the flow were pressurized water at 200°C, what velocities would give the same $\bar{h}$, the same $\overline{\text{Nu}_D}$, and the same Re_D? (c) If someone asked if you could model the water flow with an air experiment, how would you answer? [$u_\infty = 0.0156$ m/s for same Nu_D.]

8.5.   Compare the h value calculated in Example 8.3 with those calculated from the Dittus–Boelter, Colburn, and Sieder–Tate equations. Comment on the comparison.

8.6.   Water at $T_{b_{local}} = 10°$C flows in a 3-cm-I.D. pipe at 1 m/s. The pipe walls are kept at 70°C and the flow is fully developed. Evaluate h and the local value of dT_b/dx at the point of interest. The relative roughness is 0.001.

8.7.   Water at 10°C flows over a 3-cm-O.D. cylinder at 70°C. The velocity is 1 m/s. Evaluate $\bar{h}$.

8.8.   Consider the hot wire anemometer in Example 8.4. Suppose that 17.8 W/m is the constant heat input, and plot u_∞ vs. T_{wire} over a reasonable range of variables. Must you deal with any changes in the flow regime over the range of interest?

8.9.   Water at 20°C flows at 2 m/s over a 2-m length of pipe, 10 cm in diameter, at 60°C. Compare $\bar{h}$ for flow normal to the pipe with flow parallel to the pipe. What does the comparison suggest about baffling in a heat exchanger?

8.10.   A thermally fully developed flow of NAK in a 5-cm-I.D. pipe moves at $u_{av} = 8$ m/s. If $T_b = 395°$C and T_w is constant at 403°C, what is the local heat transfer coefficient? Is the flow laminar or turbulent?

8.11.   Water enters a 7-cm-I.D. pipe at 5°C and moves through it at an average speed of 0.86 m/s. The pipe wall is kept at 73°C. Plot T_b against the position in the pipe until $(T_w - T_b)/68 = 0.01$. Neglect the entry problem and consider property variations.

8.12.   Air at 20°C flows over a very large bank of 2-cm-O.D. tubes which are kept at 100°C. The air approaches at an angle 15° off normal to the tubes. The tube array is staggered with $S_L = 3.5$ cm and $S_T = 1.4$ cm. Find $\bar{h}$ on the first tubes and on the tubes deep in the array, if the air velocity is 4.3 m/s before it enters the array. [$\bar{h}_{deep} = 118$ W/m-°C.]

8.13.   Rework Problem 8.11 using a single value of $\bar{h}$ evaluated at $3(73-5)/4 = 51°$C, and treating the pipe as a heat exchanger. At what length would you judge that the pipe is no longer efficient as an exchanger? Explain.

8.14.   Go to the periodic engineering literature in your library. Find a correlation of heat transfer data. Evaluate the applicability of the correlation according to the criteria outlined in Section 8.5.

8.15. Water at 24°C flows at 0.8 m/s in a smooth 1.5-cm-I.D. tube which is kept at 30°C. The system is extremely clean and quiet, and the flow stays laminar until a noisy air compressor is turned on in the laboratory. Then it suddenly goes turbulent. Calculate the ratio of the turbulent h to the laminar h. [$h_{turb} = 4429$ W/m²-°C.]

REFERENCES

[8.1]   F. M. WHITE, *Viscous Fluid Flow*, McGraw-Hill Book Company, New York, 1974.

[8.2]   L. GRAETZ, "Über die Wärmeleitfähigkeit von Flüssigkeiten," *Ann. Phys.* vol. 25, 1885, p. 337.

[8.3]   S. R. SELLARS, M. TRIBUS, and J. S. KLEIN, "Heat Transfer to Laminar Flow in a Round Tube or a Flat Plate—The Graetz Problem Extended," *Trans. ASME*, vol. 78, 1956, pp. 441–448.

[8.4]   A. P. COLBURN, "A Method of Correlating Forced Convection Heat Transfer Data and a Comparison with Fluid Friction," *Trans. AIChE*, vol. 29, 1933, p. 174.

[8.5]   E. N. SIEDER and G. E. TATE, "Heat Transfer and Pressure Drop of Liquids in Tubes," *Ind. Eng. Chem.*, vol. 28, 1936, p. 1429.

[8.6]   B. S. PETUKHOV, "Heat Transfer and Friction in Turbulent Pipe Flow with Variable Physical Properties," *Advances in Heat Transfer*, vol. 6 (T. F. Irvine, Jr., and J. P. Hartnett, eds.) Academic Press, Inc., New York, 1970, pp. 504–564.

[8.7]   B. LUBARSKY and S. J. KAUFMAN, "Review of Experimental Investigations of Liquid-Metal Heat Transfer," NACA Tech. Note 3336, 1955.

[8.8]   R. A. SEBAN and T. T. SHIMAZAKI, "Heat Transfer to a Fluid Flowing Turbulently in a Smooth Pipe with Walls at a Constant Temperature," *Trans. ASME*, vol. 73, 1951, p. 803.

[8.9]   R. N. LYON, ed., *Liquid Metals Handbook*, 3rd ed., A.E.C. and Dept. of the Navy, Washington, D.C., 1952.

[8.10] J. H. LIENHARD, "Synopsis of Lift, Drag, and Vortex Frequency Data for Rigid Circular Cylinders," Bull. 300, Wash. State Univ., Pullman, 1966.

[8.11] W. H. GIEDT, "Investigation of Variation of Point Unit-Heat-Transfer Coefficient around a Cylinder Normal to an Air Stream," *Trans. ASME*, vol. 71, 1949, pp. 375–381.

[8.12] S. W. CHURCHILL and M. BERNSTEIN, "A Correlating Equation for Forced Convection from Gases and Liquids to a Circular Cylinder in Crossflow," *J. Heat Transfer, Trans. ASME, Ser. C*, vol. 99, 1977, pp. 300–306.

[8.13]   S. NAKAI and T. OKAZAKI, "Heat Transfer from a Horizontal Circular Wire at Small Reynolds and Grashof Numbers—1 Pure Convection," *Int. J. Heat Mass Transfer*, vol. 18, 1975, pp. 387–396.

[8.14]   A. ŽUKAUSKAS, "Heat Transfer from Tubes in Crossflow," *Advances in Heat Transfer*, vol. 8 (T. F. Irvine, Jr., and J. P. Hartnett, eds.), Academic Press, Inc., New York, 1972, pp. 93–160.

[8.15]   S. KALISH and O. E. DWYER, "Heat Transfer to NaK Flowing through Unbaffled Rod Bundles," *Int. J. Heat Mass Transfer*, vol. 10, 1967, pp. 1533–1558.

Chapter 9

Natural convection in single-phase fluids and during film condensation

9.1 Scope

The remaining convection mechanisms that we deal with are to a large degree gravity-driven. Unlike forced convection, in which the driving force is external to the fluid, these so-called "natural convection" processes are driven by body forces that are exerted directly within the fluid as the result of heating or cooling. Two such mechanisms that are rather alike, are:

- *Natural convection.* When we speak of natural convection without any qualifying words, we mean natural convection in a single-phase fluid.
- *Film condensation.* This natural convection process has much in common with single-phase natural convection.

We therefore deal with both mechanisms in this chapter. The governing equations are developed side by side in two brief opening sections. Then each mechanism is developed independently in Sections 9.3 and 9.4 and in Section 9.5, respectively.

Chapter 10 deals with other natural convection heat transfer processes that involve phase change, for example:

- *Nucleate boiling.* This heat transfer process exhibits a great deal of disorder, which distinguishes it from the processes described in Chapter 9.
- *Film boiling.* This is so similar to film condensation that it is usually treated by simply modifying film condensation predictions.
- *Dropwise condensation.* This bears some similarity to nucleate boiling.

9.2 The nature of the problems of film condensation and of natural convection

Description

The natural convection problem is sketched in its simplest form on the left-hand side of Fig. 9.1. Here we see a vertical isothermal plate that cools the fluid adjacent to it. The cooled fluid sinks downward to form a b.l. The figure would

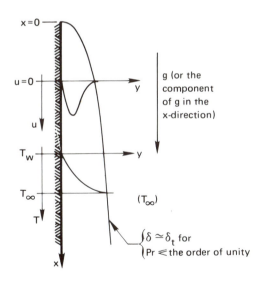

a) Natural convection

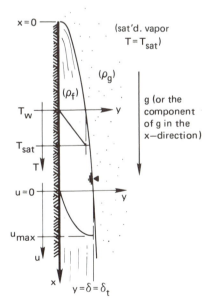

b) Film condensation

FIGURE 9.1 The convective boundary layers for natural convection and film condensation. In both sketches, but particularly in that for film condensation, the *y*-coordinate has been stretched.

be inverted if the plate were warmer than the fluid next to it. Then the fluid would buoy upward.

On the right-hand side of Fig. 9.1 is the corresponding film condensation problem in its simplest form. An isothermal vertical plate cools an adjacent vapor which condenses and forms a liquid film on the wall.[1] The film is normally very thin and it flows off as a b.l. as the figure suggests. While natural convection can carry fluid either upward or downward, a condensate film can only move downward. The temperature in the film rises from T_w at the cool wall to T_{sat} at the outer edge of the film.

In both problems, but particularly in film condensation, the b.l. and the film are normally thin enough to accommodate the b.l. assumptions [recall the discussion following eqn. (7.13)]. A second idiosyncrasy of both problems is that δ and δ_t are closely related. In the condensing film they are equal since the edge of the condensate film forms the edge of both b.l.'s. In natural convection, δ and δ_t are approximately equal when Pr is on the order of unity or less, because all cooled (or heated) fluid must buoy downward (or upward). When Pr is large, the cooled (or heated) fluid will fall (or rise) and, although it is all very close to the wall, this fluid, with its high viscosity, will also drag unheated liquid with it. In this case, δ can exceed δ_t.

Governing equations

To describe laminar film condensation and laminar natural convection, we must add a gravity term to the momentum equation. The dimensions of the terms in the momentum equation should be examined before we do this. Equation (7.13) can be written as

$$\underbrace{\left(u\frac{\partial u}{\partial x}+v\frac{\partial u}{\partial y}\right)}_{=\frac{\text{kg-m}}{\text{kg-s}^2}=\frac{\text{N}}{\text{kg}}}\frac{\text{m}}{\text{s}^2}=-\frac{1}{\rho}\frac{dp}{dx}\underbrace{\frac{\text{m}^3}{\text{kg}}\frac{\text{N}}{\text{m}^2\text{-m}}}_{\frac{\text{N}}{\text{kg}}}+\nu\frac{\partial^2 u}{\partial y^2}\underbrace{\frac{\text{m}^2}{\text{s}}\frac{\text{m}}{\text{s-m}^2}}_{=\frac{\text{m}}{\text{s}^2}=\frac{\text{N}}{\text{kg}}}$$

$$(7.13)$$

where $\partial p/\partial x\simeq dp/dx$ in the b.l. and where $\mu\simeq$ constant. Thus every term in the equation expresses acceleration (or force) per unit mass. The component of gravity in the x-direction therefore enters the momentum balance as $(+g)$. This is because x and g point in the same direction. (Gravity would enter as $-g$ if it acted opposite the x-direction.)

$$u\frac{\partial u}{\partial x}+v\frac{\partial u}{\partial y}=-\frac{1}{\rho}\frac{dp}{dx}+g+\nu\frac{\partial^2 u}{\partial y^2} \qquad (9.1)$$

---

[1]It might instead condense into individual droplets which roll off without forming into a film. This process, called dropwise condensation, is dealt with in Section 10.11.

In the two problems at hand, the pressure gradient is the hydrostatic gradient outside the b.l. Thus

$$\frac{dp}{dx} = \rho_\infty g \qquad \frac{dp}{dx} = \rho_g g \qquad (9.2)$$

$$\underbrace{\phantom{\frac{dp}{dx} = \rho_\infty g}}_{\substack{\text{natural} \\ \text{convection}}} \qquad \underbrace{\phantom{\frac{dp}{dx} = \rho_g g}}_{\substack{\text{film} \\ \text{condensation}}}$$

where ρ_∞ is the density of the undisturbed fluid and ρ_g and ρ_f are the saturated vapor and liquid densities. Equation (9.1) then becomes

$$u\frac{\partial u}{\partial x} + v\frac{\partial u}{\partial y} = \left(1 - \frac{\rho_\infty}{\rho}\right)g + v\frac{\partial^2 u}{\partial y^2} \qquad \text{for natural convection} \qquad (9.3)$$

$$u\frac{\partial u}{\partial x} + v\frac{\partial u}{\partial y} = \left(1 - \frac{\rho_g}{\rho_f}\right)g + v\frac{\partial^2 u}{\partial y^2} \qquad \text{for film condensation} \qquad (9.4)$$

Two boundary conditions, which apply to *both* problems, are

$$\left.\begin{array}{ll} u(y=0)=0 & \text{(the no-slip condition)} \\ v(y=0)=0 & \text{(no flow into the wall)} \end{array}\right\} \qquad (9.5a)$$

The third b.c. is different for the film condensation and natural convection problems.

or
$$\left.\begin{array}{ll} \dfrac{\partial u}{\partial y}\bigg|_{y=\delta} = 0 & \text{(no shear at the edge of the film)} \\[2ex] u(y \doteq \delta) = 0 & \text{(undisturbed fluid outside the b.l.)} \end{array}\right\} \qquad (9.5b)$$

The energy equation for either of the two cases is eqn. (7.40):

$$u\frac{\partial T}{\partial x} + v\frac{\partial T}{\partial y} = \alpha\frac{\partial^2 T}{\partial y^2} \qquad (7.40)$$

We leave the matter of b.c.'s for temperature until later.

The crucial thing we must recognize about the energy equation at the moment is that it is coupled to the momentum equation. Let us consider how that occurs:

In natural convection: The velocity, u, is driven by buoyancy, which is reflected in the term $(1 - \rho_\infty/\rho)g$ in the momentum equation. The density, $\rho = \rho(T)$, varies with T, so it is impossible to solve the momentum and energy equations independently of one another.

In film condensation: The third boundary condition (9.5) for the momentum equation involves the film thickness, δ. But to calculate δ we must make an energy balance on the film to find out how much latent heat—and thus how much condensate—it has absorbed. This will involve $T_{sat} - T_w$ in the solution of the momentum equation.

Recall that the boundary layer on a flat surface, during forced convection, was easy to analyze because the momentum equation could be solved completely before any consideration of the energy equation was attempted. We do not have that advantage in predicting natural convection or film condensation.

9.3 Laminar natural convection on a vertical isothermal surface

Dimensional analysis and experimental data

Before we attempt a dimensional analysis of the natural convection problem, let us simplify the buoyancy term, $(\rho - \rho_\infty)g/\rho$, in the momentum equation (9.3). The equation was derived for incompressible flow, but it has been modified by admitting a small variation of density with temperature in this term only. Now we wish to eliminate $(\rho - \rho_\infty)$ in favor of $(T - T_\infty)$ with the help of the coefficient of thermal expansion, β:

$$\beta \equiv \frac{1}{v}\frac{\partial v}{\partial T}\bigg|_P = -\frac{1}{\rho}\frac{\partial \rho}{\partial T}\bigg|_P \simeq -\frac{1}{\rho}\frac{\rho - \rho_\infty}{T - T_\infty} = -\frac{1 - \dfrac{\rho_\infty}{\rho}}{T - T_\infty} \tag{9.6}$$

where v designates the specific volume here and not a velocity component.

Figure 9.2 shows natural convection from a vertical surface that is hotter than its surroundings. In either this case or on the cold plate shown in Fig. 9.1,

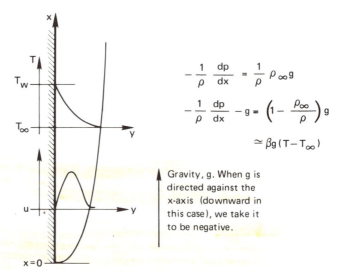

$$-\frac{1}{\rho}\frac{dp}{dx} = \frac{1}{\rho}\rho_\infty g$$

$$-\frac{1}{\rho}\frac{dp}{dx} - g = \left(1 - \frac{\rho_\infty}{\rho}\right)g$$

$$\simeq \beta g (T - T_\infty)$$

Gravity, g. When g is directed against the x-axis (downward in this case), we take it to be negative.

FIGURE 9.2   Natural convection from a vertical heated plate.

we replace $(1 - \rho_\infty/\rho)g$ with $-g\beta(T - T_\infty)$. The sign (see in Fig. 9.2) is the same in either case. Then

$$u\frac{\partial u}{\partial x} + v\frac{\partial u}{\partial y} = -g\beta(T - T_\infty) + \nu\frac{\partial^2 u}{\partial y^2} \tag{9.7}$$

This conveniently removes ρ from the equation and makes the coupling of the momentum and energy equations very clear.

The functional equation for the heat transfer coefficient, h, in natural convection is therefore (cf. Section 7.4)

$$h \quad \text{or} \quad \bar{h} = \text{fn}(k, |T_w - T_\infty|, x \text{ or } L, \nu, \alpha, g, \beta)$$

where L is a length that must be specified for a given problem. Notice that while h was assumed to be independent of ΔT in the forced convection problem (Section 7.4), the explicit appearance of $(T - T_\infty)$ in eqn. (9.7) suggests that we cannot make that assumption here. There are thus eight variables in W, m, s, and °C; so we look for $8 - 4$, or 4 pi-groups. For $\bar{h}$ and a characteristic length, L, the groups are

$$\overline{\text{Nu}_L} \equiv \frac{\bar{h}L}{k} \qquad \text{Pr} \equiv \frac{\nu}{\alpha} \qquad \Pi_3 \equiv \frac{L^3}{\nu^2}|g| \qquad \Pi_4 \equiv \beta|T_w - T_\infty| = \beta\Delta T$$

where we subsequently drop the absolute value signs. Two of these groups are new to us:

- $\Pi_3 \equiv gL^3/\nu^2$: This characterizes the importance of buoyant forces relative to viscous forces.[2]

- $\Pi_4 \equiv \beta\Delta T$: This characterizes the thermal expansion of the fluid. For an ideal gas,

$$\beta = \frac{1}{v}\frac{\partial}{\partial T}\left(\frac{RT}{p}\right)_p = \frac{1}{T_\infty}$$

where R is the gas constant. Therefore, for ideal gases

$$\beta\Delta T = \frac{\Delta T}{T_\infty} \tag{9.8}$$

It turns out that Π_3 and Π_4 (which do not bear the names of famous people) usually appear as a product. This product is called the Grashof (pronounced Gráhs-hoff) number,[3] Gr_L, where the subscript designates the length it is based

---

[2]Note that gL is dimensionally the same as a velocity squared, say u^2. Then $\sqrt{\Pi_3}$ can be interpreted as a Reynolds number: uL/ν. In a laminar b.l. we recall that $\text{Nu} \propto \text{Re}^{1/2}$; so here we expect that $\text{Nu} \propto \Pi_3^{1/4}$.

[3]Π_1, Π_2, and Gr were all suggested by Nusselt in his pioneering paper on convective heat transfer [9.1]. Grashof was a notable nineteenth-century mechanical engineering professor who was simply given the honor of having a dimensionless group named after him posthumously (see, e.g., [9.2]). He did not work with natural convection.

on:

$$\Pi_3 \Pi_4 \equiv Gr_L = \frac{g\beta \Delta T L^3}{\nu^2} \qquad (9.9)$$

Two exceptions in which Π_3 and Π_4 appear independently are rotating systems in which Coriolis forces are part of the body force, and situations in which $\beta \Delta T$ is no longer $\ll 1$ but instead approaches unity. We therefore expect to correlate data in most situations with functional equations of the form

$$Nu = fn(Gr, Pr) \qquad (9.10)$$

Another attribute of the dimensionless functional equation is that the primary independent variable is usually the product of Gr and Pr. This is called the Rayleigh number, Ra_L, where the subscript designates the length it is based on:

$$Ra_L \equiv Gr_L Pr = \frac{g\beta \Delta T L^3}{\alpha \nu} \qquad (9.11)$$

Thus most analyses and correlations of natural convection yield

$$Nu = fn(Ra, Pr) \qquad (9.12)$$

with "secondary parameter" labeling Pr and "primary (or most important) independent variable" labeling Ra.

Figure 9.3 is a careful selection of the best data available for natural convection from vertical isothermal surfaces. These data were organized by

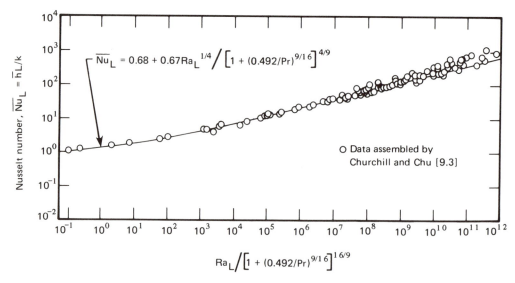

FIGURE 9.3   The correlation of $\bar{h}$ data for vertical isothermal surfaces by Churchill and Chu using $Nu_L = fn(Ra_L, Pr)$. (Applies to full range of Pr.)

Churchill and Chu [9.3] and they span 13 orders of magnitude of the Rayleigh number. The correlation of these data in the coordinates of Fig. 9.2 is exactly in the form of eqn. (9.12), and it brings to light the feature that the influence of Ra_L is large while the influence of Pr is small.

The data correlate on these coordinates within a few percent up to $Ra_L/[1 + (0.492/Pr^{9/16})]^{16/9} \simeq 10^8$. That is about where the b.l. starts exhibiting turbulent behavior. Beyond that point, the overall Nusselt number, $\overline{Nu_L}$, rises more sharply, and the data scatter increases somewhat because the heat transfer mechanisms change.

Prediction of h in natural convection on a vertical surface

The analysis of natural convection using an integral method was done independently by Squire [9.4] and by Eckert [1.7] in the 1930s. We shall refer to this important development as the Squire–Eckert formulation.

The analysis begins with the integrated momentum and energy equations. We assume $\delta = \delta_t$ and integrate both equations to the same value of δ:

$$\frac{d}{dx} \int_0^{\delta} (u^2 - \underbrace{uu_{\infty}}_{\substack{=0 \\ \text{since} \\ u_{\infty}=0}}) \, dy = -\nu \frac{\partial u}{\partial y}\Big|_{y=0} - g\beta \int_0^{\delta} (T - T_{\infty}) \, dy \qquad (9.13)$$

and

$$\frac{d}{dx} \int_0^{\delta} u(T - T_{\infty}) \, dy = \frac{q_w}{\rho c} = -\alpha \frac{\partial T}{\partial y}\Big|_{y=0} \qquad (7.47)$$

The integrated momentum equation is the same as eqn. (7.24) except that it includes the buoyancy term that was added to the differential momentum equation in eqn. (9.7).

We now must estimate the temperature and velocity profiles for use in eqns. (9.13) and (7.47). This is done here in much the same way as it was done in Sections 7.2 and 7.3 for forced convection. We write down a set of known facts about the profiles and then use these things to eliminate the constants in a power-series expression for u or T.

Since the temperature profile has a fairly simple shape, a simple quadratic expression can be used:

$$\frac{T - T_{\infty}}{T_w - T_{\infty}} = a + b\left(\frac{y}{\delta}\right) + c\left(\frac{y}{\delta}\right)^2 \qquad (9.14)$$

Notice that the thermal boundary layer thickness, δ_t, is assumed equal to δ in eqn. (9.14). This would seemingly limit the results to Prandtl numbers not too much larger than unity. Actually, the analysis will also prove useful for large Pr's because the velocity profile exerts diminishing influence on the temperature profile as Pr increases. We require the following things to be true of this profile:

- $T(y=0) = T_w$ or $\dfrac{T-T_\infty}{T_w-T_\infty}\bigg|_{y/\delta=0} = 1 = a$

- $T(y=\delta) = T_\infty$ or $\dfrac{T-T_\infty}{T_w-T_\infty}\bigg|_{y/\delta=1} = 0 = 1+b+c$

- $\dfrac{\partial T}{\partial y}\bigg|_{y=\delta} = 0$ or $\dfrac{d}{d(y/\delta)}\left(\dfrac{T-T_\infty}{T_w-T_\infty}\right)_{y/\delta=1} = 0 = b+2c$

so $a=1$, $b=-2$, and $c=1$. This gives the following dimensionless temperature profile:

$$\frac{T-T_\infty}{T_w-T_\infty} = 1-2\left(\frac{y}{\delta}\right)+\left(\frac{y}{\delta}\right)^2 = \left(1-\frac{y}{\delta}\right)^2 \tag{9.15}$$

We anticipate a somewhat complicated velocity profile (recall Fig. 9.1) and seek to represent it with a cubic function:

$$u = u_c(x)\left[\left(\frac{y}{\delta}\right)+c\left(\frac{y}{\delta}\right)^2+d\left(\frac{y}{\delta}\right)^3\right] \tag{9.16}$$

where, since there is no obvious characteristic velocity in the problem, we write u_c as an as-yet-unknown function. (u_c will have to increase with x, since u must increase with x.) We know three things about u:

- $u(y=0)=0$ we have already satisfied this condition by writing eqn. (9.16) with no lead constant

- $u(y=\delta)=0$ or $\dfrac{u}{u_c} = 0 = (1+c+d)$

- $\dfrac{\partial u}{\partial y}\bigg|_{y=\delta} = 0$ or $\dfrac{\partial u}{\partial(y/\delta)}\bigg|_{y/\delta=1} = 0 = (1+2c+3d)u_c$

These give $c=-2$ and $d=1$, so

$$\frac{u}{u_c(x)} = \frac{y}{\delta}\left(1-\frac{y}{\delta}\right)^2 \tag{9.17}$$

We could also have written the momentum equation (9.7) at the wall, where $u=v=0$, and created a fourth condition:

$$\frac{\partial^2 u}{\partial y^2}\bigg|_{y=0} = \frac{g\beta(T_w-T_\infty)}{\nu}$$

and then we could have evaluated $u_c(x)$ as $\beta g|T_w-T_\infty|\delta^2/4\nu$. A correct expression for u_c will eventually depend upon these variables, but we will not attempt to make u_c fit this particular condition. Doing so would yield two equations, (9.13) and (7.47), in a single unknown, $\delta(x)$. It would be impossible to satisfy both of them. Instead, we shall allow the velocity profile to violate this condition slightly and write

$$u_c(x) = C_1 \frac{\beta g |T_w - T_\infty|}{\nu} \delta^2(x) \qquad (9.18)$$

Then we shall solve the *two* integrated conservation equations for the two unknowns, C_1 (which should $\simeq \frac{1}{4}$) and $\delta(x)$.

The dimensionless temperature and velocity profiles are plotted in Fig. 9.4. With them are included Schmidt and Beckmann's exact calculation for air (Pr=0.7) as presented in [9.4]. Notice that the integral approximation to the temperature profile is better than the approximation to the velocity profile. That is fortunate since the temperature profile exerts the major influence in the heat transfer solution.

When we substitute eqns. (9.15) and (9.17) in the momentum equation (9.13) using eqn. (9.18) for $u_c(x)$, we get

$$C_1^2 \left(\frac{g\beta |T_w - T_\infty|}{\nu} \right)^2 \frac{d}{dx} \delta^5 \underbrace{\int_0^1 \left(\frac{y}{\delta} \right)^2 \left(1 - \frac{y}{\delta} \right)^4 d\frac{y}{\delta}}_{\frac{1}{105}} = g\beta |T_w - T_\infty| \delta \underbrace{\int_0^1 \left(1 - \frac{y}{\delta} \right)^2 d\frac{y}{\delta}}_{\frac{1}{3}}$$

$$- C_1 g\beta |T_w - T_\infty| \delta(x) \underbrace{\frac{\partial}{\partial \left(\frac{y}{\delta} \right)} \left[\frac{y}{\delta} \left(1 - \frac{y}{\delta} \right)^2 \right]_{\frac{y}{\delta} = 0}}_{1} \qquad (9.19)$$

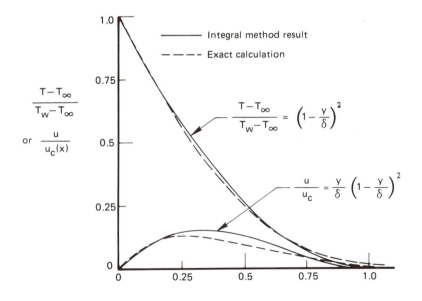

FIGURE 9.4   The temperature and velocity profiles for air (Pr=0.7) in a laminar convection b.l.

where we change the sign of the terms on left by replacing $(T_w - T_\infty)$ with its absolute value. Equation (9.19) then becomes

$$\left(\frac{1}{21} C_1^2 \frac{g\beta|T_w - T_\infty|}{\nu^2}\right) \delta^3 \frac{d\delta}{dx} = \frac{1}{3} - C_1$$

or

$$\frac{d\delta^4}{dx} = \frac{84\left(\frac{1}{3} - C_1\right)}{C_1^2 \dfrac{g\beta|T_w - T_\infty|}{\nu^2}}$$

Integrating this with the b.c., $\delta(x=0)=0$, gives

$$\delta^4 = \frac{84\left(\frac{1}{3} - C_1\right)}{C_1^2 \dfrac{g\beta|T_w - T_\infty|}{\nu^2}} x \qquad (9.20)$$

Substituting eqns. (9.15), (9.17), and (9.18) in eqn. (7.47) likewise gives

$$(T_w - T_\infty)C_1 \frac{g\beta|T_w - T_\infty|}{\nu} \frac{d}{dx} \delta^3 \underbrace{\int_0^1 \frac{y}{\delta}\left(1-\frac{y}{\delta}\right)^4 d\frac{y}{\delta}}_{\frac{1}{30}}$$

$$= -\alpha \frac{T_w - T_\infty}{\delta} \underbrace{\frac{d}{d(y/\delta)}\left[\left(1-\frac{y}{\delta}\right)^2\right]}_{-2}\bigg|_{y/\delta=0}$$

or

$$3\frac{C_1}{30}\delta^3 \frac{d\delta}{dx} = \frac{C_1}{40}\frac{d\delta^4}{dx} = \frac{2}{\Pr \dfrac{g\beta|T_w - T_\infty|}{\nu^2}}$$

Integrating this with the b.c. $\delta(x=0)=0$, we get

$$\delta^4 = \frac{80}{C_1 \Pr \dfrac{g\beta|T_w - T_\infty|}{\nu^2}} x \qquad (9.21)$$

Equating eqns. (9.20) and (9.21) for δ^4, we then obtain

$$\frac{21}{20} \frac{\frac{1}{3} - C_1}{C_1 \dfrac{g\beta|T_w - T_\infty|}{\nu^2}} x = \frac{1}{\Pr \dfrac{g\beta|T_w - T_\infty|}{\nu^2}} x$$

or

$$C_1 = \frac{\Pr}{3\left(\dfrac{20}{21} + \Pr\right)} \qquad (9.22)$$

Then from eqn. (9.21):

$$\delta^4 = \frac{240\left(\dfrac{20}{21} + \mathrm{Pr}\right)}{\mathrm{Pr}^2 \dfrac{g\beta|T_w - T_\infty|}{\nu^2}} x$$

or

$$\frac{\delta}{x} = 3.936\left(\frac{0.952 + \mathrm{Pr}}{\mathrm{Pr}^2}\right)^{1/4} \frac{1}{\mathrm{Gr}_x^{1/4}} \tag{9.23}$$

Equation (9.23) can be combined with the known temperature profile, eqn. (9.15), and substituted in Fourier's law, to find q:

$$q = -k\frac{\partial T}{\partial y}\bigg|_{y=0} = -\frac{k(T_w - T_\infty)}{\delta} \underbrace{\frac{d\left(\dfrac{T - T_\infty}{T_w - T_\infty}\right)}{d\left(\dfrac{y}{\delta}\right)}\bigg|_{y/\delta=0}}_{-2} = 2\frac{k\Delta T}{\delta} \tag{9.24}$$

so, writing $h = q/|T_w - T_\infty| \equiv q/\Delta T$, we obtain[4]

$$\mathrm{Nu}_x \equiv \frac{qx}{\Delta T k} = 2\frac{x}{\delta} = \frac{2}{3.936}(\mathrm{PrGr}_x)^{1/4}\left(\frac{\mathrm{Pr}}{0.952 + \mathrm{Pr}}\right)^{1/4}$$

or

$$\mathrm{Nu}_x = 0.508\mathrm{Ra}_x^{1/4}\left(\frac{\mathrm{Pr}}{0.952 + \mathrm{Pr}}\right)^{1/4} \tag{9.25}$$

This is the Squire–Eckert result for the local heat transfer from a vertical isothermal wall during laminar natural convection. It applies for either $T_w > T_\infty$ or $T_w < T_\infty$.

The overall heat transfer coefficient can be obtained from

$$\bar{h} = \frac{\left[\int_0^L q(x)\,dx\right]}{L\Delta T} = \frac{\left[\int_0^L h(x)\,dx\right]}{L}$$

Thus

$$\overline{\mathrm{Nu}_L} = \frac{\bar{h}L}{k} = \frac{1}{k}\int_0^L \frac{k}{x}\mathrm{Nu}_x\,dx = \frac{4}{3}(\mathrm{Nu}_x)_{x=L}$$

or

$$\boxed{\overline{\mathrm{Nu}_L} = 0.678\mathrm{Ra}_L^{1/4}\left(\frac{\mathrm{Pr}}{0.952 + \mathrm{Pr}}\right)^{1/4}} \tag{9.26}$$

[4]Recall that, in footnote 2, we anticipated that Nu would vary as $\mathrm{Gr}^{1/4}$. We now see that this is the case.

All properties in eqn. (9.26) and the preceding equations should be evaluated at $T = (T_w + T_\infty)/2$ except in gases, where β should be evaluated at T_∞.

Example 9.1

A thin-walled metal tank containing fluid at 40°C cools in air at 14°C. If the sides are 0.4 m high, compute $\bar{h}$, $\bar{q}$, and δ at the top. Are the b.l. assumptions reasonable?

SOLUTION. $\beta_{air} = 1/T_\infty = 1/(273 + 14) = 0.00348$. Then

$$\mathrm{Ra}_L = \frac{g\beta\Delta T L^3}{\nu\alpha} = \frac{9.8(0.00348)(40-14)(0.4)^3}{(1.566)(2.203)10^{-5-5}} = 1.645 \times 10^8$$

and $\mathrm{Pr} = 0.711$, where the properties are evaluated at $300°\mathrm{K} = 27°\mathrm{C}$. Then from eqn. (9.26),

$$\overline{\mathrm{Nu}}_L = 0.678(1.645 \times 10^8)^{1/4}\left(\frac{0.711}{0.952 + 0.711}\right)^{1/4} = 62.1$$

so

$$\bar{h} = \frac{62.1k}{L} = \frac{62.1(0.02614)}{0.4} = 4.06 \ \mathrm{W/m^2\text{-}°C}$$

and

$$\bar{q} = \bar{h}\Delta T = 4.06(40 - 14) = 105.5 \ \mathrm{W/m^2}$$

The b.l. thickness at the top of the tank is given by eqn. (9.23) at $x = L$:

$$\frac{x}{L} = 3.936\left(\frac{0.952 + 0.711}{0.711^2}\right)^{1/4}\frac{1}{(\mathrm{Ra}_L/\mathrm{Pr})^{1/4}} = 0.0430$$

Thus the b.l. thickness at the end of the plate is only 4% of the height, or 1.6 cm thick. This is thicker than typical forced convection b.l.'s, but it is still reasonably thin.

Example 9.2

Large thin metal sheets of length L are dipped in an electroplating bath in the vertical position. They are initially cooler than the liquid in the bath. How rapidly will they come up to bath temperature?

SOLUTION.   We can probably take $\mathrm{Bi} \ll 1$ and use the lumped-capacity response equation, (1.19). We obtain $\bar{h}$ for use in eqn. (1.19) from eqn. (9.26):

$$\bar{h} = \underbrace{0.678\frac{k}{L}\left(\frac{\mathrm{Pr}}{0.952 + \mathrm{Pr}}\right)^{1/4}\left(\frac{g\beta L^3}{\alpha\nu}\right)^{1/4}}_{\text{call this } B}\Delta T^{1/4}$$

Since $\bar{h} \propto \Delta T^{1/4}$, eqn. (1.19) becomes

$$\frac{d(T - T_b)}{dt} = -\frac{BA}{\rho cV}(T - T_b)^{5/4}$$

where $V/A = $ the half-thickness of the plate, w. Integrating this between the initial temperature of the plate, T_i, and the temperature at time, t, we get

$$\int_{T_i}^{T}\frac{d(T - T_b)}{(T - T_b)^{5/4}} = -\int_0^t\frac{B}{\rho cw}\, dt$$

so

$$T - T_b = \left[\frac{1}{(T_i - T_b)^{1/4}} + \frac{B}{4\rho c w} t \right]^{-4}$$

Before we use this result, we should check $\text{Bi} = Bw \Delta T^{1/4}/k$ to be certain that it is, in fact, less than unity. The temperature can be put in dimensionless form as

$$\frac{T - T_b}{T_i - T_b} = \left[1 + \frac{B(T_i - T_b)^{1/4}}{4\rho c w} t \right]^{-4}$$

where the coefficient of t is a kind of inverse time constant of the response. Thus the temperature dependence of $\bar{h}$ in natural convection leads to a solution quite different from the exponential response that resulted from a constant $\bar{h}$ [eqn. (1.21)].

Comparison of analysis and correlations with experimental data

Churchill and Chu have proposed two equations for the data correlated in Fig. 9.3. The simpler of the two is shown in the figure. It is

$$\overline{\text{Nu}_L} = 0.68 + 0.67 \text{Ra}_L^{1/4} \left[1 + \left(\frac{0.492}{\text{Pr}} \right)^{9/16} \right]^{-4/9} \tag{9.27}$$

This approaches to within 1.2% of the Squire–Eckert prediction as Pr and Ra_L are increased, and it only differs from the prediction by 5.5% if the fluid is a gas and $\text{Ra}_L > 10^5$. Typical Rayleigh numbers usually exceed 10^5, so we conclude that the Squire–Eckert prediction is remarkably accurate in the range of practical interest, despite the approximations upon which it is built. The additive constant of 0.68 in eqn. (9.27) is required to correct eqn. (9.27) at low Ra_L, where the b.l. assumptions are invalid and $\overline{\text{Nu}_L}$ is no longer proportional to $\text{Ra}_L^{1/4}$.

At low Prandtl numbers the Squire–Eckert prediction fails and eqn. (9.27) has to be used. In the turbulent regime ($\text{Ra}_L \gtrsim 10^8 \text{ or } {}^9$), eqn. (9.27) predicts a lower bound on the data (see Fig. 9.3). In this sense it is somewhat conservative.

In the correlation, as in eqn. (9.26), the thermal properties should all be evaluated at a mean b.l. temperature, except for β, which is to be evaluated at T_∞ if the fluid is a gas.

Example 9.3

Verify the first heat transfer coefficient in Table 1.1. This is for air at 20°C next to a 0.3-m-high wall at 50°C.

SOLUTION. At $T = 35°C = 308°K$, we find $\text{Pr} = 0.71$, $\nu = 16.45 \times 10^{-6} \text{ m}^2/\text{s}$, $\alpha = 0.2318 \times 10^{-4} \text{ m}^2/\text{s}$, and $\beta = 1/(273 + 20) = 0.00341/°K$. Then

$$\text{Ra}_L = \frac{g\beta \Delta T L^3}{\alpha\nu} = \frac{9.8(0.00341)(30)(0.3)^3}{(16.45)(0.2318)10^{-10}} = 7.10 \times 10^7$$

The Squire–Eckert prediction gives

$$\overline{\mathrm{Nu}_L} = 0.678(7.10 \times 10^7)^{1/4} \left(\frac{0.71}{0.952 + 0.71} \right)^{1/4} = 50.3$$

so

$$\bar{h} = 50.3 \frac{k}{L} = 50.3 \left(\frac{0.0267}{0.3} \right) = \underline{4.48 \ \mathrm{W/m^2\text{-}°C}}$$

And the Churchill–Chu correlation gives

$$\overline{\mathrm{Nu}_L} = 0.68 + 0.67 \frac{[7.10 \times 10^7]^{1/4}}{\left[1 + (0.492/0.71)^{9/16} \right]^{4/9}} = 47.88$$

so

$$\bar{h} = 47.88 \left(\frac{0.0267}{0.3} \right) = \underline{4.26 \ \mathrm{W/m^2\text{-}°C}}$$

The prediction is therefore within 5% of the correlation. We should use the latter result in preference to the theoretical one, although the difference is slight.

Variable-properties problem

Sparrow and Gregg [9.5] provide an extended discussion of the influence of physical property variations on predicted values of Nu. They found that while β for gases should be evaluated at T_∞, all other properties should be evaluated at T_r, where

$$T_r = T_w - C(T_w - T_\infty) \tag{9.28}$$

where $C = 0.38$ for gases. Most books recommend simply that a simple mean between T_w and T_∞ (or $C = 0.50$) be used. A simple mean does not differ much from the more precise result above, of course.

It has also been shown by Barrow and Sitharamarao [9.6] that when $\beta \Delta T$ is no longer $\ll 1$, the Squire–Eckert formula should be corrected as follows:

$$\mathrm{Nu} = \left(\mathrm{Nu}_{\mathrm{Sq-Ek}} \right) \left[1 + \tfrac{3}{5} \beta \Delta T + \mathcal{O}(\beta \Delta T)^2 \right]^{1/4} \tag{9.29}$$

This same correction can be applied to the Churchill–Chu correlation or to other expressions for Nu. Since $\beta = 1/T_\infty$ for an ideal gas, eqn. (9.29) gives only about a 1.5% correction for a 330°K plate heating 300°K air.

Note on the validity
of the boundary layer assumptions

The boundary layer assumptions are sometimes put to a rather severe test in natural convection problems. Thermal b.l. thicknesses are often fairly large, and the usual analyses that take the b.l. to be thin can be significantly in error. This is particularly true as Gr becomes small. Figure 9.5 includes three pictures which illustrate this. These pictures are interferograms (or in the case of Fig. 9.5c, data

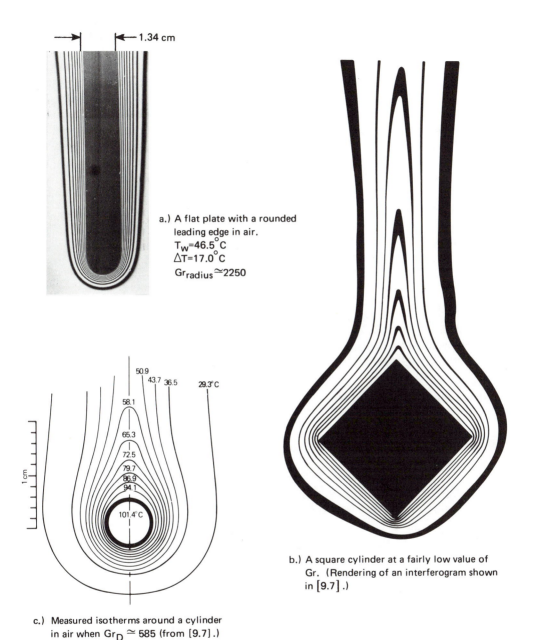

1.34 cm

a.) A flat plate with a rounded
leading edge in air.
$T_w = 46.5°C$
$\Delta T = 17.0°C$
$Gr_{radius} \simeq 2250$

50.9
43.7 36.5
29.3°C
58.1
65.3
72.5
79.7
86.9
94.1
101.4°C

1 cm

b.) A square cylinder at a fairly low value of
Gr. (Rendering of an interferogram shown
in [9.7].)

c.) Measured isotherms around a cylinder
in air when $Gr_D \simeq 585$ (from [9.7].)

FIGURE 9.5 The thickening of the b.l. during natural convection at low Gr, as
illustrated by interferograms made on two-dimensional bodies. (The dark lines in the
pictures are isotherms.)

deduced from interferograms). An interferogram is a photograph made in a kind of lighting which causes regions of uniform temperature to appear as alternating light and dark bands.

Figure 9.5a was made at the University of Kentucky by G. S. Wang and R. Eichhorn. The Grashof number based on the radius of the leading edge is 2250 in this case. This is low enough to result in a b.l. that is larger than the radius near the leading edge. Figure 9.5b and c are from Kraus's classic study of natural convection visualization methods [9.7]. Figure 9.5c shows that at Gr = 585 the b.l. assumptions are quite unreasonable, since the cylinder is small in comparison with the large region of thermal disturbance.

The analysis of free convection becomes a far more complicated problem at low Gr's, since the b.l. equations can no longer be used. We shall not discuss any of the numerical solutions of the full Navier–Stokes equations that have been carried out in this regime. We shall instead note that correlations of data using functional equations of the form

$$Nu = fn(Ra, Pr)$$

will be the first thing that we resort to in such cases. Indeed, Fig. 9.3 reveals that Churchill and Chu's equation (9.27) already serves this purpose in the case of the vertical isothermal plate, at low values of $Ra \equiv GrPr$.

9.4   Natural convection in other situations

Natural convection
from horizontal isothermal cylinders

Churchill and Chu [9.8] provide yet another comprehensive correlation of existing data. For horizontal isothermal cylinders, they find that an equation that has the same form as eqn. (9.27) correlates the data for horizontal cylinders as well. Horizontal cylinder data from a variety of sources, over about 24 orders of magnitude of the Rayleigh number based on the diameter, Ra_D, are shown in Fig. 9.6. The equation that correlates them is

$$\overline{Nu_D} = 0.36 + \frac{0.518 Ra_D^{1/4}}{\left[1 + (0.559/Pr)^{9/16}\right]^{4/9}} \tag{9.30}$$

They recommend that eqn. (9.30) be used in the range $10^{-4} \leqslant Ra_D \leqslant 10^9$.

When Ra_D is greater than 10^9, the flow becomes turbulent. The following equation is a little more complex but it gives comparable accuracy over a larger

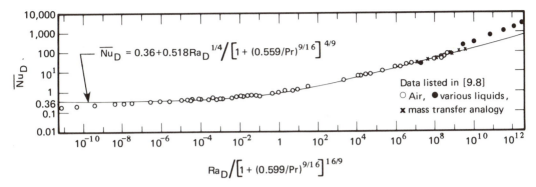

FIGURE 9.6 The data of many investigators for heat transfer from isothermal cylinders during natural convection, as correlated by Churchill and Chu [9.8].

range:

$$\overline{Nu_D} = \left\{ 0.60 + 0.387 \left[\frac{Ra_D}{\left[1 + (0.559/Pr)^{9/16} \right]^{16/9}} \right]^{1/6} \right\}^2 \qquad (9.31)$$

The recommended range of applicability of eqn. (9.31) is

$$10^{-6} \leqslant Ra_D$$

Example 9.4

The first space lab will be subject to a "g-jitter," or background variation of acceleration, on the order of 10^{-6} or 10^{-5} earth gravities. Brief periods of gravity up to 10^{-4} or 10^{-2} earth gravities can be exerted by accelerating the whole laboratory. A certain line carrying hot oil is $\frac{1}{2}$ cm in diameter and it is at 127°C. How does Q vary with g-level if $T_\infty = 27$°C in the air around the tube?

SOLUTION. The average b.l. temperature is 350°K. We evaluate properties at this temperature and write g as $g_e(g\text{-level})$, where g_e is g at the earth's surface and the g-level is the fraction of g_e in the space lab.

$$Pr = 0.697 \qquad Ra_D = \frac{g\dfrac{\Delta T}{T_\infty}D^3}{\nu\alpha} = \frac{9.8\dfrac{400-300}{300}(0.005)^3}{2.062(10)^{-5}2.92(10)^{-5}}(g\text{-level})$$

$$= (678.2)(g\text{-level})$$

From eqn. (9.31), we compute

$$\overline{Nu_D} = \left\{ 0.6 + 0.387 \left[\underbrace{\frac{678.2}{\left[1 + (0.559/0.706)^{9/16} \right]^{16/9}}}_{0.952} \right]^{1/6} (g\text{-level})^{1/6} \right\}^2$$

so

g-Level	$\overline{\mathrm{Nu}}_D$	$\bar{h}=\overline{\mathrm{Nu}}_D\dfrac{0.0297}{0.005}$	$Q=\pi D\bar{h}\Delta T$
10^{-6}	0.483	2.87 W/m²-°C	4.51 W/m of tube
10^{-5}	0.547	3.25 W/m²-°C	5.10 W/m of tube
10^{-4}	0.648	3.85 W/m²-°C	6.05 W/m of tube
10^{-2}	1.086	6.45 W/m²-°C	10.1 W/m of tube

Cooling is clearly inefficient at these low gravities.

Natural convection from vertical cylinders

The heat transfer from the wall of a cylinder with its axis running vertically is the same as that from a vertical plate as long as the thermal b.l. is thin. However, if the b.l. is thick, as is indicated in Fig. 9.7, heat transfer will be enhanced by the curvature of the thermal b.l. This correction was first considered some years ago by Sparrow and Gregg, and the analysis was recently extended with the help of more powerful numerical methods by Minkowycz and Sparrow [9.9] and by Cebeci [9.10].

Figure 9.7 includes the corrections to the vertical plate results that were calculated for many Pr's by Cebeci. The left-hand graph gives a correction that

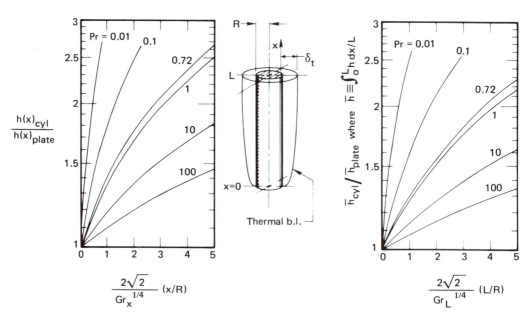

FIGURE 9.7 Corrections for h and $\bar{h}$ on vertical isothermal plates, to make them apply to vertical isothermal cylinders [9.10].

must be multiplied by the local flat-plate Nusselt number to get the vertical cylinder result. Notice that the correction increases when the Grashof number decreases. The right-hand curve gives a similar correction for the overall Nusselt number on a cylinder of height L. Notice that in either situation, the correction for all but liquid metals is less than 1% if D/x or $L < 0.02 \mathrm{Gr}_{x \text{ or } L}^{1/4}$.

Heat transfer from general submerged bodies

Spheres. The sphere is an interesting case because it has a clearly specifiable value of Nu_D as $\mathrm{Ra}_D \to 0$. We look first at this limit. When the buoyancy forces approach zero by virtue of:

- Low gravity
- Small diameter
- Very high viscosity
- A very small value of β

then heated fluid will no longer be buoyed away convectively. In that case, only conduction will serve to remove heat. Using shape factor 4 in Table 5.2, we compute in this case

$$\lim_{\mathrm{Ra}_D \to 0} \mathrm{Nu}_D = \frac{Q}{A\,\Delta T}\frac{D}{k} = \frac{k\,\Delta T(S)D}{4\pi(D/2)^2\,\Delta Tk} = \frac{4\pi(D/2)}{4\pi(D/4)} = 2 \qquad (9.32)$$

Every proper correlation of data for heat transfer from spheres therefore has the lead constant, 2, in it.[5] A typical example is that of Yuge [9.11]:

$$\boxed{\overline{\mathrm{Nu}_D} = 2 + 0.43\mathrm{Ra}_D^{1/4}}\;\Big|\;\mathrm{Ra}_D < 10^5 \qquad (9.33)$$

Rough estimate of Nu for other bodies. In 1973 Lienhard noted [9.12] that, for laminar convection, the expression

$$\boxed{\overline{\mathrm{Nu}_\tau} \simeq 0.52\mathrm{Ra}_\tau^{1/4}} \qquad (9.34)$$

would predict heat transfer from any submerged body within about 10% if Pr is not $\ll 1$. The characteristic dimension in eqn. (9.34) is the length of travel, τ, of fluid in the b.l.

In the case of spheres, for example, $\tau = \pi D/2$—the distance from the bottom to the top around the circumference. Thus for spheres, eqn. (9.34)

---

[5]It is important to note that while Nu_D approaches a limiting value at small Ra_D, no such limit exists for cylinders or vertical surfaces. The constants in eqns. (9.27) and (9.30) are not valid at extremely low values of Ra_D.

becomes

$$\frac{\bar{h}\pi D}{2k} = 0.52\left[\frac{g\beta \Delta T(\pi D/2)^3}{\nu\alpha}\right]^{1/4}$$

or

$$\frac{\bar{h}D}{k} = 0.52\left(\frac{2}{\pi}\right)\left(\frac{\pi}{2}\right)^{3/4}\left[\frac{g\beta \Delta TD^3}{\nu\alpha}\right]^{1/4}$$

or

$$\overline{Nu_D} = 0.465Ra_D^{1/4}$$

This is within 8% of Yuge's correlation if Ra_D remains fairly large.

Laminar heat transfer
from inclined and horizontal plates

In 1953, Rich [9.13] showed that heat transfer from inclined plates could be predicted by vertical plate formulas if the component of the gravity vector along the surface of the plate was used in the calculation of the Grashof number. Thus, the heat transfer rate decreases as $(\cos\theta)^{1/4}$, where θ is the angle of inclination measured from the vertical as shown in Fig. 9.8.

Subsequent studies have shown that Rich's result is substantially correct for the lower surface of a heated plate or the upper surface of a cooled plate. For the upper surface of a heated plate, or the lower surface of a cooled plate, the boundary layer becomes unstable and separates at a lower value of Gr. Experimental observations of such instability have been given by Fujii and Imura [9.14]; Vliet [9.15]; Emery, Yang, and Wilson [9.16]; and Moran and Lloyd [9.17].

In the limit $\theta \rightarrow 90$—the horizontal plate—the fluid flow above a hot plate or below a cold plate must collapse in a column as shown in Fig. 9.8c and d. The b.l. in such cases is always unstable. The unstable cases can only be represented with empirical correlations.

The following guidelines are recommended for inclined and horizontal plates:

- For small angles and large values of Gr, it is generally safe to replace g with $g\cos\theta$ in eqn. (9.27).

- For $\theta < 88°$ and $10^5 \leqslant Ra_L \leqslant 10^{11}$, eqn. (9.27) is still valid for the upper side of cold plates and the lower side of hot plates when g is replaced with $g\cos\theta$.

- For angles above 10°, we recommend that the reader consult the restrictive data cited above for the upper side of inclined hot plates and the lower side of inclined cold plates.

- For downward-facing hot plates and upward-facing cold plates—that is,

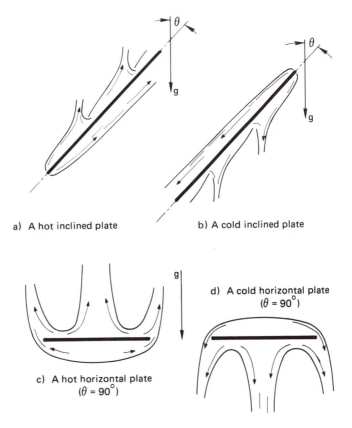

a) A hot inclined plate   b) A cold inclined plate

d) A cold horizontal plate
    ($\theta = 90°$)

c) A hot horizontal plate
    ($\theta = 90°$)

FIGURE 9.8   Natural convection b.l.'s on some inclined and horizontal surfaces.
The b.l. separation, shown here for the unstable cases in (a) and (b), only occurs at
sufficiently large values of Gr.

the stable cases—Fujii and Imura give for very shallow angles:

$$\overline{Nu_L} = 0.58 Ra_L^{1/5} \tag{9.35}$$

This is valid for $Ra_L \gg 10^9$ if $89° < \theta < 90°$, and for $Ra_L < 10^9$ if $87° < \theta < 90°$. Ra_L is based on $g\cos\theta$.

- For upward-facing hot plates and downward-facing cold plates—that is, the unstable cases—Fujii and Imura obtain

$$\overline{Nu_L} = 0.13 Ra_L^{1/5}, \; 5 \times 10^8 < Ra_L$$
$$\overline{Nu_L} = 0.16 Ra_L^{1/5}, \; 2 \times 10^8 > Ra_L \tag{9.36}$$

(There is a problem with these results in that they are based on data for only two plate sizes.) Ra_L is based on $g\cos\theta$.

- The Fujii and Imura results above are for two-dimensional plates—ones in which infinite breadth has been approximated by suppression of end effects.

- For circular plates of diameter, D, in the stable horizontal configurations, the data of Kadambi and Drake [9.18] suggest that

$$\overline{Nu_D} = 0.82 Ra_D^{1/5} Pr^{0.034} \qquad (9.37)$$

- For square plates of breadth L, the data of Saunders and Fishenden [9.19] (as correlated in [9.18]) give

$$\overline{Nu_L} = 0.82 Ra_L^{1/5} \qquad (9.38)$$

Natural convection with uniform heat flux

When q_w is specified instead of $\Delta T \equiv (T_w - T_\infty)$, in natural convection, there is a problem that did not arise in forced convection. That problem is that ΔT, which appears both in Nu on the left and Ra on the right, is now the unknown dependent variable. Since Nu usually varies as $Ra^{1/4}$, we can write

$$Nu_x = \frac{q_w}{\Delta T} \frac{x}{k} \propto Ra_x^{1/4} \propto \Delta T^{1/4} x^{3/4}$$

This can be solved for ΔT in the following way:

$$\Delta T = C \left(\frac{x}{L} \right)^{1/5} \qquad (9.39)$$

where C involves q_w, L, and the relevant physical properties. Then the average of ΔT over the length of the heater is given by

$$\frac{\overline{\Delta T}}{C} = \int_0^1 \left(\frac{x}{L} \right)^{1/5} d\left(\frac{x}{L} \right) = \frac{5}{6} \qquad (9.40)$$

We plot ΔT against x/L in Fig. 9.9. Here, $\overline{\Delta T}$ and $\Delta T(x/L = \frac{1}{2})$ are within 4% of each other. This suggests the first of two strategies for eliminating the dependent variable ΔT from the right-hand side of an equation for the Nusselt number:

1. If we are interested in *average* values of ΔT, we can use ΔT evaluated at the midpoint of the plate on the right-hand side.

2. If we want to form an equation for $Nu_x \equiv q_w x / k \Delta T(x)$, we can use a Rayleigh number, Ra*, defined as

$$Ra_x^* \equiv Ra_x Nu_x \equiv \frac{g\beta \Delta T x^3}{\nu\alpha} \frac{q_w x}{\Delta T k} = \frac{g\beta q_w x^4}{k\nu\alpha} \qquad (9.41)$$

Churchill and Chu, for example, show that their vertical plate correlation formula, eqn. (9.27), will correlate $q_w =$ constant data exceptionally well in the range $Ra_L > 1$ when Ra_L is based on ΔT at the middle of the plate. For design purposes, however, we wish to eliminate ΔT from the right-hand side, so we

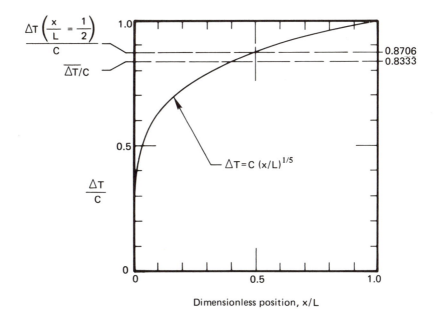

FIGURE 9.9   The mean value of $\Delta T \equiv T_w - T_\infty$ during natural convection.

replace Ra_L with $Ra_L^* / \overline{Nu_L}$. The result is

$$\overline{Nu_L} = 0.68 + 0.67 (Ra_L^*)^{1/4} \Big/ \overline{Nu_L}^{-1/4} \left[1 + \left(\frac{0.492}{Pr} \right)^{9/16} \right]^{4/9}$$

where $\overline{Nu_L} = q_w L / k \overline{\Delta T}$. This can be written in the form

$$\overline{Nu_L}^{5/4} - 0.68 \overline{Nu_L}^{-1/4} = \frac{0.67 \left[Ra_L^* \right]^{1/4}}{\left[1 + (0.492/Pr)^{9/16} \right]^{4/9}} \qquad (9.42)$$

for laminar natural convection from vertical plates with a constant wall heat flux.

The same thing can be done with eqn. (9.30) for horizontal cylinders, although the result has not been verified experimentally for very small values of Ra_L.

Some other natural convection problems

There are many natural convection situations that are beyond the scope of this book but which arise in practice.

Combined natural and forced convection.   When forced convection in, say, a duct occurs at a relatively low velocity but at a relatively high heating rate, the resulting density changes can give rise to a superimposed natural convection process. The conditions under which this can occur are mapped out in detail in many elementary heat transfer books (see, e.g., [9.20] in addition to those listed after Chapter 1). Churchill [9.21] has recently provided an extensive discussion of both the conditions that give rise to combined natural and forced convection and the prediction of heat transfer in this case.

Natural convection in enclosures.   When a natural convection process occurs within a confined space, the heated fluid buoys up and then follows the contours of the container in some way to return to the heater. This recirculation process can occur in countless different configurations, and it normally enhances heat transfer. Work on such problems, prior to 1972, is discussed in an extensive review article by Ostrach [9.22]. Much subsequent work continues to appear in the literature.

Example 9.5

A horizontal circular disk heater of diameter 0.17 m faces downward in air at 27°C. If it delivers 15 W, estimate its average surface temperature.

SOLUTION.   We have no formula for this situation, so the problem calls for some judicious guesswork. Following the lead of Churchill and Chu, we replace Ra_D with $Ra_D^*/\overline{Nu}_D$ in eqn. (9.37):

$$(\overline{Nu_D})^{6/5} = \left(\frac{q_w D}{\overline{\Delta T}k}\right)^{6/5} = 0.82(Ra_D^*)^{1/5}Pr^{0.034}$$

so

$$\overline{\Delta T} = 1.18\,\frac{q_w D/k}{\left(\dfrac{g\beta q_w D^4}{k\nu\alpha}\right)^{1/6}Pr^{0.028}}$$

$$= 1.18\,\frac{\dfrac{15}{\pi(0.085)^2}(0.17)\Big/0.02614}{\left[\dfrac{9.8\left[15/\pi(0.085)^2\right]0.17^4}{300(0.02164)(1.566)(2.203)10^{-10}}\right]^{1/6}0.711^{0.028}}$$

$$= \underline{140°C}$$

In the preceding computation all properties were evaluated at T_∞. Now we must rerun the calculation, reevaluating all properties except β at $27 + 140/2 = 97°C$:

$$\overline{\Delta T}_{corrected} = 1.18\,\frac{661(0.17)/0.03104}{\left[\dfrac{9.8\left[15/\pi(0.085)^2\right]0.17^4}{300(0.03104)(3.231)(2.277)10^{-10}}\right]^{1/6}0.99}$$

$$= \underline{142°C}$$

so the surface temperature is $27 + 142 = \underline{169°C}$.

That is rather hot. Obviously, the cooling process is quite ineffective in this case.

Dimensional analysis and experimental data

The dimensional functional equation for h (or $\bar{h}$) during film condensation is

$$\bar{h} \text{ or } h = \text{fn}\left[c_p, \rho_f, h_{fg}, g(\rho_f - \rho_g), k, \mu, (T_{sat} - T_w), L \text{ or } x \right]$$

where h_{fg} is the latent heat of vaporization. It does not appear in the differential equations (9.4) and (7.40); however, it is used in the calculation of δ [which enters in the b.c.'s (9.5)]. The film thickness, δ, depends heavily on the latent heat and slightly on the sensible heat, $c_p \Delta T$, which the film must absorb to condense. Notice, too, that $g(\rho_f - \rho_g)$ is included as a product because gravity only enters the problem as it acts upon the density difference [cf. eqn. (9.4)].

There are therefore nine variables in J, m, s, °C, and kg. It follows that we look for $9-5$, or 4 pi-groups. The ones we choose are:

$$\Pi_1 = \overline{\text{Nu}_L} \equiv \frac{\bar{h}L}{k} \qquad\qquad \Pi_2 = \text{Pr} \equiv \frac{\nu}{\alpha}$$

$$\Pi_3 = \text{Ja} \equiv \frac{c_p(T_{sat} - T_w)}{h_{fg}} \qquad \Pi_4 \equiv \frac{\rho_f(\rho_f - \rho_g)gh_{fg}L^3}{\mu k(T_{sat} - T_w)}$$

Two of these groups are new to us. The group Π_4 does not normally bear anyone's name, but it bears some similarity to the Rayleigh number. The group Π_3 is called the Jakob number, Ja, to honor Max Jakob's important pioneering work during the 1930s on problems of phase change. It compares the maximum sensible heat absorbed by the liquid to the latent heat absorbed.

Notice that if we condensed water at 1 atm on a wall 10°C below T_{sat}, then Ja would equal $4.174(10/2257) = 0.0185$. Although 10°C is a fairly large temperature difference in a condensation process, it gives a maximum sensible heat that is less than 2% of the latent heat. The Jakob number is accordingly small in most cases of practical interest, and it turns out that sensible heat can often be neglected. (There are important exceptions to this.) The same is true of the role of the Prandtl number. Therefore, during film condensation

$$\overline{\text{Nu}_L} = \text{fn}\left(\underbrace{\frac{\rho_f(\rho_f - \rho_g)gh_{fg}L^3}{\mu k(T_{sat} - T_w)}}_{\substack{\text{primary independent variable, } \Pi_4}}, \underbrace{\text{Pr}, \text{Ja}}_{\substack{\text{secondary independent} \\ \text{variables}}} \right) \qquad (9.43)$$

Equation (9.43) is not restricted to any geometrical configuration, since the same variables govern h during film condensation on any body. Figure 9.10, for

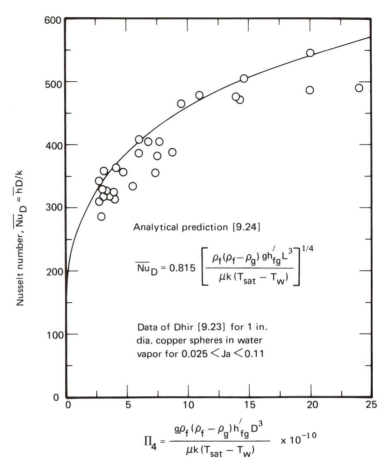

FIGURE 9.10   Correlation of the data of Dhir [9.23] for laminar film condensation on spheres, at one value of Pr and for a range of Π_4 and Ja. [Properties were evaluated at $(T_{sat} + T_w)/2$.]

example, shows laminar film condensation data given for spheres by Dhir[6] [9.23]. They have been correlated according to eqn. (9.12). The data are for only one value of Pr, but for a range of Π_4 and Ja. They generally correlate well within $\pm 10\%$, despite a broad variation of the not-very-influential variable, Ja. A predictive curve [9.23] is included in Fig. 9.10 for future reference.

Laminar film condensation on a vertical plate

Consider the following feature of film condensation. The latent heat of a liquid is normally a very large number. Therefore, even a high rate of heat transfer will

[6]Professor Dhir very kindly recalculated his data into the form shown in Fig. 9.10, for use here.

typically result in only very thin films. These films move relatively slowly, so it is safe to ignore the inertia terms in the momentum equation (9.4):

$$\underbrace{u\frac{\partial u}{\partial x} + v\frac{\partial v}{\partial y}}_{\simeq 0} = \left(1 - \frac{\rho_g}{\rho_f}\right)g + \underbrace{\nu\frac{\partial^2 u}{\partial y^2}}_{=\frac{d^2 u}{dy^2}} \tag{9.4}$$

This result will give $u = u(y, \delta)$ (where δ is the local b.l. thickness) when it is integrated. We recognize that $\delta = \delta(x)$, so that u is not really dependent on y alone. However, the y-dependence is predominant, and it is reasonable to use the approximate momentum equation

$$\frac{d^2 u}{dy^2} = -\frac{\rho_f - \rho_g}{\rho_f}g \tag{9.44}$$

This simplification was made by Nusselt in 1916 when he set down the original analysis of film condensation [9.25]. He also eliminated the convective terms from the energy equation:

$$\underbrace{u\frac{\partial T}{\partial x} + v\frac{\partial T}{\partial y}}_{\simeq 0} = \alpha\frac{\partial^2 T}{\partial y^2} \tag{7.40}$$

on the same basis. The integration of eqn. (9.44) subject to the b.c.'s

$$u(y=0)=0 \qquad \frac{\partial u}{\partial y}\bigg|_{y=\delta}=0$$

gives the parabolic velocity profile:

$$u = \frac{(\rho_f - \rho_g)g\delta^2}{2\mu}\left[2\left(\frac{y}{\delta}\right) - \left(\frac{y}{\delta}\right)^2\right] \tag{9.45}$$

And integration of the energy equation subject to the b.c.'s

$$T(y=0) = T_w \qquad T(y=\delta) = T_{sat}$$

gives the linear temperature profile:

$$T = T_w + (T_{sat} - T_w)\frac{y}{\delta} \tag{9.46}$$

To complete the analysis, we must calculate δ. This can be done in two steps. First, we express the mass flow rate in the film, $\dot{m}$, in terms of δ with the help of eqn. (9.45):

$$\dot{m} = \int_0^\delta \rho_f u\, dy = \frac{\rho_f(\rho_f - \rho_g)}{3\mu}g\delta^3 \tag{9.47}$$

Second, we neglect the sensible heat absorbed by that part of the film cooled below T_{sat} and express the local heat flux in terms of the rate of change of $\dot{m}$ (see

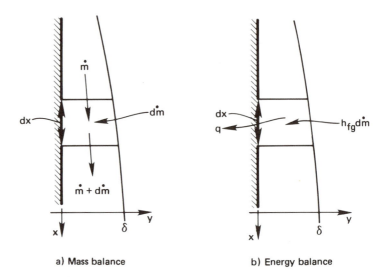

a) Mass balance   b) Energy balance

FIGURE 9.11   Heat and mass flow in an element of a condensing film.

Fig. 9.11):

$$|q| = k \frac{\partial T}{\partial y}\bigg|_{y=0} = k \frac{T_{sat} - T_w}{\delta} = h_{fg} \frac{d\dot{m}}{dx} \tag{9.48}$$

Substituting eqn. (9.47) in (9.48), we obtain a first-order differential equation for δ:

$$k \frac{T_{sat} - T_w}{\delta} = \frac{h_{fg}\rho_f(\rho_f - \rho_g)}{\mu} g\delta^2 \frac{d\delta}{dx} \tag{9.49}$$

This can be integrated directly, subject to the b.c. $\delta(x=0)=0$. The result is

$$\delta = \left[\frac{4k(T_{sat} - T_w)\mu x}{\rho_f(\rho_f - \rho_g)gh_{fg}} \right]^{1/4} \tag{9.50}$$

Both Nusselt, and subsequently Rohsenow [9.26], showed how to correct the film thickness calculation for the sensible heat that is needed to cool the inner parts of the film below T_{sat}. Rohsenow's calculation was, in part, an assessment of Nusselt's linear-temperature-profile assumption, and it led to a corrected latent heat—designated h'_{fg}—which accounted for subcooling in the liquid film:

$$h'_{fg} \equiv h_{fg} \left[1 + 0.68 \underbrace{\frac{c_p(T_{sat} - T_w)}{h_{fg}}}_{\equiv \text{ Ja, Jakob number}} \right] \tag{9.51}$$

Thus we simply must replace h_{fg} with h'_{fg} wherever it appears explicitly in the analysis, beginning with eqn. (9.48).

Finally the heat transfer coefficient is obtained from

$$h \equiv \frac{q}{T_{sat} - T_w} = \frac{1}{T_{sat} - T_w}\left[\frac{k(T_{sat} - T_w)}{\delta}\right] = \frac{k}{\delta} \tag{9.52}$$

so

$$\mathrm{Nu}_x = \frac{hx}{k} = \frac{x}{\delta} \tag{9.53}$$

Thus with the help of eqn. (9.51), we substitute eqn. (9.50) in eqn. (9.53) and get

$$\mathrm{Nu}_x = 0.707\left[\frac{\rho_f(\rho_f - \rho_g)gh'_{fg}x^3}{\mu k(T_{sat} - T_w)}\right]^{1/4} \tag{9.54}$$

This equation carries out the functional dependence that we anticipated in eqn. (9.43):

$$\mathrm{Nu}_x = \mathrm{fn}(\underbrace{\Pi_4}, \underbrace{\mathrm{Ja}}, \underbrace{\mathrm{Pr}}) \tag{9.43}$$

eliminated when we
neglected convective terms
in the energy equation

this is carried implicitly in h'_{fg}

this is clearly the dominant variable

The physical properties in Π_4, Ja, and Pr (with the exception of h_{fg}) are to be evaluated at the mean film temperature. However, if $T_{sat} - T_w$ is small—and it often is—one might approximate them at T_{sat}.

At this point we should ask just how great the missing influence of Pr is and what degree of approximation is involved in representing the influence of Ja with the use of h'_{fg}. Sparrow and Gregg [9.27] answered these questions with a complete b.l. analysis of film condensation. They did not introduce Ja in a corrected latent heat, but instead showed its influence directly.

Figure 9.12 displays two figures from the Sparrow and Gregg paper. The first shows heat transfer results plotted in the form

$$\frac{\mathrm{Nu}_x}{\sqrt[4]{\Pi_4}} = \mathrm{fn}(\mathrm{Ja}, \mathrm{Pr}) \rightarrow \text{constant as } \mathrm{Ja} \rightarrow 0 \tag{9.55}$$

Notice that the calculation approaches Nusselt's simple result for all Pr as Ja→0. It also approaches Nusselt's result, even for fairly large values of Ja, if Pr is not small. The second figure shows how the temperature deviates from the linear profile that we assumed to exist in the film in developing eqn. (9.46). If we remember that a Jakob number of 0.02 is already significant, it is clear that the linear temperature profile is a very sound assumption for nonmetallic liquids.

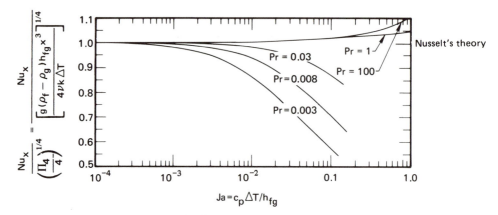

Predicted heat transfer results

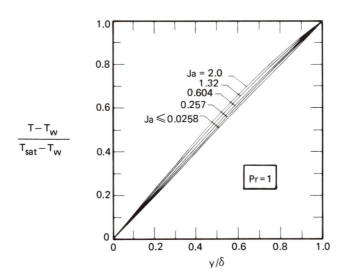

Predicted temperature profiles in condensing films

FIGURE 9.12   Results of the exact b.l. analysis of laminar film condensation on a vertical plate [9.27].

The Sparrow and Gregg analysis proves that Nusselt's analysis is quite accurate for all Prandtl numbers above the liquid-metal range. The very high Ja flows, for which Nusselt's theory requires some correction, usually result in thicker films, which become turbulent so the exact analysis no longer applies.

The average heat transfer coefficient is calculated in the usual way for $T_{wall} = \text{constant}$:

$$\bar{h} = \frac{1}{L} \int_0^L h(x)\,dx = \tfrac{4}{3} h(L) \tag{9.56}$$

so

$$\overline{\mathrm{Nu}_L} = 0.9428 \left[\frac{\rho_f(\rho_f - \rho_g)gh'_{fg}L^3}{\mu k(T_{sat} - T_w)} \right]^{1/4} \tag{9.57}$$

Example 9.6

Water at atmospheric pressure condenses on a strip 30 cm in height, which is held at 90°C. Calculate the overall heat transfer per meter, the film thickness at the bottom, and the mass rate of condensation per meter.

SOLUTION.

$$\delta = \left[\frac{4k(T_{sat} - T_w)\nu x}{(\rho_f - \rho_g)gh'_{fg}} \right]^{1/4}$$

where we have replaced h_{fg} with h'_{fg}:

$$h'_{fg} = 2257 \left[1 + 0.68 \frac{4.216(10)}{2257} \right] = 2286 \text{ kJ/kg}$$

so

$$\delta = \left[\frac{4(0.681)(10)(0.290)10^{-6}x}{(957.2 - 0.6)(9.8)(2286)(10)^3} \right]^{1/4} = 0.000139 x^{1/4}$$

Then

$$\delta(L) = 0.000103 \text{ m} = \underline{0.103 \text{ mm}}$$

Notice how thin the film is. Finally, we use eqns. (9.53) and (9.56) to compute

$$\overline{\mathrm{Nu}_L} = \frac{4}{3} \frac{L}{\delta} = \frac{4(0.3)}{3(0.000103)} = 3883$$

so

$$q = \frac{\mathrm{Nu}_L k \Delta T}{L} = \frac{3883(0.681)(10)}{0.3} = 88,155 \text{ W/m}^2$$

(This is a heat flow of over 88 kW on an area about half of a desk top. That is very high for such a small temperature difference.) Then

$$Q = 88,155(0.3) = 26,446 \text{ W/m} = \underline{26.4 \text{ kW/m}}$$

The rate of condensate, $\dot{m}$, is

$$\dot{m} = \frac{Q}{h'_{fg}} = \frac{26.4}{2286} = \underline{0.0115 \text{ kg/m-s}}$$

Condensation on other bodies

Nusselt himself extended his prediction to certain other bodies but was restricted by the lack of a digital computer from evaluating as many cases as he might have. In 1971 Dhir and Lienhard [9.24] showed how Nusselt's method could be

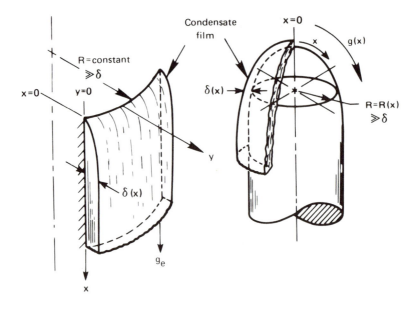

a) Vertical plate or vertical cylinder   b) Axi-symmetric body

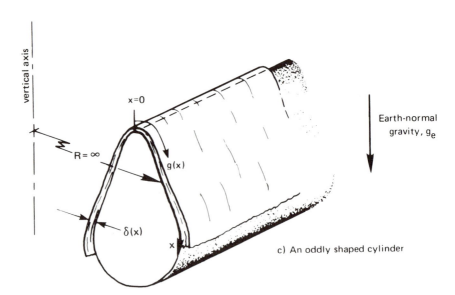

FIGURE 9.13   Condensation on various bodies. $g(x)$ is the component of gravity or other body force in the x-direction.

readily extended to a large class of problems. They showed that they needed only to replace the gravity, g, with an effective gravity, g_{eff}:

$$g_{eff} \equiv \frac{x(gR)^{4/3}}{\int_0^x g^{1/3} R^{4/3}\, dx} \qquad (9.58)$$

in eqns. (9.50) and (9.54), to predict δ and Nu_x for a variety of bodies. The terms in eqn. (9.58) are (see Fig. 9.13):

- $x =$ the distance along the liquid film measured from the upper stagnation point.
- $g = g(x)$, the component of gravity (or other body force) along x. g can vary from point to point as it does in Fig. 9.13b and c.
- $R(x)$ is a radius of curvature about the vertical axis. In Fig. 9.13a it is a constant that factors out of eqn. (9.58). In Fig. 9.13c, R is infinite and it again can be factored out of eqn. (9.58). Only in axisymmetric bodies, where R varies with x, need it be included. When it can be factored out,

$$g_{eff} \text{ reduces to } \frac{xg^{4/3}}{\int_0^x g^{1/3}\, dx} \qquad (9.59)$$

- $g_e =$ earth-normal gravity. We introduce g_e at this point to distinguish it from $g(x)$.

Example 9.7

Find Nu_x for laminar film condensation on the top of a flat surface sloping at $\theta°$ from the vertical plane.

SOLUTION. In this case $g = g_e \cos\theta$ and $R = \infty$. Therefore, eqn. (9.58) or (9.59) reduces to

$$g_{eff} = \frac{xg_e^{4/3}(\cos\theta)^{4/3}}{g_e^{1/3}(\cos\theta)^{1/3}\int_0^x dx} = g_e \cos\theta$$

as we might expect. Then for a slanting plate

$$Nu_x = 0.707 \left[\frac{\rho_f(\rho_f - \rho_g)(g_e \cos\theta)h'_{fg}x^3}{\mu k(T_{sat} - T_w)} \right]^{1/4} \qquad (9.60)$$

Example 9.8

Find the overall Nusselt number for a horizontal cylinder.

SOLUTION. There is an important conceptual hurdle here. The radius $R(x)$ is infinity as shown in Fig. 9.13c—it is not the radius of the cylinder. $g(x)$ can be shown very easily to equal $g_e \sin(2x/D)$, where D is the diameter of the cylinder. Then

$$g_{\mathrm{eff}} = \frac{xg_e^{4/3}(\sin 2x/D)^{4/3}}{g_e^{1/3}\int_0^x \sin(2x/D)\,dx}$$

and

$$\bar{h} = \frac{k}{D}\frac{2}{\pi D}\int_0^{\pi D/2}\frac{1}{\sqrt{2}}\left[\frac{\rho_f(\rho_f-\rho_g)h'_{fg}x^3}{\mu k(T_{\mathrm{sat}}-T_w)}\frac{xg_e(\sin 2x/D)^{4/3}}{\int_0^x \sin(2x/D)\,dx}\right]^{1/4}dx$$

This requires evaluation on the computer. The result, when it is put back in the form of a Nusselt number, is

$$\overline{\mathrm{Nu}_D} = 0.729\left[\frac{\rho_f(\rho_f-\rho_g)g_e h'_{fg}D^3}{\mu k(T_{\mathrm{sat}}-T_w)}\right]^{1/4} \tag{9.61}$$

for a horizontal cylinder. (Nusselt got 0.725 for the lead constant but he had to do his calculation by hand.)

Some other results of this calculation include the following cases:

Sphere of diameter, D:

$$\overline{\mathrm{Nu}_D} = 0.815\left[\frac{\rho_f(\rho_f-\rho_g)g_e h'_{fg}D^3}{\mu k(T_{\mathrm{sat}}-T_w)}\right]^{1/4} \tag{9.62}$$

This result[7] has already been compared with experimental data in Fig. 9.10.

Vertical cone with the apex on top, the bottom insulated, and a cone angle of $\alpha°$:

$$\mathrm{Nu}_x = 0.874[\cos(\alpha/2)]^{1/4}\left[\frac{\rho_f(\rho_f-\rho_g)g_e h'_{fg}x^3}{\mu k(T_{\mathrm{sat}}-T_w)}\right]^{1/4} \tag{9.63}$$

Rotating horizontal disk[8]: In this case $g = \omega^2 x$, where x is the distance from the center and ω is the speed of rotation. The Nusselt number, based on $L = (\mu/\rho_f\omega)^{1/2}$, is

$$\overline{\mathrm{Nu}} = 0.9034\left[\frac{\mu(\rho_f-\rho_g)h'_{fg}}{\rho_f k(T_{\mathrm{sat}}-T_w)}\right]^{1/4} = \mathrm{constant} \tag{9.64}$$

This result might seem strange at first glance. It says that $\mathrm{Nu} \neq \mathrm{fn}\,(x$ or $\omega)$. The reason is that δ just happens to be independent of x and hence of ω.

The Nusselt solution can thus be bent to fit many complicated geometrical figures. One of the most complicated ones that have been dealt with to date is

---

[7]There is an error in [9.24]. The constant given there is 0.785. The value of 0.815 given here is correct.

[8]This problem was originally solved by Sparrow and Gregg [9.28].

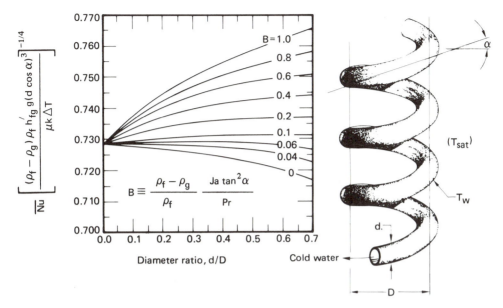

FIGURE 9.14   Fully developed film condensation heat transfer on a helical reflux condenser [9.29].

the reflux condenser shown in Fig. 9.14. In such a configuration cooling water flows through a helically wound tube and vapor condenses on the outside, running downward along the tube. As the condensate flows, centripetal forces sling the liquid outward at a downward angle. This complicated flow was analyzed by Karimi [9.29], who found that

$$\overline{Nu} \equiv \frac{\bar{h}\, d \cos \alpha}{k} = \left[\frac{(\rho_f - \rho_g)\rho_f h'_{fg}\, g (d \cos \alpha)^3}{\mu k\, \Delta T}\right]^{1/4} \mathrm{fn}\left(\frac{d}{D}, B\right) \qquad (9.65)$$

where B is a centripetal parameter:

$$B \equiv \frac{\rho_f - \rho_g}{\rho_f}\, \frac{c_p \Delta T}{h'_{fg}}\, \frac{\tan^2 \alpha}{Pr}$$

and α is the helix angle (see Fig. 9.14). The function of the tube-to-helix diameter ratio, d/D, and B must be evaluated numerically. Karimi's result is plotted in Fig. 9.14.

Laminar–turbulent transition

The mass flow rate of condensate in the film, $\dot{m}$, or Γ_c kg/m-s as it is more commonly designated, can be calculated as

$$\dot{m} \quad \text{or} \quad \Gamma_c = \rho_f \int_0^\delta u\, dy \qquad (9.66)$$

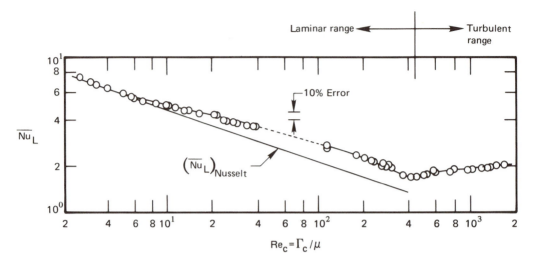

FIGURE 9.15   Film condensation on vertical plates. Data are for water [9.30].

Equation (9.45) gives $u(y)$ independently of any geometrical features. [The geometry is characterized by $\delta(x)$.] Thus we can substitute eqn. (9.45) in eqn. (9.66) and obtain

$$\Gamma_c = \frac{\rho_f(\rho_f - \rho_g)g\delta^3}{3\mu} \tag{9.67}$$

This expression is valid for any location in any film regardless of the geometry of the body. The configuration will lead to variations of $g(x)$ and $\delta(x)$, but the functional form of eqn. (9.67) will not change.

It is useful to define a Reynolds number in terms of Γ_c. This is easy to do because Γ_c is equal to $\rho u_{av}\delta$.

$$Re_c = \frac{\Gamma_c}{\mu} = \frac{\rho_f(\rho_f - \rho_g)g\delta^3}{3\mu^2} \tag{9.68}$$

It turns out that the Reynolds number dictates the onset of film instability just as it dictates the instability of a b.l. or of a pipe flow.[9] When $Re_c = 5$ or 6, scallop-shaped ripples become visible on the condensate film. When Re_c reaches 450, a full scale laminar-to-turbulent transition occurs.

Gregorig, Kern, and Turek [9.30] have reviewed existing data for the film condensation of water, and added their own measurements. Figure 9.15 shows these data in comparison with Nusselt's theory, eqn. (9.57). The comparison is almost perfect up to Re_c almost equal to 6. Then the data start yielding somewhat higher heat transfer rates than the prediction. This is because the

---

[9]Two Reynolds numbers are defined for film condensation: Γ_c/μ and $4\Gamma_c/\mu$. The latter one, which is simply four times as large as the one we use, is more common in the American literature.

ripples improve heat transfer—just a little at first, and by about 20% when the full laminar-to-turbulent transition occurs at $Re_c = 450$.

Above $Re_c = 450$, $\overline{Nu_L}$ begins to rise with Re_c. The Nusselt number begins to exhibit an increasingly strong dependence on the Prandtl number in this turbulent regime. Therefore, one can use Fig. 9.15 directly as a data correlation, to predict the heat transfer coefficient for steam condensing at 1 atm. But for other fluids with different Prandtl numbers, one should consult [9.30] or [1.13, Chap. 12].

Two final issues
in natural convection film condensation

- *Condensation in tube bundles.* Nusselt showed that if n horizontal tubes are arrayed over one another, and if the condensate leaves each one and flows directly onto the one below it without splashing, then

$$Nu_{D_{\text{for } n \text{ tubes}}} = \frac{1}{n^{1/4}} Nu_{D_{1 \text{ tube}}} \tag{9.69}$$

This is a fairly optimistic extension of the theory, of course. Other cases are briefly discussed in [1.13, Chap. 12].

- *Condensation in the presence of noncondensible gases.* When the condensing vapor is mixed with noncondensible air, uncondensed air must constantly diffuse away from the condensing film and vapor must diffuse inward toward the film. This coupled diffusion process can considerably slow condensation. The resulting h can easily be cut by a factor of five if there is as little as 5% by mass of air mixed into the steam. This effect was first analyzed in detail by Sparrow and Lin [9.31] and was treated further by Minkowycz and Sparrow [9.32] (see also [1.13, Chap. 12]).

PROBLEMS

9.1. Show that Π_4 in the film condensation problem can properly be interpreted as $PrRe^2/Ja$.

9.2. A 20-cm-high vertical plate is kept at 34°C in a 20°C room. Plot *to scale* δ and h vs. height, and the actual temperature and velocity vs. y at the top.

9.3. Redo the Squire–Eckert analysis, neglecting inertia, to get a high-Pr approximation to Nu_x. Compare your result with the Squire–Eckert formula.

9.4. Assume a linear temperature profile and a simple triangular velocity profile as shown in Fig. P9.4 for natural convection on a vertical isothermal plate. Derive

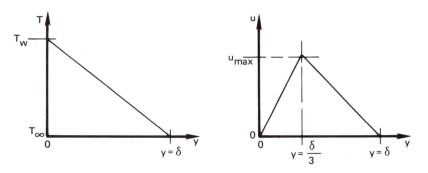

FIGURE P9.4

$Nu_x = fn(Pr, Gr_x)$, compare your result with the Squire–Eckert result, and discuss the comparison.

9.5.   A horizontal cylindrical duct of diamond-shaped cross section carries air at 35°C. Since almost all thermal resistance is in the natural convection b.l. on the outside, take T_w to be approximately 35°C. $T_\infty = 25$°C. Estimate the heat loss per meter of duct if the duct is uninsulated. [$Q = 24.0$ W/m.]

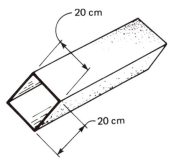

20 cm

20 cm

FIGURE P9.5

9.6.   The heat flux from a 3-m-high electrically heated panel in a wall is 75 W/m² in an 18°C room. What is the average temperature of the panel? What is the temperature at the top? at the bottom?

9.7.   Find pipe diameters and wall temperatures for which the film condensation heat transfer coefficients given in Table 1.1 are valid.

9.8.   Consider Example 9.6. What value of wall temperature (if any), or what height of the plate, would result in a laminar-to-turbulent transition at the bottom in this example?

9.9.   A plate spins as shown in Fig. P9.9, in a vapor that rotates synchronously with it. Neglect earth-normal gravity and calculate Nu_L as a result of film condensation.

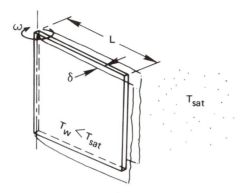

FIGURE P9.9

9.10. A laminar liquid film of temperature T_{sat} flows down a vertical wall which is also at T_{sat}. Flow is fully developed and the film thickness is δ_o. Along a particular horizontal line, the wall temperature has a lower value, T_w, and it is kept at that temperature everywhere below that position. Call the line where the wall temperature changes $x=0$. If the whole system is immersed in saturated vapor of the flowing liquid, calculate $\delta(x)$, Nu_x, and Nu_L, where $x=L$ is the bottom edge of the wall. (Neglect any transient behavior in the neighborhood of $x=0$.)

9.11. Prepare a table of formulas of the form

$$\bar{h}\,(W/m^2\text{-}°C) = C[\Delta T°C/L\ m]^{1/4}$$

for natural convection at normal gravity in air and in water at $T_\infty = 27°C$. Assume that T_w is close to $27°C$. Your table should include results for vertical plates, horizontal cylinders, spheres, and possibly additional geometries. Do not include your calculations.

9.12. For what value of Pr is the condition

$$\left.\frac{\partial^2 u}{\partial y^2}\right|_{y=0} = \frac{g\beta(T_w - T_\infty)}{\nu}$$

satisfied exactly in the Squire–Eckert b.l. solution? [Pr = 2.86.]

9.13. The overall heat transfer coefficient on the side of a particular house 10 m in height is 2.5 $W/m^2\text{-}°C$, excluding exterior convection. It is a cold, still winter night with $T_{outside} = -30°C$ and $T_{inside} = 25°C$. What is $\bar{h}$ on the outside of the house? Is external convection laminar or turbulent?

9.14. Consider Example 9.2. The sheets are mild steel 2 m long and 6 mm thick. The bath is basically water at $60°C$ and the sheets are put in it at $18°C$. (a) Plot the sheet temperature as a function of time. (b) Approximate $\bar{h}$ at $\Delta T = \left(\dfrac{60+18}{2} - 18\right)°C$ and plot the conventional exponential response on the same graph.

9.15. A vertical heater 0.15 m in height is immersed in water at 7°C. Plot $\bar{h}$ against $(T_w - T_\infty)^{1/4}$, where T_w is the heater temperature, in the range $0 < (T_w - T_\infty) < 100$°C. Comment on the result. Should the line be straight?

9.16. A 77°C vertical wall heats 27°C air. Evaluate δ_{top}/L, Ra_L, and L at the point where $Nu_L/Ra_L^{1/4}$ ceases to be a constant (see Fig. 9.3). Comment on your results. $[\delta_{top}/L \simeq 0.6.]$

9.17. A horizontal 8-cm-O.D. pipe carries steam at 150°C through a room at 17°C. The pipe has a 1.5-cm layer of 85% magnesia insulation on it. Evaluate the heat loss per meter of pipe. $[Q = 97.3 \text{ W/m.}]$

9.18. What heat rate (W/m) must be supplied to a 0.01-mm horizontal wire to keep it 30°C above the 10°C water around it?

9.19. A vertical run of copper tubing 5 mm in diameter and 20 cm long carries condensation vapor at 60°C through 27°C air. What is the total heat loss?

9.20. A body consists of two cones joined at their bases. The diameter is 10 cm and overall length of the joined cones is 25 cm. The axis of the body is vertical and the body is kept at 27°C in 7°C air. What is the rate of heat removal from the body? $[Q = 3.38 \text{ W.}]$

9.21. Consider the plate dealt with in Example 9.3. Plot $\bar{h}$ as a function of the angle of inclination of the plate as the hot side is tilted both upward and downward. Note that you must make do with discontinuous formulas in different ranges of θ.

9.22. You have been asked to design a vertical wall panel heater, 1.5 m high, for a dwelling. What should the heat flux be if no part of the wall should exceed 33°C? How much heat will be added to the room if the panel is 7 m in width?

9.23. A 14-cm-high vertical surface is heated by condensing steam at 1 atm. If the wall is kept at 30°C, how would the average heat transfer coefficient change if methanol, CCl_4, or acetone were used instead of steam to heat it? How would the heat flux change? (This problem requires that certain information be obtained from the library.)

9.24. A 1-cm-diameter tube extends 27 cm through a region of saturated steam at 1 atm. The outside of the tube can be maintained at any temperature between 50°C and 150°C. Plot the total heat transfer as a function of tube temperature.

9.25. A 2-m-high vertical plate condenses steam at atmospheric pressure. Below what temperature will Nusselt's prediction of $\bar{h}$ be in error? Below what temperature will the condensing film be turbulent?

9.26. A reflux condenser is made of copper tubing 0.8 cm in diameter with a wall temperature of 30°C. It condenses steam at 1 atm. Find $\bar{h}$ if $\alpha = 18°$ and the coil diameter is 7 cm.

9.27. The coil diameter of a helical condenser is 5 cm and the tube diameter is 5 mm. The condenser carries water at 15°C and is in a bath of saturated steam at 1 atm. Specify the number of coils and a reasonable helix angle if 6 kg/hr of steam is to be condensed. $h_{inside} = 600 \text{ W/m}^2\text{-°C}$.

9.28. A schedule 40 type 304 stainless steam pipe with a 4-in. nominal diameter carries saturated steam at 150 psia in a processing plant. Calculate the heat loss per unit length of pipe if it is bare and the surrounding air is still at 68°F. How much would this heat loss be reduced if the pipe were insulated with a 1-in. layer of 85% magnesia insulation? [$Q_{saved} \simeq 127$ W/m.]

REFERENCES

[9.1]   W. NUSSELT, "Das Grundgesetz des Wärmeüberganges," *Gesund. Ing.*, vol. 38, 1915, p. 872.

[9.2]   C. J. SANDERS and J. P. HOLMAN, "Franz Grashof and the Grashof Number," *Int. J. Heat Mass Transfer*, vol. 15, no. 3, 1972, pp. 562–563.

[9.3]   S. W. CHURCHILL and H. H. S. CHU, "Correlating Equations for Laminar and Turbulent Free Convection from a Vertical Plate," *Int. J. Heat Mass Transfer*, vol. 18, 1975, pp. 1323–1329.

[9.4]   *Modern Developments in Fluid Mechanics*, vol. 2 (S. Goldstein, ed.), Oxford University Press, New York, 1938, Chapter 14.

[9.5]   E. M. SPARROW and J. L. GREGG, "The Variable Fluid-Property Problem in Free Convection," *Recent Advances in Heat and Mass Transfer* (J. P. Hartnett, ed.), McGraw-Hill Book Company, New York, 1961, pp. 353–371.

[9.6]   H. BARROW and T. L. SITHARAMARAO, "The Effect of Variable β on Free Convection," *Brit. Chem. Eng.*, vol. 16, no. 8, 1971, p. 704.

[9.7]   W. KRAUS, *Messungen des Temperatur- und Geschwindigkeitsfeldes bei freier Konvection*, Verlag G. Braun, Karlsruhe, 1955, Chapter F.

[9.8]   S. W. CHURCHILL and H. H. S. CHU, "Correlating Equations for Laminar and Turbulent Free Convection from a Horizontal Cylinder," *Int. J. Heat Mass Transfer*, vol. 18, 1975, pp. 1049–1053.

[9.9]   W. J. MINKOWYCZ and E. M. SPARROW, "Local Nonsimilar Solutions for Natural Convection on a Vertical Cylinder," *J. Heat Transfer, Trans. ASME, Ser. C*, vol. 96, no. 2, 1974, pp. 178–183.

[9.10] T. CEBECI, "Laminar-Free-Convective-Heat Transfer from the Outer Surface of a Vertical Slender Circular Cylinder," *Proc. Fifth Int. Heat Transfer Conf., Tokyo*, September 1974, NC 1.4, pp. 15–19.

[9.11] T. YUGE, "Experiments on Heat Transfer from Spheres including Combined Forced and Natural Convection," *J. Heat Transfer, Trans. ASME, Ser. C*, vol. 82, no. 1, 1960, p. 214.

[9.12] J. H. LIENHARD, "On the Commonality of Equations for Natural Convection from Immersed Bodies," *Int. J. Heat Mass Transfer*, vol. 16, 1973, p. 2121.

[9.13] B. R. RICH, "An Investigation of Heat Transfer from an Inclined Flat Plate in Free Convection," *Trans. ASME*, vol. 75, 1953, pp. 489–499.

[9.14]   T. Fujii and H. Imura, "Natural Convection Heat Transfer from a Plate with Arbitrary Inclination," *Int. J. Heat Mass Transfer*, vol. 15, 1972, p. 755.

[9.15]   G. C. Vliet, "Natural Convection Local Heat Transfer on Constant Heat Transfer Inclined Surface," *J. Heat Transfer, Trans. ASME, Ser. C*, vol. 91, 1969, pp. 511–516.

[9.16]   A. F. Emery, A. Yang, and J. R. Wilson, "Free Convection Heat Transfer to Newtonian and Non-Newtonian High Prandtl Number Fluids from Vertical and Inclined Surfaces," ASME Prepr. 76-HT-46, 1976.

[9.17]   W. R. Moran and J. R. Lloyd, "Natural Convection Mass Transfer Adjacent to Vertical and Downward-facing Surfaces," *J. Heat Transfer, Trans. ASME, Ser. C*, vol. 97, 1975, pp. 472–474.

[9.18]   V. Kadambi and R. M. Drake, Jr., "Free Convection Heat Transfer from Horizontal Surfaces for Prescribed Variations in Surface Temperature and Mass Flow through the Surface," Tech. Rept. MECH Eng. HT-1, Princeton Univ., June 30, 1959.

[9.19]   O. Saunders and M. Fishenden, "Some Measurements of Convection by an Optical Method," *Engineering*, May 1935, pp. 483–485.

[9.20]   B. V. Karlekar and R. M. Desmond, *Engineering Heat Transfer*, West Publishing Co., St. Paul, Minn., 1977, Section 9.8.

[9.21]   S. W. Churchill, "A Comprehensive Correlating Equation for Laminar, Assisting, Forced and Free Convection," *AIChE J.*, vol. 23, no. 1, 1977, pp. 10–16.

[9.22]   S. Ostrach, "Natural Convection in Enclosures," *Advances in Heat Transfer*, vol. 8 (T. F. Irvine, Jr., and J. P. Hartnett, eds.) Academic Press, New York, 1972, pp. 161–227.

[9.23]   V. K. Dhir, "Quasi-steady Laminar Film Condensation of Steam on Copper Spheres," *J. Heat Transfer, Trans. ASME, Ser. C*, vol. 97, no. 3, 1975, pp. 347–351.

[9.24]   V. K. Dhir and J. H. Lienhard, "Laminar Film Condensation on Plane and Axi-symmetric Bodies in Non-uniform Gravity," *J. Heat Transfer, Trans. ASME, Ser. C*, vol. 93, no. 1, 1971, pp. 97–100.

[9.25]   W. Nusselt, "Die Oberflächenkondensation des Wasserdampfes," *Z. Ver. Dtsch. Ing.*, vol. 60, 1916, pp. 541 and 569.

[9.26]   W. M. Rohsenow, "Heat Transfer and Temperature Distribution in Laminar-Film Condensation," *Trans. ASME*, vol. 78, 1956, pp. 1645–1648.

[9.27]   E. M. Sparrow and J. L. Gregg, "A Boundary-Layer Treatment of Laminar-Film Condensation," *J. Heat Transfer, Trans. ASME, Ser. C*, vol. 81, 1959. pp. 13–18.

[9.28]   E. M. Sparrow and J. L. Gregg, "A Theory of Rotating Condensation," *J. Heat Transfer, Trans. ASME, Ser. C*, vol. 81, 1959, pp. 113–120.

[9.29] A. KARIMI, "Laminar Film Condensation on Helical Reflux Condensers and Related Configurations," *Int. J. Heat Mass Transfer*, vol. 20, 1977, pp. 1137–1144.

[9.30] R. GREGORIG, J. KERN, and K. TUREK, "Improved Correlation of Film Condensation Data Based on a More Rigorous Application of Similarity Parameters," *Wärme- und Stoffübertragung*, vol. 7, 1974, pp. 1–13.

[9.31] E. M. SPARROW and S. H. LIN, "Condensation in the Presence of a Non-condensible Gas," *J. Heat Transfer, Trans. ASME, Ser. C*, vol. 86, 1963, p. 430.

[9.32] W. J. MINKOWYCZ and E. M. SPARROW, "Condensation Heat Transfer in the Presence of Non-condensibles, Interfacial Resistance, Superheating, Variable Properties, and Diffusion," *Int. J. Heat Mass Transfer*, vol. 9, 1966, pp. 1125–1144.

Chapter 10

Heat transfer in boiling
and other phase-change configurations

For a charm of powerful trouble,
Like a Hell-broth boil and bubble, ...
Cool it with a baboon's blood,
Then the charm is firm and good.

Macbeth, Wm. Shakespeare

"A watched pot never boils"—the water in a teakettle takes a long time to get hot enough to boil because natural convection initially warms it rather slowly. Once boiling begins, the water is heated the rest of the way to the saturation point very quickly. Boiling is of interest to us because it is remarkably effective in carrying heat from a heater into a liquid. The heater in question might be a redhot horseshoe quenched in a bucket, or the core of a nuclear reactor with coolant flowing through it. Our aim is to learn enough about the boiling process to design systems that use boiling for cooling. We begin by considering pool boiling—the boiling that occurs when a stationary heater transfers heat to an otherwise stationary liquid.

10.1 Nukiyama's experiment
and the pool boiling curve

Hysteresis in the q vs. ΔT
relation for pool boiling

In 1934, Nukiyama [10.1] did the experiment described in Fig. 10.1. He boiled saturated water on a horizontal wire that functioned as both an electric resistance heater and as a resistance thermometer. By calibrating the resistance of a

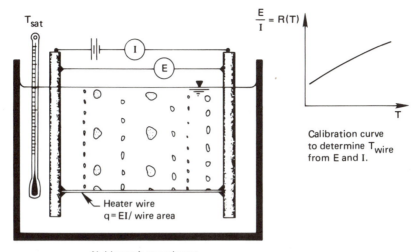

Nukiyama's experiment

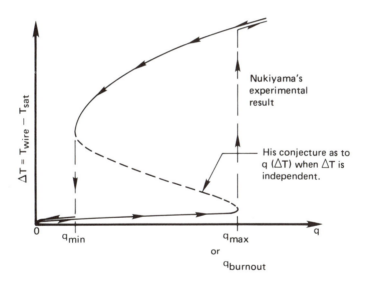

FIGURE 10.1 Nukiyama's boiling hysteresis loop.

Nichrome wire as a function of temperature before the experiment, he was able to obtain both the heat flux and the temperature using the observed current and voltage. He found that as he increased the power input to the wire, the heat flux rose sharply but the temperature of the wire increased relatively little. Suddenly, at a particular high value of the heat flux, the wire abruptly melted. Nukiyama then obtained a platinum wire and tried again. This time the wire reached the same limiting heat flux, but then it turned almost white-hot without melting.

As he reduced the power input to the white-hot wire, the temperature dropped in a continuous way, as shown in Fig. 10.1, until the heat flux was far below the value where the first temperature jump occurred. Then the temperature dropped abruptly to the original q vs. $\Delta T = (T_{wire} - T_{sat})$ curve, as shown. Nukiyama suspected that the hysteresis would not occur if ΔT could be specified as the independent controlled variable. He conjectured that such an experiment would result in the connecting line shown between the points where the temperatures jumped.

In 1937, Drew and Mueller [10.2] succeeded in making an independent ΔT experiment by boiling organic liquids outside a tube. Steam was allowed to condense inside the tube at an elevated pressure. The steam saturation tem-

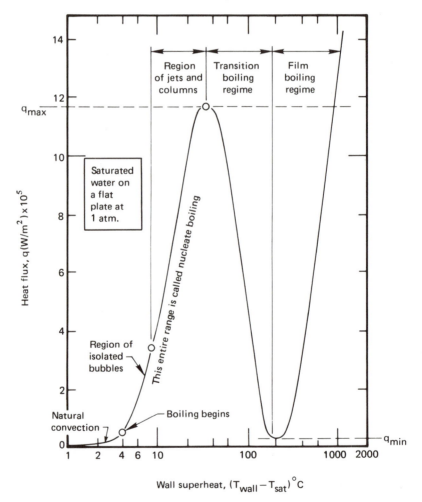

FIGURE 10.2   Typical boiling curve and regimes of boiling for an unspecified heater surface.

perature—and hence the tube-wall temperature—were controlled independently by varying the steam pressure. This permitted them to verify Nukiyama's conjecture and complete the boiling curve.

Figure 10.2 is a completed boiling curve for saturated water at atmospheric pressure on a particular flat horizontal heater. It displays the behavior shown in Fig. 10.1, but it has been rotated to place the independent variable, ΔT, on the abscissa.

Modes of pool boiling

The boiling curve in Fig. 10.2 has been divided into five regimes of behavior. These regimes, and the transitions that divide them, are discussed next.

Natural convection. Water that is not in contact with its own vapor does not boil at the so-called normal boiling point,[1] T_{sat}. Instead, it continues to rise in temperature until bubbles finally do begin to form. On conventional machined metal surfaces this occurs when the surface is a few degrees above T_{sat}. Below the bubble inception point, heat is removed by natural convection and it can be predicted by the methods laid out in Chapter 9.

Nucleate boiling. The nucleate boiling regime embraces the two distinct regimes that lie between bubble inception and Nukiyama's first transition point:

1. *The region of isolated bubbles*. In this range, bubbles rise from isolated nucleation sites, more or less as they are sketched in Fig. 10.1. As q and ΔT increase, more and more sites are activated. Figure 10.3a is a photograph of this regime as it appears on a horizontal cylinder.

2. *The region of slugs and columns*. When the active sites become very numerous, the bubbles start to merge into one another, and an entirely different kind of vapor escape path comes into play. Vapor formed at the surface merges immediately into jets which feed into large overhead bubbles of "slugs" of vapor. This process is pictured in Fig. 10.3b.

Peak heat flux. Clearly, it is very desirable to be able to operate heat exchange equipment at the upper end of the region of slugs and columns. Then the temperature difference is low while the heat flux is very high. Heat transfer coefficients in this range are enormous. However, it is very dangerous to run equipment near q_{max} in systems in which q is given independently of ΔT (as in nuclear reactors). If q is raised beyond the upper limit of the nucleate boiling regime, such a system will suffer a sudden and damaging increase of temperature. This transition[2] is known by a variety of names: the *burnout* point

---

[1]This notion might be new to some readers. It is explained in Section 10.2.
[2]We defer a proper physical explanation of the transition to Section 10.3.

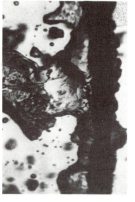

c) Two views of transitional boiling in acetone on a 0.32 cm diam. tube.

2.54 cm

d) Film boiling of acetone on a 22 gage wire at earth-normal gravity.

a) Isolated bubble regime—water.

3.45 cm length of 0.0322 cm diam. wire in methanol at 10 earth-normal gravities. $q = 0.35 \times 10^6$ W/m²

3.75 cm length of 0.164 cm diam. wire in benzene at earth-normal gravity. $q = 0.35 \times 10^6$ W/m²

b) Two views of the regime of slugs and columns.

FIGURE 10.3 Typical photographs of boiling in the four regimes identified in Fig. 10.2.

(although a complete burning up or melting away does not always accompany it); the *peak heat flux* (a modest descriptive term); the *boiling crisis* (a Russian term); the *DNB* or *departure from nucleate boiling* and the *CHF* or *critical heat flux* (terms more often used in flow boiling); and the *first boiling transition* (which term ignores previous transitions). We designate the peak heat flux as q_{max}.

Transitional boiling regime.   It is a curious fact that the heat flux actually diminishes with ΔT after q_{max} is reached. In this regime the effectiveness of the vapor escape process becomes worse and worse. Furthermore, the hot surface becomes completely blanketed in vapor and q reaches a minimum heat flux which we call q_{min}. Figure 10.3c shows a typical instance of transitional boiling.

Film boiling.   Once a stable vapor blanket is established, q will only increase with increasing ΔT. The mechanics of the heat removal process during film boiling, and the regular removal of bubbles, has a great deal in common with film condensation, but the heat transfer coefficients are much lower because heat must conduct through a vapor film instead of through a liquid film. We see an instance of film boiling in Fig. 10.3d.

Experiment 10.1

Set an open pan of cold tap water on your stove, to boil. Observe the following stages as you watch:

- At first nothing appears to happen; then you notice that numerous small stationary bubbles have formed over the bottom of the pan. These bubbles have nothing to do with boiling—they contain air that has been driven out of solution as the temperature has risen.

- Suddenly the pan will begin to "sing." There will be a somewhat high-pitched buzzing–humming sound as the first vapor bubbles are triggered. They grow at the heated surface and condense very suddenly when their tops encounter the still-cold water above them. This cavitation collapse is accompanied by a small "ping" or "click," over and over, as the process is repeated at a fairly high frequency.

- As the liquid bulk increases in temperature, the singing is increasingly muted. You may then look in the pan and see a number of points on the bottom where a feathery blur appears to be affixed. These blurred images are bubble columns emanating scores of bubbles per second. The bubbles in these columns condense completely at some distance above the surface. Notice that the air bubbles are all gradually being swept away.

- The "singing" finally gives way to a full rolling boil, accompanied by a gentle burbling sound. Bubbles no longer condense but now reach the surface, where they break.

- A full rolling-boil process, in which the liquid bulk is saturated, is a form of isolated-bubble process, as plotted in Fig. 10.2. No kitchen stove supplies energy

fast enough to boil water in the slugs-and-columns regime. You might, therefore, reflect on the relative intensity of the slugs-and-columns process.

Experiment 10.2

Repeat Experiment 10.1 with a glass beaker instead of a kitchen pan. Place a strobe light,
blinking about 6 to 10 times per second, behind the beaker with a piece of frosted glass or
tissue paper between it and the beaker. You can now see the evolution of bubble columns
from the first singing mode up to the rolling boil. You will also be able to see natural
convection in the refraction of the light before boiling begins.

10.2 Nucleate boiling

Inception of boiling

Figure 10.4 shows a highly enlarged sketch of a heater face. Most metal-finish-
ing operations score tiny grooves on the surface, but they also typically involve
some *chattering* or bouncing action, which hammers small holes into the surface.
When a surface is wetted, liquid is prevented by surface tension from entering
these holes, so small gas or vapor pockets are formed. These little pockets are
the sites at which bubble nucleation occurs.

To see why vapor pockets serve as nucleation sites, consider Fig. 10.5. Here
we see the problem in highly idealized form. Suppose that a spherical bubble of
pure saturated steam is at equilibrium with an infinite superheated liquid. To
determine the size of such a bubble, we impose the conditions of mechanical
and thermal equilibrium.

The bubble will be in *mechanical* equilibrium when the pressure difference
between the inside and the outside of the bubble is balanced by the forces of
surface tension, σ, as indicated in the cutaway sketch in Fig. 10.5. Since *thermal*
equilibrium requires that the temperature must be the same inside and outside
the bubble, and since the vapor inside must be saturated at T_{sup} because it is in
contact with its liquid, the force balance takes the form

$$R_b = \frac{2\sigma}{p_{sat} \text{ at } T_{sup} - p_{ambient}} \qquad (10.1)$$

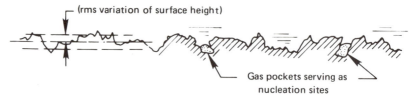

FIGURE 10.4   Enlarged sketch of a typical metal surface.

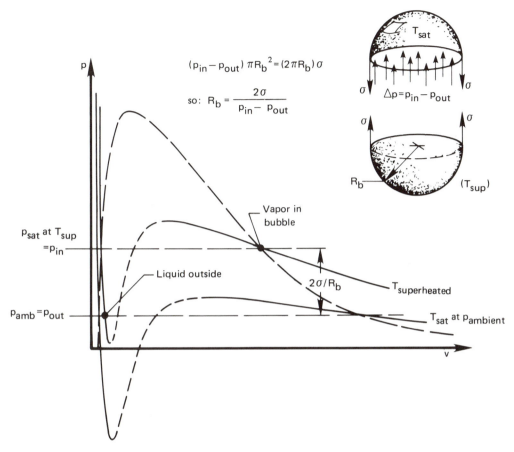

FIGURE 10.5 The conditions required for simultaneous mechanical and thermal equilibrium of a vapor bubble.

The $p-v$ diagram in Fig. 10.5 shows the state points of the internal vapor and external liquid for a bubble at equilibrium. Notice that the external liquid is superheated to $(T_{sup} - T_{sat})°C$ above its boiling point at the ambient pressure; but the vapor inside, being held at just the right elevated pressure by surface tension, is just saturated.

Physical digression 10.1

The surface tension of water in contact with its vapor is given with great accuracy by [10.3]:

$$\sigma_{water} = 235.8\left(1 - \frac{T_{sat}}{T_c}\right)^{1.256}\left[1 - 0.625\left(1 - \frac{T_{sat}}{T_c}\right)\right]\frac{dyn}{cm} \qquad (10.2)$$

where both T_{sat} and the thermodynamical critical temperature, $T_c = 647.2°K$, are expressed in °K. Table 10.1 gives additional values of σ for several substances.

Table 10.1   Surface tension for various substances from the collection of Jasper [10.4][a]

Substance	Temperature Range (°C)	σ (dyn/cm)	$\sigma = a - bT$ (°C) a(dyn/cm)	b(dyn/cm-°C)
Acetone	25 to 50		26.26	0.112
Ammonia	−70	42.39		
	−60	40.25		
	−50	37.91		
	−40	35.38		
Aniline	15 to 90		44.83	0.1085
Benzene	10	30.21		
	30	27.56		
	50	24.96		
	70	22.40		
Butyl alcohol	10 to 100		27.18	0.08983
Carbon dioxide	−30	10.08		
	−10	6.14		
	10	2.67		
	30	0.07		
Carbon tetrachloride (CCl_4)	15 to 105		29.49	0.1224
Cyclohexanol	20 to 100		35.33	0.0966
Ethyl alcohol	10 to 100		24.05	0.0832
Ethylene glycol	20 to 140		50.21	0.089
Hydrogen	−258	2.80		
	−255	2.29		
	−253	1.95		
Isopropyl alcohol	10 to 100		22.90	0.0789
Mercury	5 to 200		490.6	0.2049
Methane	90	18.877		
	100	16.328		
	115	12.371		
Methyl alcohol	10 to 60		24.00	0.0773
Napthalene	100 to 200		42.84	0.1107
Nicotine	−40 to 90		41.07	0.1112
Nitrogen	78 to 90		26.42	0.2265
Octane	10 to 120		23.52	0.09509
Pentane	10 to 30		18.25	0.11021
Toulene	10 to 100		30.90	0.1189
Water	10 to 100		75.83	0.1477

[a]The function $\sigma = \sigma(T)$ is not really linear, but Jasper was able to linearize it over modest ranges of temperature [e.g., compare the water equation above with eqn. (10.2)].

It is easy to see that the equilibrium bubble, whose radius is described by eqn. (10.1), is unstable. If its radius is less than this value, surface tension will overbalance $p_{sat}(T_{sup}) - p_{amb}$. The vapor inside will condense at this higher pressure and the bubble will collapse. If the bubble radius is slightly larger than the equation specifies, liquid at the interface will evaporate and the bubble will begin to grow.

Thus, as the surface temperature is increased, higher and higher values of $p_{sat}(T_{sup}) - p_{amb}$ will result and the equilibrium radius, R_b, will decrease in accordance with eqn. (10.1). It follows that smaller and smaller vapor pockets will be triggered into active bubble growth as the temperature is increased. As an approximation we can use eqn. (10.1) to specify the radius of those vapor pockets which become active nucleation sites. More accurate estimates can be made using: Hsu's [10.5] bubble inception theory, the more recent work by Rohsenow and his coworkers (see, e.g., [1.13]. Chap. 13]), or the still more recent technical literature.

Example 10.1

Estimate the approximate size of active nucleation sites in water at 1 atm on a wall superheated to 8 and to 16°C. (This is roughly in the regime of isolated bubbles indicated in Fig. 10.2.)

SOLUTION. $p_{sat} = 1,203,000$ dyn/cm^2 at 108°C and 1,769,000 dyn/cm^2 at 116°C, and σ at $T_{sat} = 100$°C is given as 58.89 dyn/cm by eqn. (10.2). Then

$$R_b = \frac{2(58.89) \text{ dyn/cm}}{(1,203,000 \text{ or } 1,769,000 - 1,013,000) \text{ dyn/cm}^2}$$

so the radius of active nucleation sites is on the order of

$$R_b = 0.00062 \text{ cm at } T = 108°C \quad \text{or} \quad 0.00016 \text{ cm at } 116°C$$

This means that active nucleation sites would be holes with diameters very roughly on the order of magnitude of 0.005 mm or 5 μm—at least on the heater represented by Fig. 10.2. That is within the range of roughness of commercially finished surfaces.

Region of isolated bubbles

The mechanism of heat transfer enhancement in the isolated bubble regime was hotly argued in the years following World War II. A few conclusions have emerged from that debate, and we shall attempt to identify them. There is little doubt that bubbles act in some way as small pumps that keep replacing liquid heated at the wall with cool liquid. The question is that of specifying the correct mechanism. Figure 10.6 shows the way bubbles probably act to remove hot liquid from the wall and introduce cold liquid to be heated.

It is apparent that the number of active nucleation sites generating bubbles will strongly influence q. On the basis of his experiments, Yamagata showed in 1955 (see, e.g., [10.6]) that

$$q \propto \Delta T^a n^b \tag{10.3}$$

where $\Delta T \equiv T_w - T_{sat}$ and n is the site density or number of active sites per square meter. A great deal of subsequent work has been done to fix the constant of proportionality and the constant exponents, a and b. The exponents turn out to be approximately $a = 1.2$ and $b = \frac{1}{3}$.

A bubble growing and departing in saturated liquid.

A bubble growing in subcooled liquid.

The bubble grows, absorbing heat from the superheated liquid on its periphery. As it leaves, it entrains cold liquid onto the plate which then warms up until nucleation occurs and the cycle repeats.

When the bubble protrudes into cold liquid, steam can condense on the top while evaporation continues on the bottom. This provides a short-circuit for cooling the wall. Then, when the bubble caves in, cold liquid is brought to the wall.

FIGURE 10.6   Heat removal by bubble action during boiling. Dark regions denote locally superheated liquid.

The problem with eqn. (10.3) is that it introduces what engineers call a "nuisance variable." A nuisance variable is one that varies from system to system and cannot easily be evaluated—the site density, n, in this case. Normally, n increases with ΔT in some way, but how? If all sites were identical in size, all sites would be activated simultaneously, and q would be a discontinuous function of ΔT. When the sites have a typical distribution of sizes, n (and hence q) can increase very strongly with ΔT.

It is a lucky fact that, for a large class of factory-finished materials, n varies approximately as $\Delta T^{5 \text{ or } 6}$, so q varies roughly as ΔT^3. This has made it possible for various authors to correlate q approximately for a large variety of materials. One of the first and most useful correlations for nucleate boiling was that of Rohsenow [10.7] in 1952. It is

$$\frac{c_p(T_w - T_{sat})}{h_{fg} \mathrm{Pr}^s} = C_{sf} \left[\frac{q}{\mu h_{fg}} \sqrt{\frac{\sigma}{g(\rho_f - \rho_g)}} \right]^{0.33} \tag{10.4}$$

where all properties, unless otherwise noted, are for liquid at T_{sat}. The constant C_{sf} is an empirical correction for typical surface conditions. Table 10.2 includes a set of values of C_{sf} for common surfaces (taken from [10.7]) as well as the Prandtl number exponent, s.

We noted, initially, that there are two nucleate boiling regimes, and the Yamagata equation (10.3) applies only to the first of them. Rohsenow's equation is frankly empirical and does not depend on the rational analysis of either nucleate boiling process. It turns out that it represents $q(\Delta T)$ in both regimes, but it is not terribly accurate in either one. Figure 10.7 shows Rohsenow's

Table 10.2 Selected values of the surface correction factor for use with eqn. (10.4) [10.7]

Surface–Fluid Combination	C_{sf}	s
Water–nickel	0.006	1.0
Water–platinum	0.013	1.0
Water–copper	0.013	1.0
Water–brass	0.006	1.0
CCl_4–copper	0.013	1.7
Benzene–chromium	0.010	1.7
n-Pentane–chromium	0.015	1.7
Ethyl alcohol–chromium	0.0027	1.7
Isopropyl alcohol–copper	0.0025	1.7
35% K_2CO_3–copper	0.0054	1.7
50% K_2CO_3–copper	0.0027	1.7
n-Butyl alcohol–copper	0.0030	1.7

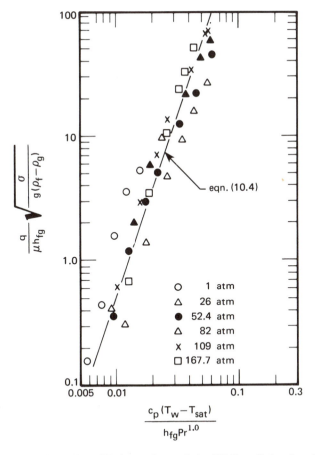

FIGURE 10.7 Illustration of Rohsenow's correlation [10.7] applied to data for water boiling on 0.61 mm diam. platinum wire.

original comparison of eqn. (10.4) with data for water over a large range of conditions. It shows typical errors in heat flux of 100%, and typical errors in ΔT of about 25%.

Thus our ability to predict the nucleate pool boiling heat flux is poor. Our ability to predict ΔT is better because, with $q \sim \Delta T^3$, a large error in q gives a much smaller error in ΔT. It appears that any substantial improvement in this situation will have to wait until someone has managed to deal realistically with the nuisance variable, n. Current research efforts are dealing with this matter, and we can simply hope that such work will produce a method for achieving heat transfer design relationships for nucleate boiling within a few years.

It is indeed fortunate that we do not often have to calculate q, given ΔT, in the nucleate boiling regime. More often, the major problem is to avoid exceeding q_{max}. We turn our attention in the next section to predicting this limit.

Example 10.2

What is C_{sf} for the heater surface in Fig. 10.2?

SOLUTION.   From eqn. (10.4) we obtain

$$\frac{q}{\Delta T^3} C_{sf}^3 = \frac{\mu c_p^3}{h_{fg}^2 \mathrm{Pr}^3} \sqrt{\frac{g(\rho_f - \rho_g)}{\sigma}}$$

where we note that $s = 1.0$. Then for water at $T_{sat} = 100°C$:

$$c_p = 4.22 \text{ kJ/kg}°\text{-C}$$

$$\mathrm{Pr} = 1.72$$

$$\rho_f - \rho_g = 957 \text{ kg/m}^3$$

$$\sigma = 0.0589 \text{ N/m or kg/s}^2$$

$$h_{fg} = 2257 \text{ kJ/kg}$$

$$\mu = 0.000277 \text{ kg/m-s}$$

so

$$\frac{q}{\Delta T^3} C_{sf}^3 = 3.20 \times 10^{-7} \frac{\text{kW}}{\text{m}^2\text{-}°\text{C}^3}$$

At $q = 800 \text{ kQ/m}^2$-°C we read $\Delta T = 22°C$ from Fig. 10.2. This gives

$$C_{sf} = \left[\frac{3.20 \times 10^{-7}(22)^3}{800} \right]^{1/3} = 0.016$$

This value compares favorably with C_{sf} for a platinum or copper surface under water.

Transitional boiling regime and Taylor instability

It will help us to understand the peak heat flux if we first consider the process that connects the peak and the minimum heat fluxes. During transitional boiling, a large amount of vapor is glutted about the heater. It wants to buoy upward, but it has no clearly defined escape route. The jets that carry vapor away from the heater in the region of slugs and columns are unstable and cannot serve that function in the transitional boiling regime. Therefore, vapor buoys up in big slugs—then liquid falls in, touches the surface briefly, and a new slug begins to form. Figure 10.3c shows part of this process.

The transitional boiling regime is of relatively little practical interest because one seldom operates equipment in it, and we offer no formulas to describe it. However, it holds a conceptual key both to predicting q_{max} and to understanding film boiling as well. Notice that during transitional boiling the heater is almost blanketed with vapor, and in film boiling it is completely blanketed. In both cases we must contend with the unstable configuration of a liquid on top of a vapor.

Figure 10.8 shows two commonplace examples of such behavior. In either an inverted honey jar or in the water condensing from a cold water pipe, we have seen how a heavy fluid falls into a light one (water or honey, in this case, collapses into air). The heavy phase falls down at one node of a wave and the light fluid rises into the other node.

The collapse process is called *Taylor instability* after G. I. Taylor, who first predicted it. The so-called Taylor wavelength, λ_d, is the length of the wave that grows fastest, and therefore predominates during the collapse of an infinite plane horizontal interface. It can be predicted using dimensional analysis. The dimensional functional equation for λ_d is

$$\lambda_d = \text{fn}\left[\sigma, g(\rho_f - \rho_g)\right] \tag{10.5}$$

since the wave is formed as a result of the balancing forces of surface tension against inertia and gravity. There are three variables involving m and kg/s^2, so we look for just one dimensionless group:

$$\lambda_d \sqrt{\frac{(\rho_f - \rho_g)g}{\sigma}} = \text{constant}$$

This relationship was derived analytically by Bellman and Pennington [10.8] for one-dimensional waves and by Sernas [10.9] for the two-dimensional waves that

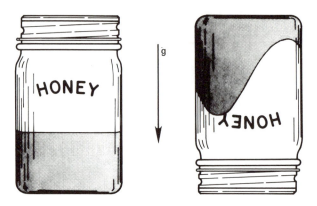

Taylor instability in the surface of the honey in an inverted honey jar

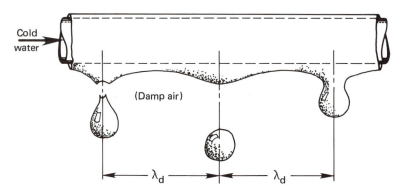

Taylor instability in the interface of the water condensing
on the underside of a small cold water pipe

FIGURE 10.8   Two examples of Taylor instabilities that one might commonly
experience.

actually occur in a plane horizontal interface. The results were

$$\lambda_d \sqrt{\frac{(\rho_f - \rho_g)g}{\sigma}} = \begin{cases} 2\pi\sqrt{3} & \text{for one-dimensional waves} \\ 2\pi\sqrt{6} & \text{for two-dimensional waves} \end{cases} \tag{10.6}$$

Experiment 10.3

Hang a metal rod in the horizontal position by threads at both ends. The rod should be
about 30 cm in length and perhaps 1 to 2 cm in diameter. Pour motor oil or glycerin in a
narrow cake pan and lift the pan up under the rod until it is submerged. Then lower the
pan and watch the liquid drain into it. Take note of the wave action on the underside of
the rod. The same thing can be done in an even more satisfactory way by running cold

water through a horizontal copper tube above a beaker of boiling water. The condensing liquid will also come off in a Taylor wave such as is shown in Fig. 10.8. In either case, the waves will approximate λ_{d_1} (the length of a one-dimensional wave, since they are arrayed on a line), but the wavelength will be influenced by the curvature of the rod.

Throughout the transitional boiling regime, vapor rises into liquid on the nodes of Taylor waves, and at q_{max} this rising vapor forms into jets. These jets arrange themselves on a staggered square grid as shown in Fig. 10.9. The basic

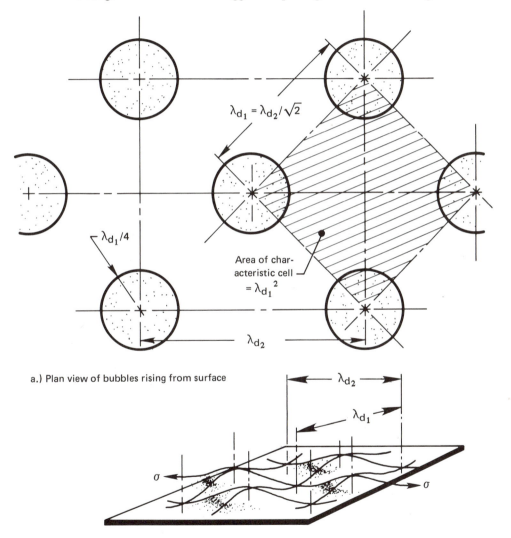

a.) Plan view of bubbles rising from surface

b.) Wave-form underneath the bubbles shown in (a)

FIGURE 10.9   The array of vapor jets as seen on an infinite horizontal heater surface.

spacing of the grid is λ_{d_2} (the two-dimensional Taylor wavelength). Since

$$\lambda_{d_2} = \sqrt{2}\, \lambda_{d_1} \tag{10.7}$$

[recall eqn. (10.6)], the spacing of the most basic module of jets is actually λ_{d_1}, as shown in Fig. 10.9.

Next we must consider how the jets become unstable at the peak, to bring about burnout.

Helmholtz instability of vapor jets

Figure 10.10 shows a commonplace example of what is called *Helmholtz instability*. This is the phenomenon that causes the vapor jets to cave in when the vapor velocity in them reaches a critical value. Any flag in a breeze will constantly be in a state of collapse as the result of relatively high pressures where the velocity is low and relatively low pressures where the velocity is high, as is indicated in the top view.

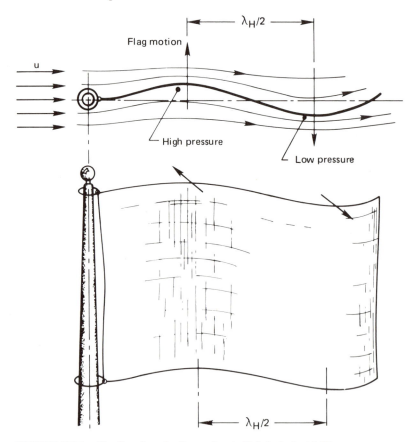

FIGURE 10.10   The flapping of a flag owing to Helmholtz instability.

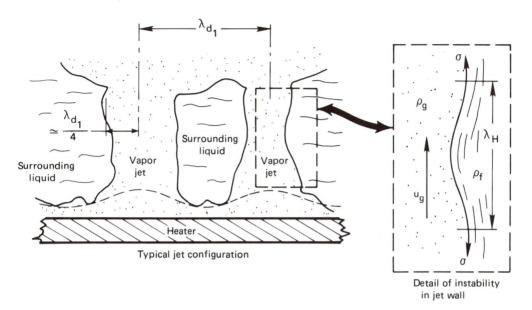

FIGURE 10.11   Helmholtz instability of vapor jets.

This same instability is shown as it occurs in a vapor jet wall in Fig. 10.11. This situation differs from the flag in one important particular. There is surface tension in the jet walls, It tends to balance the flow-induced pressure forces that bring about collapse. Thus while the flag is unstable in *any* breeze, the vapor velocity in the jet must reach a limiting value, u_g, before the jet becomes unstable.

Lamb [10.10] gives the following relation between the vapor flow u_g, shown in Fig. 10.11, and the wavelength of a disturbance in the jet wall, λ_H:

$$u_g = \sqrt{\frac{2\pi\sigma}{\rho_g \lambda_H}} \tag{10.8}$$

[This result, like eqn. (10.6), can be predicted within a constant using dimensional analysis. See Problem 10.21.] A real liquid–vapor interface will usually be irregular, and therefore it can be viewed as containing all possible sinusoidal wavelengths superposed on one another. One problem we face is that of guessing whether or not one of those wavelengths will be better developed than the others and therefore more liable to collapse.

Example 10.3

Saturated water at 1 atm flows down the periphery of the inside of a 10-cm-I.D. tube. Steam flows upward in the center. The wall of the pipe has circumferential corrugations in it, with a 4-cm wavelength in the axial direction. Neglect problems raised by curvature and the finite depth of the liquid, and estimate the steam velocity required to destabilize the liquid flow over these corrugations, assuming the liquid moves slowly.

SOLUTION. The flow will be Helmholtz stable until the steam velocity reaches the value given by eqn. (10.8).

$$u_g = \sqrt{\frac{2\pi(0.0589)}{0.577(0.04\ m)}}$$

Thus the maximum stable steam velocity would be $u_g = 4\ m/s$. Beyond that the liquid will form whitecaps and be blown back upward.

Example 10.4

Mercury is held in place between two parallel steel plates by capillary forces. The plates are slowly pulled apart until the mercury interface collapses. Approximately what is the maximum spacing?

SOLUTION. The mercury is most susceptible to Taylor instability when the spacing reaches the wavelength given by eqn. (10.6):

$$\lambda_{d_1} = 2\pi\sqrt{3}\ \sqrt{\frac{\sigma}{g(\rho_f - \rho_g)}} = 2\pi\sqrt{3}\ \sqrt{\frac{0.487}{9.8(13600)}} = 0.021\ m = 2.1\ cm$$

(Actually, this spacing would give the maximum *rate* of collapse. It can be shown that collapse would begin at $1/\sqrt{3}$ times this value, or at 1.2 cm.)

Prediction of q_{max}

General expression for q_{max}. The heat flux must be balanced by the latent heat carried away in the jets when the liquid is saturated. Thus we can write immediately

$$q_{max} = \rho_g h_{fg} u_g \left(\frac{A_j}{A_h}\right) \tag{10.9}$$

where A_j is the cross-sectional area of a jet and A_h is the heater area that supplies each jet.

For any heater configuration there are two things that must be determined. One is the length of the particular disturbance in the jet wall, λ_H, which will trigger Helmholtz instability and fix u_g in eqn. (10.8) for use in eqn. (10.9). The other thing is the ratio A_j/A_h. The prediction of q_{max} in any pool boiling configuration always comes down to these two problems.

q_{max} on an infinite horizontal plate. The original analysis this type was done by Zuber in his doctoral dissertation at UCLA in 1958 (see [10.11]). He first guessed that the jet radius was $\lambda_{d_1}/4$. This guess has received corroboration by subsequent investigators and (with reference to Fig. 10.9) it gives

$$\frac{A_j}{A_h} = \frac{\text{cross-sectional area of circular jet}}{\text{area of the square portion of the heater that feeds the jet}} = \frac{\pi(\lambda_{d_1}/4)^2}{(\lambda_{d_1})^2} = \frac{\pi}{16} \qquad (10.10)$$

Lienhard and Dhir ([10.12] to [10.14]) assumed that the Helmholtz unstable wavelength was equal to λ_{d_1}, so eqn. (10.9) becomes

$$q_{max} = \rho_g h_{fg} \sqrt{\frac{2\pi\sigma}{\rho_g} \frac{1}{2\pi\sqrt{3}}} \sqrt{\frac{g(\rho_f - \rho_g)}{\sigma}} \frac{\pi}{16}$$

or

$$\boxed{q_{max} = 0.149 \rho_g^{1/2} h_{fg} \sqrt[4]{g(\rho_f - \rho_g)\sigma}} \qquad (10.11)$$

Equation (10.11) is compared with available data for large flat heaters, with vertical sidewalls to prevent any liquid sideflow, in Fig. 10.12. As long as the

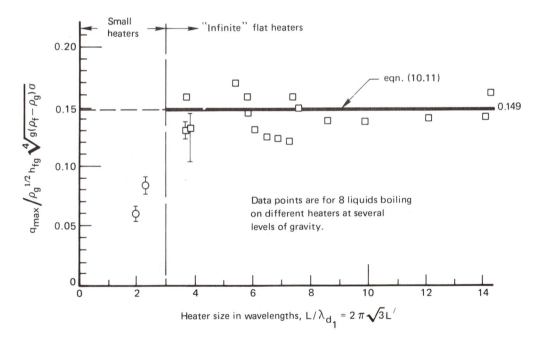

FIGURE 10.12 Comparison of the q_{max} prediction for infinite horizontal heaters with data reported in [10.12].

diameter or width of the heater is more than about $3\lambda_{d_1}$, the prediction is quite accurate. When the width or diameter is less than this, there is a small integral number of jets on a plate which may be larger or smaller in area than $16/\pi$ per jet. When this is the case, the actual q_{max} may be larger or smaller than that predicted by eqn. (10.11) (see Problem 10.13).

The preceding prediction is usually credited to Kutateladze [10.15] and Zuber [10.11]. Kutateladze (then working in Leningrad, and later director of the Heat Transfer Laboratory near Novosibirsk, Siberia) recognized that burnout resembled the flooding of a distillation column. At any level in a distillation column, alcohol-rich vapor (for example) rises while water-rich liquid flows downward in counterflow. If the process is driven too far, the flows become Helmholtz-unstable and the process collapses. The liquid then cannot move downward and the column is said to "flood."

Kutateladze did the dimensional analysis of q_{max} based on the flooding mechanism and obtained the following relationship, which, lacking a characteristic length and being of the same form as eqn. (10.11), is really only valid for an infinite horizontal plate:

$$q_{max} = C\rho_g^{1/2}h_{fg}\sqrt[4]{g(\rho_f - \rho_g)\sigma}$$

He then suggested that C was equal to 0.131 on the basis of data from configurations other than infinite flat plates (horizontal cylinders, for example). Zuber's analysis yielded $C = \pi/24 = 0.1309$, which was quite close to Kutateladze's value but lower by 14% than eqn. (10.11). We therefore designate the Zuber–Kutateladze prediction as q_{max_z}. However, we shall not use it directly, since it does not predict any actual physical configuration.

$$q_{max_z} \equiv 0.131\rho_g^{1/2}h_{fg}\sqrt[4]{g(\rho_f - \rho_g)\sigma} \qquad (10.12)$$

It is very interesting that C. F. Bonilla, whose q_{max} experiments in the early 1940s are included in Fig. 10.12, also suggested that q_{max} should be compared with the column flooding mechanism. He presented these ideas in a paper, but A. P. Colburn wrote to him: "A correlation [of the flooding velocity plots with] boiling data would not serve any great purpose and would perhaps be very misleading." And T. H. Chilton—another eminent chemical engineer of that period—wrote to him: "I venture to suggest that you delete from the manuscript ... the relationship between boiling rates and loading velocities in packed towers." Thus the technical conservativism of the period prevented the idea from gaining acceptance for another decade.

Example 10.5

Predict the peak heat flux for Fig. 10.2.

SOLUTION. We use eqn. (10.11) to evaluate q_{max} for water at 100°C on an infinite flat plate.

$$q_{max} = 0.149 \rho_g^{1/2} h_{fg} \sqrt[4]{g(\rho_f - \rho_g)\sigma}$$

$$= 0.149(0.597)^{1/2}(2,257,000)\sqrt[4]{9.8(958.2 - 0.6)(0.0589)}$$

$$= 1,260,000 \text{ W/m}^2$$

Figure 10.2 shows $q_{max} \simeq 1,160,000$ W/m^2, which is less by only about 6%.

Example 10.6

What is q_{max} in mercury at 1 atm?

SOLUTION. The normal boiling point of mercury is 355°C. At this temperature, $h_{fg} = 292,500$ J/kg, $\rho_f = 13,400$ kg/m^3, $\rho_g = 4.0$ kg/m^3, and $\sigma \simeq 0.418$ kg/s^2, so

$$q_{max} = 0.149(4.0)^{1/2}(292,500)\sqrt[4]{9.8(13,400 - 4)(0.418)}$$

$$= 1,334,000 \text{ W/m}^2$$

The result is very close to that for water. The increases in density and surface tension have been compensated by a much lower latent heat.

Peak heat flux
in other pool boiling configurations

The prediction of q_{max} in configurations other than an infinite flat heater will involve a characteristic length, L. Thus the dimensional functional equation for q_{max} becomes

$$q_{max} = \text{fn}\left[\rho_g, h_{fg}, \sigma, g(\rho_f - \rho_g), L\right]$$

which involves six variables and four dimensions: J, m, s, and kg. There are thus two pi-groups. The first group can arbitrarily be multiplied by $24/\pi$ to give

$$\Pi_1 = \frac{q_{max}}{(\pi/24)\rho_g^{1/2} h_{fg}\sqrt[4]{\sigma g(\rho_f - \rho_g)}} = \frac{q_{max}}{q_{max_z}} \tag{10.13}$$

Notice that the factor of $24/\pi$ has served to make the denominator equal to Zuber's expression for q_{max}. Thus for q_{max} on a flat plate, Π_1 equals $0.149/0.131$, or 1.14. The second pi-group is

$$\Pi_2 = \frac{L}{\sqrt{\sigma/g(\rho_f - \rho_g)}} = 2\pi\sqrt{3}\,\frac{L}{\lambda_{d_1}} \equiv L' \tag{10.14}$$

The latter group, Π_2, is the square root of the *Bond number*, Bo, which is used to compare buoyant force with capillary forces.

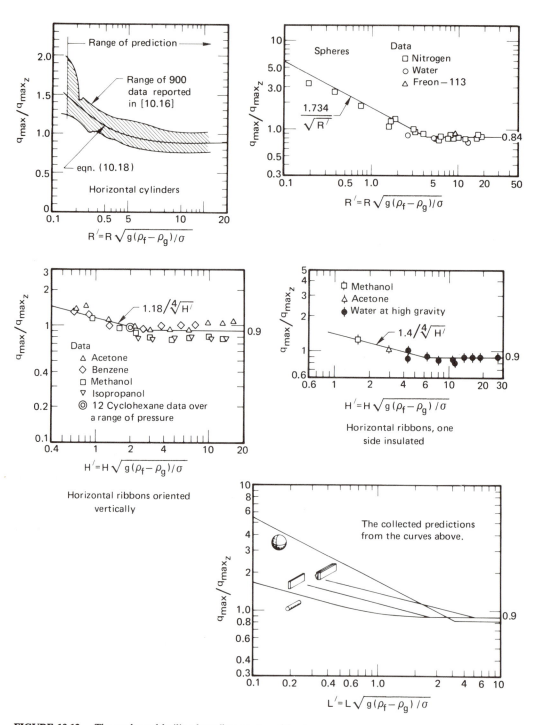

FIGURE 10.13   The peak pool boiling heat flux on several heaters.

Predictions and correlations of q_{max} have been made for several finite geometries in the form

$$\frac{q_{max}}{q_{max_z}} = \text{fn}(L') \tag{10.15}$$

The dimensionless characteristic dimension in eqn. (10.15) might be a dimensionless radius (R'), dimensionless diameter (D'), or dimensionless height (H'). The graphs in Fig. 10.13 are comparisons of several of the existing predictions and correlations with experimental data. These predictions and others are listed in Table 10.3. Notice that the last three items in Table 10.3 (10, 11, and 12) are general expressions from which several of the preceding expressions in the table can be obtained.

The equations in Table 10.3 are all valid within ± 15 or 20%, which is very little more than the inherent scatter of q_{max} data. However, they are subject to the following conditions:

• The bulk liquid is saturated.

• There are no pathological surface imperfections.

• There is no forced convection.

Another limitation on all the equations in Table 10.3 is that neither the size of the heater nor the relative force of gravity can be too small. When $L' < 0.15$ in

Table 10.3   Predictions of the peak pool boiling heat flux

Situation	q_{max}/q_{max_z}	Basis for L'	Range of L'	Source	Equation Numbers
1. Infinite flat heater	1.14	Heater width or diameter	$L' \gtrsim 2.7$	[10.13]	(10.16)
2. Small flat heater	$1.14 A_{heater}/\lambda_{d_1}^2$	Heater width or diameter	$0.07 \gtrsim L' < 0.2$	[10.13]	(10.17)
3. Horizontal cylinder	$0.89 + 2.27e^{-3.44\sqrt{R'}}$	Cylinder radius, R	$R' \gtrsim 0.15$	[10.16]	(10.18)
4. Large horizontal cylinder	0.90	Cylinder radius, R	$R' \gtrsim 1.2$	[10.14]	(10.19)
5. Small horizontal cylinder	$0.94/(R')^{1/4}$	Cylinder radius, R	$0.15 \lesssim R' \lesssim 1.2$	[10.14]	(10.20)
6. Large sphere	0.84	Sphere radius, R	$0.15 \lesssim R' \lesssim 4.26$	[10.17]	(10.21)
7. Small sphere	$1.734/(R')^{1/2}$	Sphere radius, R	$4.26 \lesssim R'$	[10.17]	(10.22)
8, 9. Small horizontal ribbon oriented vertically } plain: 1 side	$1.18/(H')^{1/4}$	Height of side, H	$0.15 \lesssim H' \lesssim 2.96$	[10.14]	(10.23)
insulated	$1.4/(H')^{1/4}$	Height of side, H	$0.15 \lesssim H' \lesssim 5.86$	[10.14]	(10.24)
10. Any large finite body	~0.90	Characteristic length, L	(cannot specify generally; $L' \gtrsim 4$)	[10.14]	(10.25)
11. Small slender cylinder of any cross section	$1.4/(P')^{1/4}$	Transverse perimeter, P	$0.15 \lesssim P' \lesssim 5.86$	[10.14]	(10.26)
12. Small bluff body	Constant$/(L')^{1/2}$	Characteristic length, L	(cannot specify generally; $L' \gtrsim 4$)	[10.14]	(10.27)

most configurations, the Bond number

$$\mathrm{Bo} \equiv L'^2 = \frac{g(\rho_f - \rho_g)L^3}{\sigma L} = \frac{\text{buoyant force}}{\text{capillary force}} < \frac{1}{40}$$

In this case, the process becomes completely dominated by surface tension and the Taylor–Helmholtz wave mechanisms no longer operate. As L' is reduced, the peak and minimum heat fluxes cease to occur and the boiling curve becomes monotonic. When nucleation occurs on a small wire, the wire is immediately enveloped in vapor and the mechanism of heat removal passes directly from natural convection to film boiling.

Example 10.7

A spheroidal metallic body of surface area 400 cm^2 and volume 600 cm^3 is quenched in saturated water at 1 atm. What is the most rapid rate of heat removal during the quench?

SOLUTION.   As the cooling process progresses, the object goes through the boiling curve from film boiling, through q_{min}, up the transitional boiling regime, through q_{max}, down the nucleate boiling curve, and cooling is finally completed by natural convection. One who has watched the quenching of a red-hot horseshoe will recall the great gush of bubbling that occurs as q_{max} is reached. We therefore calculate the required heat flow as $Q = q_{max}A_{spheroid}$, where q_{max} is given by eqn. (10.25) in Table 10.3:

$$q_{max} = 0.9 q_{max_z} = 0.9(0.131)\rho_g^{1/2}h_{fg}\sqrt[4]{g\sigma(\rho_f - \rho_g)}$$

so

$$Q = \left[0.9(0.131)(0.597)^{1/2}(2,257,000)\sqrt[4]{9.8(0.0589)(958)} \right](400 \times 10^{-4})$$

or

$$Q = 39{,}900 \text{ W} \quad \text{or} \quad \underline{39.9 \text{ kW}}$$

This is a startlingly large rate of energy removal for such a small object.

To complete the calculation it is necessary to check whether or not R' is large enough to justify the use of eqn. (10.25):

$$R' = \frac{V/A}{\sqrt{g(\rho_f - \rho_g)/\sigma}} = \frac{0.0006}{0.04}\sqrt{\frac{9.8(958)}{0.0589}} = 6.0$$

This is larger than the specified lower bound of about 4.

10.4 Film boiling

Film boiling bears an uncanny similarity to film condensation. The similarity is so great that in 1950 Bromley [10.18] was able to use the eqn. (9.55) for condensation on cylinders—almost directly—to predict film boiling from cylinders. He simply changed k and ν from liquid to vapor properties, and altered the lead constant from 0.729 to 0.62 to fit the available boiling data. The resulting $\overline{Nu}_D$ (based on k_g) was

$$\overline{Nu}_D = 0.62 \left[\frac{(\rho_f - \rho_g)gh'_{fg}D^3}{\nu_g k_g(T_w - T_{sat})} \right]^{1/4} \qquad (10.28)$$

where vapor and liquid properties should be evaluated at $T_{sat} + \Delta T/2$, and at T_{sat}, respectively, and where h'_{fg} is approximately $h_{fg}(1 + 0.34\text{Ja})$.

Dhir and Lienhard [9.5] did the same thing for film boiling from *spheres*, 20 years later. Their result [cf. eqn. (9.56)] was

$$\overline{Nu}_D = 0.67 \left[\frac{(\rho_f - \rho_g)gh'_{fg}D^3}{\nu_g k_g(T_w - T_{sat})} \right]^{1/4} \qquad (10.29)$$

The preceding expressions are based on heat transfer by convection through the vapor film, alone. However when film boiling occurs much beyond q_{min} in water, the heater glows dull cherry-red to white-hot. Radiation in such cases can be enormous. One's first temptation might be to simply add a radiation heat transfer coefficient, $\overline{h}_{rad}$ to $\overline{h}_{boiling}$ as obtained from eqn. (10.28) or (10.29), where

$$\overline{h}_{rad} = \frac{q_{rad}}{T_w - T_{sat}} = \frac{\varepsilon\sigma(T_w^4 - T_{sat}^4)}{T_w - T_{sat}} \qquad (10.30)$$

and where ε is a surface radiation property called the emittance (see Section 11.1).

Unfortunately, such addition is not correct because the additional radiative heat transfer will increase the vapor blanket thickness reducing the convective contribution. Bromley [10.18] suggested for cylinders the approximate relation

$$\overline{h}_{total} = \overline{h}_{boiling} + \tfrac{3}{4}\overline{h}_{rad} \qquad \overline{h}_{rad} < \overline{h}_{conv} \qquad (10.31)$$

An accurate correction would be considerably more complex than this. Other suggested forms have subsequently been offered for the radiation correction. One of the most comprehensive is that of Pitschmann and Grigull [10.19]. Their correlation, which is fairly intricate, brings together an enormous range of heat transfer data for cylinders, within 20%. It is worth noting that radiation is seldom important when the heater temperature is less than 300°C.

The use of the analogy between film condensation and film boiling is somewhat questionable during film boiling on a vertical surface. In this case the liquid–vapor interface becomes Helmholtz-unstable at a short distance from the leading edge. However, Leonard, Sun, and Dix [10.20] have shown that, by using $\lambda_{d_1}/\sqrt{3}$ in place of D in eqn. (10.28), one obtains a very satisfactory prediction of $\overline{h}$ for rather tall vertical plates.

The analogy between film condensation and film boiling also deteriorates when it is applied to small curved bodies. The reason is that the thickness of the vapor film in boiling is far greater than the liquid film during condensation. Consequently, a curvature correction, which could be ignored in film condensation, must be included during film boiling from small cylinders, spheres, and other curved bodies. The first curvature correction to be made was an empirical one given by Westwater and Breen in 1962 [10.21]. They showed that the equation

$$\overline{Nu_D} = \left[\left(0.661 + \frac{0.243}{R'} \right)(R')^{1/4} \right] \overline{Nu_{D_{Bromley}}}$$ (10.32)

applies when $R' < 3.52$. Otherwise, Bromley's equation should be used directly.

10.5 Minimum heat flux

Zuber also provided a prediction of the minimum heat flux, q_{min}, along with his prediction of q_{max} [10.11]. He assumed that as $T_w - T_{sat}$ is reduced in the film boiling regime, the rate of vapor generation eventually becomes too small to sustain the Taylor wave action that characterizes film boiling. Zuber's q_{min} prediction, based on this assumption, has to include an arbitrary constant. The result for horizontal heaters is

$$q_{min} = C \rho_g h_{fg} \sqrt[4]{\frac{\sigma g (\rho_f - \rho_g)}{(\rho_f + \rho_g)^2}}$$ (10.33)

Zuber guessed a value of C which Berenson [10.22] subsequently corrected on the basis of experimental data. Berenson used measured values of q_{min} on horizontal heaters to get

$$q_{min} = 0.09 \rho_g h_{fg} \sqrt[4]{\frac{\sigma g (\rho_f - \rho_g)}{(\rho_f + \rho_g)^2}}$$ (10.34)

Lienhard and Wong [10.23] did the parallel prediction for horizontal wires and found that

$$q_{min} = 0.515 \left[\frac{18}{R'^2 (2R'^2 + 1)} \right]^{1/4} q_{min_{Berenson}}$$ (10.35)

where the constant 0.515 is susceptible to large variation as the result of a "nuisance variable"—namely, the end mounting of the cylinder.

Example 10.8

Check the value of q_{min} shown in Fig. 10.2.

SOLUTION. The heater is a flat surface, so we use equation (10.34) and the physical properties given in Example 10.5.

$$q_{min} = 0.09(0.597)(2,257,000)\sqrt[4]{\frac{9.8(0.0589)(958)}{(959)^2}}$$

or

$$q_{min} = \underline{18,990 \text{ W}/\text{m}^2}$$

From Fig. 10.2 we read 20,000 W/m², which is the same, within the accuracy of the graph.

10.6 Parametric Influences

There are many system variables that can change the pool boiling behavior we have discussed thus far. These variables include forced convection, subcooling, gravity, surface roughness, vibration, heater configuration, and liquid impurities. Let us consider some of these influences.

Forced convection

The influence of superposed flow on the pool boiling curve for a given heater (e.g., Fig. 10.2) is generally to improve heat transfer everywhere. But flow is particularly effective in raising q_{max}. We discuss such influence in detail in Section 10.7.

Subcooling

A stationary pool will normally not remain below its saturation temperature over an extended period of time. When heat is transferred to the pool, the liquid soon becomes saturated—as it does in a teakettle (recall Experiment 10.1). However, before a liquid comes up to temperature, or if a very small rate of forced convection continuously replaces warm liquid with cool liquid, we can justly ask what the effect of a cool liquid bulk might be.

Figure 10.14 shows how a typical boiling curve might be changed if $T_{bulk} < T_{sat}$: We know that in *natural convection* q will increase as $(T_w - T_{bulk})^{5/4}$ or as $[(T_w - T_{sat}) + \Delta T_{sub}]^{5/4}$, where $\Delta T_{sub} \equiv T_{sat} - T_{bulk}$. During *nucleate boiling*, the influence of subcooling on q is known to be small. The *peak and minimum heat fluxes* are known to increase linearly with ΔT_{sub}. These increases are quite significant. The *film boiling* heat flux increases rather strongly, especially at

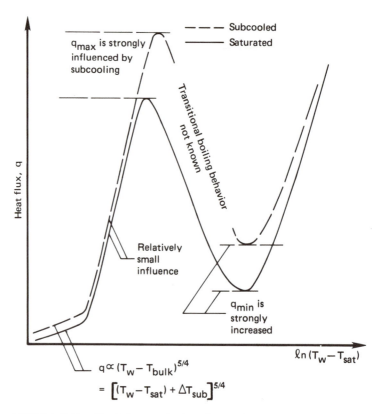

FIGURE 10.14   The influence of subcooling on the boiling curve.

lower heat fluxes. The influence of ΔT_{sub} on transitional boiling is not well documented.

Gravity

The influence of gravity (or any other such body force) is of concern because boiling processes frequently take place in rotating or accelerating systems. The reduction of gravity is a serious concern in boiling processes on-board space vehicles. Since g appears explicitly in the q_{max} and q_{min} equations, we know what its influence is. Both q_{max} and q_{min} increase directly as $g^{1/4}$ in finite bodies, and there is a secondary gravitational influence which enters through the parameter L'. Although Rohsenow's equation suggests that q is proportional to $g^{1/2}$ in the nucleate boiling regime, other evidence suggests that the influence of gravity is very slight.

Surface roughness

In 1960, Berenson [10.22] provided us with the definitive study of the influence of surface roughness during pool boiling from a flat plate. He varied the surface

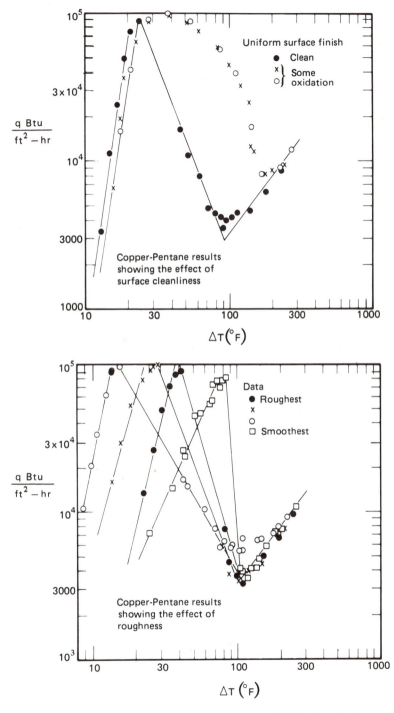

FIGURE 10.15 Typical data from Berenson's study [10.22] of the influence of surface condition on the boiling curve.

condition of heaters both mechanically (by sanding them to different roughnesses) and chemically (by treating them with various chemicals). Figure 10.15
shows typical boiling curves from Berenson's work. They show that q_{max} is
hardly affected by either roughness or by those surface-chemistry changes that
alter the contact angle. It is also no surprise that surface condition does not
influence film boiling, because there is *no* liquid–solid contact.

On the other hand, the influence of surface condition on nucleate boiling is
dramatic. This was predictable by the Yamagata equation (10.4). When the
surface is roughened, the site density, n, is far higher at a given temperature
difference, ΔT, and q is much greater. It is thus easy to increase q tenfold or
more on a given surface by, for example, sandpapering it. The transitional
boiling region is also strongly influenced by surface condition, since physical
contact between the liquid and the heater is an important part of the heat
transfer process. Although the minimum heat flux is subject to less influence,
there are kinds of surface chemistry influences that can increase q_{min} greatly.

10.7 Forced convection boiling in external flows

Influences of forced convection on nucleate boiling

Figure 10.16 shows nucleate boiling during the forced convection of water over a
flat plate. Bergles and Rohsenow (see, e.g., [1.13, Chap. 13]) offer an empirical
strategy for predicting the heat flux during nucleate flow boiling when the net
vapor generation is still relatively small. (The photograph in Fig. 10.16 shows
how a substantial buildup of vapor can radically alter flow boiling behavior.)
They suggest that

$$q = q_{FC}\sqrt{1 + \frac{q_B}{q_{FC}}\left(1 - \frac{q_i}{q_B}\right)^2} \tag{10.36}$$

where

- q_{FC} is the single-phase forced convection heat transfer for the heater, as
  one might calculate using the methods of Chapters 7 and 8.

- q_B is the *pool* boiling heat flux for that liquid and that heater.

- q_i is the heat flux from the pool boiling curve evaluated at the value of
  $(T_w - T_{sat})$ where boiling begins during flow boiling (see Fig. 10.16).

Notice that as q_B increases, eqn. (10.36) suggests that

$$q \rightarrow \sqrt{q_{FC}q_B} = \text{a geometric mean } q$$

Equation (10.36) will provide a first approximation in most boiling configurations, but it is restricted to subcooled flows or other situations in which vapor
generation is not too great.

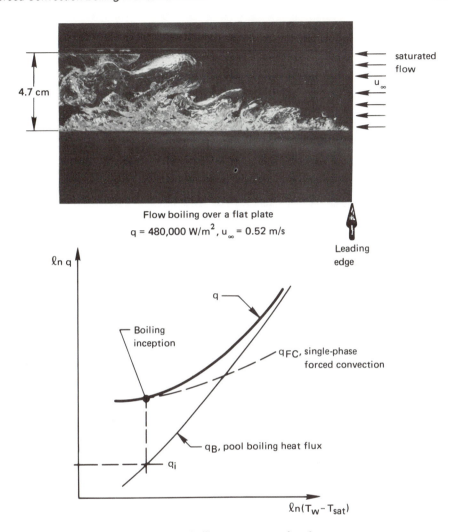

FIGURE 10.16   Forced convection boiling on an external surface.

Peak heat flux in external flows

Consider a heated body moving through a saturated liquid at a speed equal to u_∞, as shown in Fig. 10.17, or consider liquid flowing past a stationary heater at a speed, u_∞. As the heat flux approaches q_{max}, a tubelike (or, in some cases, sheetlike) jet of vapor leaves the body as shown. This jet, which has a cross-sectional area comparable to that of the body, periodically breaks off in large vapor bubbles. These bubbles are stationary with respect to the flow.

It turns out that a Helmholtz stability analysis of this system cannot easily be completed, but it can be circumvented with a simple mechanical energy argument. Notice that the motion of the body through the fluid requires an

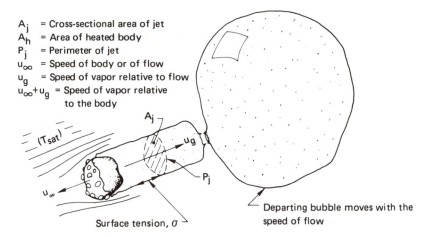

A_j = Cross-sectional area of jet
A_h = Area of heated body
P_j = Perimeter of jet
u_∞ = Speed of body or of flow
u_g = Speed of vapor relative to flow
$u_\infty + u_g$ = Speed of vapor relative
         to the body

A_j

(T_{sat})

u_g

u_∞

P_j

Surface tension, σ

Departing bubble moves with the
speed of flow

FIGURE 10.17   A heated body moving relative to a saturated liquid and transferring heat to it at a rate just below q_{max}.

amount of work equal to $u_\infty P_j \sigma$ (where P_j is the perimeter of the jet) to be done by the wake of the body. If the rate at which the kinetic energy of the vapor is delivered to the wake is less than $u_\infty P_j \sigma$, the system will be Helmholtz-stable. But when the rate of kinetic energy of the vapor exceeds the surface work, the vapor escape wake is asked to store mechanical energy. It has no means for doing this and it therefore collapses. This is rather like lifting a window shade: As long as you allow the shade to pull upon your hand it stays stable, but if you try to hasten its rise by pushing it faster than the spring can wind it in, it will collapse.

To predict q_{max}, we require that burnout will occur after the vapor kinetic energy rate first exceeds the surface energy rate. This occurs at u_g, given by

$$\sigma u_\infty P_j = \rho_g u_g A_j \left(\frac{u_g^2}{2} \right)$$

or

$$\frac{u_g}{u_\infty} = \sqrt[3]{\frac{2\sigma P_j}{u_\infty^2 A_j}} \tag{10.37}$$

The energy balance equation which gives q_{max} is the same as eqn. (10.9). However, we must now add u_∞ to u_g, since u_g is only the *relative* vapor speed. The result is

$$q_{max} = \frac{(u_g + u_\infty)\rho_g h_{fg} A_j}{A_h} \tag{10.38}$$

Substituting eqn. (10.37) in eqn. (10.38), we obtain

$$\frac{q_{max}}{\rho_g h_{fg} u_\infty} = \frac{A_j}{A_h}\left[1 + \frac{\sqrt[3]{P_j/A_j}}{\sqrt[3]{\rho_g u_\infty^2/2\sigma}}\right] \qquad (10.39)$$

Under the definition of a Weber number, We_L,

$$We_L \equiv \frac{\rho_g u_\infty^2 L}{\sigma} = \frac{\text{inertia force}/L}{\text{surface force}/L}$$

where L is any characteristic length, eqn. (10.39) can be written as

$$\boxed{\frac{q_{max}}{\rho_g h_{fg} u_\infty} = \frac{A_j}{A_h}\left[1 + \frac{\sqrt[3]{2P_j L/A_j}}{We_L^{1/3}}\right]} \qquad (10.40)$$

for any external flow.

It is possible to approximate A_j/A_h and $\sqrt[3]{P_j L/A_j}$ for a given flow, but to get precise values is difficult because the jet size does not necessarily conform to the heater size. A more detailed pursuit of the problem reveals that these ratios can depend on the density ratio, ρ_f/ρ_g. This dependence is quite hard to predict.

One such problem has been solved fairly completely to date with equations in the form of eqn. (10.40). Lienhard and Eichhorn [10.24] showed that q_{max} on a cylinder normal to a crossflow is given by

$$\boxed{\frac{q_{max}}{\rho_g h_{fg} u_\infty} = \frac{1}{\pi}\left(1 + \frac{4^{1/3}}{We_D^{1/3}}\right)} \qquad \text{low velocity} \qquad (10.41)$$

and

$$\boxed{\frac{q_{max}}{\rho_g h_{fg} u_\infty} = \frac{(\rho_f/\rho_g)^{3/4}}{169\pi} + \frac{(\rho_f/\rho_g)^{1/2}}{19.2\pi We_D^{1/3}}} \qquad \text{high velocity} \qquad (10.42)$$

where the low- and high-velocity regimes are defined in terms of the heat flux by the empirical criterion

$$\frac{q_{max}}{\rho_g h_{fg} u_\infty} \begin{cases} < \dfrac{0.275}{\pi}\sqrt{\rho_f/\rho_g} + 1 & \text{high velocity} \qquad (10.43) \\[2mm] > \dfrac{0.275}{\pi}\sqrt{\rho_f/\rho_g} + 1 & \text{low velocity} \qquad (10.44) \end{cases}$$

The low-velocity expression is derived directly from eqn. (10.40) on the presumption that A_j is the projected area of the cylinder (Problem 10.17). At higher liquid flow velocities, a transition occurs and the jet becomes thinner than the

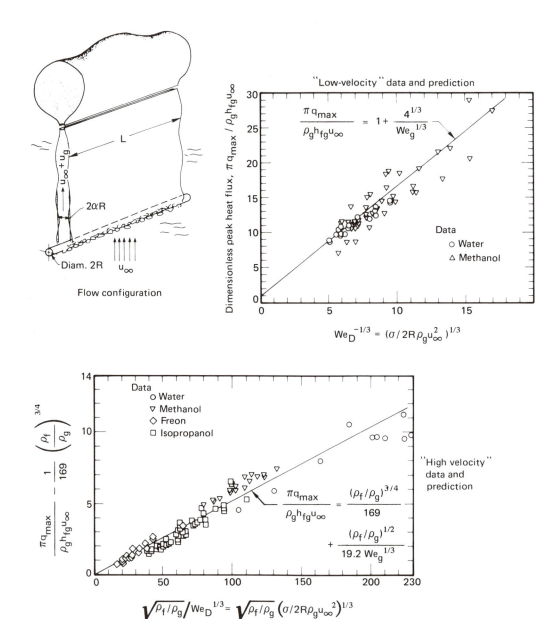

FIGURE 10.18 Peak heat flux during flow boiling over cylinders [10.24].

wire. Equations (10.41) and (10.42) predict q_{max} within $\pm 20\%$ of the experimental data, in most cases, as we see in Fig. 10.18.

Example 10.9

Saturated water at 1 atm flows over a 0.2-cm-diameter heating rod. Plot the burnout heat flux as a function of free stream velocity.

SOLUTION. The physical properties of the water are as follows:

$$h_{fg} = 2,257,000 \text{ J/kg} \qquad \rho_f = 958.3 \text{ kg/m}^3$$

$$\sigma = 0.0589 \text{ kg/s}^2 \qquad \rho_g = 0.597 \text{ kg/m}^3$$

so $R' = 0.001\sqrt{9.8(958)/0.0589} = 0.4$. We therefore should use eqn. (10.20) to predict the pool boiling limit at $u_\infty = 0$.

$$q_{max} = (0.94/(R')^{1/4}) q_{max_z}$$

$$= \frac{0.94}{(0.4)^{1/4}} \frac{\pi}{24} (0.597)^{1/2} (2,257,000) \sqrt[4]{9.8(958)(0.0589)}$$

$$= 1.31 \times 10^6$$

In the "low-velocity" range we use eqn. (10.41) to calculate

$$q_{max} = \frac{\rho_g h_{fg}}{\pi} \left[1 + \left(\frac{4\sigma}{\rho_g u_\infty^2 D} \right)^{1/3} \right] u_\infty$$

$$= \frac{0.597(2,257,000)}{\pi} \left\{ 1 + \left[\frac{4(0.0589)}{0.597(0.002)} \right]^{1/3} u_\infty^{-2/3} \right\} u_\infty$$

$$= 429,000 u_\infty + 2,497,000 u_\infty^{1/3}$$

and in the "high-velocity" range we use eqn. (10.42):

$$q_{max} = \frac{\rho_g h_{fg}}{\pi} \left[\frac{1}{169} \left(\frac{\rho_f}{\rho_g} \right)^{3/4} + \frac{1}{19.2} \sqrt{\frac{\rho_f}{\rho_g}} \left(\frac{\sigma}{\rho_g u_\infty^2 D} \right)^{1/3} \right] u_\infty$$

$$= \frac{0.597(2,257,000)}{\pi} \left[\frac{1}{169} \left(\frac{958.3}{0.597} \right)^{3/4} u_\infty \right.$$

$$\left. + \frac{1}{19.2} \sqrt{\frac{958.3}{0.597}} \left(\frac{0.0589}{0.597(0.002)} \right)^{1/3} u_\infty^{1/3} \right]$$

$$= 644,000 u_\infty + 3,282,000 u_\infty^{1/3}$$

The transition from low to high velocity occurs when

$$q_{max} = \left(\frac{0.275}{\pi} \sqrt{\frac{958.3}{0.597}} + 1 \right) (0.597)(2,257,000) u_\infty$$

$$= 6.073,000 u_\infty$$

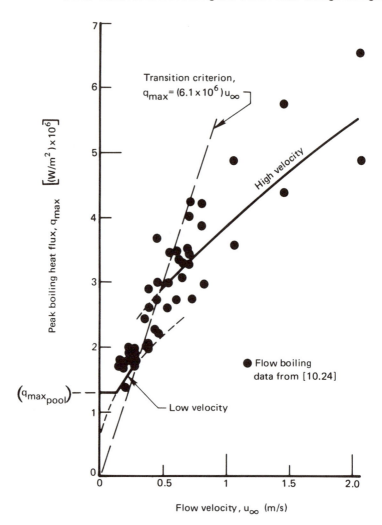

FIGURE 10.19   The peak heat flux during boiling from a 2 mm diam. rod in a cross-flow of saturated water at 1 atm.

These results are plotted in Fig. 10.19, where we see that there is a short transition range in which we do not know whether burnout occurs in the high- or the low-velocity mode. The data of many investigators (as reported in [10.23]) are included in Fig. 10.19. We see that they generally fall within ±20% of the predictive equations.[3]

Another kind of external boiling flow is that which occurs when a jet is directed onto a heated plate to cool it. Katto and his co-workers (see, e.g.,

---

[3]Actually, the data were obtained at pressures slightly higher than 1 atm, so the theory underpredicts them by a few percent.

[10.25]) have done a great deal of experimental work with such systems; and Lienhard, Eichhorn, and Hasan ([10.26] and [10.27]) have extended the mechanical energy stability criterion to rationalize the data. For a jet of diameter, d, impinging on a plate of diameter, D, they showed that

$$\frac{q_{max}}{\rho_g h_{fg} u_{jet}} = \left(0.744 + 0.0087 \frac{\rho_f}{\rho_g}\right)\left(\frac{D}{d}\right)^{3A-1}\left(\frac{\rho_g/\rho_f}{We_D}\right)^A \qquad (10.45)$$

where A depends on ρ_f/ρ_g in the following way:

$$A = 0.4346 + 0.1027 \ell n \frac{\rho_f}{\rho_g} - 0.0474 \left(\ell n \frac{\rho_f}{\rho_g}\right)^2 + 0.00426\left(\ell n \frac{\rho_f}{\rho_g}\right)^3 \qquad (10.46)$$

This correlation is generally accurate within $\pm 7\%$ for $\rho_f/\rho_g < 50$, and within $\pm 20\%$ for larger values of ρ_f/ρ_g.

10.8 Forced convection boiling in tubes

Relationship between heat transfer and temperature difference

Forced convection boiling in a tube or duct is a process that becomes very hard to delineate because it takes so many forms. In addition to the usual system variables that must be considered in pool boiling, the formation of many regimes of boiling requires that we understand several boiling mechanisms and the transitions between them, as well.

Collier's excellent book *Convective Boiling and Condensation* [1.22] provides a comprehensive discussion of the issues involved in forced convection boiling. Figure 10.20 is his representation of the fairly simple case of flow of liquid in a *uniform wall heat flux* tube in which body forces can be neglected. This situation is representative of a fairly low heat flux at the wall. The vapor fraction, or "quality," of the flow increases steadily until the wall "dries out." Then the wall temperature rises rapidly. With a very high wall heat flux, the pipe could burn out before dryout occurs.

Figure 10.21, also provided by Collier, shows how the regimes shown in Fig. 10.20 are distributed in heat flux and in position along the tube. Notice that at high enough heat fluxes, burnout can be made to occur at any station in the pipe. In the nucleate boiling regimes the heat transfer can be predicted fairly well using the methods described in Section 10.7. But in the annular flow regimes (E and F in Fig. 10.20) the heat transfer mechanism is radically altered, and one of the best methods for predicting q is that of Chen [10.28].

Chen developed a fairly complex—but fairly accurate—method for computing h for water in an annular pipe flow. It is best explained in the form of a

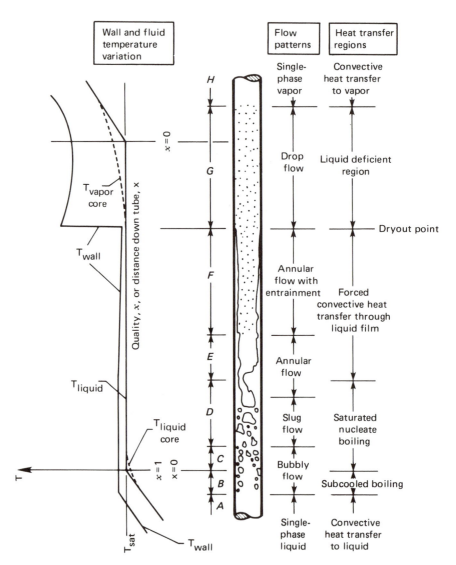

FIGURE 10.20 The development of a two-phase flow in a tube with a constant wall heat flux (not to scale).

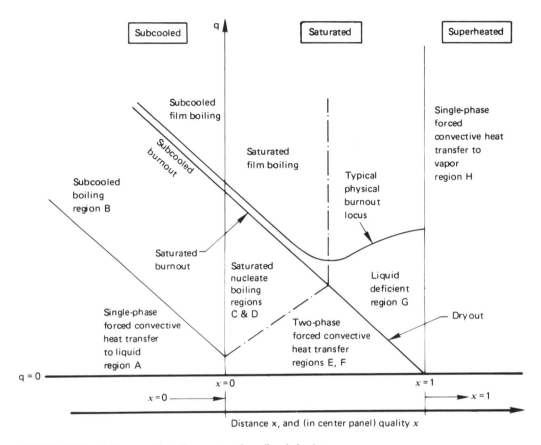

FIGURE 10.21 Influence of heat flux on two-phase flow behavior.

recipe:

* Compute the Martinelli parameter,[4] X_{tt}, for the flow:

$$X_{tt} \cong \left(\frac{1-x}{x}\right)^{0.9} \left(\frac{\rho_g}{\rho_f}\right)^{0.5} \left(\frac{\mu_f}{\mu_g}\right)^{0.1} \qquad (10.47)$$

where x is the "quality" of the flow at the point of interest. The Martinelli parameter is defined as

$$X_{tt} \equiv \sqrt{\left(\frac{dp}{dx}\right)_f \bigg/ \left(\frac{dp}{dx}\right)_g} \qquad (10.48)$$

[4]R. C. Martinelli was an important figure in American heat transfer for a few brief years in the 1940s before he died of leukemia at an early age. He contributed to the famous Berkeley *Heat Transfer Notes* [1.2] and he set down the foundations for predicting heat transfer in two-phase flows, among other accomplishments.

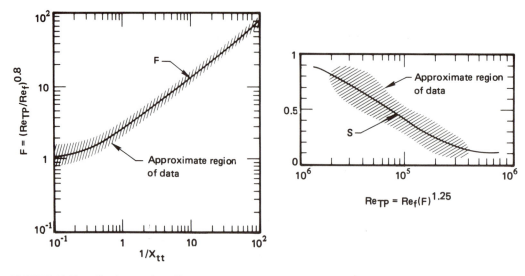

FIGURE 10.22   Chen's two-phase flow parameters [10.28].

and eqn. (10.47) is a correlation that approximates X_{tt} as it is defined by eqn. (10.48). Thus X_{tt}^2 is the ratio of the single-phase pressure drop for the liquid component, to that for the vapor.

- Obtain the empirical function, F, at this X_{tt} from Fig. 10.22. $F^{1/0.8}$ is the ratio of the "two-phase Reynolds number," $\mathrm{Re_{TP}}$ (defined below), to the conventional liquid-phase Reynolds number, $\mathrm{Re_f}$.

- Calculate the superficial mass flux, G, through the pipe:

$$G \equiv \frac{\dot{m}}{A_{\text{pipe}}}$$

- Calculate the single-phase heat transfer coefficient, h_c, from the Dittus–Boelter equation, eqn. (8.24), using saturated liquid properties and the Reynolds number, $\mathrm{Re_{TP}}$:

$$\mathrm{Re_{TP}} \equiv F^{1.25}\left[G(1-x)D/\mu_f\right] \equiv F^{1.25}\mathrm{Re_f} \qquad (10.49)$$

- Obtain the empirical factor, S, from Fig. 10.22 at the known value of $\mathrm{Re_{TP}}$.

- Calculate a "nucleate boiling heat transfer coefficient," h_{NB}, from

$$h_{\text{NB}} = 0.00122\left[\frac{k_f^{0.79}c_{Pf}^{0.45}\rho_f^{0.49}}{\sigma^{0.5}\mu_f^{0.29}h_{fg}^{0.24}\rho_g^{0.24}}\right](\Delta T_{\text{sat}})^{0.24}(\Delta p_{\text{sat}})^{0.75} \qquad (10.50)$$

where Δp_{sat} is p_{sat} at T_w minus p_{sat} at T_{sat}, ΔT_{sat} is $(T_w - T_{\text{sat}})$, and any consistent units may be used.

- Calculate h_{TP} from

$$h_{TP} = S h_{NB} + F h_c \qquad (10.51)$$

for a range of values of ΔT_{sat}.

- Plot $q = h_{TP} \Delta T_{sat}$ against ΔT_{sat} and read ΔT_{sat}, for the case of interest, where this curve intersects q_w; or solve eqn. (10.51) for ΔT_{sat} by trial and error, using the steam tables to get Δp_{sat}.

Example 10.10

0.6 kg/s of H_2O at 200°C flows in a 5-cm-diameter tube heated by 184,000 W/m². Find the wall temperature at a point where the quality x is 20%.

SOLUTION.

$$X_{tt} = \left(\frac{1 - 0.20}{0.2} \right)^{0.9} (0.0091)^{0.5} \left(\frac{0.000139}{0.00001607} \right)^{0.1} = 0.411,$$

so from Fig. 10.22 we read $F = 5.1$. Then, since

$$G = \frac{\dot{m}}{A_{pipe}} = \frac{0.6}{0.00196} = 306 \text{ kg/m}^2\text{-s}$$

we calculate

$$\text{Re}_{TP} = F^{1.25} \left[G(1-x) \frac{D}{\mu_t} \right] = \frac{7.66(306)(1-0.2)(0.05)}{0.00139}$$

$$= 67,500$$

Then from eqn. (8.24),

$$h_c = 0.0246 \frac{k}{D} \text{Pr}^{0.4} \text{Re}_{TP}^{0.8}$$

$$= 0.0246 \frac{0.658}{0.05} (0.915)^{0.4} (67,500)^{0.8}$$

$$= 2281 \text{ W/m}^2\text{-°C}$$

and from Fig. 10.22, we read $S = 0.51$. Finally, we calculate

$$h_{NB} = 0.00122 \left[\frac{(0.658)^{0.79} (4505)^{0.45} (865)^{0.49}}{(0.0377)^{0.5} (0.000139)^{0.29} (1,941,000)^{0.24} (0.597)^{0.24}} \right] \Delta T_{sat}^{0.24} \Delta p_{sat}^{0.75}$$

$$= 2.52 \Delta T_{sat}^{0.24} \Delta p_{sat}^{0.75}$$

so

$$h_{TP} = S h_{NB} + F h_c = 1.284 \Delta T_{sat}^{0.24} \Delta p_{sat}^{0.75} + 11,633$$

and

$$q_w = 25,000 = 1.284 \Delta T_{sat}^{1.24} \Delta p_{sat}^{0.75} + 11,633 \Delta T_{sat}$$

Then, using a steam table to evaluate Δp_{sat}, we solve for ΔT_{sat} by trial and error. The first trial goes like this: first guess:

$$\Delta T_{sat} = 10°C \qquad \text{so } T_w = 210°C$$

then

$$\Delta p_{sat} = p_{sat}(210°C) - p_{sat}(200°C) = 352,900 \text{ N/m}^2$$

and

$$184,000 \neq 323,075 + 116,330 = 439,405$$

so we try a lower ΔT. After two more tries we get

$$\Delta T \simeq 6°C \qquad \text{so } \underline{T_w = 206°C}$$

This is a very low temperature difference because the heat transfer process is very efficient. In this case

$$h = \frac{184,000}{6} = 30,650 \text{ W/m}^2\text{-}°C$$

Peak heat flux

We have seen that there are two limiting heat fluxes in flow boiling in a tube, dryout and burnout. The latter is the more dangerous of the two since it occurs at higher heat fluxes and gives rise to more catastrophic temperature rises. A great deal of work continues to be done on this problem, but the matter is far from resolved. Collier provides a two-chapter discussion of attempts to predict burnout prior to 1972. Hsu and Graham [1.23] include a useful catalog of restrictive empirical burnout formulas.

A promising development in the prediction of the burnout heat flux has recently been given by Katto [10.29]. Katto employed some advanced strategies of dimensional analysis to show that

$$\frac{q_{max}}{Gh_{fg}} = \text{fn}\left(\frac{\rho_g}{\rho_f}, \frac{\sigma\rho_f}{G^2L}, \frac{L}{D}\right)$$

where L is the length of the tube and D its diameter. Since $G^2L/\sigma\rho_f$ is a Weber number, we can see that this equation is of the same form as the burnout equations in Section 10.7. Katto identifies several regimes of flow boiling with both saturated and subcooled liquid entering the pipe. For each of these regions he fits a successful correlation of this form to existing data.

10.9 Two-phase flow in horizontal tubes

The preceding discussion of flow boiling in tubes is restricted to vertical tubes. Several of the flow regimes in Fig. 10.20 will be altered as shown in Fig. 10.23 if the tube is oriented horizontally. The reason is that, especially at low quality, liquid will tend to flow along the bottom of the pipe and vapor along the top. The pattern shown in Fig. 10.23, by the way, will be observed during boiling, during the reverse process—condensation—or during adiabatic two-phase flow.

Many methods have been suggested to predict what flow patterns will result for a given set of conditions in the pipe. Figure 10.24 shows a so-called

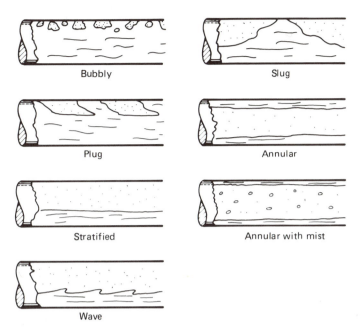

Bubbly

Slug

Plug

Annular

Stratified

Annular with mist

Wave

FIGURE 10.23 The discernible flow regimes during boiling, condensation, or adiabatic flow from left to right in horizontal tubes.

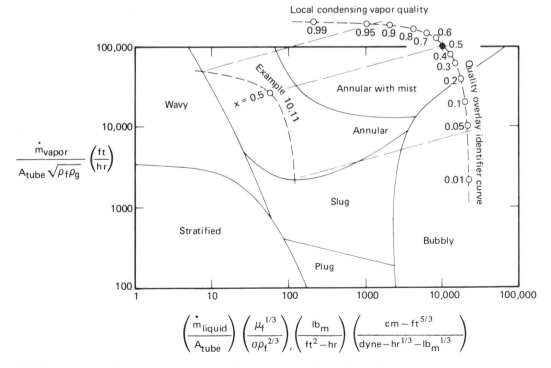

FIGURE 10.24 Modified Baker plot for identifying two-phase flow regimes (after [10.30]).

"modified Baker plot" given by Bell, Taborek, and Fenoglio [10.30]. This graph gives the approximate flow regime as a function of the liquid and vapor flow rates in the tube. The precision of such a representation is not high, since transitions themselves are not sharply defined. The coordinates, which involve other variables as well as the flow rates, are in mixed English and metric units.

In the upper right-hand corner of the flow regime plot (Fig. 10.24) is shown a quality overlay curve. By translating this dashed curve so that it overlays one point of known quality on Fig. 10.24, it is possible to read off any other quality directly with no additional computation. We illustrate its use with an example.

Example 10.11

Water vapor is condensing in a 4-cm-I.D. horizontal tube at 1 atm. The total mass flow rate is 0.2 kg/s. Estimate how much heat transfer will occur in the annular flow regime.

SOLUTION.   We first identify the point of 50% quality. This will be the point at which $\dot{m}_{vapor} = \dot{m}_{liquid} = 0.1$ kg/s.

$$\frac{\dot{m}_{vapor}}{A_{tube}\sqrt{\rho_f\rho_g}} = \frac{0.1}{(\pi/4)(0.04)^2\sqrt{958(0.597)}} = 3.33 \text{ m/s} = 39{,}331 \text{ ft/hr}$$

and

$$\frac{\dot{m}_{liquid}}{A_{tube}}\left(\frac{\mu_f^{1/3}}{\sigma\rho_f^{2/3}}\right) = \frac{0.1}{\frac{\pi}{4}(0.04)^2}\left(\frac{0.000277^{1/3}}{0.0589(958)^{2/3}}\right)$$

$$= 0.906 \frac{\text{kg}}{\text{m}^2\text{-s}}\frac{\text{m}^{8/3}}{\text{N-s}^{1/3}\text{-kg}^{1/3}}$$

$$= 0.906 \frac{\text{kg}}{\text{m}^2\text{-s}}\frac{\text{m}^{8/3}}{\text{N-s}^{1/3}\text{-kg}^{1/3}}\left(\frac{\text{N/m}}{10^3 \text{ dyn/cm}}\right)\left(2.205\frac{\text{lb}_m}{\text{kg}}\right)^{2/3}\left(0.3048\frac{\text{m}}{\text{ft}}\right)^{1/3}\left(3600\frac{\text{s}}{\text{hr}}\right)^{4/3}$$

$$= 57.0 \frac{\text{lb}_m}{\text{ft}^2\text{-hr}}\frac{\text{cm-hr}^{5/3}}{\text{dyn-hr}^{1/3}\text{-lb}_m^{1/3}}$$

Now we identify the point with these coordinates on Fig. 10.24 and slide the dashed curve over so that the point at which $x = 0.5$ lies on top of it. Then we note where the curve crosses the boundaries of the annular flow regime. This can easily be done by connecting the calculated point with the $x = 0.5$ point on the dashed line, and by locating the parallel line segments of equal length that connect the dashed line to the annular flow region boundaries. These line segments intersect the overlay line at $x = 0.94$ and 0.043.

The heat transfer required to change the quality of 0.2 kg/s of steam/water from 0.94 to 0.043 is $\dot{m}_{total}h_{fg}(x_{initial} - x_{final})$, or

$$Q = 0.2\frac{\text{kg}}{\text{s}}\left(2257\frac{\text{kJ}}{\text{kg}}\right)(0.94 - 0.043) = \underline{405 \text{ kW}}$$

When vapor is blown or forced past a cool wall it exerts a shear stress on the condensate film. If the direction of forced flow is downward, it will drag the condensate film along, thinning it out and enhancing heat transfer. It is not hard to show (see Problem 10.24) that

$$\frac{4\mu k(T_{sat}-T_w)x}{gh'_{fg}\rho_f(\rho_f-\rho_g)}=\delta^4+\frac{4}{3}\left[\frac{\tau_\delta\delta^3}{(\rho_f-\rho_g)g}\right] \tag{10.52}$$

where τ_δ is the shear stress exerted by the vapor flow on the condensate film.

Equation (10.52) is the starting point for any analysis of forced convection condensation on an external surface. Notice that if τ_δ is negative—if the shear opposes the direction of gravity—then it will have the effect of thickening δ and reducing heat transfer. Indeed, if for any value of δ,

$$\tau_\delta=-\frac{3g(\rho_f-\rho_g)}{4}\delta \tag{10.53}$$

the shear stress will have the effect of halting the flow of condensate completely for a moment until δ grows to a larger value.

Heat transfer solutions based upon eqn. (10.52) are complex because they require that one solve the boundary layer problem in the vapor in order to evaluate τ_δ; and this solution must be matched with the velocity at the outside of the condensate film. Collier [1.22, Sec. 10.5] discussed such solutions in some detail. One explicit result has been obtained in this way for condensation on the outside of a horizontal cylinder, by Shekriladze and Gomelauri [10.31]:

$$\overline{Nu_D}=0.64\left\{\frac{\rho_f u_\infty D}{\mu_f}\left[1+\left(1+1.69\frac{gh'_{fg}\mu_f D}{u_\infty^2 k_f(T_{sat}-T_w)}\right)^{1/2}\right]\right\}^{1/2} \tag{10.54}$$

where u_∞ is the free stream velocity and $\overline{Nu_D}$ is based on the liquid conductivity. Equation (10.54) is valid up to $Re_D\equiv\rho_f u_\infty D/\mu_f=10^6$. Notice too that under appropriate flow conditions (large values of u_∞, for example) gravity becomes unimportant and

$$\overline{Nu_D}\rightarrow0.64\sqrt{Re_D} \tag{10.55}$$

The prediction of heat transfer during forced convective condensation in tubes becomes a different problem for each of the many possible flow regimes. The reader is referred to [1.22, Sec. 10.5] or [10.30] for details.

An automobile windshield normally is covered with droplets during a rainfall. They are hard to see through, and one must keep the windshield wiper moving constantly to achieve any kind of visibility. A glass windshield is normally quite clean and it is free of any natural oxides, so the water forms a contact angle on it and any film will be unstable. The water tends to pull into droplets which intersect the surface at the contact angle. Visibility can be improved by mixing a surfactant chemical into the window-washing water to reduce surface tension. It can also be improved by preparing the surface with a "wetting agent" to reduce the contact angle.[5]

Such behavior also occurs on a metallic condensing surface, but there is an important difference. Such surfaces do not retain the level of surface perfection required to break a film into drops. Wetting can be temporarily suppressed, and dropwise condensation can be encouraged, by treating an otherwise clean surface (or the vapor) with oil, kerosene, or a fatty acid. But the liquid condensed in a heat exchanger almost always forms a film. It would take very specialized kinds of fouling, or special surface preparations, to cause the liquid to condense in droplets.

It is regrettable that this is the case, because what is called *dropwise* condensation is an extremely effective heat removal mechanism. Figure 10.25 shows how it works. Droplets grow from active nucleation sites on the surface, and in this sense there is a great similarity between nucleate boiling and dropwise condensation. The similarity persists as the droplets grow, touch, and merge with one another until one is large enough to be pulled away from its position by gravity. It then slides off, wiping away the smaller droplets in its path and leaving a dry swathe in its wake. New droplets immediately begin to grow at the nucleation sites in the path.

The repeated re-creation of the early droplet growth cycle creates a very efficient heat removal mechanism. It is typically 10 times more effective than is film condensation under the same temperature difference. Indeed, condensing heat transfer coefficients as high as 200,000 W/m²-°C can be obtained with water at 1 atm. Were it possible to sustain dropwise condensation, we would certainly design equipment in such a way as to make use of it. Unfortunately, laboratory experiments are almost always done on surfaces that have been prepared with oleic, stearic, or other fatty acids. These nonwetting agents, or "promoters" as they are called, are discussed in [1.13, Chap. 12B]. While promoters are normally impractical for industrial use, experienced plant engineers have sometimes added rancid butter through the cup valves of commercial condensers to get at least temporary dropwise condensation.

---

[5]A way in which you can accomplish one of these ends is by wiping the wet window with a cigarette. It is hard to tell which of the two effects the many chemicals in the cigarette achieve.

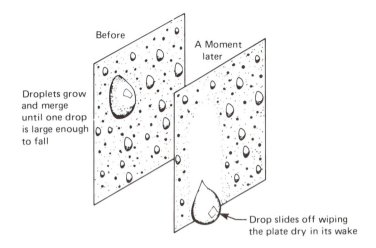

Before

A Moment later

Droplets grow and merge until one drop is large enough to fall

Drop slides off wiping the plate dry in its wake

The process of liquid removal during dropwise condensation.

Typical photograph of dropwise condensation provided by Professor Bora Mikic. Notice the dry paths on the left and in the wake of the middle drop.

FIGURE 10.25　Dropwise condensation.

One of the more significant recent advances in heat transfer technology is a device that combines the high efficiencies of boiling and condensation. The device, called a *heat pipe*, is aptly named because it literally pipes heat from a hot region to a cold one.

The operation of the heat pipe is shown in Fig. 10.26. The pipe is a tube that can be bent or turned in any way that is convenient. The inside of the tube is lined with a layer of wicking material. The wick is wetted with an appropriate liquid. One end of the tube is exposed to a heat source that evaporates the liquid, drying out the wick. Capillary action quickly replenishes the evaporated fluid and moves liquid axially along the wick. Vapor likewise flows from the hot end of the tube to the cold end, where it is condensed.

Placing a heat pipe between a hot region and a cold one is thus similar to connecting the regions with a material of extremely high thermal conductivity—orders of magnitude higher than any known substance (other than helium II). Such devices are not only used for achieving high heat transfer between a source and a sink but for a variety of less obvious purposes. They are used, for example, to level out temperature hot spots in systems, since they function almost isothermally and require enormous heat transfer to sustain any temperature difference.

Design considerations in the specification of a given heat pipe for a given application center on the following issues:

- *Selection and installation of the wick.* The wick is normally made of stainless steel, copper, or another metallic mesh. Many ingenious schemes have been created for bonding it to the inside of the pipe and keeping it at optimum density.

- *Selection of the right liquid.* The liquid can be a cryogen, water, liquid metal, or almost any substance, depending on the operating temperature of the device. The following physical property characteristics make a fluid desirable for heat pipe application:

  (a) High latent heat.
  (b) High thermal conductivity.
  (c) High surface tension.
  (d) Low viscosity.
  (e) It should wet the wick material.
  (f) It should have a suitable boiling point.

  Two liquids which meet the first four criteria admirably are water and mercury.

- *Operating limits of the heat pipe.* The heat flux through a heat pipe is

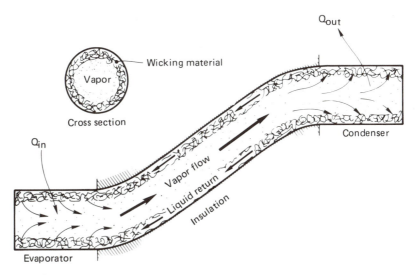

FIGURE 10.26   A typical heat pipe configuration.

restricted by

(a)   Viscous drag in the wick at low temperature.

(b)   Ability of the wick to move the liquid through the required head.

(c)   Drag of the vapor on the returning liquid.

(d)   The sonic or choking speed of the vapor.

(e)   The burnout heat flux during boiling in the evaporation section.

•   *Control of the pipe performance.* Often a given heat pipe will be called upon to function over a range of conditions—under varying evaporator temperatures, for example, or under varying heat loads. One way to vary its performance is to "spike" its effectiveness by injecting more-or-less noncondensible gas into the pipe with an automatic control system.

The reader interested in designing or selecting a heat pipe will find a broad discussion of such devices in the review by Winter and Barsch [10.32]. Tien [10.33] has provided a useful review of the fluid mechanics problems involved in heat pipes. Since the field is advancing rapidly the *Engineering Index* is also a useful source of information.

PROBLEMS

10.1.   A large square tank with insulated sides has a copper base 1.27 cm thick. The base is heated to 650°C and saturated water is suddenly poured in the tank. Plot the temperature of the base as a function of time on the basis of Fig. 10.2 if the bottom of the base is insulated. In your graph, indicate the regimes of boiling and note the temperature at which cooling is most rapid.

10.2. Predict q_{max} for the two heaters in Fig. 10.3b. At what percentage of q_{max} is each one operating?

10.3. A very clean glass container of water at 70°C is depressurized until it is subcooled 30°C. Then it suddenly and explosively "flashes" (or boils). What is the pressure at which this happens? Approximately what diameter of gas bubble, or other disturbance in the liquid, caused it to flash?

10.4. Plot the unstable bubble radius as a function of liquid superheat for water at 1 atm. Comment on the significance of your curve.

10.5. In chemistry class you have probably witnessed the phenomenon of "bumping" in a test tube (the explosive boiling that blows the contents of the tube all over the ceiling). Yet you have never seen this happen in a kitchen pot. Explain why not.

10.6. Use van der Waals' equation of state to approximate the highest reduced temperature to which water can be superheated at low pressure. How many degrees of superheat does this suggest that water can sustain at the low pressure of 1 atm? (It turns out that this calculation is accurate within about 10%.) What would R_b be at this superheat?

10.7. Use Yamagata's equation to determine how nucleation site density increases with ΔT for Berenson's curves in Fig. 10.15. (That is, find c in the relation $n = \text{constant } \Delta T^c$.)

10.8. Suppose that C_{sf} for a given surface is high by 50%. What will be the percentage error in q calculated for a given value of ΔT? [Low by 70%.]

10.9. Water at 100 atm boils on a nickel heater whose temperature is 6°C above T_{sat}. Find h and q.

10.10. Water boils on a large flat plate at 1 atm. Calculate q_{max} if the plate is operated on the surface of the moon (at $\frac{1}{6}$ of $g_{earth-normal}$). What would q_{max} be in the NASA space lab at 10^{-4} of $g_{earth-normal}$?

10.11. Water boils on a 0.002-m-diameter horizontal copper wire. Plot as much of the boiling curve on $\ell n\, q$ vs. $\ell n\, \Delta T$ coordinates as you can. The system is at 1 atm.

10.12. Redo Problem 10.11 for a 0.03-m-diameter sphere in water at 10 atm.

10.13. Verify eqn. (10.17).

10.14. Make a sketch of the q vs. $(T_w - T_{sat})$ relation for a pool boiling process, and invent a graphical method for locating the points where h is maximum and minimum.

10.15. Saturated water at 1 atm flows across a 6-mm-diameter heater rod at 2 m/s. The rod is 10 cm long. How many watts can safely be dissipated from the wire? [7.83 kW.]

10.16. A 2-mm-diameter jet of methanol is directed normal to the center of a 1.5-cm-diameter disk heater at 1 m/s. How many watts can safely be supplied by the heater?

10.17. Verify eqn. (10.41).

10.18. Saturated water at 1 atm boils on a $\frac{1}{2}$-cm-diameter platinum rod. Estimate the temperature of the rod at burnout.

10.19. Plot $(T_w - T_{sat})$ and the quality x as a function of position for the conditions in Example 10.6. Set $x = 0$ where $x = 0$.

10.20. Plot $(T_w - T_{sat})$ and the quality x as a function of position in an 8-cm-I.D. pipe if 0.3 kg/s of water at 100°C passes through it and $q_w = 200{,}000$ W/m². Explain how you would use Fig. 10.21 to set the range of the calculation if it were plotted to scale.

10.21. Use dimensional analysis to verify the form of eqn. (10.8).

10.22. Compare the peak heat flux calculated from the data given in Problem 5.6, with the appropriate prediction. [The prediction is within 11%.]

10.23. Find the highest and lowest mass flow rates for which the annular flow region would not occur (except at extremely high qualities) in Example 10.10.

10.24. Verify eqn. (10.52) by repeating the analysis following eqn. (9.44) but using the b.c. $(\partial u / \partial y)_{y=\delta} = \tau_\delta / \mu$ in place of $(\partial u / \partial y)_{y=\delta} = 0$. Verify the statement involving eqn. (10.53).

10.25. A cool-water-carrying pipe 7 cm in outside diameter has an outside temperature of 40°C. Saturated steam at 80°C flows across it. Plot $\bar{h}_{condensation}$ over the range of Reynolds numbers $0 \leqslant \mathrm{Re}_D \leqslant 10^6$. Do you get the value at $\mathrm{Re}_D = 0$ that you would anticipate from Chapter 9?

REFERENCES

[10.1]   S. NUKIYAMA, "The Maximum and Minimum Values of the Heat Q Transmitted from Metal to Boiling Water under Atmospheric Pressure," *J. Jap. Soc. Mech. Eng.*, vol. 37, 1934, pp. 367–374 (transl.: *Int. J. Heat Mass Transfer*, vol. 9, 1966, pp. 1419–1433).

[10.2]   T. B. DREW and C. MUELLER, "Boiling," *Trans. AIChE*, vol. 33, 1937, p. 449.

[10.3]   "Release of Surface Tension Substance," The International Association for the Properties of Steam, Dec. 1976; available from the Executive Secretary, IAPS, Office of Standard Reference Data, National Bureau of Standards, Washington, D.C.

[10.4]   J. J. JASPER, "The Surface Tension of Pure Liquid Compounds," *J. Phys. Chem. Ref. Data*, vol. 1, no. 4, 1972, pp. 841–1010.

[10.5]   Y. Y. HSU, "On the Size Range of Active Nucleation Cavities on a Heating Surface," *J. Heat Transfer, Trans. ASME, Ser. C*, vol. 84, 1962, pp. 207–216.

[10.6]   K. YAMAGATA, F. KIRANO, K. NISHIWAKA, and H. MATSUOKA, "Nucleate Boiling of Water on the Horizontal Heating Surface," *Mem. Fac. Eng. Kyushu*, vol. 15, 1955, p. 98.

[10.7]   W. M. ROHSENOW, "A Method of Correlating Heat Transfer Data for Surface Boiling of Liquids," *Trans. ASME*, vol. 74, 1952, p. 969.

[10.8]   R. BELLMAN and R. H. PENNINGTON, "Effects of Surface Tension and Viscosity on Taylor Instability," *Quart. Appl. Math.*, vol. 12, 1954, p. 151.

[10.9]   V. SERNAS, "Minimum Heat Flux in Flim Boiling—A Three Dimensional Model," *Proc. 2nd Can. Cong. Appl. Mech.*, Univ. of Waterloo, Canada, 1969, pp. 19–23.

[10.10]   H. LAMB, *Hydrodynamics*, 6th ed., Dover Publications, Inc., New York, 1945.

[10.11]   N. ZUBER, "Hydrodynamic Aspects of Boiling Heat Transfer," AEC Report AECU-4439, Physics and Mathematics, 1959.

[10.12]   J. H. LIENHARD and V. K. DHIR, "Extended Hydrodynamic Theory of the Peak and Minimum Pool Boiling Heat Fluxes," NASA CR-2270, July 1973.

[10.13]   J. H. LIENHARD, V. K. DHIR, and D. M. RIHERD, "Peak Pool Boiling Heat-Flux Measurements on Finite Horizontal Flat Plates," *J. Heat Transfer, Trans. ASME, Ser. C*, vol. 95, 1973, pp. 477–482.

[10.14]   J. H. LIENHARD and V. K. DHIR, "Hydrodynamic Prediction of Peak Pool-Boiling Heat Fluxes from Finite Bodies," *J. Heat Transfer, Trans. ASME, Ser. C*, vol. 95, 1973, pp. 152–158.

[10.15]   S. S. KUTATELADZE, "On the Transition to Film Boiling under Natural Convection," *Kotloturbostroenie*, no. 3, 1948, p. 10.

[10.16]   K. H. SUN and J. H. LIENHARD, "The Peak Pool Boiling Heat Flux on Horizontal Cylinders," *Int. J. Heat Mass Transfer*, vol. 13, 1970, pp. 1425–1439.

[10.17]   J. S. DED and J. H. LIENHARD, "The Peak Pool Boiling Heat Flux from a Sphere," *AIChE J.*, vol. 18, no. 2, 1972, pp. 337–342.

[10.18]   A. L. BROMLEY, "Heat Transfer in Stable Film Boiling," *Chem. Eng. Progr.*, vol. 46, 1950, pp. 221–227.

[10.19]   P. PITSCHMANN and U. GRIGULL, "Filmverdampfung an waagerechten Zylindern," *Wärme- und Stoffübertragung*, vol. 3, 1970, pp. 75–84.

[10.20]   J. E. LEONARD, K. H. SUN, and G. E. DIX, "Low Flow Film Boiling Heat Transfer on Vertical Surfaces: Part II: Empirical Formulations and Application to BWR-LOCA Analysis," ASME–AIChE Nat. Heat Transfer Conf., St. Louis, Mo., August 1976.

[10.21]   J. W. WESTWATER and B. P. BREEN, "Effect of Diameter of Horizontal Tubes on Film Boiling Heat Transfer," *Chem. Eng. Progr.*, vol. 58, 1962, pp. 67–72.

[10.22]   P. J. BERENSON, "Transition Boiling Heat Transfer from a Horizontal Surface," M.I.T. Heat Transfer Lab. Tech. Rep. 17, 1960.

[10.23]   J. H. LIENHARD and P. T. Y. WONG, "The Dominant Unstable Wavelength and Minimum Heat Flux during Film Boiling on a Horizontal Cylinder," *J. Heat Transfer, Trans. ASME, Ser. C*, vol. 86, no. 2, 1964, pp. 220–226.

[10.24]   J. H. LIENHARD and R. EICHHORN, "Peak Boiling Heat Flux on Cylinders in a Cross Flow," *Int. J. Heat and Mass Transfer*, vol. 19, 1966, pp. 1135–1142.

[10.25]   Y. KATTO and M. SHIMIZU, "Upper Limit of CHF in the Saturated Forced Convection Boiling on a Heated Disk with a Small Impinging Jet," *J. Heat Transfer, Trans. ASME, Ser. C*, vol. 101, no. 2, 1979, pp. 265–269.

[10.26]   J. H. LIENHARD and R. EICHHORN, "On Predicting Boiling Burnout for Heaters Cooled by Liquid Jets," *Int. J. Heat Mass Transfer*, vol. 22, 1979, pp. 774–776.

[10.27]   J. H. LIENHARD and M. Z. HASAN, "Correlation of Burnout Data for Disc Heaters Cooled by Liquid Jets," *J. Heat Transfer, Trans. ASME, Ser. C*, vol. 101, no. 2, 1979, pp. 276–279.

[10.28]   J. C. CHEN, "A Correlation for Boiling Heat Transfer to Saturated Fluids in Convective Flow," ASME Prepr. 63-HT-34, 6th ASME–AIChE Heat Transfer Conf., Boston, August 1963.

[10.29]   Y. KATTO, "A Generalized Correlation of Critical Heat Flux for the Forced Convection Boiling in Vertical Uniformly Heated Round Tubes," *Int. J. Heat Mass Transfer*, vol. 21, no. 12, 1978, pp. 1527–1542.

[10.30]   K. J. BELL, J. TABOREK, and F. FENOGLIO, "Interpretation of Horizontal In-Tube Condensation Heat Transfer Correlations with a Two-Phase Flow Regime Map," *Chem. Eng. Symp. Ser.*, vol. 66, no. 102, 1970, pp. 150–163.

[10.31]   I. G. SHEKRILADZE and V. I. GOMELAURI, "Theoretical Study of Laminar Film Condensation of Flowing Vapour," *Int. J. Heat Mass Transfer*, vol. 9, 1966, pp. 581–591.

[10.32]   E. R. F. WINTER and W. O. BARSCH, "The Heat Pipe," *Advances in Heat Transfer*, vol. 7, (T. F. Irvine, Jr., and J. P. Hartnett, eds.), Academic Press, Inc., New York, 1971, pp. 219–320e.

[10.33]   C. L. TIEN, "Fluid Mechanics of Heat Pipes," *Annu. Rev. Fluid Mech.*, vol. 7, 1975, pp. 167–185.

Part IV

Thermal
Radiation

Chapter 11

Radiative heat transfer

11.1 The problem of radiative exchange

Chapter 1 described the elementary mechanisms of heat radiation. Before we proceed, you should reflect upon what you remember about the following key ideas from Chapter 1:

- Electromagnetic wave spectrum
- Black body
- Hohlraum
- Infrared (and other) radiation
- Heat radiation
- Transmittance
- Reflectance
- Absorptance
- $\alpha + \rho + \tau = 1$
- The Stefan–Boltzmann law

- The Stefan–Boltzmann constant
- Planck's law
- F_{1-2} and $\mathscr{F}_{1-2}$
- $e(T)$ and $e_\lambda(T)$ for black bodies
- Radiation shielding

We presume that the reader understands these concepts.

The heat exchange problem

Figure 11.1 shows two arbitrary surfaces radiating energy to one another. The net heat exchange, Q_{net}, from the hotter surface (1) to the cooler surface (2) depends upon the following influences:

- T_1 and T_2.
- The areas of (1) and (2).
- The configurations of (1) and (2) and the spacing between them.
- The radiative characteristics of the surfaces.
- Additional surfaces in the environment.
- The medium between (1) and (2). (If the medium is air, we can probably neglect its influence.)

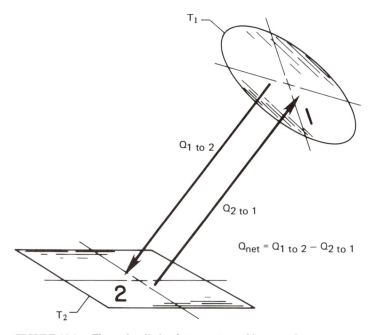

FIGURE 11.1   Thermal radiation between two arbitrary surfaces.

If surfaces (1) and (2) are black, if they are surrounded by air, if the surfaces in the environment are black, and if no heat flows between (1) and (2) by conduction or convection, then only the first three considerations are involved in determining Q_{net}. We saw some elementary examples of how this could be done in Chapter 1. In this case

$$Q_{net} = F_{1-2} A_1 \sigma \left(T_1^4 - T_2^4 \right) \qquad (11.1)$$

The last three considerations lead to great complications of the problem. In Chapter 1 we saw that these nonideal factors were customarily included in a "real body view factor," $\mathcal{F}_{1-2}$, such that

$$Q_{net} = \mathcal{F}_{1-2} A_1 \sigma \left(T_1^4 - T_2^4 \right) \qquad (11.2)$$

Before we undertake the problem of evaluating $\mathcal{F}_{1-2}$, it is necessary to set down several definitions.

Some definitions

Emittance. A real body at temperature, T, does not emit with the black body emissive power, $e_b = \sigma T^4$, but rather with some fraction, ε, of e_b. Thus we define either the monochromatic emittance, ε_λ:

$$\varepsilon_\lambda \equiv \frac{e_\lambda(\lambda, T)}{e_{\lambda_b}(\lambda, T)} \qquad (11.3)$$

or the total emittance, ε:

$$\varepsilon \equiv \frac{e(T)}{e_b(T)} = \frac{\int_0^\infty e_\lambda(\lambda, T)\, d\lambda}{\sigma T^4} \qquad (11.4)$$

The emittance is determined entirely by the properties of the surface of the particular body and its temperature. It is independent of the environment of the body.

Table 11.1 lists typical values of the total emittance for a variety of real substances. (These were summarized from [1.20].) Notice that most metals have quite low emittances, unless they are oxidized. Most nonmetals have emittances that are quite high—approaching the black body limit of unity. Notice that among the "blackest" surfaces in the table are white paint, paper, and ice.

One particular kind of surface behavior is that for which ε_λ is independent of λ. We call such a surface a *gray body*. The emittance of a gray body is a constant fraction of e_b, as is indicated in the inset of Fig. 11.2. No real body is gray, but many exhibit approximately gray behavior. We see in Fig. 11.2, for example, that the sun appears to us on earth as an approximately gray body with an emittance of approximately 0.6. We shall often use the gray body simplification in this chapter to avoid the formidable difficulties of considering the variation of ε_λ with λ.

Table 11.1   Total emittances for a variety of surfaces

Metals			Nonmetals		
Surface	Temperature (°C)	ε	Surface	Temperature (°C)	ε
Aluminum			Asbestos	40	0.93–0.97
Polished, 98% pure	200–600	0.04–0.06	Brick		
Commercial sheet	90	0.09	Red, rough	40	.093
Heavily oxidized	90–540	0.20–0.33	Silica	980	0.80–0.85
Brass			Fireclay	980	0.75
Highly polished	260	0.03	Ordinary refractory	1090	0.59
Dull plate	40–260	0.22	Magnesite refractory	980	0.38
Oxidized	40–260	0.46–0.56	White refractory	1090	0.29
Copper			Carbon		
Highly polished electrolytic	90	0.02	Filament	1040–1430	0.53
Slightly polished, to dull	40	0.12–0.15	Lampsoot	40	0.95
Black oxidized	40	0.76	Concrete, rough	40	0.94
Gold: pure, polished	90–600	0.02–0.035	Glass		
Iron and Steel			Smooth	40	0.94
Mild steel, polished	150–480	0.14–0.32	Quartz glass (2 mm)	260–540	0.96–0.66
Steel, polished	40–260	0.07–0.10	Pyrex	260–540	0.94–0.74
Sheet steel, rolled	40	0.66	Gypsum	40	0.80–0.90
Sheet steel, strong rough oxide	40	0.80	Ice	0	0.97–0.98
Cast iron, oxidized	40–260	0.57–0.66	Limestone	400–260	0.95–0.83
Iron, rusted	40	0.61–0.85	Marble	40	0.93–0.95
Wrought iron, smooth	40	0.35	Mica	40	0.75
Wrought iron, dull oxidized	20–360	0.94	Paints		
Stainless, polished	40	0.07–0.17	Black gloss	40	0.90
Stainless, after repeated heating	230–900	0.50–0.70	White paint	40	0.89–0.97
Lead			Lacquer	40	0.80–0.95
Polished	40–260	0.05–0.08	Various oil paints	40	0.92–0.96
Oxidized	40–200	0.63	Red lead	90	0.93
Mercury: pure, clean	40–90	0.10–0.12	Paper		
Platinum			White	40	0.95–0.98
Pure, polished plate	200–590	0.05–0.10	Other colors	40	0.92–0.94
Oxidized at 590°C	260–590	0.07–0.11	Roofing	40	0.91
Drawn wire and strips	40–1370	0.04–0.19	Plaster, rough lime	40–260	0.92
Silver	200	0.01–0.04	Quartz	100–1000	0.89–0.58
Tin	40–90	0.05	Rubber	40	0.86–0.94
Tungsten			Snow	10–20	0.82
Filament	540–1090	0.11–0.16	Water, 0.1 mm or more thick	40	0.96
Filament	2760	0.39	Wood	40	0.80–0.90

Diffuse and specular emittance and reflection.   The energy emitted by a surface, or that portion of an incoming ray of energy that is reflected from a surface which is not black, may leave the body *diffusely* or *specularly*. It may also be emitted or reflected in a way that lies between these limits. Figure 11.3 shows how radiation might be reflected in these various ways. A mirror reflects visible radiation in an almost perfectly *specular* fashion. The "reflection" of a billiard

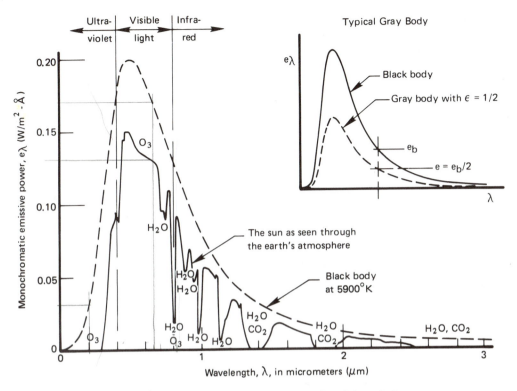

FIGURE 11.2 Comparison of the energy emitted by the sun (as viewed through the earth's atmosphere) with a black body at the same mean temperature. (Notice that the effective e_{λ_b}, just outside the earth's atmosphere, is far less than it is on the surface of the sun because the radiation has spread out.)

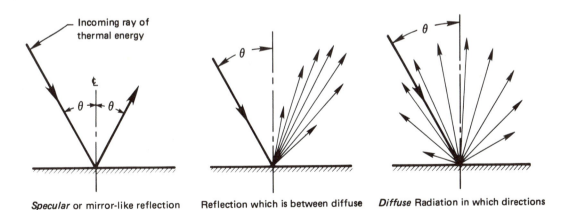

Specular or mirror-like reflection of incoming ray

Reflection which is between diffuse and specular. (a real surface)

Diffuse Radiation in which directions of departure are uninfluenced by incoming ray angle, θ.

FIGURE 11.3 Specular and diffuse reflection of radiation. (Arrows indicate magnitude of the heat flux in the directions indicated.)

ball from the side of a table is also specular. When reflection or emission is diffuse, equal rates of radiation leave in all directions.

The character of the emittance or reflectance of a surface will normally change with the wavelength of the radiation. We shall rather frequently assume diffuse behavior on the part of the surface, but this will be strictly true only if the surface is black.

Experiment 11.1

Obtain a flashlight with as narrow a spot focus as you can find. Direct it at an angle, onto a mirror, onto the surface of a bowl filled with sugar, and onto a variety of other surfaces, all in a darkened room. In each case, move the palm of your hand around the surface of an imaginary hemisphere centered on the point where the spot touches the surface. Notice how your palm is illuminated and categorize the kind of reflectance of each surface—at least in the range of visible wavelengths.

Intensity of radiation.   Consider radiation from a circular surface element, dA, as shown at the top of Fig. 11.4. If the element is black, the radiation that it emits is indistinguishable from the radiation that would be emitted from a black cavity at the same temperature. Thus the rate at which energy is emitted in any direction is proportional to the projected area of dA normal to the direction of view, as shown in the upper left corner of Fig. 11.4.

If an aperture of area dA_a is placed at a radius, r, from dA, and normal to the radius, it will intercept a fraction of the energy emitted by dA. The magnitude of that fraction is equal to the ratio of the solid angle,[1] ω, subtended by dA_a to the solid angle subtended by the entire hemisphere. We define a quantity called the ***intensity of radiation,*** i (W/m²-steradian), which is defined by an energy balance statement:

$$dq = i\,d\omega \cos\theta = \text{fraction of heat transfer from}$$
$$dA \text{ that is intercepted by } dA_a \qquad (11.5)$$

Notice that while the heat flux from dA decreases with θ (as indicated on the right side of Fig. 11.3), the intensity of energy from a diffuse surface is uniform in all directions.

Finally, we compute i in terms of q by integrating $i\,d\omega$ over the entire unit hemisphere and noting (see Fig. 11.4) that $d\omega = \sin\theta\,d\theta\,d\phi$.

$$q = \int_{\phi=0}^{2\pi} \int_{\theta=0}^{\pi/2} i \cos\theta (\sin\theta\,d\theta\,d\phi) = \pi i \qquad (11.6a)$$

---

[1]The unit of solid angle is the steradian. One steradian is the solid angle subtended by a spherical segment whose area equals the square of its radius. A full sphere therefore subtends $4\pi r^2 / r^2 = 4\pi$ steradians.

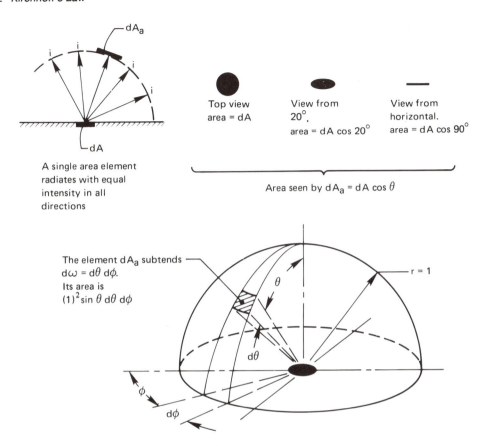

FIGURE 11.4   Radiation intensity through a unit sphere.

Thus for a black body

$$i_b = \frac{\sigma T^4}{\pi} = \frac{e_b}{\pi} = \text{fn}(T \text{ only}) \tag{11.6b}$$

and for any particular wavelength, we define the monochromatic intensity

$$i_\lambda = \frac{e_\lambda}{\pi} = \text{fn}(T, \lambda) \tag{11.6c}$$

11.2 Kirchhoff's law

The problem of predicting α

The total emittance, ε, of a surface is uniquely determined by the characteristics of the surface. But the absorptance, α, while it is surface-dependent, is also influenced by the environment from which the surface receives energy. The reason is that α depends on the way in which incoming energy is distributed in wavelength. That distribution is determined by the characteristics of the surfaces from which the surface of interest receives radiation. Furthermore, if the temperatures of those bodies from which radiation is received are changed, the energy distribution in wavelength will generally change as well.

We are thus faced with the problem that α depends on the surface characteristics and the temperatures of *all* bodies involved in a given heat exchange process. Kirchhoff's law[2] is a theoretical relation which can be used to predict α under certain restrictions. Next, we shall derive the law and state the restrictions on it.

Simple heat exchange problem

Figure 11.5 shows two surfaces that exchange heat by radiation. Our objective is to trace the way in which each "packet" of energy, leaving each of the two surfaces, is emitted, reflected, absorbed, reemitted—back and forth—ad infinitum. In so doing we shall relate α and ε to one another through the net heat transfer. To do this we consider e_1 and e_2 to be the flux of energy packets emitted by each surface. All the energy emitted by one of the surfaces in any instant will eventually come to rest either back in that surface, or in the other surface. Let us trace the process:

surface (1) emits e_1	surface (2) emits e_2
surface (2) absorbs $e_1\alpha_2$	surface (1) absorbs $e_2\alpha_1$
surface (2) reflects $e_1(1-\alpha_2)$	surface (1) reflects $e_2(1-\alpha_1)$
surface (1) absorbs $e_1(1-\alpha_2)\alpha_1$	etc.
surface (1) reflects $e_1(1-\alpha_2)(1-\alpha_1)$	
surface (2) absorbs $e_1(1-\alpha_2)(1-\alpha_1)\alpha_2$	
surface (2) reflects $e_1(1-\alpha_2)(1-\alpha_1)(1-\alpha_2)$	
surface (1) absorbs $e_1(1-\alpha_2)(1-\alpha_1)(1-\alpha_2)\alpha_1$	
etc.	

The next step is to add up all the energy e_1 that finally comes back to be absorbed in surface (1), and all the energy e_2 that finally is absorbed in surface (1). This addition will be greatly simplified if we define the recurring factor $(1-\alpha_1)(1-\alpha_2)$ as γ. Then the fraction of the energy e_1 that surface (1) finally absorbs is

$$e_1(1+\gamma+\gamma^2+\ldots)(1-\alpha_2)\alpha_1$$

[2]Gustav Robert Kirchhoff (1824–1887) was a very important German physicist of the nineteenth century. He presented this "Kirchhoff's law" when he was only 25 years old. But he is also known for a great deal of basic work in the thermodynamics of phase change and in electric theory.

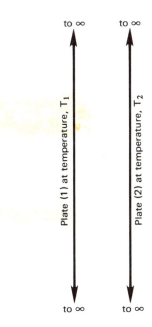

FIGURE 11.5   Heat transfer between two infinite parallel plates.

The binomial expansion

$$\frac{1}{1-\gamma} = (1-\gamma)^{-1} = 1 + \gamma + \gamma^2 + \ldots$$

allows us to summarize the preceding equation as

$$\frac{e_1(1-\alpha_2)\alpha_1}{1-\gamma}$$

We can likewise write the fraction of e_2 that is absorbed by surface (1) as

$$\frac{e_2\alpha_1}{1-\gamma}$$

Thus the net heat flux from (1) to (2) is

$$q_{\text{net}_{1 \text{ to } 2}} = e_1 - \frac{e_1(1-\alpha_2)\alpha_1}{1-\gamma} - \frac{e_2\alpha_1}{1-\gamma} \qquad (11.7)$$

Substituting $\gamma = (1-\alpha_1)(1-\alpha_2)$ back into eqn. (11.7), we get

$$q_{\text{net}_{1 \text{ to } 2}} = e_1 \left[\frac{1 - 1 + \alpha_2 + \alpha_1 - \alpha_1\alpha_2 - \alpha_1 + \alpha_1\alpha_2 - (e_2/e_1)\alpha_1}{1 - 1 + \alpha_2 + \alpha_1 - \alpha_1\alpha_2} \right]$$

or

$$q_{1 \text{ to } 2} = \frac{e_1\alpha_2 - e_2\alpha_1}{\alpha_2 + \alpha_1 - \alpha_1\alpha_2} \qquad (11.8)$$

In the figure, the labels read: "to ∞" (top and bottom of both plates), "Plate (1) at temperature, T_1", "Plate (2) at temperature, T_2".

Finally, we note that *if* $T_1 = T_2$, $q_{1 \text{ to } 2}$ must equal zero, and

$$e_1 \alpha_2 = e_2 \alpha_1$$

All the quantities here depend on the common temperature $T_1 = T_2 = T$. Thus we can rearrange this expression in the form

$$\frac{e_1}{\alpha_1} = \frac{e_2}{\alpha_2} = \text{fn}(T) \tag{11.9}$$

This result is *Kirchhoff's law.* The most important consequence of Kirchoff's law is obtained by allowing, say, body (2) to be black. Then $\alpha_2 = 1$, $e_2 = \sigma T^4$. Equation (11.9) then becomes

$$\frac{\varepsilon_1 \sigma T^4}{\alpha_1} = \frac{\sigma T^4}{1}$$

so $\varepsilon_1 = \alpha_1$. The subscripts are superfluous and we can write

$$\boxed{\varepsilon = \alpha}$$

which is subject to restrictions discussed below $\tag{11.10a}$

We can also write for each wavelength:

$$\boxed{\varepsilon_\lambda = \alpha_\lambda}$$

which is generally true $\tag{11.10b}$

Limitations on the result $\varepsilon = \alpha$

Equation (11.10) was derived, subject to an important tacit assumption: that both surfaces were at the same temperature. We have noted that when radiation from a hot surface falls onto a cooler one, the wavelength distribution of the incoming energy will differ from that of the reemitted energy. Under these conditions we could hardly expect eqn. (11.10) to remain valid. Consequently, we must be wary of using it when the temperatures of the body of interest differ significantly from those around it. Strictly speaking, eqn. (11.10) is true only for the following circumstances:

*   The body is gray. Then $\alpha = \varepsilon \neq \text{fn}(\lambda)$.
*   The surroundings are black, so that $\alpha_\lambda = \varepsilon_\lambda \neq f(T)$.
*   The trivial case in which the body and its surroundings are at the same temperature.

It can also be shown for metallic surfaces that, if the surroundings are black or gray, $\alpha = \varepsilon(\overline{T})$, where

$$\overline{T} \equiv \sqrt{(T_{\text{surroundings}})(T_{\text{surface}})}$$

As a typical example of the failure of eqn. (11.10), consider solar radiation incident on a roof, painted black. From Table 11.1 we see that ε is on the order of 0.94. It turns out that α is just about the same. If we repaint the roof white, ε will not change noticeably. However, much of the energy arriving from the sun is carried in visible wavelengths. Our eyes tell us that white paint reflects sunlight very strongly in these wavelengths, and indeed this is the case. The absorptance of white paint to energy from the sun is only on the order of 0.10—much less than ε.

11.3 Simple radiant heat exchange between two surfaces

One body enclosed by another

Parallel plates. Equation (11.8) is not a useful design equation in its present form. But when we substitute $e_1 = \varepsilon_1 \sigma T_1^4$ and $e_2 = \varepsilon_2 \sigma T_2^4$ and use $\varepsilon = \alpha$, we get

$$q_{1 \text{ to } 2} = \sigma \frac{\varepsilon_1 \varepsilon_2 T_1^4 - \varepsilon_2 \varepsilon_1 T_2^4}{\varepsilon_1 + \varepsilon_2 - \varepsilon_1 \varepsilon_2}$$

or

$$q_{1 \text{ to } 2} = \frac{1}{\dfrac{1}{\varepsilon_1} + \dfrac{1}{\varepsilon_2} - 1} \sigma\left(T_1^4 - T_2^4\right) \qquad (11.11)$$

Comparing eqn. (11.11) with eqn. (11.2), we see that

$$\mathcal{F}_{1 \text{ to } 2} = \frac{1}{\dfrac{1}{\varepsilon_1} + \dfrac{1}{\varepsilon_2} - 1} \qquad (11.12)$$

for infinite parallel plates. Notice too, that if the surfaces are both black, $\varepsilon_1 = \varepsilon_2 = 1$ and

$$\mathcal{F}_{1 \text{ to } 2} = 1 = F_{1 \text{ to } 2} \qquad (11.13)$$

which, of course, is what we would expect.

Example 11.1

A stainless steel plate at 100°C faces a firebrick wall at 500°C. Estimate the heat flux and radiant "heat transfer coefficient," h_r (recall Section 6.2).

SOLUTION. From Table 11.1 we read the emittances of stainless steel and firebrick as approximately 0.6 and 0.75. Thus

$$q_{1 \text{ to } 2} = \frac{1}{\dfrac{1}{0.75} + \dfrac{1}{0.6} - 1} 5.67 \times 10^{-8}\left[(773°\text{K})^4 - (373°\text{K})^4\right]$$

$$= 9573 \ \text{W/m}^2$$

This can be put in the form of a "radiation heat transfer coefficient:"

$$h_r = \frac{q_{1 \text{ to } 2}}{T_1 - T_2} = \frac{9573}{500 - 100} = \underline{24 \text{ W/m}^2\text{-°C}}$$

This heat transfer coefficient is rather low. If we had done the calculation for a brick wall at 1500°C, we would have found that $q = 280,000$ and $h_r = 200$. Thus we see that h_r rises in a fairly dramatic nonlinear way with T_1.

General case in which one body surrounds another

A pair of parallel plates is a special case of the general situation in which one body surrounds another. The general situation is suggested in Fig. 11.6, and it includes such configurations as concentric cylinders or a sphere within a sphere. The factor $\mathcal{F}_{1 \text{ to } 2}$ in this case takes either of two limiting forms, which we state, for the moment, without proof.

$$\mathcal{F}_{1 \text{ to } 2} = \begin{cases} \left[\dfrac{1}{\dfrac{1}{\varepsilon_1} + \dfrac{A_1}{A_2}\left(\dfrac{1}{\varepsilon_2} - 1\right)} \right] & \text{for diffusely re-} \\[2em] & \text{flecting bodies} \\[3em] \left[\dfrac{1}{\dfrac{1}{\varepsilon_1} + \dfrac{1}{\varepsilon_2} - 1} \right] & \text{for specularly} \\[1em] & \text{reflecting bodies} \end{cases} \tag{11.14}$$

The latter result is interestingly identical to eqn. (11.12).

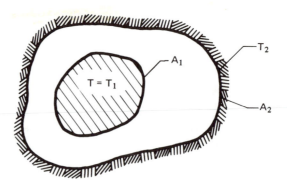

FIGURE 11.6   Heat transfer between an enclosed body and the body surrounding it.

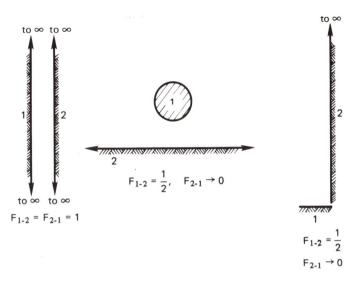

FIGURE 11.7 Some configurations for which the value of the view factor is immediately apparent.

Radiant heat exchange between two finite black bodies

Some evident results. Let us now return to the purely geometric problem of evaluating the view factor, F_{1-2}. Although the evaluation of F_{1-2} is relevant to the calculation of $\mathscr{F}_{1-2}$ for nonblack bodies, it is the only correction of the Stephan–Boltzmann law that we need for black bodies, and therefore it is of immediate use to us.

Figure 11.7 shows three elementary situations in which the value of F_{1-2} is evident under the definition:

$$F_{1-2} = \text{fraction of energy emitted by (1) that reaches (2)}$$

A second apparent result in regard to the view factor is that all the energy leaving a body (1) reaches something else. Thus the first law gives

$$1 = F_{1-1} + F_{1-2} + F_{1-3} + \ldots + F_{1-n} \tag{11.15}$$

where $(2), (3), \ldots, (n)$ are all of the bodies in the neighborhood of (1). Figure 11.8 shows a representative situation in which a body (1) is surrounded by three other bodies. It sees all three bodies but it also views itself, in part. This accounts for the inclusion of the view factor F_{1-1} in eqn. (11.15).

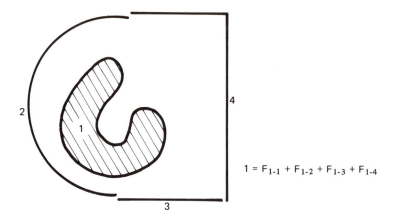

$$1 = F_{1-1} + F_{1-2} + F_{1-3} + F_{1-4}$$

FIGURE 11.8   A body, 1, that views three other bodies and itself as well.

Example 11.2

A jet of liquid metal at 2000°C pours from a crucible. It is 3 mm in diameter. A 5-cm-diameter cylindrical radiation shield surrounds the jet through an angle of 330°, but there is a 30° slit in it. The jet is otherwise surrounded by a large cubic room at 30°C. How much radiant energy reaches the room per meter of length of the shield if it is legitimate to assume that the jet and the shield are both black under these conditions?

SOLUTION.   There are two paths by which radiant energy can reach the room: directly through the slit, and from the shield to the room. To calculate the former, we need $F_{\text{jet–room}}$:

$$1 = F_{\text{jet–room}} + F_{\text{jet–shield}}$$

but $F_{\text{jet–shield}} = 330/360 = 0.917$, so $F_{\text{jet–room}} = 1 - 0.917 = 0.0833$. Then

$$Q_{\text{jet–room}} = F_{\text{jet–room}} A_{\text{jet}} \sigma \left(T_{\text{jet}}^4 - T_{\text{room}}^4 \right)$$

$$= 0.0833 \left[\frac{\pi(0.003) \text{ m}^2}{\text{m length}} \right] (5.67 \times 10^{-8})(2273^4 - 303^4)$$

$$= \underline{1188 \text{ W/m}}$$

The shield temperature is obtained by balancing the heat that it receives with that which it emits:

$$F_{\text{jet–shield}} \sigma A_{\text{jet}} \left(T_{\text{jet}}^4 - T_{\text{shield}}^4 \right) = F_{\text{shield–room}} \sigma A_{\text{shield}} \left(T_{\text{shield}}^4 - T_{\text{room}}^4 \right)$$

The view factor $F_{\text{shield–room}}$ can be approximated as unity. (Notice that we neglect heat transfer from the *inside* of the shield to the room.) Then

$$\frac{0.917(0.003)}{1(0.050)} \left(2273^4 - T_{\text{shield}}^4 \right) = T_{\text{shield}}^4 - 303^4$$

or

$$T_{shield} = \left[\frac{303^4 + \dfrac{0.917(0.003)}{1(0.050)} 2273^4}{1 + \dfrac{0.917(0.003)}{1(0.050)}} \right]^{1/4} = \underline{1088\,°C}$$

It is now possible to calculate $Q_{shield-room}$:

$$Q_{shield-room} = F_{shield-room} \sigma A_{shield} \left(T_{shield}^4 - T_{room}^4 \right)$$

$$= 1(5.67)(10)^{-8} \left[\frac{\pi(0.05)\mathrm{m}^2}{\mathrm{m\ length}} \right] (1088^4 - 303^4)$$

$$= \underline{12{,}400\ \mathrm{W/m}}$$

so the total heat transfer is

$$Q_{shield-room} + Q_{jet-room} = \underline{13{,}490\ \mathrm{W/m}}$$

most of which comes through the shield.

Notice that the unshielded jet would have transferred

$$\frac{1}{0.0833}(1188) = 14{,}260\ \mathrm{W/m}$$

to the room. Therefore, this particular shield has accomplished only a 5.4% reduction of heat transfer. To be effective, either the shield would have to be made much larger or it would have to have a low emittance.

Calculation of F_{1-2} for black bodies. Consider two elements, dA_1 and dA_2, of larger black bodies (1) and (2), as shown in Fig. 11.9. the entire body (1), and the entire body (2), are isothermal. Since element dA_2 subtends a solid angle $d\omega_1$, we use eqn. (11.5) to write

$$dQ_{1\ to\ 2} = (i_1 d\omega_1)(dA_1 \cos\beta_1)$$

But from eqn. (11.6)

$$i_1 = \frac{\sigma T_1^4}{\pi}$$

Furthermore,

$$d\omega_1 = \frac{dA_2 \cos\beta_2}{s^2}$$

where s is the distance from (1) to (2). Thus:

$$dQ_{1\ to\ 2} = \frac{\sigma T_1^4}{\pi} \left(\frac{\cos\beta_1 \cos\beta_2 dA_1 dA_2}{s^2} \right)$$

By the same token

$$dQ_{2\ to\ 1} = \frac{\sigma T_2^4}{\pi} \left(\frac{\cos\beta_2 \cos\beta_1 dA_2 dA_1}{s^2} \right)$$

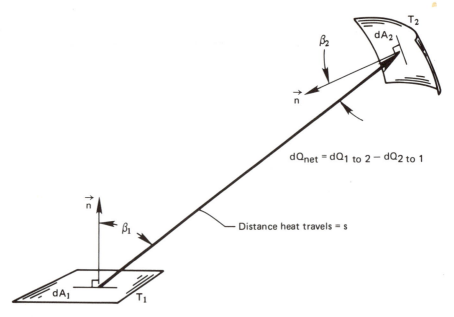

FIGURE 11.9   Radiant exchange between two black elements which are part of the bodies, 1 and 2.

Then

$$Q_{net} = \sigma(T_1^4 - T_2^4) \int_{A_1} \int_{A_2} \frac{\cos\beta_1 \cos\beta_2}{\pi s^2} dA_1 dA_2 \qquad (11.16)$$

The view factors, F_{1-2} and F_{2-1}, are immediately obtainable from eqn. (11.16). If we compare this result with $Q_{net} = F_{1-2}A_1\sigma(T_1^4 - T_2^4)$, we get

$$F_{1-2} = \frac{1}{A_1} \int_{A_1}\int_{A_2} \frac{\cos\beta_1 \cos\beta_2}{\pi s^2} dA_1 dA_2 \qquad (11.17a)$$

From the inherent symmetry of the problem, we can also write

$$F_{2-1} = \frac{1}{A_2} \int_{A_2}\int_{A_1} \frac{\cos\beta_2 \cos\beta_1}{\pi s^2} dA_2 dA_1 \qquad (11.17b)$$

It follows from eqns. (11.17a) and (11.17b) that

$$\int_{A_1}\int_{A_2} \frac{\cos\beta_1 \cos\beta_2}{\pi s^2} dA_1 dA_2 = \boxed{A_1 F_{1-2} = A_2 F_{2-1}} \qquad (11.18)$$

This reciprocity relation will prove to be very useful in subsequent work.

The direct evaluation of F_{1-2} from eqn. (11.17a) becomes fairly involved, even for the simplest configurations. Siegel and Howell [1.21] provide an

Table 11.2 View factors for a variety of two-dimensional configurations (infinite in extent normal to the paper)

Configuration	Equation
1.	$$F_{1\text{-}2} = F_{2\text{-}1} = \sqrt{1 + \left(\frac{h}{w}\right)^2} - \left(\frac{h}{w}\right)$$
2.	$$F_{1\text{-}2} = F_{2\text{-}1} = 1 - \sin(\alpha/2)$$
3.	$$F_{1\text{-}2} = \frac{1}{2}\left[1 + H - \sqrt{1 + H^2}\right], \quad H \equiv \frac{h}{w}$$
4.	$$F_{1\text{-}2} = (A_1 + A_2 - A_3)/2A_1$$
5.	$$F_{1\text{-}2} = \frac{r}{b - a}\left[\tan^{-1}\frac{b}{c} - \tan^{-1}\frac{a}{c}\right]$$
6.	let $X = 1 + s/D$. Then: $$F_{1\text{-}2} = F_{2\text{-}1} = \frac{1}{\pi}\left[\sqrt{X^2 - 1} + \sin^{-1}\frac{1}{X} - X\right]$$
7.	$$F_{1\text{-}2} = 1, \quad F_{2\text{-}1} = \frac{r_1}{r_2}, \text{ and}$$ $$F_{2\text{-}2} = 1 - F_{2\text{-}1} = 1 - \frac{r_1}{r_2}$$

Table 11.3   View factors for some three-dimensional configurations

Configuration	Equation
1. 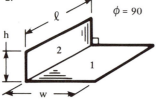	Let $X = a/c$ and $Y = b/c$. Then: $$F_{1\text{-}2} = \frac{2}{\pi XY} \left\{ \ell n \left[\frac{(1 + X^2)(1 + Y^2)}{1 + X^2 + Y^2} \right]^{\frac{1}{2}} \right.$$ $$+ X\sqrt{1 + Y^2}\ \tan^{-1} \frac{X}{\sqrt{1 + Y^2}}$$ $$+ Y\sqrt{1 + X^2}\ \tan^{-1} \frac{Y}{\sqrt{1 + X^2}}$$ $$\left. - X \tan^{-1} X - Y \tan^{-1} Y \right\}$$
2. 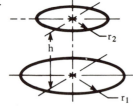 $\phi = 90$	Let $H = b/\ell$ and $W = w/\ell$. Then: $$F_{1\text{-}2} = \frac{1}{\pi W} \left(W \tan^{-1} \frac{1}{W} + H \tan^{-1} \frac{1}{H} - \sqrt{H^2 + W^2}\ \tan^{-1}(H^2 + W^2)^{-\frac{1}{2}} \right.$$ $$\left. + \frac{1}{4} \ell n \left\{ \left[\frac{(1+W^2)(1+H^2)}{1+W^2+H^2} \right] \left[\frac{W^2(1+W^2+H^2)}{(1+W^2)(W^2+H^2)} \right]^{W^2} \left[\frac{H^2(1+H^2+W^2)}{(1+H^2)(H^2+W^2)} \right]^{H^2} \right\} \right)$$
3. 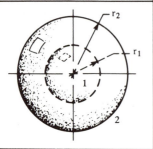	Let $R_1 = r_1/h$, $R_2 = r_2/h$, and $$X = 1 + (1 + R_2{}^2)/R_1{}^2$$ $$F_{1\text{-}2} = \frac{1}{2} \left[X - \sqrt{X^2 - 4(R_2/R_1)^2} \right]$$
4.	Concentric spheres: $$F_{1\text{-}2} = 1; \qquad F_{2\text{-}1} = (r_1/r_2)^2$$ and $F_{2\text{-}2} = 1 - (r_1/r_2)^2$

unusually comprehensive discussion of such calculations and a large catalog of their results. We shall not actually use eqns. (11.17a) and (11.17b) directly but shall instead refer the interested reader to Siegel and Howell for the results of such calculations.

We list some typical results of the calculation in Tables 11.2 and 11.3. Table 11.2 gives calculated values of F_{1-2} for two-dimensional bodies—various configurations of cylinders and strips that approach being infinite in length. Table 11.3 gives F_{1-2} for some three-dimensional configurations.

Many of these and other results have been evaluated numerically and presented in graphical form for easy reference. Figure 11.10, for example, includes the solutions for configurations (1), (2), and (3) from Table 11.3. The reader should study these results and be sure that the tendencies that they show make sense: Is it clear, for example, that $F_{1-2} \rightarrow$ constant, which is <1 in each case, as the abscissa becomes large? Can the right-hand element of Fig. 11.7 be located in Fig. 11.10? And so forth?

Figure 11.11 shows view factors for another kind of configuration—one in which one area is very small in comparison with the other one. Many such solutions exist because they are somewhat less difficult to calculate, and they can often be very useful in practice.

Example 11.3

A heater (h) as shown in Fig. 11.12 radiates to the partially conical shield (s) that surrounds it. If the heater and shield are black, calculate the net heat transfer from the heater to the shield.

SOLUTION. First imagine a plane (i) laid across the open top of the shield.

$$F_{h-s} + F_{h-i} = 1$$

But F_{h-i} can be obtained from Fig. 11.10 or the equation in case 3, for $R_1 = r_1/h = 5/20 = 0.25$ and $R_2 = r_2/h = 10/20 = 0.5$. The result is $F_{h-i} = 0.192$. Then

$$F_{h-s} = 1 - 0.192 = 0.808$$

Then

$$Q_{net} = F_{h-s} A_h \sigma (T_h^4 - T_s^4)$$

$$= 0.808 \left(\frac{\pi}{4} 0.1^2 \right) 5.67 \times 10^{-8} [(1200 + 273)^4 - 373^4] = \underline{1687 \text{ W}}$$

Example 11.4

Suppose that the shield were heating the region where the heater is presently located. What would F_{s-h} be?

SOLUTION. From eqn. (11.18) we have

$$A_s F_{s-h} = A_h F_{h-s}$$

But the frustrum-shaped shield has an area of

$$A_s = \frac{\pi}{2} (0.2 + 0.1) \sqrt{0.05^2 + 0.20^2} = 0.0971 \text{ m}^2$$

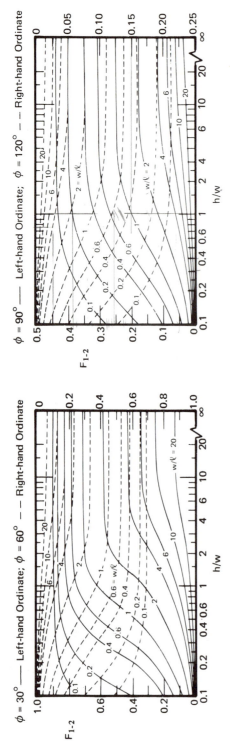

$\phi = 30°$ —— Left-hand Ordinate; $\phi = 60°$ – – – Right-hand Ordinate

$\phi = 90°$ —— Left-hand Ordinate; $\phi = 120°$ – – – Right-hand Ordinate

Intersecting rectangles, case 2., Table 11.3

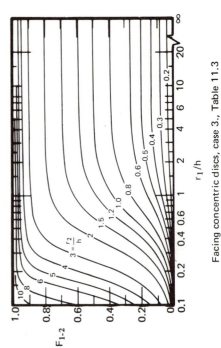

Facing concentric discs, case 3., Table 11.3

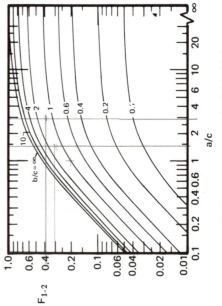

Facing rectangles, case 1., Table 11.3

FIGURE 11.10 The view factors for configurations shown in Table 11.3.

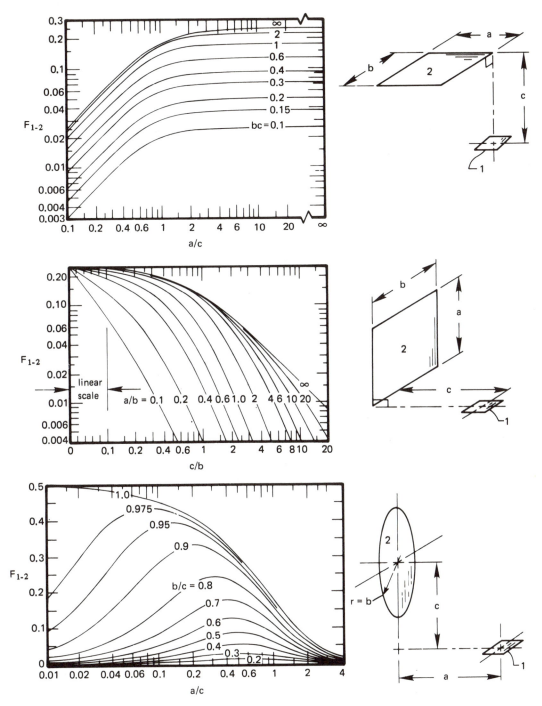

FIGURE 11.11 The view factor for three very small surfaces "looking at" three large surfaces. $(A_1 \ll A_2)$

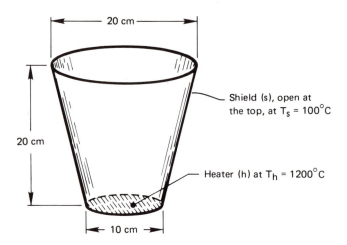

20 cm

20 cm

Shield (s), open at
the top, at $T_s = 100^\circ$C

Heater (h) at $T_h = 1200^\circ$C

10 cm

FIGURE 11.12   Heat transfer from a disc heater to its radiation shield.

and

$$A_h = \frac{\pi}{4}(0.1)^2 = 0.007854 \text{ m}^2$$

so

$$F_{s-h} = \frac{0.007854}{0.0971}(0.808) = \underline{0.0654}$$

Example 11.5

Verify F_{1-2} for case 4 in Table 11.2.

SOLUTION.   Multiply $F_{1-2} + F_{1-3} = 1$ by A_1:

$$A_1 F_{1-2} + A_1 F_{1-3} = A_1$$

Likewise:

$$A_2 F_{2-i} + A_2 F_{2-3} = A_2$$

and

$$A_3 F_{3-1} + A_3 F_{3-2} = A_3$$

Then use $A_2 F_{2-1} = A_1 F_{1-2}$, $A_3 F_{3-1} = A_1 F_{1-3}$, and $A_3 F_{3-2} = A_2 F_{2-3}$ to get

$$A_1 F_{1-2} + A_1 F_{1-3} = A_1$$
$$A_1 F_{1-2} + A_2 F_{2-3} = A_2$$
$$A_1 F_{1-3} + A_2 F_{2-3} = A_3$$

This is easily solved for F_{1-2}:

$$F_{1-2} = \frac{A_1 + A_2 - A_3}{2A_1}$$

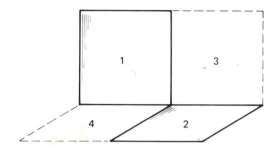

FIGURE 11.13   Radiation between two offset perpendicular squares.

Example 11.6

Find F_{1-2} for the configuration of two offset squares of area A, shown in Fig. 11.13.

SOLUTION.

$$2AF_{(1+3)-(4+2)} = AF_{1-4} + AF_{1-2} + AF_{3-4} + AF_{3-2}$$

$$2F_{(1+3)-(4+2)} = 2F_{1-4} + 2F_{1-2}$$

And $F_{(1+3)-(4+2)}$ can be read from Fig. 11.10 (at $\Phi = 90$, $w/l = \frac{1}{2}$, and $h/w = 1$) as 0.29 and $F_{1-4} = 0.20$. Thus

$$F_{1-2} = (0.29 - 0.20) = \underline{0.09}$$

11.4 Evaluation of $\mathcal{F}_{1-2}$

Electrical analogy for gray body heat exchange

There exists a rather clever adaptation of the electric analogy for calculating heat exchange among black bodies. It was developed by Oppenheim [11.1] in 1956, and it requires the definition of two new quantities. These are:

$$H \frac{W}{m^2} \equiv irradiance = \text{flux of energy that irradiates the surface}$$

and

$$B \frac{W}{m^2} \equiv radiosity = \text{total flux of radiative energy away from the surface}$$

The radiosity can be expressed as the sum of the irradiated energy that reflects away from the surface and the radiation emitted from it. Thus

$$B = \rho H + \varepsilon e_b \tag{11.19}$$

We can immediately write the net heat flux from the surface as the difference between B and H. Then with the help of eqn. (11.19), we get

$$q_{net} = B - H = B - \frac{B - \varepsilon e_b}{\rho} \tag{11.20}$$

This can be rearranged as

$$q_{\text{net}} = \frac{\varepsilon}{\rho} e_{\text{b}} - \frac{1-\rho}{\rho} B \qquad (11.21)$$

As long as the surface is opaque ($\tau = 0$), $\varepsilon = \alpha = 1 - \rho$, eqn. (11.21) gives

$$q_{\text{net}} A = Q_{\text{net}} = \frac{e_{\text{b}} - B}{\rho / \varepsilon A} = \frac{e_{\text{b}} - B}{(1-\varepsilon)/\varepsilon A} \qquad (11.22)$$

Equation (11.22) is a form of Ohm's law which tells us that $(e_{\text{b}} - B)$ can be viewed as a driving potential for transferring heat away from a surface through an effective resistance, $(1 - \varepsilon)/\varepsilon A$. Now consider heat transfer from one infinite gray plane to another parallel with it, in these terms. Radiant energy flows past an imaginary surface, parallel with the first infinite plate in Fig. 11.14 but quite close to it. There is no way of telling whether this energy comes from a real surface or a black body. Therefore, we can isolate radiation at the surface, and treat radiation from just above the surface as though it were from a black body. Thus

$$Q_{\text{net}} = A_1 F_{1-2}(B_1 - B_2) = \frac{B_1 - B_2}{\dfrac{1}{A_1 F_{1-2}}} \qquad (11.23)$$

which is again a form of Ohm's law. The radiosity difference $(B_1 - B_2)$ can be imagined to drive heat through the resistance $1/A_1 F_{1-2}$. An electric circuit, shown in Fig. 11.14, expresses the analogy and gives us means for calculating Q_{net} in accordance with Ohm's law. Recalling that $e_{\text{b}} = \sigma T^4$, we obtain

$$Q_{\text{net}} = \frac{e_{b_1} - e_{b_2}}{\Sigma \text{ resistances}} = \frac{1}{\left(\dfrac{1-\varepsilon}{\varepsilon A}\right)_1 + \dfrac{1}{A_1 F_{1-2}} + \left(\dfrac{1-\varepsilon}{\varepsilon A}\right)_2} \sigma\left(T_1^4 - T_2^4\right) \quad (11.24)$$

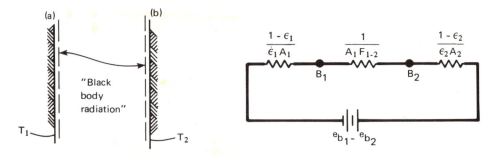

FIGURE 11.14   The electrical analogy for radiation between two gray infinite plates.

or, if we remember that $F_{1-2}=1$ and $A_1=A_2$ for infinite parallel plates,

$$q_{1-2}= \underbrace{\frac{1}{\dfrac{1}{\varepsilon_1}+\dfrac{1}{\varepsilon_2}-1}}_{\mathcal{F}_{1-2}} \sigma(T_1^4-T_2^4) \qquad (11.11)$$

This result is one that we arrived at during the derivation of Kirchhoff's law. But the method we have used to develop it here can quickly be extended to develop other results as well.

Example 11.7

Derive $\mathcal{F}_{1-2}$ for one gray body enclosed by another as shown in Fig. 11.6.

SOLUTION. The electric circuit analog is exactly the same as that shown in Fig. 11.14, and F_{1-2} is still unity. Therefore,

$$Q_{net}=q_{net}A_1= \frac{1}{\dfrac{1-\varepsilon_1}{\varepsilon_1 A_1}+\dfrac{1}{A_1}+\dfrac{1-\varepsilon_2}{\varepsilon_2 A_2}} \sigma(T_1^4-T_2^4) \qquad (11.25)$$

Therefore,

$$\mathcal{F}_{1-2}= \frac{1}{\dfrac{1}{\varepsilon_1}+\dfrac{A_1}{A_2}\left(\dfrac{1}{\varepsilon_2}-1\right)}$$

which is the result we anticipated in eqn. (11.14) for diffusely reflecting bodies.

Example 11.8

Derive $\mathcal{F}_{2-1}$ for the enclosed bodies shown in Fig. 11.6.

SOLUTION. By the same rationale as we used in Example 11.7, but replacing the center resistance with $1/A_2 F_{2-1}$, we get

$$\mathcal{F}_{2-1}= \frac{1}{\dfrac{1}{\varepsilon_2}+\dfrac{A_2}{A_1}\left(\dfrac{1}{\varepsilon_1}-1\right)+\left(\dfrac{1}{F_{2-1}}-1\right)}$$

To eliminate the unknown view factor, F_{2-1}, from this result we use the reciprocity relation, eqn. (11.18):

$$F_{2-1}=\frac{A_1}{A_2}\underbrace{F_{1-2}}_{=1}=\frac{A_1}{A_2}$$

so

$$\mathcal{F}_{2-1}= \frac{1}{\dfrac{A_2}{A_1}\dfrac{1}{\varepsilon_1}+\left(\dfrac{1}{\varepsilon_2}-1\right)} \qquad (11.26)$$

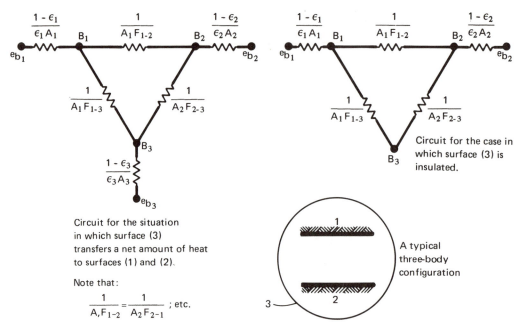

FIGURE 11.15   The electrical analogy for radiation among three gray surfaces.

Use of the electrical analogy when more than one gray body is involved in heat exchange

Let us first consider a three-body transaction as pictured in Fig. 11.15. Body (3) might either be insulated or it might exchange a net amount of heat with bodies (1) and (2). But in either case it absorbs and reemits energy. If it is insulated, there is no net surface heat transfer and we can eliminate one leg of the circuit, as shown in Fig. 11.15.

The circuit for such an exchange is not as easy to analyze as the in-line circuits that we have previously analyzed. In this case one must sum the energy exchanges at each of the interior nodes:

node B_1:
$$\frac{e_{b_1} - B_1}{\dfrac{1 - \epsilon_1}{\epsilon_1 A_1}} = \frac{B_1 - B_2}{\dfrac{1}{A_1 F_{1-2}}} + \frac{B_1 - B_3}{\dfrac{1}{A_1 F_{1-3}}} \qquad (11.27)$$

node B_2:
$$\frac{e_{b_2} - B_2}{\dfrac{1 - \epsilon_2}{\epsilon_2 A_2}} = \frac{B_2 - B_1}{\dfrac{1}{A_1 F_{1-2}}} + \frac{B_2 - B_3}{\dfrac{1}{A_2 F_{2-3}}} \qquad (11.28)$$

node B_3:
$$\left[\frac{e_{b_3} - B_3}{\dfrac{1 - \epsilon_3}{\epsilon_3 A_3}} \quad \text{or} \quad 0 \right] = \frac{B_3 - B_1}{\dfrac{1}{A_1 F_{1-3}}} + \frac{B_3 - B_2}{\dfrac{1}{A_2 F_{2-3}}} \qquad (11.29)$$

These equations must be solved simultaneously for the three unknowns, B_1, B_2, and B_3. When they are solved, one can compute the net heat transfer to or from any body (a), as a result of all surrounding bodies, (i), as

$$Q_{net} = \sum_i \frac{B_i - B_a}{\dfrac{1}{A_a F_{a-i}}} \qquad (11.30)$$

Notice that when the third body is insulated, the right-hand circuit in Fig. 11.15 can be treated as a series–parallel circuit, since all heat flows from (1) to (2). In this case

$$Q_{net} = \frac{e_{b_1} - e_{b_2}}{\dfrac{1 - \varepsilon_1}{\varepsilon_1 A_1} + \cfrac{1}{\cfrac{1}{\cfrac{1}{A_1 F_{1-3}} + \cfrac{1}{A_2 F_{2-3}}} + \cfrac{1}{\cfrac{1}{A_1 F_{1-2}}}} + \dfrac{1 - \varepsilon_2}{\varepsilon_2 A_2}} \qquad (11.31)$$

However when the third body participates in the transaction, the three equations (11.27), (11.28), and (11.29) have to be used and this simplification cannot be used.

Example 11.9

Two very long strips 1 m wide and 2.33 m apart face each other as shown in Fig. 11.16. (a) Find Q_{net} W/m from one to the other if the surroundings are cold and black. (b) Find Q_{net} W/m if they are connected by an insulated diffuse reflector between the edges on both sides. (c) Evaluate the temperature of the reflector in part (b). Assume $(T_{surroundings})^4 \ll T_1^4$ or T_2^4.

SOLUTION. From Fig. 11.10 we read $F_{1-2} = 0.2 = F_{2-1}$. Then the three nodal equations [(11.27), (11.28), and (11.29)] become

$$\frac{1451 - B_1}{2.333} = \frac{B_1 - B_2}{5} + \frac{B_1 - B_3}{\dfrac{1}{1 - 0.2}}$$

$$\frac{459.3 - B_2}{1} = \frac{B_2 - B_1}{5} + \frac{B_2 - B_3}{\dfrac{1}{1 - 0.2}}$$

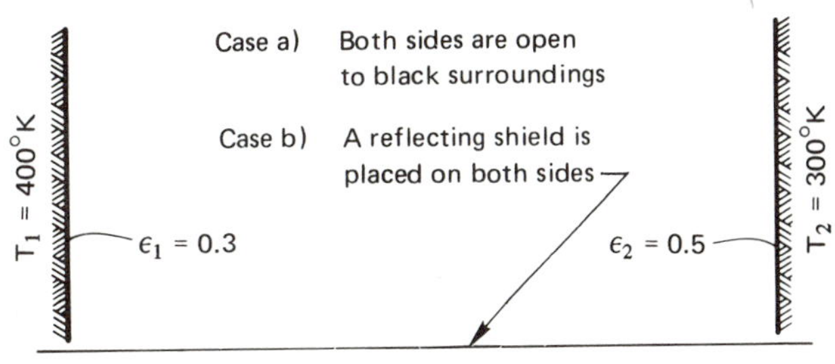

Case a) Both sides are open to black surroundings

Case b) A reflecting shield is placed on both sides

$T_1 = 400°K$ $\varepsilon_1 = 0.3$ $\varepsilon_2 = 0.5$ $T_2 = 300°K$

FIGURE 11.16   Example 11.9.

$$0 = \frac{B_3 - B_1}{\frac{1}{1-0.2}} + \frac{B_3 - B_2}{\frac{1}{1-0.2}}$$

or

$$B_1 - 0.14B_2 - 0.56B_3 = 435$$
$$-B_1 + 10B_2 - 4B_3 = 2296.5$$
$$-B_1 - B_2 + 2B_3 = 0$$

Without the reflecting shield, we neglect B_3 since the surroundings are very cold. Then the first two equations reduce to

$$B_1 - 0.14B_2 = 435 \qquad\qquad b_1 = 473.78 \text{ W/m}$$
$$\text{so}$$
$$-B_1 + 10B_2 = 2296.5 \qquad\qquad B_2 = 277.03 \text{ W/m}$$

Thus the net flow from 1 to 2 is quite small:

$$Q_{1-2_{\text{no shield}}} = \frac{B_1 - B_2}{\frac{1}{A_1 F_{1-2}}} = \underline{39.35 \text{ W/m}}$$

When the shield is in place we must solve the full set of nodal equations. This can be done manually, by the use of determinants, or with matrix algebra methods that have been packaged as computer subroutines (see, e.g., Example 6.10). The result is

$$B_1 = 987.7 \qquad B_2 = 657.4 \qquad B_3 = 822.6$$

Then from eqn. (11.30) we get

$$Q_{\text{net}} = 1 \frac{\text{m}^2}{\text{m}} \left(\frac{987.7 - 657.4}{\frac{1}{0.2}} + \frac{822.6 - 657.4}{\frac{1}{0.8}} \right) \frac{\text{W}}{\text{m}^2} = \underline{198.5 \text{ W/m}}$$

Notice that, because node 3 is insulated, we could also have used eqn. (11.31) to get Q_{net}. The result would be

$$Q_{\text{net}} = \frac{5.67 \times 10^{-8}(400^3 - 300^3)}{\dfrac{0.7}{0.3} + \dfrac{1}{\dfrac{1}{\dfrac{1}{0.2} + \dfrac{1}{0.8}} + \dfrac{1}{0.2}} + \dfrac{0.5}{0.5}} = \underline{198.5 \text{ W/m}}$$

The result, of course, is the same. We note that the presence of the reflector increases the net heat flow from (1) to (2).

(c) The temperature of the reflector (3) is obtained from

$$Q_{3 \text{ to (1 or 2)}} = 0 = e_{b_3} - B_3 = 5.67(10)^{-8} T_3^4 - 822.6$$

so

$$T_3 = \underline{347.06°\text{K}}$$

Holman [1.3] gives a very nice discussion of the application of the electric analog to more complicated problems, and he provides a number of useful examples. However, the digital computer now makes it more feasible to approach complicated problems directly with numerical methods. Sparrow and Cess [1.20] provide an excellent discussion of these methods. While they generally lie beyond the scope of this text, it is instructive to treat one important class of such solutions in the following section.

Algebraic solution of compound radiation problems

Radiant heat exchange in gray, diffuse enclosures. An enclosure can consist of any number of surfaces participating in radiant energy exchange. For example, the case shown in Fig. 11.16 could have been treated as a rectangular enclosure if, in addition to the two walls and the shield, we had assumed a fictitious surface of $0°K$ to make up the fourth side.

An enclosure formed by n surfaces is shown in Fig. 11.17. We assume that:

- Each surface emits or reflects diffusely and it is gray and opaque ($\varepsilon = \alpha$, $\rho = 1 - \varepsilon$).
- Each surface is either at a uniform temperature, or its heat flux is a uniform known value, and its emittance is known.
- The view factor, F_{i-j}, between any two surfaces i and j is known.
- Conduction and convection can be neglected.

We are interested in determining the heat fluxes at the surfaces where temperatures are specified, and vice versa.

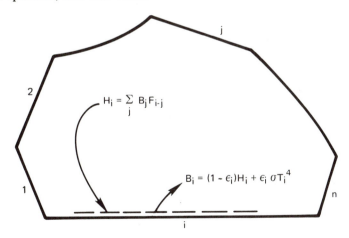

FIGURE 11.17   An enclosure surrounded by gray and diffuse, isothermal and constant-heat-flux segments.

The rate of heat loss from the ith surface of the enclosure can conveniently be written in terms of the radiosity, B_i, and the incident surface heat flux, H_i [see eqn. (11.22)].

$$q_i = B_i - H_i = \frac{\varepsilon_i}{1 - \varepsilon_i}\left(\sigma T_i^4 - B_i\right) \tag{11.32}$$

where

$$B_i = \rho_i H_i + \varepsilon_i e_{b_i} = (i - \varepsilon_i)H_i + \varepsilon_i \sigma T_i^4 \tag{11.33}$$

However, $A_i H_i$, the incident radiant heat transfer rate to the surface, i, is the sum of energies reaching i from all other surfaces, including itself:

$$A_i H_i = \sum_{j=1}^{n} A_j B_j F_{j-i} = \sum_{j=1}^{n} B_j A_i F_{i-j}$$

where we have used the reciprocity rule, $A_j F_{j-i} = A_i F_{i-j}$. Then

$$H_i = \sum_{j=1}^{n} B_j F_{i-j} \tag{11.34}$$

It follows from eqns. (11.33) and (11.34) that

$$B_i = (1 - \varepsilon_i)\sum_{j=1}^{n} B_j F_{i-j} + \varepsilon_i \sigma T_i^4 \tag{11.35}$$

When all the surface temperatures are specified, eqn. (11.35) can be written for each surface. This yields n algebraic equations that can be solved for the n unknown B's. The rate of heat loss, Q_i, from the ith surface ($i = 1, 2, \ldots, n$) can then be obtained from eqn. (11.32).

For those surfaces where heat fluxes are prescribed, we can eliminate the $\varepsilon_i \sigma T_i^4$ term in eqn. (11.35) using eqn. (11.32). The B's can still be calculated as before, and eqn. (11.32) can be solved for the unknown temperature of that particular surface.

Example 11.10

Two sides of a long triangular duct as shown in Fig. 11.18 are made of stainless steel ($\varepsilon = 0.5$) and are maintained at 500°C. The third side is of copper ($\varepsilon = 0.15$) and is at a uniform temperature of 100°C. Calculate the rate of heat transferred to the copper base per meter of length of the duct.

SOLUTION. Assume the duct walls to be gray and diffuse, the fluid in the duct to be radiatively inactive, and convection to be negligible. The view factors can be calculated from configuration 4 of Table 11.2 or Example 11.5:

$$F_{1-2} = \frac{A_1 + A_2 - A_3}{2A_1} = \frac{5 + 3 - 4}{10} = 0.4$$

Similarly, $F_{2-1} = 0.67$, $F_{1-3} = 0.6$, $F_{3-1} = 0.75$, $F_{2-3} = 0.33$, and $F_{3-2} = 0.25$. The surfaces

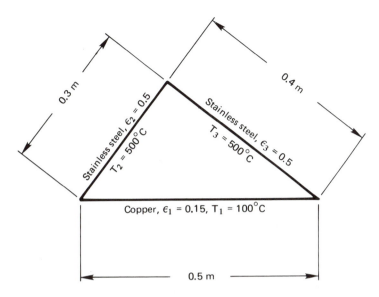

FIGURE 11.18   Example 11.10.

cannot "see" themselves, so $F_{1-1}=F_{2-2}=F_{3-3}=0$. We therefore use eqn. (11.35) to write the three algebraic equations for the three unknowns B_1, B_2, and B_3.

$$B_1 = \underbrace{(1-\varepsilon_1)}_{0.85}\left(\underbrace{F_{1-1}B_1}_{0} + \underbrace{F_{1-2}B_2}_{0.4} + \underbrace{F_{1-3}B_3}_{0.6}\right) + \underbrace{\varepsilon_1 \sigma T_1^4}_{0.15}$$

$$B_2 = \underbrace{(1-\varepsilon_2)}_{0.5}\left(\underbrace{F_{2-1}B_1}_{0.67} + \underbrace{F_{2-2}B_2}_{0} + \underbrace{F_{2-3}B_3}_{0.33}\right) + \underbrace{\varepsilon_2 \sigma T_2^4}_{0.5}$$

$$B_3 = \underbrace{(1-\varepsilon_3)}_{0.5}\left(\underbrace{F_{3-1}B_1}_{0.75} + \underbrace{F_{3-2}B_2}_{0.25} + \underbrace{F_{3-3}B_3}_{0}\right) + \underbrace{\varepsilon_3 \sigma T_3^4}_{0.5}$$

If there were more surfaces, it would be easy to solve this system numerically using matrix methods. In this case we can obtain B_1 algebraically:

$$B_1 = 0.232\sigma T_1^4 - 0.319\sigma T_2^4 + 0.447\sigma T_3^4$$

Equation (11.32) gives the rate of heat lost by surface 1 as

$$Q_1 = A_1\frac{\varepsilon_1}{1-\varepsilon_1}\left(\sigma T_1^4 - B_1\right)$$

$$= A_1\frac{\varepsilon_1}{1-\varepsilon_1}\sigma\left(T_1^4 - 0.232T_1^4 + 0.319T_2^4 - 0.447T_3^4\right)$$

$$= (0.5)\left(\frac{0.15}{0.85}\right)(5.67\times10^{-8})\left[(373)^4 - 0.232(373)^4 + 0.319(773)^4 - 0.447(773)^4\right] \text{ W/m}$$

$$= \underline{-154.3 \text{ W/m}}$$

The negative sign indicates that the copper base is gaining heat.

11.5 Gaseous radiation

Absorptance, transmittance, and emittance

Every radiation problem that we have considered thus far has been treated as though heat flow in the space separating the surfaces of interest were completely unobstructed. However, all gases and liquids impede the radiation of heat through them to some extent. We have ignored this effect in air because it is generally quite minor. We now turn our attention briefly to those problems in which we must consider the role of gases (or liquids, for that matter) as participants in the heat exchange process.

The photons of radiant energy passing through a gaseous region can be impeded in two ways. Some can be "scattered," or deflected in various directions, as they pass through, and some can be absorbed into the molecules. Scattering is a fairly minor influence in most gases unless they contain foreign particles, such as dust or fog. In cloudless air, for example, we are only aware of the scattering of sunlight when it passes through many miles of the atmosphere. Then the short-wavelength light (which, as it happens, is far more susceptible to scattering) is scattered, giving the sky its blue hues.

At sunset, sunlight passes through the atmosphere at a shallow angle for hundreds of miles. Radiation in the blue wavelengths has all been scattered out before it can be seen. Thus we see only the unscattered red hues, just before dark.

Radiant energy can be absorbed by molecules only if the appropriate quantum mechanical conditions prevail. For practical purposes we can assert that monatomic, and symmetrical diatomic, molecules are transparent to thermal radiation. Thus the major components of air—N_2 and O_2—are nonabsorbing; so too are H_2 and such monatomic gases as argon. Two particularly important absorbing molecules are CO_2 and H_2O, which are usually present in air. Other absorbing gases include ammonia, ozone, CO, and SO_2.

Figure 11.19 shows radiant energy flowing through an absorbing gas, with a monochromatic intensity, i_λ. As it passes through an element of thickness dx, the intensity will be reduced by an amount di_λ:

$$di_\lambda = -\kappa_\lambda i_\lambda \, dx \qquad (11.36)$$

where κ_λ is called the *monochromatic absorption coefficient*. If the gas scatters radiation, we replace κ_λ with γ_λ, the *monochromatic scattering coefficient*. If it both absorbs and scatters radiation, we replace κ_λ with $\beta_\lambda \equiv \kappa_\lambda + \gamma_\lambda$, the *monochromatic extinction coefficient*.[3] The dimensions of κ_λ, β_λ, and γ_λ are all m^{-1}.

---

[3]All three coefficients, κ_λ, γ_λ, and β_λ, are expressed on a volumetric basis. They could, alternatively, have been expressed on a mass basis.

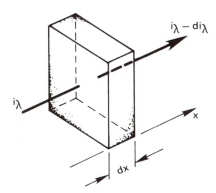

FIGURE 11.19 The attenuation of radiation through an absorbing (and/or scattering) gas.

Equation (11.36) can be integrated between $i_\lambda(x=0)=i_{\lambda_0}$ and $i_\lambda(x)=i_\lambda$. The result is

$$\frac{i_\lambda}{i_{\lambda_0}} = e^{-\kappa_\lambda x} \tag{11.37}$$

This result is called Beer's (pronounced Bayr's or Baer's) law. The ratio

$$\frac{i_\lambda}{i_{\lambda_0}} \equiv \text{monochromatic transmittance, } \tau_\lambda, \text{ of the gas}$$

as we saw in Chapter 1. Since gases do not normally reflect radiant energy, $\tau_\lambda + \alpha_\lambda = 1$. Thus eqn. (11.37) gives the monochromatic absorptance, α_λ, as

$$\alpha_\lambda = 1 - e^{-\kappa_\lambda x} \tag{11.38}$$

The dependence of α_λ on λ is normally very strong. It arises from the fact that, in certain narrow bands of wavelength, radiation will interact with certain molecules and be absorbed, while radiation with somewhat higher or lower wavelengths might pass almost unhindered. Figure 11.20 shows the absorptance of steam as a function of wavelength for a vapor layer of a particular depth.

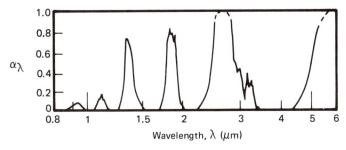

FIGURE 11.20 The monochromatic absorptance of a 1.09 m thick layer of steam at 127°C.

A comparison of Fig. 11.20 with Fig. 11.2 readily shows why the apparent emittance of the sun, as viewed from the earth's surface, shows a number of spiked indentations at certain wavelengths. Several of these indentations occur at those wavelengths at which water vapor in the air absorbs the incoming radiation of the sun in accordance with Fig. 11.20. The other indentations in Fig. 11.2 occur where ozone and CO_2 absorb radiation. The sun does not exhibit these regions of low emittance. It is just that much of the radiation in certain wavelength ranges is blocked from our view and trapped in the upper atmosphere.

Just as α and ε are equal to one another for a given surface, under certain restrictions, the monochromatic absorption coefficient, κ_λ, and the monochromatic emittance of a gas, ε_{g_λ}, are also related. However, while ε_{g_λ} is dimensionless, κ_λ has the dimensions of inverse length. To better see why that should be, consider Fig. 11.21. Figure 11.21a shows a slab of thickness, l, in which molecules at various depths are emitting energy. If the gas is isothermal and at steady state, the emittance will be balanced uniformly by absorption. Thus if we consider a sufficiently thin slice, as shown in Fig. 11.21b, it will be accurate to conceive all of the absorbing and emitting molecules being located in its center, as shown in Fig. 11.21c. Then an energy balance gives, for $l \to 0$:

$$q_{in} = 2\left[\kappa_\lambda \left(\frac{1}{m} \right) \right] \left(\frac{l}{2} \frac{m^3}{m^2} \right) \left(e_b \frac{W}{m^2} \right) = 2\varepsilon_{g_\lambda} e_b = q_{out}$$

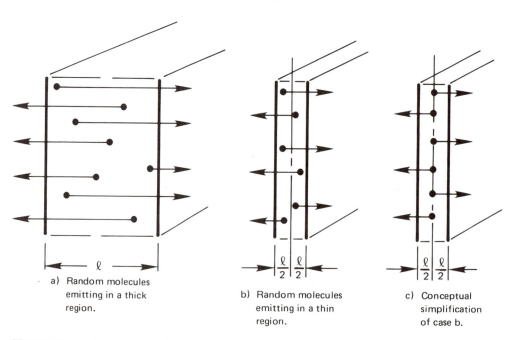

a) Random molecules emitting in a thick region.

b) Random molecules emitting in a thin region.

c) Conceptual simplification of case b.

FIGURE 11.21 One dimensional emission of radiant energy from within a gas.

or

$$\kappa_\lambda = \lim_{l \to 0} \frac{\varepsilon_{g_\lambda}}{l/2} \qquad (11.39)$$

For a gas that is kept at a temperature different from the surroundings to or from which it radiates, Hottel and Sarofim [1.19] quote:

$$\alpha_g = \left(\frac{T_{\text{gas}}}{T_{\text{surroundings}}} \right)^{0.65} \varepsilon_g \qquad (11.40)$$

where ε_g is the total emittance, defined as

$$\varepsilon_g \equiv \frac{\int_0^\infty \varepsilon_{g_\lambda} e_{g_\lambda} d\lambda}{e_b} \qquad (11.41)$$

Notice that for a thin slice of gas of thickness, $l/2$, in equilibrium with its surroundings, $\alpha_g = \varepsilon_g$, and eqn. (11.38) gives

$$\alpha_g = 1 - e^{-\kappa(l/2)} \simeq 1 - 1 + l\kappa/2 = \varepsilon_g$$

which is consistent with eqn. (11.39).

It is therefore clear that ε_g for an emitting gas depends on the thickness of the emitting layer. Notice too that ε_g also increases if the molecules are packed more closely by virtue of an increase in pressure. Thus ε_g is a fairly complicated function of temperature, pressure, size, and configuration of a gaseous region.

Hottel and Sarofim provide empirical correlations of ε_g, using a single parameter, $L_e \equiv$ *mean beam length*, to represent both the size and configuration of gaseous region. The mean beam length is defined as

$$L_e \equiv \frac{4(\text{volume of gas})}{\text{boundary area that is irradiated}} \qquad (11.42)$$

Thus for two infinite parallel plates a distance l apart, $L_e = 4Al/2A = 2l$. Some other values of L_e for volumes radiating to all points on their boundaries (unless otherwise noted) are

- For a sphere of diameter, D $L_e = 2D/3$
- For an infinite cylinder of diameter, D $L_e = D$
- Cube of side, L, radiating to one face $L_e = 2L/3$
- A cylinder with height $= D$ $L_e = 2D/3$

We then provide empirical correlations of the form

$$\varepsilon_g = f_1 [\text{total pressure, } (L)(\text{partial pressure of}$$
$$\text{absorbing component})] \cdot f_2 [T, (L)(\text{partial pressure})] \qquad (11.43)$$

where the experimental functions f_1 and f_2 are plotted in Figs. 11.22 and 11.23 for CO_2 and H_2O, respectively.

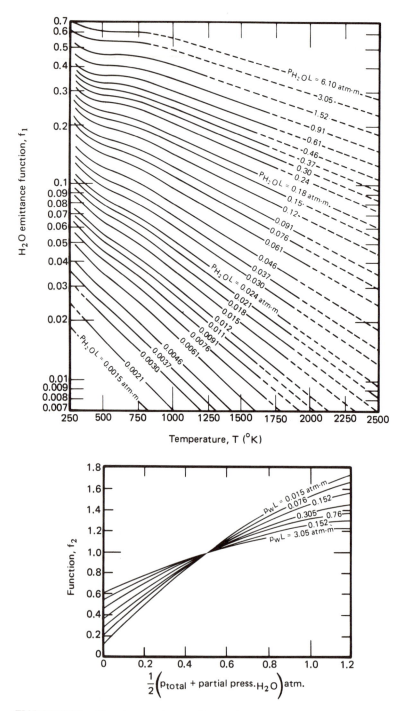

FIGURE 11.22 Functions used to predict $\varepsilon_g = f_1 f_2$, for water vapor in air.

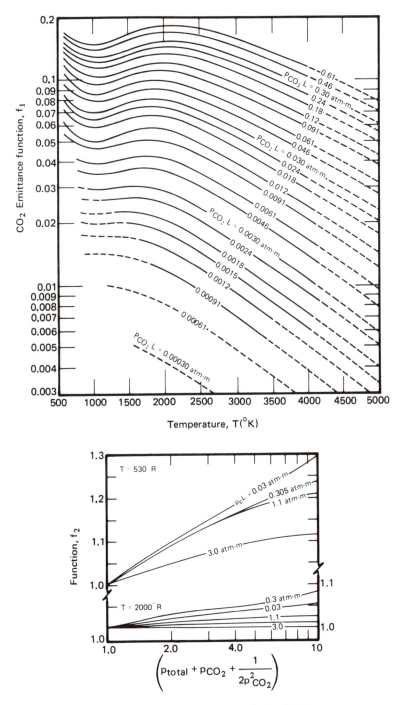

FIGURE 11.23 Functions used to predict $\varepsilon_g = f_1 f_2$, for CO_2 in air.

Consider the problem of a hot gas—say the products of combustion—in a black container. We are now in position to calculate the net heat flow from the gas to the container in such circumstances.

Example 11.11

A long cylindrical combustor 40 cm in diameter contains a gas at 1200°C consisting of 80% N_2 and 20% CO_2 at 1 atm. How much heat must we remove from the walls to keep them at 300°C?

SOLUTION.   First calculate $q_{gas \text{ to wall}}$. To do this we note that $L_e = D = 0.4$ m and $p_{CO_2} = 0.2$ atm. Then Fig. 11.23a gives f_1 as 0.098 and Fig. 11.23b gives f_2 as 1, so $\varepsilon_g = 0.098$. The view factor is unity, so

$$q_{gas \text{ to wall}} = \sigma F_{g-w} \varepsilon_g T_g^4 = 5.67 \times 10^{-8}(0.098)(1200+273)^4 = 26{,}160 \text{ W}/\text{m}^2$$

Next we need α_g to calculate $q_{wall \text{ to gas}}$. Using eqn. (11.40), we get

$$\alpha_g = \left(\frac{1200+273}{300+273} \right)^{0.65}(0.098) = 0.181$$

so, since the wall "sees" itself *through* gas with this absorptance, we use $F_{w-g} = 1$ and obtain

$$q_{wall \text{ to gas}} = \sigma F_{w-g} \alpha_g T_w^4 = 5.67(10)^{-8}(0.181)(573)^4 = 1106 \text{ W}/\text{m}^2$$

Thus

$$q_{net} = \underline{25{,}054 \text{ W}/\text{m}^2}$$

or

$$Q_{net/m \text{ length}} = \pi(0.4)(25{,}054) = \underline{31{,}483 \text{ W}/\text{m}}$$

The problem of heat transfer among gases and gray bodies will be left as being beyond the scope of this course. Sparrow and Cess [1.20] provide a more advanced treatment of analytical methods for treating the problem. Holman's undergraduate text [1.3] shows how to apply the electric analogy to problems of gaseous radiation.

Finally, it is worth noting that gaseous radiation is frequently less important than one might imagine. Consider for example, two flames: a bright orange candle flame and a "cold-blue" hydrogen flame. Both have a great deal of water vapor in them, as a result of oxidizing H_2. But the candle will warm your hands if you place them near it and the hydrogen flame will not. Yet the temperature in the hydrogen flame is *higher*. It turns out that what is radiating both heat and light from the candle are small solid particles of almost thermally black carbon. The CO_2 and H_2O in the flame actually contribute relatively little to radiation.

11.6 Solar energy

The sun as an energy source

The sun bestows energy on the earth at a rate[4] just over $1.7(10)^{14}$ kW. We absorb most of it by day, and that which is absorbed is radiated away by night. If the world population is 4 billion people, each of us has a renewable energy birthright of about 43,000 kW. Of course, we can use very little of this. Most of it must go to sustaining those processes that make the earth a fit place to live—to creating weather and to supplying the flora and fauna we live with.

In the United States alone, we consume energy at the rate of about 2×10^9 kW. The interesting thing about this enormous consumption is that almost none of it comes from our renewable energy birthright. Instead, we are burning up the planet to get it. It is interesting to notice that if we price energy at 5 cents/kWh, the average American could steadily buy a little more than 10 kW by investing all earnings in nothing but energy. This is rather close to our per capita rate of energy consumption in this country—a fact that probably reflects the intimate connection between energy and money.

There is little doubt that our short-term needs—during the next century or so—can be met by coal and, perhaps nuclear power, combined with a less wasteful attitude than most of us have been raised with. But our long-term hope for an adequate energy supply probably lies in the sun.[5] Solar energy can be made useful in many different forms; some possibilities include:

- Hydroelectric power. (There is no hope for a drastic increase in this source because much of the available rainfall runoff has already been harnessed.)

- The combustion of renewable organic matter. (Wood has been used in this way for years, and now we recognize at least the possibility of replacing gasoline with methanol.)

- OTEC, or offshore thermal energy conversion. (This involves the potential use of large floating heat engines operating offshore in tropical ocean waters.)

- Direct solar heating.

- Beaming of energy collected in space to the earth's surface by microwave transmission.

- Photovoltaic collection.

- The energy of ocean waves.

- And so on.

[4]This and other numbers are from [11.2].

[5]Nuclear fusion—that process by which we might manage to create mini-suns upon the earth —might also be a hope of the future.

Notice that some of these sources lend themselves to heat production and some lend themselves to work production. Any time we turn thermal energy to electricity or any other form of work, the Second Law of Thermodynamics exacts a severe "tax" on the energy. Usually, we can only recover about one-third of the total thermal energy as work. Electrical heating, for example, is inherently wasteful because we first sacrifice two-thirds of the energy present in the fuel, or from the sun, in producing electricity. Then we degrade the electricity back to heat.

Distribution of the sun's energy

Figure 11.24 shows what becomes of the solar energy that impinges on the earth if we average it over all kinds of weather. Only 47% of it actually reaches the earth's surface. The lower left-hand portion of the figure shows how this energy is, in turn, returned to the atmosphere and to space. That portion absorbed by the earth's atmosphere (from both the sun and the earth's surface) is eventually reradiated to space.

The heat flux from the sun to the outer edge of the atmosphere is 1353 W/m^2 when the sun is at a mean distance from the earth. We have seen that 47% of this, or 636 W/m^2, reaches the earth's surface. The solar radiation that is felt at the earth's surface includes direct radiation that has passed through the atmosphere; diffuse radiation from the sky; and reflected radiation from snow, water, or other features on the surface. There are some interesting problems in connection with these arriving and departing flows of solar energy.

A large fraction of the sun's energy arrives at the earth's surface in the ultraviolet and visible wavelengths. However, it is reradiated from the relatively cool surface of the earth in wavelengths that are generally far longer. We have already noted that α and ε for objects that are subject to solar radiation might differ greatly as a consequence of this.

Another important consequence of the difference between incoming and outgoing radiation wavelengths is called the *greenhouse effect*. We know that a greenhouse admits shortwave energy from the sun. This energy is absorbed and reradiated at a much lower temperature—a temperature at which the major heat radiation is accomplished in wavelengths above 3 or 4 μm. But this, in turn, is the wavelength range above which glass is virtually *opaque*. The heat is therefore trapped inside.

If we look again at Fig. 11.2, we see that our own sky creates a partial greenhouse effect if it is heavily loaded with CO_2, H_2O, and to a lesser extent, ozone. The escape of long-wavelength reradiated energy from the earth's surface will be reduced in the neighborhood of $\lambda = 1.4$, 1.9, and 2.7 μm. But it will be even more strongly impeded at certain higher-wavelength bands not shown in Fig. 11.2. Water, of course, will condense out in rain, or snow, but CO_2 must be removed by photosynthesis, and it can build up without limit.

A major objection to the continued use of fossil fuels, or renewable organic fuels, is that we are loading the atmosphere with CO_2 faster than our flora can

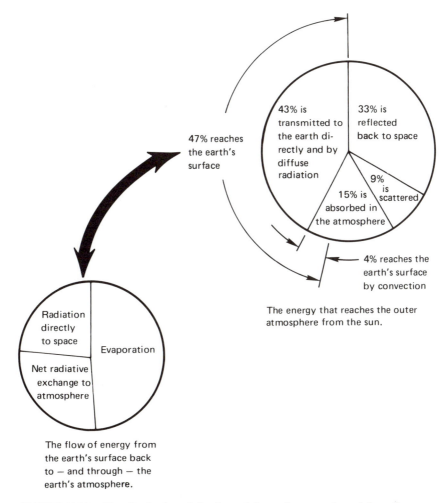

47% reaches
the earth's
surface

43% is
transmitted to
the earth di-
rectly and by
diffuse
radiation

33% is
reflected
back to space

9%
is
scattered

15% is
absorbed in
the atmosphere

4% reaches the
earth's surface
by convection

The energy that reaches the outer
atmosphere from the sun.

Radiation
directly
to space

Evaporation

Net radiative
exchange to
atmosphere

The flow of energy from
the earth's surface back
to — and through — the
earth's atmosphere.

FIGURE 11.24   The distribution of the flow of the sun's energy to and from the earth's surface.

remove it. The long-range effect of this buildup could be a significant rise in the average temperature of the earth's atmosphere, with accompanying climatic changes. These changes are hard to predict accurately but remain potentially dangerous.

The potential for solar power

With so much solar energy falling upon all parts of the world, and with the apparent safety, reliability, and cleanliness of most—but not all—schemes for utilizing solar energy, one might ask why we do not generally use solar power already. The reason is that solar power involves many serious heat transfer and thermodynamics design problems. We shall discuss the problems qualitatively, and refer the reader to [11.3], [11.4], or [11.5] for detailed discussions of the design of solar energy systems.

Solar energy reaches the earth with very low intensity. We began this discussion in Chapter 1 by noting that human beings can only interface with a few hundred watts of energy. We could not live on earth if the sun were not very gentle. It follows that any large solar power source must concentrate the energy that falls on a very large area. By way of illustration, suppose that we sought to convert 636 W/m^2 of solar energy into electric power with a 3% thermal efficiency (which is not pessimistic) during 8 hr of each day. This would correspond with less than 0.6 W/ft^2, on the average, and we would need 50 square miles of collector area to match the steady output of an 800-MW power plant.

Hydroelectric power also requires a large collector area, in the form of the watershed and reservoir behind it. The burning of organic matter requires a large forest to be fed by the sun, and so forth. Any energy supply that is served by the sun must draw from a large area of the earth's surface. This, in turn, means that solar power systems inherently involve very high capital investments, and they introduce their own kinds of environmental complications.

A second problem stems from the intermittent nature of solar devices. To provide steady power—day and night, rain or shine—requires thermal storage systems which are often complex and expensive.

These problems are minimal when one uses solar power merely to heat air or water to low temperatures (50 to 90°C). In this case the efficiency will improve from just a few percent to as high as 70%. Such heating can be used for such industrial processes as crop drying, or it can be used on a small scale for domestic heating of air and/or water.

Figure 11.25 shows a typical configuration of a domestic solar collector of the flat-plate type. Solar radiation passes through one to four glass plates and impinges on a nearly black absorber. The absorber plate might be copper painted with a high-absorptance green paint. The glass plates, of course, are almost transparent in the visible range, and each one admits about 90% of the solar energy that reaches it. Once the energy is absorbed, it is reemitted as long-wavelength infrared radiation. Glass is almost opaque in this range, and energy is retained in the collector by a greenhouse effect.

Water flowing through tubes, which are held in close contact with the absorbing plate, carries the energy away for use. The flow rate is adjusted to give an appropriate temperature rise.

When the working fluid is to be brought to fairly high temperature, it is necessary to focus the direct radiation from the sun from a large area down to a very small region, using reflecting mirrors. Collectors equipped with a small parabolic reflector, focused on a water or air pipe, can raise the fluid to between 100 and 200°C. Any scheme intended to produce electrical power with a conventional thermal cycle needs to focus energy in an area ratio on the order of 1000:1 if it is to achieve a practical efficiency.

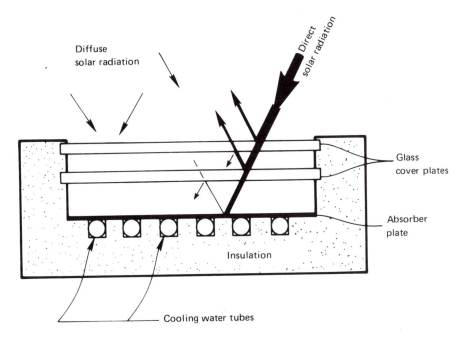

Diffuse
solar radiation

Direct
solar radiation

Glass
cover plates

Absorber
plate

Insulation

Cooling water tubes

FIGURE 11.25   A typical flat-plate solar collector.

PROBLEMS

11.1. What will ε_λ of the sun appear to be to an observer on the earth's surface at $\lambda = 0.2$ μm and at 0.65 μm? How do these emittances compare with the real emittances of the sun? [At 0.2 μm ε is 54.3%.]

11.2. Plot e_{λ_b} against λ for $T = 300°$K and 10,000°K with the help of eqn. (1.29). About what fraction of energy from each black body is visible?

11.3. A 0.6-mm-diameter wire is drawn out through a mandril at 950°C. It's emittance is 0.85. It then passes through a long cylindrical shield of commercial aluminum sheet, 7 cm in diameter. The shield is horizontal in still air at 25°C. What is the temperature of the shield? Is it reasonable to neglect natural convection inside and radiation outside? [$T_{shield} = 153°$C.]

11.4. A 1-ft^2 shallow pan with adiabatic sides is filled to the brim with water at 32°F. It radiates to a night sky whose temperature is 360°R, while a 50°F breeze blows over it at 1.5 ft/s. Will the water freeze or warm up?

11.5. A thermometer is held vertically in a room with air at 10°C and walls at 27°C. What temperature will the thermometer read if everything can be considered black? State your assumptions.

11.6. Rework Problem 11.5, taking the room to be wall-papered and considering the thermometer to be nonblack.

11.7. Two thin aluminum plates, the first polished and the second painted black, are placed horizontally outdoors, where they are cooled by air at 10°C. The heat transfer coefficient is 5 W/m²-°C on both the top and bottom. The top is irradiated with 750 W/m²-°C and it radiates to the sky at 170°K. The earth below the plates is black at 10°C. Find the equilibrium temperature of each plate.

11.8. A sample holder of 99% pure aluminum 1 cm in diameter and 16 cm in length protrudes from a small housing on an orbital space vehicle. The holder "sees" almost nothing but outer space at an effective temperature of 30°K. The base of the holder is 0°C and you must find the temperature of the sample at its tip. It will help if you note that aluminum is used so that the temperature of the tip stays quite close to that of the root. [$T_{end} = -0.7$°C.]

11.9. There is a radiant heater in the bottom of the box shown in Fig. P11.9. What percentage of the heat goes out the top? What fraction impinges on each of the four sides? (Remember that the percentages must add up to 100.)

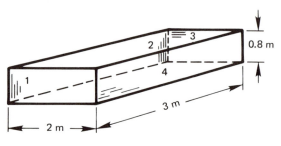

FIGURE P11.9

11.10. With reference to Fig. 11.13, find $F_{1-2,4}$ and $F_{2,4-1}$.

11.11. Find F_{2-4} for the surfaces shown in Fig. P11.11. [0.315.]

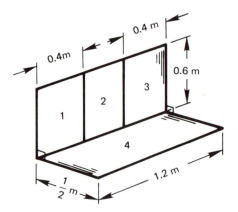

FIGURE P11.11

11.12. What is F_{1-2} or the squares shown in Fig. P11.12?

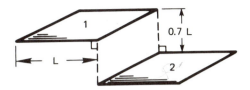

FIGURE P11.12

11.13. A particular internal combustion engine has an exhaust manifold at 600°C running parallel to a water cooling line at 20°C. If both the manifold and the cooling line are 4 cm in diameter, their centers are 7 cm apart, and both are approximately black, how much heat will be transferred to the cooling line by radiation? [239 W/m.]

11.14. Prove that F_{1-2} for any pair of two-dimensional plane surfaces as shown in Fig. P11.14 is equal to $[(a+b)-(c+d)]/2L_1$. This is called the *string method* because we can imagine that the numerator equals the sum of the lengths of a set of crossed strings (a and b) and a set of uncrossed strings (c and d).

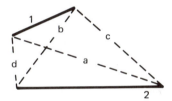

FIGURE P11.14

11.15. Find F_{1-5} for the surfaces shown in Fig. P11.15.

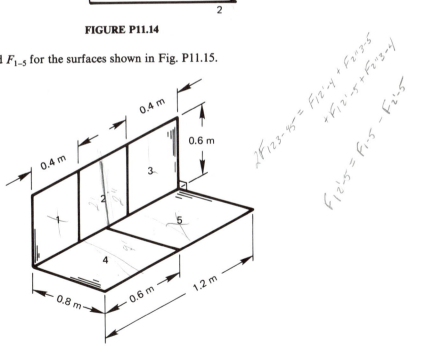

FIGURE P11.15

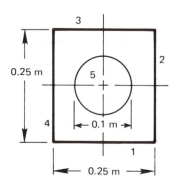

FIGURE P11.16

11.16. Find F_{1-3} for the surfaces shown in Fig. P11.16.

11.17. A cubic box 1 m on the side is black except for one side, which has an emittance of 0.2 and is kept at 300°C. An adjacent side is kept at 500°C. The other sides are insulated. Find Q_{net} inside the box. [2494 W.]

11.18. Rework Problem 11.17, but this time set the emittance of the insulated walls equal to 0.6. Compare the insulated wall temperature with the value you would get if the walls were black.

11.19. An insulated black cylinder 10 cm in length with an inside diameter of 5 cm has a black cap on one end and a cap with an emittance of 0.1 on the other. The black end is kept at 100°C and the reflecting end is kept at 0°C. Find Q_{net} inside the cylinder, and $T_{cylinder}$.

11.20. Rework Example 11.3 if the shield has an inside emittance of 0.34 and the room is at 20°C. How much cooling must be provided to keep the shield at 100°C?

11.21. A 0.8-m-long cylindrical burning chamber is 0.2 m in diameter. The hot gases within it are at a temperature of 1500°C and a pressure of 1 atm, and the absorbing components consist of 12% by volume of CO_2 and 18% H_2O. Neglect end effects and determine how much cooling must be provided the walls to hold them at 750°C if they are black.

11.22. A 30-ft by 40-ft house has a conventional 30° sloping roof with a peak running in the 40-ft direction. Calculate the temperature of the roof in 20°C still air when the sun is overhead (a) if the roofing is of wooden shingles; (b) if it is commercial aluminum sheet. Assume that the incident solar energy = 670 W/m², Kirchhoff's law applies for both roofs, and T_{eff} for the sky is 22°C.

11.23. Calculate the radiant heat transfer from a 0.2-m-diameter stainless steel hemisphere ($\varepsilon_{ss} = 0.4$) to a copper floor ($\varepsilon_{Cu} = 0.15$) which forms its base. The hemisphere is kept at 300°C and the base at 100°C. Use the algebraic method. [21.24 W.]

11.24. A hemispherical indentation in a smooth wrought iron plate has a 0.008 m radius. How much heat radiates from the 40°C dent to the −20°C surroundings?

11.25. A conical hole in a block of metal, for which $\varepsilon = 0.5$, is 5 cm-in-diameter at the surface and 5 cm deep. By what factor will the radiation from the area of the hole be changed by the presence of the hole? (This problem can be done to a close approximation using the methods in this chapter if the cone does not become very deep and slender. If it does, then the fact that the apex is receiving far less radiation makes it incorrect to use the network analogy.)

REFERENCES

[11.1]   A. K. OPPENHEIM, "Radiation Analysis by the Network Method," *Trans. ASME*, vol. 78, 1956, pp. 725–735.

[11.2]   *Scientific American*, "Energy and Power," vol. 224, no. 3, September 1971. (See especially M. K. Hubbert, "The Energy Resources of the Earth," pp. 60–70.)

[11.3]   J. A. DUFFIE and W. A. BECKMAN, *Solar Energy Thermal Processes*, John Wiley & Sons, Inc., New York, 1974.

[11.4]   F. KREITH and J. F. KREIDER, *Principles of Solar Engineering*, Hemisphere Publishing Corp./McGraw-Hill Book Company, Washington, D.C., 1978.

[11.5]   *Solar Heating and Cooling of Residential Buildings* (2 vols.), U.S. Dept. of Commerce, Washington, D.C., October 1977.

Appendix A

Some thermophysical properties of selected materials

A *primary source* of thermophysical properties is a document in which the experimentalist who obtained the data reports the details and results of his or her measurements. The term *secondary source* generally refers to a document, based on primary sources, which presents other peoples' data, and does so critically. This appendix is neither a primary nor a secondary source, since it has been assembled from a variety of secondary and even tertiary sources.

We attempted to cross-check the data against different sources, and this often led to contradictory values. Such contradictions are usually the result of differences between the experimental samples that are reported, or of differences in the accuracy of experiments themselves. We resolved such differences by judging the source, by reducing the number of significant figures to accommodate the conflict, or by omitting the substance from the table. The resulting numbers will suffice for most calculations. However, the reader who needs high accuracy should be sure of the physical constitution of the material and then should seek out one of the more reliable secondary data sources.

The format of these tables is quite close to that established by R. M. Drake in his excellent appendix on thermophysical data [1.7]. However, while we use some of Drake's numbers directly in Table A.4, many of his other values have been superseded by more recent measurements. The two secondary sources from which most of the data here were obtained were the Purdue University series on the *Thermophysical Properties of Matter* [A.1] and the G.E. *Heat Transfer Data Book* [A.2]. The Purdue series is the result of an enormous property-gathering effort carried out under the direction of Y. S. Touloukian and several coworkers. The various volumes in the series are dated since 1970, and addenda continue to be issued.

Numbers that did not come directly from [A.1], [A.2], or [1.7] were obtained from a large variety of manufacturers' tables, handbooks, other textbooks, and so on. No attempt has been made to document all these diverse sources and the various compromises that were made in quoting them.

Table A.1 gives the density, specific heat, thermal conductivity, and thermal diffusivity for various metallic solids. These values were obtained from volumes 1 and 4 of [A.1] whenever it was possible to find them there. Most thermal conductivity values in the table have been rounded off to two significant figures. The reason is that k is sensitive to very minor variations in physical structure that cannot be detailed fully here. Notice, for example, the significant differences between pure silver and 99.9% pure silver, or between pure aluminum and 99% pure aluminum.

Table A.2 gives the same properties as Table A.1 (where they are available) but for nonmetallic substances. Volumes 2 and 5 of [A.1] provided many of the data here and they revealed even greater variations in k than the metallic data did. For the various sands reported, k varied by a factor of 500, and graphite varied by a factor of 50, for example. The sensitivity of k to small variations in the packing of fibrous materials, or to the water content of hygroscopic materials, forced us to restrict many of the k values to a single significant figure.

Table A.3 gives ρ, c_p, k, α, ν, Pr, and β for several liquids. Volumes 3, 6, 10, and 11 of [A.1] gave most of the needed values of c_p, k, and $\mu = \rho\nu$, and occasional independently measured values of α. Other values came from [A.3]. Densities came from [3.5], [A.3], and a variety of other sources. Values of α that disagreed only slightly with $k/\rho c_p$ were allowed to stand.

Table A.4 was drawn from [1.7], with five exceptions: air and argon data came from [A.2], except for density, which was calculated using the ideal gas law and a compressibility factor. Water data were taken from [A.4]. Nitrogen data and a few high-temperature Hydrogen data are from [A.3].

REFERENCES

[A.1]   Y. S. TOULOUKIAN, *Thermophysical Properties of Matter*, vols. 1–6, 10, and 11, Purdue University, West Lafayette, Ind., 1970 to 1975.

[A.2]   *Heat Transfer Data Book* (R. H. Norris, F. F. Buckland, N. D. Fitzroy, R. H. Roecker, D. A. Kaminski, eds.), General Electric Co., Schenectady, N.Y., 1977.

[A.3]   N. B. VARGAFTIK, *Tables on the Thermophysical Properties of Liquids and Gases*, 2nd ed., Hemisphere Publishing Corp., Washington, D.C., 1975.

[A.4]   *Steam and Air Tables in S.I. Units* (T. F. Irvine, Jr., and J. P. Hartnett, eds.), Hemisphere Publishing Corp./McGraw-Hill Book Company, Washington, D.C., 1976.

Table A.1 Properties of metallic solids

Metal	Properties at 20°C				Thermal Conductivity, k (W/m-°K)									
	ρ (kg/m³)	c_p (J/kg-°K)	k (W/m-°K)	α (10^{-5} m²/s)	−170°C	−100°C	0°C	100°C	200°C	300°C	400°C	600°C	800°C	1000°C
Aluminum														
Pure	2707	905	237	9.61	302	242	236	240	238	234	228	215	~95 (liq)	
99% pure			211		220	206	209							
Duralumin (~4% Cu)	2787	883	164	6.66		126	164	182	194					
Chromium	7190	453	90	2.77	158	120	95	88	85	82	77	69	64	62
Copper and Cu alloys														
Pure	8954	384	398	11.57	483	420	401	391	389	384	378	366	352	336
Brass (30% Zn)	8522	385	109	3.32	73	89	106	133	143	146	147			
Bronze (25% Sn)	8666	343	26	0.86	(Data on this and other bronzes vary by a factor of about 2)									
Constantan (40% Ni)	8922	410	22	0.61	17	19	22	26	35					
German silver (15% Ni; 22% Zn)	8618	394	25	0.73	18	19	24	31	40	45	48			
Gold	19,320	129	315	12.64			318		309					
Ferrous metals														
Pure iron	7897	447	80	2.26	132	98	84	72	63	56	50	39	30	29.5
Cast iron (0.4% C)	7272	420	52	1.70										
Steels (C < 1.5%)														
0.5% carbon (mild)	7833	465	54	1.47			55	52	48	45	42	35	31	29
1.0% carbon	7801	473	43	1.17			43	43	42	40	36	33	29	28
1.5% carbon	7753	486	36	0.97			36	36	36	35	33	31	28	28

k and α for carbon steels can vary greatly, owing to trace elements

Material														
Stainless steel, type:														
304	8000	400	13.8	0.4				15	17+	19−	21	25	24	26+
316	8000	460	13.5	0.37				15	16	17+	19−	21+	26	28
347	8000	420	15	0.44				16+	18−	19	20	23		
410	7700	460	25	0.7		12		25+	26	27	27+	28+		
414	7700	460	25	0.7		13					29	29		
Lead	11,373	130	35	2.34	40	37	36	34	33	31	17 (liq.)	20 (liq.)		
Magnesium	1746	1023	156	8.76	17	16	157	154	152	150	148		90 (liq.)	
Mercury (polycrystalline)					32	30	7.8 (liq.)							
Nickel														
Pure	8906	445	91	2.30	156	114	94	83	74	67	64	69	73	78
Nichrome (24% Fe, 16% Cr)	8250	448		0.34				13						
Nichrome V (20% Cr)	8410	466	13	0.33			12	14	15	17	19			
Platinum	21,450	133	71	2.50	78	73	72	72	72	73	74	77	80	84
Silver														
99.99+% pure	10,524	236	427	17.19	449	431	428	422	417	409	401	386	370	176 (liq.)
99.9% pure	10,524	236	411	16.55		422	405		373	367	364			
Tin (polycrystalline)	7304	~220	67	4.17	85	76	68	63						
Titanium (polycrystalline)	4540	523	22	0.93	31	26	22	21	20	20	19	21	21	22
Tungsten	19,350	133	178	6.92	235	223	182	166	153	141	134	125	122	114
Uranium	18,700	116	28	1.29	22	24	27	29	31	33	36	41	46	
Zinc	7144	388	121	4.37	124	122	122	117	110	106	100	60 (liq.)		

Material	Temperature Range (°C)	Density, ρ (kg/m^3)	Specific Heat, c (J/kg-°C)	Thermal Conductivity, k (W/m-°C)	Thermal Diffusivity, α (m^2/s)
Asbestos					
Cement board	20			0.6	
Fiber ⎫ properties vary	20	1930		0.8	
Fiber ⎭ with packing	20	980		0.14	
Asphalt	20–25			0.75	
Beef	25				1.35×10^{-7}
Brick					
B&W, K-28 insulating	300			0.3	
B&W, K-28 insulating	1000			0.4	
Cement	10	720		0.34	
Common	0–1000			0.7	
Chrome	100			1.9	
Facing	20			1.3	
Firebrick	300	2000	960	0.1	5.4×10^{-8}
Firebrick	1000			0.2	
Carbon					
Diamond (type II b)	20	~3250	510	1350.	8.1×10^{-4}
Graphite	20	~2100	~2090	Highly variable structure	
Cardboard	0–20	790		0.14	
Clay					
Fireclay	500–750			1.	
Sandy clay	20	1780		0.9	
Coal					
Anthracite	900	~1500		~0.2	
Brown coal	900			~0.1	
Bituminous in situ		~1300		0.5–0.7	3 to 4×10^{-7}
Concrete					
Limestone gravel	20	1850		0.6	
Portland cement	90	2300		1.7	
Sand : cement (3 : 1)	230			0.1	
Sand and gravel	20			1.8	
Slag cement	20			0.14	
Corkboard (medium ρ)	30	170		0.04	
Egg white	20				1.37×10^{-7}
Glass					
Lead	36			1.2	
Plate	20			1.3	
Pyrex	60–100	2210	753	1.3	7.8×10^{-7}
Soda	20			0.7	
Window	46			1.3	
Glass wool	20	64–160		0.04	
Ice	0	917	2100	2.215	1.15×10^{-6}
Ivory	80			0.5	
Kapok	30			0.035	
Magnesia (85%)	38			0.067	
	93			0.071	
	150			0.074	
	204			0.08	

Material	Temperature Range (°C)	Density, ρ (kg/m³)	Specific Heat, c (J/kg-°C)	Thermal Conductivity, k (W/m-°C)	Thermal Diffusivity, α (m²/s)
Lunar surface dust (high vacuum)	250	1500 ± 300	~600	~0.0006	~7 × 10⁻¹⁰
Rock wool	−5	~130		0.03	
	93			0.05	
Rubber (hard)	0	1200	2010	0.15	6.2 × 10⁻⁸
Silica aerogel	0	140		0.024	
	120	136		0.022	
Silo-cel (diatomaceous earth)	0	320		0.061	
Soil (mineral)					
Dry	15	1500	1840	1.	4 × 10⁻⁷
Wet	15	1930		2.	
Stone					
Granite (NTS)	20	~2640	~820	1.6	~7.4 × 10⁻⁷
Limestone (Indiana)	100	2300	~900	1.1	~5.3 × 10⁻⁷
Sandstone (Berea)	25			~3	
Slate	100			1.5	
Wood (perpendicular to grain)					
Ash	15	740		0.15–0.3	
Balsa	15	100		0.05	
Cedar	15	480		0.11	
Fir	15	600	2720	0.12	7.4 × 10⁻⁸
Mahogany	20	700		0.16	
Oak	20	600	2390	0.1–0.4	(0.7–2.8) × 10⁻⁷
Pitch pine	20	450		0.14	
Sawdust (dry)	17	128		0.05	
Sawdust (dry)	17	224		0.07	
Spruce	20	410		0.11	
Wool (sheep)	20	145		0.05	

Table A.3 Thermophysical properties of saturated liquids

Temp. °K	°C	$\rho(kg/m^3)$	$c_p(J/kg\text{-}°K)$	$k(W/m\text{-}°K)$	$\alpha(m^2/s)$	$\nu(m^2/s)$	Pr	$\beta(°K^{-1})$
\multicolumn{9}{Ammonia (there is considerable disagreement among sources)}
220	−53	706	4426	0.66	2.11×10^{-7}			
240	−33	682	4484	0.61	2.00	4.17×10^{-7}	2.09	
260	−13	656	4547	0.57^-	1.91	3.27	1.71	
280	7	629	4625	0.52	1.79	2.68	1.50	0.00025
300	27	600	4736	0.470	1.65	2.32	1.41	
320	47	568	4962	0.424	1.50	2.06	1.37	
340	67	533	5214	0.379	1.36	1.79	1.32	
360	87	490	5635	0.335	1.21	1.55	1.28	
380	107	436		0.289		1.34		
400	127	345		0.245		1.19		

Temp. °K	°C	$\rho(kg/m^3)$	$c_p(J/kg\text{-}°K)$	$k(W/m\text{-}°K)$	$\alpha(m^2/s)$	$\nu(m^2/s)$	Pr	$\beta(°K^{-1})$
\multicolumn{9}{CO_2}
250	−23	1046	1990	0.135	6.49×10^{-8}			
260	−13	998	2110	0.123	5.84	1.15×10^{-7}	1.97	
270	−3	944	2390	0.113	5.09	1.08	2.12	
280	7	883	2760	0.102	4.19	1.04	2.48	
290	17	805	3630	0.090	3.08	0.99	3.20	0.014
300	27	676	7690	0.076	1.46	0.88	6.04	
303	30	604						

Temp. °K	°C	$\rho(kg/m^3)$	$c_p(J/kg\text{-}°K)$	$k(W/m\text{-}°K)$	$\alpha(m^2/s)$	$\nu(m^2/s)$	Pr	$\beta(°K^{-1})$
\multicolumn{9}{D_2O (heavy water)}
589	316	740	2034	0.0509	0.978×10^{-7}	1.23×10^{-7}	1.257	

Temp. °K	°C	$\rho(kg/m^3)$	$c_p(J/kg\text{-}°K)$	$k(W/m\text{-}°K)$	$\alpha(m^2/s)$	$\nu(m^2/s)$	Pr	$\beta(°K^{-1})$
\multicolumn{9}{Freon 11 (trichlorofluoromethane)}
220	−53		829	0.110				
240	−33	1607	841	0.105	7.8×10^{-8}	4.78×10^{-7}	6.1	
260	−13	1564	855	0.099	7.4	4.10	5.5	
280	7	1518	871	0.093	7.0	3.81	5.4	
300	27	1472	888	0.088	6.7	2.82	4.2	0.00154
320	47	1421	906	0.082	6.4			0.00163
340	67	1369	927	0.076	6.0			

Temp. °K	°C	$\rho(kg/m^3)$	$c_p(J/kg\text{-}°K)$	$k(W/m\text{-}°K)$	$\alpha(m^2/s)$	$\nu(m^2/s)$	Pr	$\beta(°K^{-1})$
\multicolumn{9}{Freon 12 (dichlorodifluoromethane)}
160	−113			0.133				
180	−93	1664	834	0.124	8.935×10^{-8}			
200	−73	1610	856	0.1148	8.33			
220	−53	1555	873	0.1057	7.79	3.2×10^{-7}	4.11	0.00263
240	−33	1498	892	0.0965	7.22	2.60	3.60	
260	−13	1438	914	0.0874	6.65	2.26	3.40	
280	7	1374	942	0.0782	6.04	2.06	3.41	
300	27	1305	980	0.0690	5.39	1.95	3.62	
320	47	1229	1031	0.0599	4.72	1.9	4.03	
340	67		1097	0.0507				

Temp. °K	°C	$\rho(kg/m^3)$	$c_p(J/kg\text{-}°K)$	$k(W/m\text{-}°K)$	$\alpha(m^2/s)$	$\nu(m^2/s)$	Pr	$\beta(°K^{-1})$
\multicolumn{9}{Freon 113 (CCl_2FCClF_2)}
376	103	1496	9620	0.571	3.97×10^{-8}	2.08×10^{-7}	5.24	

Temp. °K	°C	$\rho(\text{kg/m}^3)$	$c_p(\text{J/kg-}°\text{K})$	$k(\text{W/m-}°\text{K})$	$\alpha(\text{m}^2/\text{s})$	$\nu(\text{m}^2/\text{s})$	Pr	$\beta(°\text{K}^{-1})$
				Glycerin (or glycerol)				
273	0	1276	2200	0.282	1.00×10^{-7}	0.0083	83,000	
293	20	1261	2350	0.285	0.962	0.001120	11,630	0.00048
303	30	1255	2400	0.285	0.946	0.000488	5,161	0.00049
313	40	1249	2460	0.285	0.928	0.000227	2,451	0.00049
323	50	1243	2520	0.285	0.910	0.000114	1,254	0.00050
				20% glycerin, 80% water				
293	20	1047	3250	0.519	1.53×10^{-7}	1.681×10^{-6}	11.0	0.00031
303	30	1043	3230	0.532	1.58	1.294	8.19	0.00036
313	40	1039	3300	0.540	1.57	1.030	6.56	0.00041
323	50	1035	3360	0.553	1.59	0.849	5.34	0.00046
				40% glycerin, 60% water				
293	20	1073	3030	0.448	1.38×10^{-7}	3.467×10^{-6}	23.7	0.00041
303	30	1069	3010	0.452	1.40	2.484	17.1	0.00045
313	40	1090	2900	0.461	1.46	1.900	13.0	0.00048
323	50	1085	3140	0.469	1.38	1.493	10.2	0.00051
				60% glycerin, 40% water				
293	20	1154	2590	0.381	1.27×10^{-7}	9.36×10^{-6}	73.4	0.00048
303	30	1148	2610	0.381	1.27	6.89	49.3	0.00050
313	40	1143	2660	0.385	1.27	4.44	35.1	0.00052
323	50	1137	2700	0.389	1.27	3.31	26.1	0.00053
				80% gylcerin, 20% water				
293	20	1209	2430	0.327	1.11×10^{-7}	4.97×10^{-5}	447	0.00051
303	30	1203	2470	0.327	1.10	2.82	259	0.00052
313	40	1197	2520	0.327	1.08	1.74	161	0.00053
323	50	1191	2570	0.331	1.08	1.14	106	0.00053

Helium I and helium II

- k for He I is about 0.020 W/m-°K near the λ-transition (~2.17°K).
- k for He II below the λ-transition is hard to measure. It appears to be about 80,000 W/m-°K between 1.4 and 1.75°K and it might go as high as 340,000 W/m-°K at 1.92°K. These are the highest conductivities known (cf. copper, silver, and diamond).

Temp. °K	°C	$\rho(\text{kg/m}^3)$	$c_p(\text{J/kg-}°\text{K})$	$k(\text{W/m-}°\text{K})$	$\alpha(\text{m}^2/\text{s})$	$\nu(\text{m}^2/\text{s})$	Pr	$\beta(°\text{K}^{-1})$
				Lead				
644	371	10,540	159	16.1	1.084×10^{-5}	2.276×10^{-7}	0.024	
755	482	10,442	155	15.6	1.223	1.85	0.017	
811	538	10,348	145	15.3	1.02	1.68	0.017	
				Mercury				
234	−39		141.5	6.97	3.62×10^{-6}			
250	−23		140.5	7.32	3.83			
300	27	13,611	139.1	8.34	4.41	1.2×10^{-7}	0.027	
350	77	13,489	137.7	5.29	4.91	1.0	0.020	

Temp. °K	°C	$\rho(kg/m^3)$	$c_p(J/kg\text{-}°K)$	$k(W/m\text{-}°K)$	$\alpha(m^2/s)$	$\nu(m^2/s)$	Pr	$\beta(°K^{-1})$
\multicolumn{9}{c}{Mercury (continued)}								
400	127	13,367	136.7	5.69	5.83×10^{-6}	0.95×10^{-7}	0.016	
500	277	13,128	135.6	6.36	6.00	0.80	0.013	
600	327		135.4	6.93	6.55	0.68	0.010	
700	427		136.1	7.34				
800	527			7.40				
\multicolumn{9}{c}{Methyl alcohol (methanol)}								
260	−13	823	2336	0.2164	1.126×10^{-7}	$\sim 1.3 \times 10^{-6}$	~ 11.5	
280	7	804	2423	0.2078	1.021	~ 0.9	~ 8.8	0.00114
300	27	785	2534	0.2022	1.016	~ 0.7	~ 6.9	
320	47	767	2672	0.1965	0.959	~ 0.6	~ 6.3	
340	67	748	2856	0.1908	0.893	~ 0.44	~ 4.9	
360	87	729	3036	0.1851	0.836	~ 0.36	~ 4.3	
380	107	710	3265	0.1794	0.774	~ 0.30	~ 4.1	
\multicolumn{9}{c}{NAK (eutectic mixture of sodium and potassium)}								
366	93	849	946	24.4	3.05×10^{-5}	5.8×10^{-7}	0.019	
672	399	775	879	26.7	3.92	2.67	0.0068	
811	538	743	872	27.7	4.27	2.24	0.0053	
1033	760	690	883			2.12		
\multicolumn{9}{c}{Oils (some approximate viscosities)}								
273	0	\multicolumn{3}{c}{MS-20}				0.0076	100,000	
339	66	\multicolumn{3}{c}{California crude (heavy)}				0.00008		
289	16	\multicolumn{3}{c}{California crude (light)}				0.00005		
339	66	\multicolumn{3}{c}{California crude (light)}				0.000010		
289	16	\multicolumn{3}{c}{Light machine oil}				0.0007		
339	66	\multicolumn{3}{c}{Light machine oil}				0.00004		
289	16	\multicolumn{3}{c}{Light machine oil ($\rho=907$)}				0.00016		
339	66	\multicolumn{3}{c}{Light machine oil ($\rho=907$)}				0.000013		
289	16	\multicolumn{3}{c}{SAE 30}				0.00044	$\sim 5,000$	
339	66	\multicolumn{3}{c}{SAE 30}				0.00003		
289	16	\multicolumn{3}{c}{SAE 30 (Eastern)}				0.00011		
339	66	\multicolumn{3}{c}{SAE 30 (Eastern)}				0.00001		
289	16	\multicolumn{3}{c}{Spindle oil ($\rho=885$)}				0.00005		
339	66	\multicolumn{3}{c}{Spindle oil ($\rho=885$)}				0.000007		
\multicolumn{9}{c}{Oxygen}								
54		1276	1648	0.191	9.08×10^{-8}	6.5×10^{-7}	7.15	
60	−213		1649	0.185				
80	−193		1653	0.1623				
90	−183	1114	1655	0.1501	8.14×10^{-8}	1.75×10^{-7}	2.15	
120	−153			0.1096				
150	−123			0.061				

Temp. °K	°C	$\rho(kg/m^3)$	$c_p(J/kg\text{-}°K)$	$k(W/m\text{-}°K)$	$\alpha(m^2/s)$	$\nu(m^2/s)$	Pr	$\beta(°K^{-1})$
				Water				
273	0	999.8	4205	0.5750	1.368×10^{-7}	1.753×10^{-6}	12.81	
280	7	999.9	4196	0.5818	1.386	1.422	10.26	
300	27	996.6	4177	0.6084	1.462	0.826	5.65	0.000275
320	47	989.3	4177	0.6367	1.541×10^{-7}	0.566×10^{-6}	3.67	0.000435
340	67	979.5	4187	0.6587	1.606	0.420	2.61	
360	87	967.4	4206	0.6743	1.657	0.330	1.99	
373	100	957.2	4219	0.6811	1.683	0.290	1.72	
400	127	937.5	4241	0.6864	1.726	0.229	1.33	
420	147	919.9	4306	0.6836	1.726×10^{-7}	2.000×10^{-7}	1.16	
440	167	900.5	4391	0.6774	1.713	1.786	1.04	
460	187	879.5	4456	0.6672	1.703	1.626	0.955	
480	207	856.6	4534	0.6530	1.681	1.504	0.894	
500	227	831.5	4647	0.6348	1.463	1.412	0.859	
520	247	803.9	4831	0.6123	1.577×10^{-7}	1.345×10^{-7}	0.853	
540	267	773.0	5099	0.5857	1.486	1.298	0.873	
560	287	738.2	5487	0.555	1.370	1.269	0.926	
580	307	697.6	6010	0.520	1.240	1.240	1.000	
600	327	648.8	6691	0.481	1.108	1.215	1.097	
620	347	586.3				1.213×10^{-7}		
640	367	482.1				1.218		
647.3	374.2	306.8				1.356		

Table A.4   Thermophysical properties of gases at atmospheric pressure[a]

T(°K)	ρ(kg/m³)	c_p(J/kg-°K)	μ(kg/m-s)	ν(m²/s)	k(W/m-°K)	α(m²/s)	Pr
				Air			
100		1009	0.706×10^{-5}	$\sim 0.2 \times 10^{-5}$	0.00922		
150		1005	1.038	~ 0.4	0.01375		
200	1.79	1003	1.336	0.746	0.01810	1.01×10^{-5}	0.74
250	1.43	1003	1.606	1.123	0.02226	1.55	0.724
300	1.183	1003	1.853	1.566	0.02614	2.203	0.711
350	1.009	1008	2.081	2.062	0.02970	2.920	0.706
400	0.8826	1013	2.294	2.599	0.03305	3.697	0.703
450	0.7846	1020	2.493	3.177	0.03633	4.540	0.700
500	0.7061	1029	2.682×10^{-5}	3.798×10^{-5}	0.03951	5.438×10^{-5}	0.698
550	0.6419	1039	2.860	4.456	0.0426	6.387	0.698
600	0.5884	1051	3.030	5.150	0.0456	7.374	0.698
650	0.5432	1063	3.193	5.878	0.0484	8.382	0.701
700	0.5044	1075	3.349	6.64	0.0513	9.461	0.702
750	0.4707	1087	3.498	7.43	0.0541	10.57	0.703
800	0.4413	1099	3.643	8.26	0.0569	11.73	0.704
850	0.4154	1110	3.783	9.11	0.0597	12.95	0.704
900	0.3923	1121	3.918	9.99	0.0625	14.22	0.703
950	0.3716	1131	4.049	10.90	0.0649	15.44	0.706
1000	0.3531	1142	4.177×10^{-5}	11.83×10^{-5}	0.0672	16.67×10^{-5}	0.710
1100	0.3210	1159	4.42	13.8	0.0717	19.27	0.716
1200	0.2942	1175	4.65	15.8	0.0759	29.96	0.720
1300	0.2716	1189	4.88	18.0	0.0797	24.7	0.729
1400	0.2522		5.09	20.2	0.0835	27.5	0.734
1500	0.2354		5.30	22.5	0.0870	30.3	0.74
				Argon			
100		550			0.00590		0.700
144		528			0.00922		0.683
200	2.51	523.6	1.62×10^{-5}	0.647×10^{-5}	0.01246	0.948×10^{-5}	0.683
255	1.955	522.1	2.01	1.028	0.01542	1.511	0.679
311	1.59	521.5	2.36	1.49	0.01821	2.20	0.676
366.5	1.34	521.1	2.68	2.00	0.02080	2.98	0.672
422	1.154	520.9	2.982	2.58	0.02322	3.86	0.669
477.6	1.020	520.8	3.265	3.20	0.02553	4.81	0.666
533	0.913	520.8	3.53	3.87	0.02771	5.83	0.664
811	0.6007	520.5	4.71	7.85	0.03714	11.88	0.660
1089	0.4474		5.73	12.82	0.0451	19.37	0.663
1366.5	0.3565		6.66	18.69	0.0519	28.0	0.668
1500	0.3243		7.08	21.82	0.0550	32.5	0.671
				Ammonia (NH_2)			
220	0.3828	2.198×10^3	7.255×10^{-6}	1.90×10^{-5}	0.0171	0.2054×10^{-4}	0.93
273	0.7929	2.177	9.353	1.18	0.0220	0.1308	0.90
323	0.6487	2.177	11.035	1.70	0.0270	0.1920	0.88
373	0.5590	2.236	12.886	2.30	0.0327	0.2619	0.87
423	0.4934	2.315	14.672	2.97	0.0391	0.3432	0.87
473	0.4405	2.395	16.49	3.74	0.0467	0.4421	0.84

$T(°K)$	$\rho(kg/m^3)$	$c_p(J/kg\text{-}°K)$	$\mu(kg/m\text{-}s)$	$\nu(m^2/s)$	$k(W/m\text{-}°K)$	$\alpha(m^2/s)$	Pr
			Carbon dioxide				
220	2.4733	0.783×10^3	11.105×10^{-6}	4.490×10^{-6}	0.010805	0.05920×10^{-4}	0.818
250	2.1657	0.804	12.590	5.813	0.012884	0.07401	0.793
300	1.7973	0.871	14.958	8.321	0.016572	0.10588	0.770
350	1.5362	0.900	17.205	11.19	0.02047	0.14808	0.755
400	1.3424	0.942	19.32	14.39	0.02461	0.19463	0.738
450	1.1918	0.980	21.34	17.90	0.02897	0.24813	0.721
500	1.0732	1.013	23.26	21.67	0.03352	0.3084	0.702
550	0.9739	1.047	25.08	25.74	0.03821	0.3750	0.685
600	0.8938	1.076	26.83	30.02	0.04311	0.4483	0.668
			Carbon monoxide				
220	1.55363	1.0429×10^3	13.832×10^{-6}	8.903×10^{-6}	0.01906	0.11760×10^{-4}	0.758
250	0.8410	1.0425	15.40	11.28	0.02144	0.15063	0.750
300	1.13876	1.0421	17.843	15.67	0.02525	0.21280	0.737
350	0.97425	1.0434	20.09	20.62	0.02883	0.2836	0.728
400	0.85363	1.0484	22.19	25.99	0.03226	0.3605	0.722
450	0.75848	1.0551	24.18	31.88	0.0436	0.4439	0.718
500	0.68223	1.0635	26.06	38.19	0.03863	0.5324	0.718
550	0.62024	1.0756	27.89	44.97	0.04162	0.6240	0.721
600	0.56850	1.0877	29.60	52.06	0.04446	0.7190	0.724
			Helium				
3		5.200×10^3	8.42×10^{-7}		0.0106		
33	1.4657	5.200	50.2	3.42×10^{-6}	0.0353	0.04625×10^{-4}	0.74
144	3.3799	5.200	125.5	37.11	0.0928	0.5275	0.70
200	0.2435	5.200	156.6	64.38	0.1177	0.9288	0.694
255	0.1906	5.200	181.7	95.50	0.1357	1.3675	0.70
366	0.13280	5.200	230.5	173.6	0.1691	2.449	0.71
477	0.10204	5.200	275.0	269.3	0.197	3.716	0.72
589	0.08282	5.200	311.3	375.8	0.225	5.215	0.72
700	0.07032	5.200	347.5	494.2	0.251	6.661	0.72
800	0.06023	5.200	381.7	634.1	0.275	8.774	0.72
900	0.05286	5.200	413.6	781.3	0.298	10.834	0.72
			Hydrogen				
30	0.84722	10.840×10^3	1.606×10^{-6}	1.805×10^{-6}	0.0228	0.0249×10^{-4}	0.759
50	0.50955	10.501	2.516	4.880	0.0362	0.0676	0.721
100	0.24572	11.229	4.212	17.14	0.0665	0.2408	0.712
150	0.16371	12.602	5.595	34.18	0.0981	0.475	0.718
200	0.12270	13.540	6.813	55.53	0.1282	0.772	0.719
250	0.09819	14.059	7.919	80.64	0.1561	1.130	0.713
300	0.08185	14.314	8.963	109.5	0.182	1.554	0.706
350	0.07016	14.436	9.954	141.9	0.206	2.031	0.697
400	0.06135	14.491	10.864	177.1	0.228	2.568	0.690
450	0.05462	14.499	11.779	215.6	0.251	3.164	0.682
500	0.04918	14.507	12.636	257.0	0.272	3.817	0.675
550	0.04469	14.532	13.475	301.6	0.292	4.516	0.668

$T(°K)$	$\rho(kg/m^3)$	$c_p(J/kg\text{-}°K)$	$\mu(kg/m\text{-}s)$	$\nu(m^2/s)$	$k(W/m\text{-}°K)$	$\alpha(m^2/s)$	Pr
			Hydrogen (continued)				
600	0.04085	14.537×10^{-3}	14.285×10^{-6}	349.7×10^{-6}	0.315	5.306×10^{-4}	0.664
700	0.03492	14.574	15.89	455.1	0.351	6.903	0.659
800	0.03060	14.675	17.40	569	0.384	8.563	0.664
900	0.02723	14.821	18.78	690	0.412	10.21	0.675
1000	0.02451	14.968	20.16	822	0.445	12.13	0.678
1100	0.02227	15.165	21.46	965	0.488	14.45	0.668
1200	0.02050	15.366	22.75	1107	0.528	16.76	0.661
1300	0.01890	15.575	24.08	1273	0.568	19.3	0.660
1333	0.01842	15.638	24.44	1328	0.58	20.1	0.661
			Nitrogen				
100	3.439	1.0722×10^3	6.862×10^{-6}	1.995×10^{-6}	0.00958	0.026×10^{-4}	0.767
200	1.688	1.0429	12.947	7.67	0.0183	0.104	0.738
300	1.1233	1.0408	17.84	15.88	0.0259	0.222	0.715
400	0.8425	1.0459	21.98	26.1	0.0327	0.371	0.704
500	0.6739	1.0555	25.70	38.1	0.0389	0.547	0.696
600	0.5615	1.0756	29.11	51.8	0.0446	0.738	0.702
700	0.4812	1.0969	32.13	66.8	0.0499	0.945	0.707
800	0.4211	1.1225	34.84	82.7	0.0548	1.16	0.713
900	0.3743	1.1464	37.49	100.2	0.0597	1.39	0.721
1000	0.3368	1.1677	40.00	119.	0.0647	1.65	0.723
1100	0.3062	1.1857	42.28	138.	0.0700	1.93	0.716
1200	0.2807	1.2037	44.50	158.	0.0758	2.24	0.704
			Oxygen				
100	3.9918	0.9479×10^2	7.768×10^{-6}	1.946×10^{-6}	0.00903	0.0239×10^{-4}	0.815
150	2.6190	0.9178	11.490	4.387	0.01367	0.0569	0.773
200	1.9559	0.9131	14.850	7.593	0.01824	0.1021	0.745
250	1.5618	0.9157	17.87	11.45	0.02259	0.1579	0.725
300	1.3007	0.9203	20.63	15.86	0.02676	0.2235	0.709
350	1.1133	0.9291	23.16	20.80	0.03070	0.2968	0.702
400	0.9755	0.9420	25.54	26.18	0.03161	0.3768	0.695
450	0.8682	0.9567	27.77	31.99	0.03828	0.4609	0.694
500	0.7801	0.9722	29.91	38.34	0.04173	0.5502	0.697
550	0.7096	0.9881	31.97	45.05	0.04517	0.6441	0.700
600	0.6504	1.0044	33.92	52.15	0.04832	0.7399	0.704
			Steam (H_2O vapor)				
373.15	0.597	2030	12.28×10^{-6}	21.28×10^{-6}	0.0237	2.023×10^{-5}	1.052
393.15	0.547	1997	13.04	23.85	0.0251	2.298	1.038
413.15	0.520	1980	13.81	26.56	0.0265	2.574	1.032
433.15	0.494	1972	14.59	29.53	0.0280	2.874	1.027
453.15	0.473	1963	15.38	32.52	0.0294	3.166	1.027
473.15	0.452	1963	16.18	35.80	0.0309	3.483	1.029
493.15	0.433	1968	17.00	39.25	0.0323	3.790	1.036
513.15	0.416	1972	17.81	42.82	0.0338	4.120	1.039
533.15	0.400	1976	18.63	46.58	0.0354	4.479	1.040

$T(°K)$	$\rho(kg/m^3)$	$c_p(J/kg\text{-}°K)$	$\mu(kg/m\text{-}s)$	$\nu(m^2/s)$	$k(W/m\text{-}°K)$	$\alpha(m^2/s)$	Pr
			Steam (H_2) vapor)				
553.15	0.386	1985	19.46×10^{-6}	50.42×10^{-6}	0.0369	4.816×10^{-5}	1.047
573.15	0.372	1997	20.29	54.54	0.0385	5.183	1.052
593.15	0.359	2010	21.12	58.84	0.0401	5.557	1.059
613.15	0.348	2022	21.95	63.09	0.0416	5.912	1.067

[a]Portions of this table were adapted from material appearing in Eckert and Drake, *Analysis of Heat and Mass Transfer*, © 1972 by McGraw-Hill Book Company.

Appendix B

Units and
conversion factors

The reader is presumably familiar with the *Système International d'Unités* (the "S.I. system") and will probably make primary use of it in his or her subsequent work. But the need to deal with English units will remain with us for many years to come. We therefore list S.I. definitions, and conversion factors to and from the English system, in this appendix.

The dimensions that are used consistently in the subject of heat transfer are length, mass, force, energy, temperature, and time. Of course, we generally avoid using both force and mass dimensions in the same equation, since force is always expressible in dimensions of mass, length, and time, and vice versa. We do not make a practice of eliminating energy in terms of force times length because the accounting of work and heat is often kept separate in heat transfer problems. The text makes occasional reference to electrical units; however, these are conventional and do not have counterparts in the English system. Hence we do not discuss them here.

All conversions are presented in the form of conversion factors whose numerical worth is unity in any equation. Thus, for example, the conversion factor

$$1 = 0.0001663 \frac{\text{m/s}}{\text{furlong/fortnight}}$$

Table B.1 Elementary units and conversion factors

Quantity	Units		Conversion Factor $\equiv 1$		S.I. or Metric Equivalent
	S.I.	English	S.I. to English	English to S.I.	
Length	meter (m)	foot (ft)	$3.2808 \dfrac{\text{ft}}{\text{m}}$	$0.3048 \dfrac{\text{m}}{\text{ft}}$	$\text{m} = 10^2 \text{ cm} = 10^3 \text{ mm} = 10^{10} \text{ Å}$
Time		second (s)			
Mass	kilogram (kg)	pound mass (lb_m)	$2.2046 \dfrac{\text{lb}_m}{\text{kg}}$	$0.4539 \dfrac{\text{kg}}{\text{lb}_m}$	$\text{kg} = 10^3 \text{ g}$
Force	newton (N)	pound force (lb_f)	$0.2248 \dfrac{\text{lb}_f}{\text{N}}$	$4.4482 \dfrac{\text{N}}{\text{lb}_f}$	$\text{N} = \text{kg-m/s}^2 = 10^5 \text{ dyn}$
Energy	kilojoule (kJ)	British thermal unit (Btu)	$0.94783 \dfrac{\text{Btu}}{\text{kJ}}$	$1.05504 \dfrac{\text{kJ}}{\text{Btu}}$	$\text{kJ} = 10^3 \text{ J} = 238.8 \text{ cal}$
Temperature	degree Celsius (°C)	degree Fahrenheit (°F)	(see note [a])		$T°\text{C} = T°\text{K} - 273.15$
	degree Kelvin (°K)	degree Rankine (°R)	$1.8 \dfrac{°\text{R}}{°\text{K}}$	$\dfrac{5}{9} \dfrac{°\text{K}}{°\text{R}}$	$T°\text{K} = T°\text{C} + 273.15$

[a]Note that while an increment of 1°C and an increment of 1°K are identical in magnitude, $T°\text{C} \neq T°\text{K}$. Thus the conversion factor of $\frac{5}{9}$ °C/°F will convert an *increment* ΔT of temperature, but to convert $T°\text{F}$ to $T°\text{C}$ we must use $T°\text{C} = \frac{5}{9}(T°\text{F} - 32)$.

could be multiplied by a velocity, on just one side of an equation, to convert it from furlongs per fortnight[1] to meters per second.

Table B.1 provides conversion factors in this form for the elementary units. Notice that the S.I. prefix k is used to indicate multiplication by a factor of 1000 (e.g., 1 kg = 1000 g). Other standard prefixes that appear in the text include:

- μ, which stands for *micro*. It designates multiplication by 10^{-6}.
- m, which stands for *milli*. It designates multiplication by 10^{-3}.
- c, which stands for *centi*. It designates multiplication by 0.01.
- M, which stands for *mega*. It designates multiplication by 10^6.

Table B.2 provides conversion factors in the same form for the derived units that appear in the text.

[1]Shortly after World War II, a group of staff physicists at Boeing Airplane Co. answered angry demands by engineers that calculations be presented in English units, with a report translated entirely into such dimensions as these.

Table B.2   Conversion factors for derived units

Quantity	Symbol	Conversion Factor ≡ 1		S.I. or Metric Equivalent
		S.I. to English	English to S.I.	
Area	A	$10.764 \dfrac{\text{ft}^2}{\text{m}^2}$	$0.092903 \dfrac{\text{m}^2}{\text{ft}^2}$	$\text{m}^2 = 10^4 \text{ cm}^2$
Volume	V	$35.3134 \dfrac{\text{ft}^3}{\text{m}^3}$	$0.028317 \dfrac{\text{m}^3}{\text{ft}^3}$	$\text{m}^3 = 1000 \text{ liters}$ $= 10^6 \text{ cm}^3$
Velocity	u	$3.2808 \dfrac{\text{ft/s}}{\text{m/s}}$	$0.3048 \dfrac{\text{m/s}}{\text{ft/s}}$	$\text{m/s} = 3.6 \text{ km/hr}$
Acceleration	g	$0.3048 \dfrac{\text{ft/s}^2}{\text{m/s}^2}$	$3.2808 \dfrac{\text{m/s}^2}{\text{ft/s}^2}$	$\text{m/s}^2 = 100 \text{ cm/s}^2$
Density	$\rho \equiv 1/v$	$0.06243 \dfrac{\text{lb}_\text{m}\text{-ft}^3}{\text{kg/m}^3}$	$16.018 \dfrac{\text{kg/m}^3}{\text{lb}_\text{m}/\text{ft}^3}$	kg/m^3 $= 1000 \text{ g/cm}^3$
Specific volume	v	$16.018 \dfrac{\text{ft}^3/\text{lb}_\text{m}}{\text{m}^3/\text{kg}}$	$0.06243 \dfrac{\text{m}^3/\text{kg}}{\text{ft}^3/\text{lb}_\text{m}}$	$\text{m}^3/\text{kg} = 0.001 \text{ cm}^3/\text{g}$
Pressure	p	$0.00014504 \dfrac{\text{psi}}{\text{N/m}^2}$	$6894.8 \dfrac{\text{N/m}^2}{\text{psi}}$	$\text{N/m}^2 \equiv \text{pascal}$ $= 10^{-5} \text{ bar}$ $= 0.98692 \times 10^{-5} \text{ atm}$
Heat rate	Q	$3.4121 \dfrac{\text{Btu/hr}}{\text{W}}$	$0.29307 \dfrac{\text{W}}{\text{Btu/hr}}$	$\text{W} = \text{J/s}$
Heat rate per unit length	Q/L	$1.0403 \dfrac{\text{Btu/ft-hr}}{\text{W/m}}$	$0.9613 \dfrac{\text{W/m}}{\text{Btu/ft-hr}}$	$\text{W/m} = \text{J/m-s}$
Heat flux	$q = Q/A$	$0.3170 \dfrac{\text{Btu/ft}^2\text{-hr}}{\text{W/m}^2}$	$3.154 \dfrac{\text{W/m}^2}{\text{Btu/ft}^2\text{-hr}}$	$\text{W/m}^2 = \text{J/m}^2\text{-s}$
Volumetric heat generation	$\dot{q}$	$0.096623 \dfrac{\text{Btu/ft}^3\text{-hr}}{\text{W/m}^3}$	$10.35 \dfrac{\text{W/m}^3}{\text{Btu/ft}^3\text{-hr}}$	$\text{W/m}^3 = \text{J/m}^3\text{-s}$
Energy per unit mass	(many forms)	$0.4299 \dfrac{\text{Btu/lb}_\text{m}}{\text{kJ/kg}}$	$2.326 \dfrac{\text{kJ/kg}}{\text{Btu/lb}_\text{m}}$	$\text{kJ/kg} = \text{J/g}$ $= 0.2388 \text{ cal/g}$
Specific heat	$c, c_\text{p}, c_\text{v}$	$0.23884 \dfrac{\text{Btu/lb}_\text{m}\text{-}^\circ\text{F}}{\text{kJ/kg-}^\circ\text{C}}$	$4.1869 \dfrac{\text{kJ/kg-}^\circ\text{C}}{\text{Btu/lb}_\text{m}\text{-}^\circ\text{F}}$	$\text{kJ/kg-}^\circ\text{C} \equiv \text{kJ/kg-}^\circ\text{K} = 0.2388 \text{ cal/g-}^\circ\text{C}$
Entropy	s	Same as c_p	Same as c_p	Should only be used with $^\circ\text{K}$
Thermal conductivity	k	$0.5778 \dfrac{\text{Btu/ft-hr-}^\circ\text{F}}{\text{W/m-}^\circ\text{C}}$	$1.7307 \dfrac{\text{W/m-}^\circ\text{C}}{\text{Btu/ft-hr-}^\circ\text{F}}$	
Convective, or overall, heat transfer coefficient	$h, \bar{h}, U$	$0.1761 \dfrac{\text{Btu/ft}^2\text{-hr-}^\circ\text{F}}{\text{W/m}^2\text{-}^\circ\text{C}}$	$5.6786 \dfrac{\text{W/m}^2\text{-}^\circ\text{C}}{\text{Btu/ft}^2\text{-hr-}^\circ\text{F}}$	

Quantity	Symbol	Conversion Factor ≡ 1		S.I. or Metric Equivalent
		S.I. to English	English to S.I.	
Dynamic viscosity	μ	$0.672 \dfrac{lb_m/ft\text{-}s}{kg/m\text{-}s}$	$1.4881 \dfrac{kg/m\text{-}s}{lb_m/ft\text{-}s}$	$kg/m\text{-}s = N\text{-}s/m^2$ $\equiv 10 \text{ poise}$ $= 10^3 \text{ centipoise}$
Kinematic viscosity	ν			
Thermal diffusivity	α	$10.764 \dfrac{ft^2/s}{m^2/s}$	$0.092903 \dfrac{m^2/s}{ft^2/s}$	$m^2/s \equiv 10^4 \text{ stokes}$
Coefficient of diffusion	$\mathcal{D}$			
Stefan–Boltzmann constant	σ	$0.03023 \dfrac{Btu/ft^2\text{-}hr\text{-}°R^4}{W/m^2\text{-}°K^4}$	$33.08 \dfrac{W/m^2\text{-}°K^4}{Btu/ft^2\text{-}hr\text{-}°R^4}$	$W/m^2\text{-}°K^4$ $= 10^{-3} \text{ ergs/cm}^2\text{-s-}°K^4$
Surface tension	σ	$0.06852 \dfrac{lb_f/ft}{N/m}$	$14.594 \dfrac{N/m}{lb_f/ft}$	$N/m = 10^3 \text{ dyn/cm}$ $= kg/s^2$

Index[1]

[1]Reference-list entries are italicized.